SOTHIS PRESS

Sothis Press aims to completely reimagine the public domain book with an emphasis on universal access to captivatingly designed print editions. We believe that technology will enable every single book ever published to be perpetually accessible in beautiful print form. Sothis is beauty in ubiquity.

www.sothis.press
1 Waterhouse Square, London, ECIN 2ST

Creams

A Field Scale Model for Chemicals, Runoff, and Erosion From Agricultural Management Systems; Model Documentation

Walter G. Knisel Jr.

SOTHIS PRESS

ISBN 978-1-0365-0215-7

This edition is published by
Sothis Press © 2023. The company
is registered in England and Wales.
The Company Number is 14900389.

CREAMS

A Field Scale Model for Chemicals, Runoff, and Erosion From Agricultural Management Systems

VOLUME I. MODEL DOCUMENTATION

CONTENTS

ABSTRACT

Knisel, Walter G., editor. CREAMS: A Field-Scale Model for Chemicals, Runoff, and Erosion from Agricultural Management Systems. U.S. Department of Agriculture, Conservation Research Report No. 26, 640 pp., illus.

This publication describes a mathematical model developed to evaluate nonpoint source pollution from field-sized areas. CREAMS consists of three components: hydrology, erosion/sedimentation, and chemistry. The publication is presented in three volumes: Volume I, model documentation, describes the concepts of each of the model components; Volume II, user manual, describes the model application and selection of parameter values; Volume III, supporting documentation, provides additional data and parameter information.

Keywords: hydrology, erosion, sediment transport, plant nutrient transport, pesticide transport, mathematical model, nonpoint source pollution, agricultural management.

PREFACE

Section 208 of PL 92-500, the 1972 amendments of the Clean Water Act, placed emphasis on nonpoint-source pollution. Planning required by this legislation needed methods to assess nonpoint-source pollution under various management practices for selecting Best Management Practices (BMP's) to reduce nonpoint pollution to acceptable levels.

Expertise by the staff of the Science and Education Administration-Agricultural Research (SEA-AR) in soil and water management research, along with the high priority needs of action agencies, prompted SEA's National Program Staff (NPS) to develop plans for a concerted national effort to assemble mathematical models for evaluating nonpoint-source pollution. Staff scientists met with planners in action agencies to determine their needs for such models. A. R. Robinson, D. A. Farrell, and J. Lunin, all of the NPS, and J. C. Lance, temporarily assigned to this staff, planned the mechanism for a national project. The plans were approved by C. W. Carlson, Associate Administrator of Agricultural Research, and a request was made to T. W. Edminster, Administrator of Agricultural Research, for unassigned program funds to initiate the project. T. W. Edminster and R. J. McCracken, then the Assistant Administrator, made funds available for this important project.

The project coordinator and Technical Work Group met with the Steering Committee (NPS scientists) at Beltsville, Md., in October 1977, to initiate a national project to develop mathematical models for evaluating nonpoint-source pollution. On February 14-16, 1978, a workshop was held at Arlington, Tex., to assemble SEA-AR scientists interested in participating in this project. The workshop was to:

> ...(1) review, refine, and adopt an approach, (2) select group leaders (lead scientists), (3) plot the course of action and set the time table..., (4) assign tasks (to) investigate specific components of the system to be considered...

To develop a model quickly, participants at the workshop determined that existing physically based models, or those that could be readily modified and improved, would be assembled into a package to estimate runoff, sediment, plant nutrient, and pesticide movement in a field.

Lead scientists identified for the four components are:

Hydrology---------A. D. Nicks, Chickasha, Okla.
Erosion-----------G. R. Foster, Lafayette, Ind.
Plant nutrients---M. H. Frere, Chickasha, Okla. (New Orleans, La.)
Pesticides--------R. A. Leonard, Athens, Ga.

The hydrology component was further represented by two options lead by R. E. Smith, Fort Collins, Colo., and J. R. Williams, Temple, Tex. J. D. Nowlin,

CONTENTS

Issued May 1980

Purdue University, Lafayette, Ind., working with SEA-AR under cooperative agreement, programmed the model concepts.

The lead scientists drew upon material provided by other contributors to develop and document the model. These contributors are acknowledged throughout the publication.

Scientists in SEA-AR worked together to assemble state-of-the-art mathematical models to evaluate nonpoint-source pollution for field-scale areas. pesults of these efforts have culminated in an operational continuous simulation model. This publication documents and provides a user manual for the model named CREAMS - a field scale model for Chemicals, Runoff, and Erosion from Agricultural Management Systems.

The CREAMS model was developed using units common for the individual components. That is, customary units in the hydrology and erosion/sedimentation fields are English units, whereas metric units are common in chemistry. Rainfall data available from the National Weather Service and SEA-AR are reported in inches. Temperature data are generally in degrees Fahrenheit. Runoff, percolation, soil water, and evapotranspiration are generally reported in inches. Erosion/sedimentation data are generally reported in pounds per acre or tons per acre. Plant nutrient and pesticide losses are reported in milligrams per liter and kilograms per hectare. Model input, output, and operations were structured accordingly. Although the CREAMS model has the potential for international use, the principal users will be action agencies and consulting firms in the United States and the model, therefore, contains mixed or customary units. Users are cautioned, however, against indiscriminately modifying model components without a complete understanding of the units of operations. This version will be improved over the next several months to provide a more comprehensive model and will incorporate consistent English or metric units for user option specification.

The purpose of this publication is to provide a complete package for potential users of the model. It is divided into three main divisions. Volume I, model documentation, presents the concepts of model components. Volume II, user manual, provides information on selection of parameter values and model operation. Volume III, supporting documentation, provides the user additional information to obtain parameter values.

Results of sensitivity analysis for each component are included in the publication to indicate effects of errors in parameter estimation. This enables the user to be aware of potential difficulties resulting from inaccuracies in individual parameters.

The results of considerable testing of the components using data available from SEA-AR research locations are included. Users should be aware of two significant points: (1) Statements of model accuracy in the publication are made realistically based upon the scientist's evaluation of the mathematical representation of the real-world system and his scientific knowledge of the range and confidence in parameter estimation, and (2) ranges of conditions considered appropriate for application of the model are given in the publication.

Magnetic tapes of the computer model can be furnished to anyone interested in using the model. A user, in turn, can send a magnetic tape to the project coordinator and the program will be taped along with a set of test data, parameter values, and summary output. These data will enable users to be sure the model is operating properly on their respective computer systems. A tape can be generated for CDC or IBM computers and the user should specify the system when requesting the program.

Walter G. Knisel, Jr.
Project Coordinator
USDA-SEA-AR
442 East Seventh Street
Tucson, Ariz. 85705

ACKNOWLEDGMENTS

The authors acknowledge T. W. Edminster, R. J. McCracken, and C. W. Carlson for their confidence in and support of the scientists working on the modeling project.

National Program Staff scientists who make up the Steering Committee for the project, D. A. Farrell, J. Lunin, J. C. Lance, and A. R. Robinson, are recognized for their technical support.

The Technical Work Group, D. G. DeCoursey, E. T. Engman, L. D. Meyer, M. H. Frere, and R. A. Leonard, is recognized for planning and conducting a workshop at Arlington, Tex., to initiate the modeling effort.

K. G. Renard is acknowledged for providing support staff assistance to the project. Virginia Ferreira, Karen Mellor, Sue Schield, John Rocha, and Bob Wilson are recognized for their respective assistance in computer terminal operations, typing, and drafting.

W. E. Modenhauer, SEA-AR, and G. W. Isaacs, Agricultural Engineering Department, Purdue University, Lafayette, Ind., are recognized for providing J. D. Nowlin for computer programming assistance.

The following scientists participated in the workshop at Arlington, Tex., in 1978 and contributed ideas, direction, and data as well as completed task assignments that led to the model development.

C. V. Alonso, Oxford, Miss.
A. P. Barnett, Watkinsville, Ga.
J. V. Bonta, Coshocton, Ohio
A. J. Bowie, Oxford, Miss.
D. L. Chery, Athens, Ga.
K. R. Cooley, Phoenix, Ariz.
D. G. DeCoursey, Oxford, Miss.
E. T. Engman, Beltsville, Md.
G. R. Foster, Lafayette, Ind.
M. H. Frere, Chickasha, Okla.
W. R. Hamon, Coshocton, Ohio
C. L. Hanson, Boise, Idaho
A. T. Hjelmfelt, Columbia, Mo.
D. E. Kissel, Temple, Tex.
W. G. Knisel, Tucson, Ariz.
J. M. Laflen, Ames, Iowa
J. C. Lance, Beltsville, Md.
L. J. Lane, Tucson, Ariz.
W. E. Larson, St. Paul, Minn.
R. A. Leonard, Athens, Ga.
C. E. Murphree, Oxford, Miss.
C. K. Mutchler, Oxford, Miss.
R. G. Nash, Beltsville, Md.
E. L. Neff, Sidney, Mont.
A. D. Nicks, Chickasha, Okla
C. A. Onstad, Morris, Minn.
L. B. Owens, Coshocton, Ohio
R. F. Piest, Columbia, Mo.
H. B. Pionke, Univ. Park, Penn.
W. J. Rawls, Beltsville, Md.
C. W. Richardson, Temple, Tex.
J. C. Ritchie, Beltsville, Md.
M. J. M. Romkens, Oxford, Miss.
K. E. Saxton, Pullman, Wash.
E. H. Seely, Chickasha, Okla.
D. E. Smika, Akron, Colo.
R. E. Smith, Fort Collins, Colo.
S. J. Smith, Durant, Okla.
W. F. Spencer, Riverside, Calif.
D. R. Timmons, Morris, Minn.

D. K. McCool, Pullman, Wash.	R. D. Wauchope, Stoneville, Miss.
L. L. McDowell, Oxford, Miss.	J. R. Williams, Temple, Tex.
R. G. Menzel, Durant, Okla.	G. H. Willis, Baton Rouge, La.
L. D. Meyer, Oxford, Miss.	R. A. Young, Morris, Minn.

J. B. Burford and Jane DeLashmutt of the Water Data Laboratory, Beltsville, Md., helped obtain and format data for testing the hydrologic components of the model.

USDA-Soil Conservation Service personnel critiqued the model in a technology transfer workshop at the South Technical Service Center, Fort Worth, Tex. John Burt, Gary Margheim, E. C. Nicholas, and S. J. Robbins helped arrange the workshop and provided input to improve the model. Margheim obtained SCS funds for the SEA-AR model testing and technology transfer.

B. C. Dysart, III, and R. C. Warner, Environmental Systems Engineering Department, Clemson University, Clemson, S.C., are acknowledged for their comments and suggestions on the erosion component. Their application in South Carolina enabled improvements of the erosion component.

CREAMS: A FIELD SCALE MODEL FOR CHEMICALS, RUNOFF, AND EROSION FROM AGRICULTURAL MANAGEMENT SYSTEMS

VOLUME I. MODEL DOCUMENTATION

Chapter 1. INTRODUCTION

W. G. Knisel and A. D. Nicks[1/]

Under the Federal Water Pollution Control Act Amendments of 1972, Public Law No. 92-500, the Administrator of the Environmental Protection Agency (EPA), in cooperation with other agencies, provides guidelines for identifying and evaluating nonpoint sources of pollutants. The U.S. Department of Agriculture (USDA) is one of the cooperating agencies. The USDA-Soil Conservation Service (USDA-SCS) has the technical responsibility for evaluating nonpoint source pollution and implementing Best Management Practices (BMP's) to limit nonpoint source pollution to an acceptable level. The Science and Education Administration-Agricultural Research (SEA-AR), as the research agency of USDA, has obligations for research to meet the needs of SCS and EPA.

Scientists in SEA-AR (formerly Agricultural Research Service) by request of EPA prepared a two-volume document on control of potential water pollutants from cropland. The two volumes, published in 1976, include information on the basic principles of control of specific pollutants (28, 29). A list of BMP's was included in volume I (28). Although simple models for estimating annual values of runoff, percolation, erosion, plant nutrient, and pesticide losses were given in volume II (29), the BMP's were not quantified. Management practices are site-specific. Stewart and others, (28) stated: "Because of the variation of climate, soils, and agricultural practices throughout the United States, no single group of control measures can be used for every region, nor will the regional information printed herein be accurate for all areas within the region."

SEA-AR recognized the need for development of physically based mathematical models to make the next logical step beyond the Stewart and others reports (28, 29), and in 1978 scientists began a concerted effort to assemble such models. Since management practices are applied on a farm or field basis, it was thought that the size range to be considered should be the field scale. Figure I-1 shows a schematic representation of a field with natural and management input and the associated water, sediment, and chemical output.

1/ Hydraulic engineer, USDA-SEA-AR, Tucson, Ariz., and hydraulic engineer, USDA-SEA-AR, Chickasha, Okla., respectively.

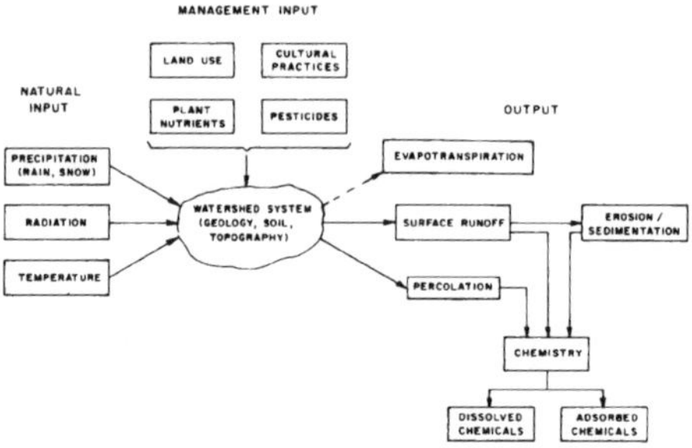

Figure I-1.—Flow chart of system for evaluating nonpoint source pollution.

A question arose immediately: What size is a field? The physical size of farm fields varies from a few acres in ridge and valley provinces to a few tens of acres in the Corn Belt to a few hundreds of acres in the Wheat Belt and western rangelands. Such a size range required some arbitrarily imposed constraints. Thus, a field herein is defined as a management unit having (1) a single land use, (2) relatively homogeneous soils, (3) spatially uniform rainfall, and (4) single management practices, such as conservation tillage or terraces. This definition allows different physical sizes in different climatic regions and Land Resource Areas (LRA's).

To achieve the goal of model assembly in a year, state-of-the-art models were assembled and/or modified. Criteria for the model were: (1) the model must be physically based and not require calibration for each specific application, (2) the model must be simple, easily understood with as few parameters as possible and still represent the physical system relatively accurately, (3) the model must estimate runoff, percolation, erosion, and dissolved and adsorbed plant nutrients and pesticides, and (4) the model must distinguish between management practices.

Although hydrology is only one component of the total system, water is the principle element; it causes erosion, carries chemicals, and is an uncontrolled natural input. Each climatic region and physiographic area has its own characteristics that affect the response of the system. These varied conditions must be kept in mind when considering wide-scale applicabilty of a model. Figure I-2 is a generalized schematic representation of the water balance for different areas of the United States. The width of bars in the figure are drawn to scale to show the relative magnitude of each component among the five regions. In the Southeast Coastal Plain, rainfall averages about 50 inches (1,270 mm), and evapotranspiration is about 35 inches (890 mm). Approximately 80 percent of the water that ultimately reaches streamflow has at one time been subsurface flow. That is, about 12 inches (305 mm) of total streamflow comes from subsurface flow (24). Only 3 inches (76 mm) comes from direct overland

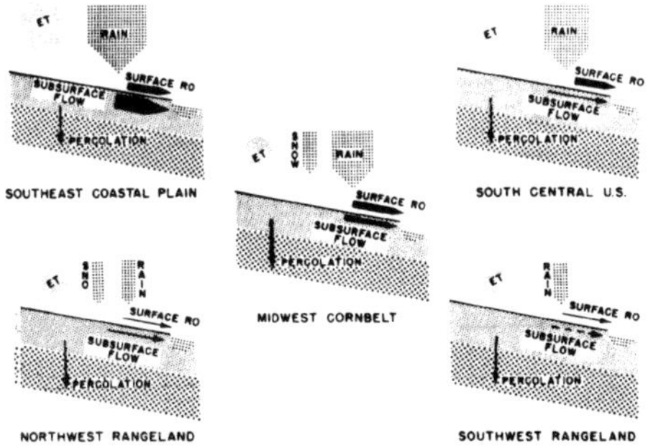

Figure I-2.—Schematic representation of water balance for selected locations in the United States.

flow. Deep percolation to regional groundwater is negligible. In the south central United States, average rainfall is about 34 inches (864 mm), and runoff is about 8 inches (203 mm). For all practical purposes, there is no subsurface flow or deep percolation, and evapotranspiration is about 26 inches (661 mm). In the semiarid Southwest, precipitation is about 13 inches (330 mm), with only 0.5 inch (13 mm) of surface runoff and negligible groundwater recharge or subsurface flow. Snow is a part of the precipitation input for the Corn Belt and Northwest rangelands. Subsurface flow and groundwater recharge are significant components in the Corn Belt. Dissolved chemicals may be an important potential nonpoint source pollutant in the Coastal Plain and in the Corn Belt. Sediment and adsorbed chemicals may be the major pollutants in the Western rangelands and the South Central areas, as well as the Corn Belt. Although these representations are generalized, they indicate the varied conditions that a nonpoint source pollution model must be capable of considering.

The system represented in figure I-1, the conditions represented in figure I-2, the model criteria, and the constraints of field size were guidelines used in the development and testing of CREAMS. This publication documents the model and provides a user manual to aid in selection of parameter values to run the model. CREAMS is the first step beyond the Stewart and others reports (28, 29), and is preliminary to a basin scale model.

The general logic of the model is that hydrologic processes provide the transport medium for sediment and agricultural chemicals. Therefore, the hydrologic component provides input to the other model components. The erosion/sediment yield component in turn provides estimates of sediment yield and silt/clay/organic matter enrichment to be used in the chemical transport components. The documentation generally follows this logic, with evaluation included in each section. A separate section gives results of sensitivity analysis of the model parameters.

DEVELOPMENT OF NONPOINT SOURCE POLLUTION MODELS

Hydrologists have long used models to depict relationships between such hydrologic variables as rainfall, runoff, evapotranspiration, and infiltration. These were generally graphical representations or regression equations that could be solved easily with desk calculators. The relationships like the rational formula Q = CIA (13) often were gross simplifications of complex processes.

In the 1950's, the USDA-SCS recognized the need for a more comprehensive model to estimate runoff from rainfall as a function of soil, vegetation, and antecedent moisture, and developed the SCS curve number (SCSCN) model. The basic model is still being used at present by SCS (30). This model related storm runoff to storm rainfall, and was used to estimate runoff in the report of Stewart and others (29).

Wischmeier and Smith (33) analyzed many years of plot data to develop the Universal Soil Loss Equation (USLE) for estimating gross erosion by water. The USLE, a relatively simple regression equation, is presently being used by many agencies and consultants, and it was updated recently (34).

Development of electronic data processing equipment eliminated the time-consuming repetitive hand calculations necessary for analyses of large volumes of data. Also, scientists may now formulate more complex conceptual models and solve more complex equations with the computer than was possible earlier. Model proliferation began in the late 1950's, and continues to the present. These models cover a range of sophistication and mathematical complexity. Models range from deterministic to stochastic, with various combinations in between. The models were all developed for specific purposes that range from analyses of data to extrapolation of data to some future condition. These specific purposes include prediction of runoff from rainfall, estimation of erosion from rainfall, projection of downstream sediment yield from field erosion processes within a watershed, and so forth.

In 1962, Crawford and Linsley (4) published one of the earliest computer hydrologic simulation models. The model became widely known as the Stanford watershed model. It uses conceptual simplifications for physical processes of overland flow, interflow, upper zone soil water storage, lower zone soil water storage, deep percolation, groundwater storage, and evapotranspiration to estimate streamflow from rainfall records (5). The model requires calibration to specific watershed conditions and was primarily intended to show effects of watershed changes on streamflow.

The Stanford watershed model (SWM) became the basis for numerous studies, and several scientists have made revisions, particularly in optimization procedures for calibration (20, 25). More recently, the Stanford model has been used as the basic hydrologic component for field-scale, water-quality models (3). The basic concepts of the model were retained with internal revisions, but calibration of the model to specific fields is still required.

Glymph and Holtan (15) developed an infiltration-based hydrologic model, known as the USDAHL (U.S. Department of Agriculture Hydrograph Laboratory) model, to estimate streamflow using a concept of soil zones on the watershed landscape. Snowmelt, separation of flow regimes, and ground water contributions to streamflow have been incorporated recently (17, 18).

Passage of the Federal Water Pollution Control Act Amendments, Public Law 92-500, (commonly known as the Clean Waters Act) in 1972 created an awareness by many agencies and consultants for models to simulate processes affecting water quality. More specifically, Section 208 of the Clean Waters Act specified that by October 1978 the States would have completed plans for limiting streamflow pollution from nonpoint sources, particularly agriculture. This specification emphasized the need for mathematical models to evaluate nonpoint-source pollution and consider BMP's to reduce the pollution (6). All these models, that is, SCSCN, USLE, USDAHL, and the SWM, were used later as basic components, with or without modification, for water-quality models. There was little precedent for chemical transport models, especially for upland areas, although diffusion models had been applied in river-channel systems. Since water is the carrier of sediment and chemicals, most water quality models were developed by selecting a hydrologic model, and "piggy-backing" sediment and chemistry components to produce a model package.

Hydrocomp, under contract with the Environmental Protection Agency (EPA), developed the Pesticide Transport and Runoff (PTR) model (3). A revision of the SWM (5) became the hydrologic component of the PTR model. The sediment loss component of PTR consists of a part of Negev's equation for sediment detachment and transport (23). Although Negev simulated the entire sheet, rill, and channel erosion, the PTR model only uses the sheet and rill erosion components which include the detachment and transport of soil particles by overland flow. Pesticide simulation includes the process of adsorption/desorption to determine the division between the sediment and water phases of runoff. Volatilization of pesticides is considered along with degradation, which is represented by a first-order, decay-type relation with time. Plant nutrients were not considered in the PTR model.

Frere, Onstad, and Holtan (12) developed an agricultural chemical transport model (ACTMO) based on the USDAHL model (18). The erosion/sediment transport component of ACTMO is a modification of the USLE to reflect both rainfall and runoff erosivity and transport processes (11). The erosion component estimates the contribution of rill and interrill sources to sediment load. The chemical component of ACTMO included pesticide and nitrate options. The pesticide option treated adsorption, breakdown, and movement processes. Very little field data were available to validate the proposed relationships. The nitrate option considered mineralization, plant uptake, and movement processes.

Bruce and others (2), developed a parametric model for water-sediment-chemical (WASCH) runoff for single storm events. The hydrologic component consists of three functions: a retention function, a characteristic function, and a variable state function (26). Two-stage convolution is used to produce nonlinear watershed response. The sediment component of WASCH considers the rill/interrill erosion concepts developed by Foster and Meyer (10), but uses erosion and routing functions for both rill and interrill erosion. Sediment transport capacity in the WASCH model is a function of overland flow discharge rather than velocity. The chemical component of WASCH considers only pesticides and does not treat plant nutrients. The pesticide model is a single mathematical expression relating pesticide runoff to rill and interrill sediment with extraction and enrichment factors.

Donigian and Crawford (7) modified, tested, and further developed the PTR model, and these revisions resulted in the Agricultural Runoff Management (ARM) model. Although the model was revised, the original basic components were the same, that is, SWM for the hydrology component, and Negev's equations for the sediment component. A plant nutrient component was incorporated into the new version.

Donigian and Crawford (8) developed a Nonpoint Source Pollutant loading (NPS) model to simulate pollutant contributions to stream channels from nonpoint sources. NPS considers a maximum of five pollutants from each of a maximum of five separate land use categories. The hydrology and erosion components are identical with those in ARM (7). The water quality component relates pollutants to sediment by specifying pollutant strength or potency factors. NPS does not have a component for channel processes, but simulates loads of pollutants reaching the stream channels.

Williams and Hann (32) developed a basin scale model to consider surface runoff, sedimentation, and plant nutrients. The hydrologic component is a modification of the SCS curve number model. The USLE was modified for the erosion component by replacing the rainfall energy term with a product of storm runoff volume and peak rate of discharge raised to a power. Subsurface or baseflow is not considered by the model. The plant nutrient component of the model considers both organic and inorganic nitrogen and denitrification, immobilization, and mineralization processes. Nitrogen fertilization, nitrogen in rainfall, and nitrogen from crop residue are inputs to the basin soils, while plant uptake and nitrate leaching were simulated to remove nitrogen from the soils. The phosphorus component of the nutrient model considered only that portion adsorbed to soil particles. Both the nitrogen and phosphorus components use enrichment ratios to develop loading functions. The model routes runoff, sediment, nitrogen, and phosphorus to the basin outlet. Linear programming techniques are used to select an alternate management practice.

Gianessi, Pleskin, and Young (14) developed a water pollution network model, referred to as the RFF model (Resources for the Future), to link sources of pollutants to concentrations in water bodies throughout the Nation. The water network identifies 1,051 node points along rivers of the United States to correspond with U.S. Geological Survey (USGS) gaging station locations. Each county in the United States is assigned to at least one node. The average distance between nodes is 66 miles. Streams were classed by ranges of mean discharges, and USGS periodic stream gaging measurements at the nodes are used to determine velocity at the nodes. The RFF model emphasizes pollutants, including sediment, which are input at node points and are assumed input uniformly between nodes. Loading functions, on a county basis, are obtained from McElroy and others (21). Sediment from construction, forestry, and mining activities is obtained by prorating national estimates to each county based on the county's share of employment in these activities and weighted by an estimate of runoff. The RFF model is basically a routing technique for 66-mile river reaches with generalized loadings of pollutants without identification of conservation systems on less than a county basis.

The CREAMS is a physically based, daily simulation model that estimates runoff, erosion/sediment transport, plant nutrient, and pesticide yield from field-sized areas. The hydrologic component consists of two options. When only daily rainfall data are available to the user, the SCS curve number model is used to estimate surface runoff. If hourly or breakpoint rainfall data are available, an infiltration-based model is used to simulate runoff. Both methods estimate percolation through the root zone of the soil. The erosion component maintains elements of the USLE, but includes sediment transport capacity for overland flow. A channel erosion/deposition feature of the model permits consideration of concentrated flow within a field. Impoundments are treated in the erosion component also. The plant nutrient submodel of CREAMS has a nitrogen component that considers mineralization, nitrification, and denitrification processes. Plant uptake is estimated, and nitrate leached by percolation out of the root zone is calculated. Both the nitrogen and phosphorus parts of the nutrient component use enrichment ratios to estimate that portion of the two nutrients transported with sediment. The pesticide component considers foliar interception, degradation, and washoff, as well as adsorption, desorption, and degradation in the soil. This method, like the nutrient model, uses enrichment ratios and partitioning coefficients to calculate the separate sediment and water phases of pesticide loss.

These models are compared in table I-1. In addition to these models for predicting runoff, erosion, and agricultural chemicals, several models were developed to estimate runoff and erosion for relatively large basins. Representative features are given in table I-2.

Beasley, Monke, and Huggins (1) developed a distributed deterministic model. (ANSWERS) for predicting runoff and erosion/sediment transport for different agricultural management systems. The basic hydrologic component from Huggins and Monke (19) describes surface runoff, subsurface flow, and channel flow in a system of square grids laid over the watershed. The infiltration element of the model is basically the infiltration function of the USDAHL model (18). When the water content of the control zone exceeds field capacity, infiltrated water becomes subsurface drainage. The erosion component of ANSWERS consists of modifications of the USLE (33). Two soil detachment processes were included: (a) rainfall detachment, described by Meyer and Wischmeier (22), and (b) overland flow detachment, described by Foster (9). Sediment transport of both overland and channel flow is based on transport capacity. Channel erosion is assumed to be negligible, and only deposition is allowed in channel flow.

Simons, Li, and Ward (26) developed an event model to predict runoff and sediment yield from small basins. The hydrologic component consists of the kinematic wave model for overland flow and channel flow with infiltration approximated by the Green and Ampt (16) infiltration equation. The sediment component considers erosion by raindrop splash and shear stress of overland flow. Raindrop erosion is expressed as a power function of rainfall intensity and an empirically determined erodibility factor. Erosion by overland flow uses a detachment coefficient that requires calibration for specific soils. Sediment transport in the model considers transport capacity for individual sediment sizes. Bed load transport and suspended load transport are estimated.

Wade and Heady (29) developed an economic model based on agricultural crop production considering sediment as a pollutant. The model, referred to as the

Table I-1.--Water quality models, basic components and scale of application

Model	Date	Hydrology component	Erosion/ sedimentation component	Pesticide component[1/]	Nutrient component[2/]	Scale of application
PRT	1973	SWM	Negev	As,Ds,Vo,De	None	Field.
ACTMO	1975	USDAHL	Modified USLE.	As,Ds,Vo,De	M,N,NL	Basin.
WASCH	1975	Parametric	Parametric	Parametric	None	Field.
ARM	1976	SWM	Negev	As,Ds,De	M,D,N,I,NL,AP,SP	Field.
NPS	1976	SWM	Negev	None	None	Basin.
Williams	1978	SCSCN	Williams-Modified USLE.	None	M,D,N,I,NL,AP,SP	Basin.
RFF	1978	Mean river flow(?).	Loading functions.	Loading functions.	Loading functions.	Basin.
CREAMS	1979	SCSCN, infiltration.	Interrill-rill detachment; overland flow transport capacity; concentrated flow detachment and transport capacity; impoundment deposition.	As,Ds,Vo,De	M,N,NL,D, AP,SP.	Field.

1/ No precedent for pesticide model; symbols for processes are: As - adsorption; Ds - desorption; Vo - volatilization; De - degradation.
2/ No precedent for nutrient model; symbols for process are: M - mineralization; D - denitrification; N - nitrification; I - imobilization; NL - nitrate leaching; AP - adsorbed phosphorus; SP - solution phosphorus.

Table I-2.--Hydrology-sedimentation models, basic components, and scale of application

Model	Date	Hydrology component	Erosion component	Scale
ANSWERS	1977	USDAHL infiltration; kinematic flow; channel routing.	Interrill-rill detachment, overland and channel flow transport capacity.	Basin.
Simons and others	1977	Infiltration, kinematic flow, channel routing.	Raindrop and overland flow detachment and transport capacity, channel flow detachment and transport capacity.	Basin.
Wade and Heady	1978	None	Sediment delivery ratios, sediment transport ratios.	Basin.

National Water Assessment (NWA) model, does not contain a hydrologic component but estimates average annual erosion with the USLE (33) for 105 Producing Areas (PA) covering the United States. Sediment delivery ratios, estimated for each PA by using measured and computed data, are used to estimate sediment delivery. River basin sediment accounting is made by sediment ratios estimated for the rivers of the PA's. River flow apparently is not used in the accounting system, and the transport ratios are determined subjectively to give river sediment yields. Where lakes were involved in the river systems, estimated trap efficiencies were used in determining transport ratios. Linear programming was used with the NWA model to consider 5 sediment control alternatives to calculate the associated sediment yield to the oceans from 18 river basins of the United States.

REFERENCES

(1) Beasley, D. B., E. J. Monke, and L. F. Huggins.
1977. ANSWERS: A model for watershed planning. Purdue Agricultural Experiment Station Journal Paper No. 7038. 34 pp.

(2) Bruce, R. R., L. A. Harper, R. A. Leonard, W. M. Snyder, and A. W. Thomas.
1975. A model for runoff of pesticides from small upland watersheds. Journal of Environmental Quality 4(4):541-548.

(3) Crawford, N. H., and A. S. Donigian, Jr.
1973. Pesticide transport and runoff model for agricultural lands. U.S. Environmental Protection Agency, EPA-660/2-74-013. 211 pp. Washington, D.C.

(4) ------, and R. K. Linsley.
1962. The synthesis of continuous streamflow hydrographs on a digital computer. Stanford University, Department of Civil Engineering, Technical Report No. 12. Stanford, Calif.

(5) ------, and R. K. Linsley.
1966. Digital simulation in hydrology: Stanford Watershed Model IV. Stanford University, Department of Civil Engineering, 210 pp. Stanford, Calif.

(6) Davey, W. B.
1975. Conservation districts and 208 water quality management. U.S. Environmental Protection Agency and National Association of Conservation Districts, 349 pp. U.S. Government Printing Office, Washington, D.C.

(7) Donigian, A. S., Jr., and N. H. Crawford.
1976. Modeling pesticides and nutrients on agricultural lands. U.S. Environmental Protection Agency, Environmental Protection Technology Series, EPA-600/2-76-043, 317 pp. Washington, D.C.

(8) Donigian, A. S., Jr., and N. H. Crawford.
1976. Modeling nonpoint pollution from the land surface. U.S. Environmental Protection Agency, Ecological Research Series, EPA-600/3-76083, 279 pp. Washington, D.C.

(9) Foster, G. R.
1976. Sedimentation, general. Proceedings National Symposium on Urban Hydrology, Hydraulics, and Sediment Control, University of Kentucky, Lexington, Ken. July 26-29.

(10) ------, and L. D. Meyer.
1975. Mathematical simulation of upland erosion using fundamental erosion mechanics. In: Present and prospective technology for predicting sediment yields and sources. U.S. Department of Agriculture, Agricultural Research Servcice, Southern Region, ARS-S-40, pp. 190-207. (Series discontinued; Agricultural Research Service is now Science and Education Administration-Agricultural Research.)

(11) ------, L. D. Meyer, and C. A. Onstad.
1973. Erosion equation derived from modeling principles. Paper 73-2550. Presented at 1973 Winter Meeting ASAE, December 11-14, 1973, Chicago, Ill.

(12) Frere, M. H., C. A. Onstad, and H. N. Holtan.
1975. ACTMO, an agricultural chemical transport model. U.S. Department of Agriculture, Agricultural Research Service, Headquarters, ARS-H-3, 54 pp. (Series discontinued; Agricultural Research Service is now Science and Education Administration-Agricultural Research.)

(13) Frevert, R. K., G. O. Schwab, T. W. Edminster, and K. K. Barnes.
1955. Soil and Water Conservation. John Wiley and Sons, Inc., New York, N.Y. 479 pp.

(14) Gianessi, L. P., H. M. Peskin, and G. K. Young.
1978. A national water pollution network model. [Unpublished] 21 pp.

(15) Glymph, L. M., and H. N. Holtan.
1969. Land treatment in agricultural watershed hydrology research. Effects of watershed changes on stream flow, Water Resources Symposium No. 2, University of Texas, Austin, Tex., pp. 44-68.

(16) Green, W. H., and G. A. Ampt.
1911. Studies on soil physics, Part I: The flow of air and water in soils. Journal of Agricultural Science.

(17) Holtan, H. N., and N. C. Lopez.
1971. USDAHL-70 model of watershed hydrology. U.S. Department of Agriculture, Technical Bulletin No. 1435, 84 pp. Washington, D.C.

(18) ------, G. J. Stiltner, W. H. Henson, and N. C. Lopez.
1975. USDAHL-74 revised model of watershed hydrology, a United States contribution to the International Hydrological Decade. U.S. Department of Agriculture, Technical Bulletin No. 1518, 99 pp. Washington, D.C.

(19) Huggins, L. F., and E. J. Monke.
1966. The mathematical simulation of the hydrology of small watersheds. Purdue University, Water Resources Research Center, Technical Report No. 1. 130 pp.

(20) James, L. D.
1970. An evaluation of relationships between streamflow patterns and watershed characteristics through the use of OPSET: A self-calibrating version of the Stanford watershed model. University of Kentucky, Water Resources Institute, Research Report No. 36. Lexington, Ken.

(21) McElroy, A. D., S. Y. Chiu, J. W. Nebgen, A. Aleti, and F. W. Bennett.
1976. Loading functions for assessment of water pollution from nonpoint sources. U.S. Environmental Protection Agency, Environmental Protection Technology Series, EPA-600/2-76-151, Washington, D.C. 444 pp.

(22) Meyer, L. D., and W. H. Wischmeier.
1969. Mathematical simulation of the processes of soil erosion by water. Transactions of the American Society of Agricultural Engineers 12(6):754-758.

(23) Negev, M. A.
1967. Sediment model on a digital computer. Department of Civil Engineering, Stanford University, Technical Report No. 76, Stanford Calif. 109 pp.

(24) Rawls, W. J., and L. E. Asmussen.
1973. Subsurface flow in the Georgia Coastal Plain. Journal of Irrigation and Drainage, Proceedings of the American Society of Civil Engineers Paper No. 10013, 99(IR3):375-385.

(25) Ross, G. A.
1970. The Stanford watershed model: The correlation of parameter values selected by a computerized procedure with measurable physical characteristics of the watershed. Water Resources Institute Research Report No. 35. University of Kentucky, Lexington, 178 pp.

(26) Simons, D. B., R. M. Li, and T. J. Ward.
1977. A simple procedure for estimating on-site soil erosion. Proceedings of the International Symposium on Urban Hydrology, Hydraulics, and Sediment Control, University of Kentucky, Lexington, July 18-21. pp. 95-102.

(27) Snyder, W. M.
1974. Development of a parametric hydrologic model useful for sediment yield. In: Present and prospective technology for predicting sediment yields and sources. U.S. Department of Agriculture, Agricultural Research Service, Southern Region, ARS-S-40. pp. 220-230. (Series discontinued; Agricultural Research Service is now Science and Education Administration-Agricultural Research.)

(28) Stewart, B. A., D. A. Woolhiser, W. H. Wischmeier, J. H. Caro, and M. H. Frere.
1975. Control of water pollution from cropland. Vol. I - A manual for guideline development. U.S. Department of of Agriculture, Agricultural Research Service, Headquarters, Report No. ARS-H-5-1. 111 pp. (Series discontinued; Agricultural Research Service is now Science and Education Administration-Agricultural Research.)

(29) ------, D. A. Woolhiser, W. H. Wischmeier, J. H. Caro, and M. H. Frere.
1976. Control of water pollution from cropland. Vol. II: An overview. U.S. Department of Agriculture-Agricultural Research Service, Headquarters, ARS-H-5-2. 187 pp. (Series discontinued; Agricultural Research Service is now Science and Education Administration-Agricultural Research.)

(30) U.S. Soil Conservation Service.
1972. SCS National Engineering Handbook, Sec. 4, Hydrology. 548 pp.

(31) Wade, J. C., and E. O. Heady.
1978. Measurement of sediment control impacts on agriculture. Water Resources Research 14(1):1-8.

(32) Williams, J. R., and R. W. Hann, Jr.
1978. Optimal operation of large agricultural watersheds with water quality constraints. Texas A&M University, Texas Water Resources Institute, Technical Report No. 96, College Station. 152 pp.

(33) Wischmeier, W. H., and D. D. Smith.
1965. Predicting rainfall-erosion losses from cropland east of the Rocky Mountains--Guide for selection of practices for soil and water conservation. U.S. Department of Agriculture, Agriculture Handbook No. 282. 47 pp.

(34) ------, and D. D. Smith.
1978. Predicting rainfall erosion losses--a guide to conservation planning. U.S. Department of Agriculture, Agriculture Handbook No. 537. 58 pp.

Chapter 2. SIMULATION OF THE SURFACE WATER HYDROLOGY

R. E. Smith and J. R. Williams[1/]

INTRODUCTION

Central to the simulation of pollutant movement on and from a field site is the simulation of the amount and rate of water movement on the surface and through the soil. All major hydraulic processes which occur during a rainstorm — such as rainfall infiltration, soil water movement, and surface water flow — can be simulated in detail with current knowledge of hydraulics and the capabilities of modern computers. The constraint in the construction of this model, however, is to approximate the complexity of these processes and their interrelations with a model whose sophistication is appropriate to the detail of data expected to be available in its intended use.

The field-scale hydrologic response simulation includes models for infiltration, soil water movement, and soil/plant evapotranspiration between storms. It is a continuous simulation model using a day as the time step for evaporation and soil water movement between storms, and using shorter time increments dictated by available rainfall records during storms. The between-storm simulation provides prediction of amount of seepage below the root zone and gives an initial soil water content at the beginning of a storm, which is an important initial condition for storm runoff simulation. When storm rainfall records are not available, runoff is estimated by the SCS curve number procedure (7).

INFILTRATION

Infiltration From Daily Rainfall (SCS Curve Number Model)

The SCS curve number technique (7) was selected for predicting runoff from daily rainfall because (1) it is a familiar procedure that has been used for many years in the United States; (2) it is computationally efficient; (3) the required inputs are generally available; and (4) it relates runoff to soil type, land use, and management practices. The use of readily available daily rainfall is a particularly important attribute of the curve number technique. For many locations, rainfall data with time increments of less than 1 day are not available. Also, daily rainfall data manipulation and runoff computation are more efficient than similar operations with shorter time increments.

1/ Hydraulic engineer, USDA-SEA-AR, Fort Collins, Colo., and hydraulic engineer, USDA-SEA-AR, Grassland-Soil and Water Research Laboratory, Temple, Tex., respectively.

Traditionally, the SCS has used an antecedent rainfall index to estimate antecedent moisture as one of three conditions (I - dry, II - normal, and III - wet). The relation between rainfall and runoff for these three conditions is expressed as a curve number (CN). Each storm in a rainfall series is assigned one of the three curve numbers according to antecedent rainfall. In reality, CN varies continuously with soil moisture, and thus has many values instead of only three. Runoff prediction accuracy was increased by using a soil moisture accounting procedure to estimate the curve number for each storm (9). Although the soil moisture accounting model was found to be superior to the antecedent rainfall method, it did not contain a percolation component or a physically based water balance. Also, the model required calibration with measured runoff data.

Here the curve number technique was linked with evapotranspiration and percolation models to form a model capable of maintaining a continuous water balance. Calibration is not necessary, because the new model is more physically based. Besides predicting daily runoff volumes, an equation was also developed for predicting peak runoff rates. Tests with data from watersheds in Texas, Nebraska, Georgia, Ohio, Oklahoma, Arizona, New Mexico, West Virginia, Mississippi, Iowa, and Montana indicate that the model simulates runoff volumes and peak rates realistically (tables I-3 and I-4).

Model Description

Runoff is predicted for daily rainfall using the SCS equation

$$Q = \frac{(P - 0.2s)^2}{P + 0.8s} \qquad [I\text{-}1]$$

where Q is the daily runoff; P is the daily rainfall; and s is a retention parameter, all having dimensions of length. The retention parameter s is related to soil water content with the equation

$$s = s_{mx}\left(\frac{UL - SM}{UL}\right) \qquad [I\text{-}2]$$

where SM is the soil water content in the root zone, UL is upper limit of soil water storage in the root zone, and s_{mx} is the maximum value of s. The maximum value of s is estimated with the I moisture condition CN using the SCS (7) equation

$$s_{mx} = \frac{1000}{CN_I} - 10 \qquad [I\text{-}3]$$

where CN_I is the moisture condition I CN. An estimate of the moisture condition II CN can be obtained easily for any watershed using the SCS Hydrology Handbook (7). The corresponding CN_I values are also tabulated. For computing purposes CN_I was related to CN_{II} with the polynomial

$$CN_I = -16.91 + 1.348(CN_{II}) - 0.01379(CN_{II})^2 + 0.0001177(CN_{II})^3. \qquad [I\text{-}4]$$

If soil water is distributed uniformly in the soil profile, equation [I-2] should give a good estimate of the retention parameter, and thus the runoff. However, if the soil water content is greater near the surface, equation [I-2] would tend to give low runoff predictions. Conversely, runoff would be over-predicted if the soil water content was greater in the lower root zone. To account for the soil water distribution, a weighting technique was developed. The root zone was divided into seven layers and weighting factors (decreasing with depth) were applied. The depth-weighted retention parameter is computed with the equation

$$s = s_{mx}\left[1.0 - \sum_{i=1}^{N} W_i\left(\frac{SM_i}{UL_i}\right)\right] \qquad [I-5]$$

where W_i is the weighting factor, SM_i is the water content, and UL_i is the upper limit of water storage in storage i. The weighting factors decrease with depth according to the equation

$$W_i = 1.016\left[e^{-4.16\left(\frac{D_{i-1}}{RD}\right)} - e^{-4.16\left(\frac{D_i}{RD}\right)}\right] \qquad [I-6]$$

where D_i is the depth to the bottom of storage i, and RD is the root zone depth. Equation [I-6] assures that $\sum_{i=1}^{N} W_i = 1$.

The evapotranspiration and percolation components of the model are described below. Since the model maintains a continuous water balance, mixed land use watersheds are subdivided to reflect differences in ET for various crops. Thus, runoff is predicted separately for each subarea and combined to obtain the total runoff for the watershed. Division by land use increases accuracy and gives a much better physical description of the water balance.

Peak runoff rate is predicted with the equation

$$q_p = 200(DA)^{0.7}(CS)^{0.159}\,(Q)^{(0.917DA^{0.0166})}\,(LW)^{-0.187} \qquad [I-7]$$

where q_p is the peak runoff rate in ft^3/s; DA is the drainage area in mi^2; CS is the mainstem channel slope in ft/mi; Q is the daily runoff volume in in; and LW is the length-width ratio of the watershed. Data from 304 storms that occurred on 56 watersheds located in 14 states were used to develop equation [I-7]. Watershed areas ranged from 0.275 to 24 mi^2. Since these areas are larger than what is usually considered field-scale, the equation has variable exponents for DA and Q to accommodate areas down to 1 acre or less. These variable exponents simply prevent unreasonably high predictions for small areas.

Model Testing and Evaluation

The runoff model based on the SCS curve number technique has been tested on basins in Texas, Ohio, Georgia, Oklahoma, Nebraska, Arizona, New Mexico, West Virginia, Mississippi, Iowa, and Montana. Results of the tests are shown in tables I-3 through I-6. Table I-3 shows that the model generally approximates long-term water yield (average annual runoff) well. Also, average ET and percolation predictions seem realistic. The monthly R^2 values shown in table I-3 were obtained by comparing measured and predicted monthly runoff. Table I-4 contains statistics obtained by comparing measured and predicted individual runoff events. Although some of the R^2 values are lower than desirable, the standard deviations of the measured and predicted runoff are similar. This indicates that the model simulates runoff with a frequency distribution similar to that of the measured runoff, although the measured record is not duplicated precisely. There are many reasons for prediction errors. Some more important reasons are (1) the curve number system's inability to consider rainfall intensity, duration, or distribution; (2) the use of average values for temperature, solar radiation, and leaf area index instead of actual values; (3) lack of information on planting and tillage dates and incomplete soils descriptions; and (4) errors in rainfall and runoff data.

Table I-5 shows a comparison of measured and predicted percolation for watershed Z at Tifton, Ga. The measured values are actually subsurface flow measured at the watershed outlet. Of course, the predicted percolation is the amount of water that flows downward below the root zone. Considering these differences, the test can only indicate that the percolation model gave reasonable results.

Table I-6 contains measured and predicted percolation and evapotranspiration for watershed 115 and lysimeter Y103A near Coshocton, Ohio. The measured values were obtained from the lysimeter. Both the watershed and the lysimeter had the same crop each year. Close comparisons between measured and predicted values indicate satisfactory test results. Limited data prohibit percolation and ET model tests as extensive as those of the runoff model.

Infiltration Simulation from Breakpoint Rainfall Data

Whenever rainfall information is available in terms of actual time pattern of rainfall intensity or rate, the present understanding of soil water dynamics allows a significantly improved prediction of infiltration and runoff as compared with predictions based on amount of rain alone, such as the SCS curve number method discussed above.

Infiltration during rainfall is composed of two phases, as illustrated in figure I-3. At the beginning of a rain, the soil has an initial saturation not necessarily uniform with depth, but here assumed to be uniform in the (usually) small upper region which most affects infiltration. The saturation, S_i, is defined as

$$S_i = \frac{\theta_i}{\Phi} \qquad [I\text{-}8]$$

where θ_i is initial water content by volume, and Φ is porosity.

In the early stages of rainfall, the surface saturation increases from S_i to a maximum value S_0 (theoretically, $S_0 \rightarrow 1$), if the rainfall lasts long enough. For $S \geq S_0$, the soil controls surface flux, and the time when this begins is

Table I-3.—Runoff model test results (annual-monthly)

Watershed location	Drainage area	Length of record	Average measured P	Average measured Q	Annual predicted Q	Annual predicted ET	Annual predicted Percolation	Monthly R^2
	(mi^2)	(yr)	(in)	(in)	(in)	(in)	(in)	
SW-2, Riesel, Tex.	0.004	4	36.94	6.28	7.73	27.82	1.14	0.75
SW-12, Do.	.005	9	38.08	9.05	6.45	30.63	1.05	.72
Y-6, Do.	.025	9	38.08	5.72	7.34	30.10	.80	.86
Y-8, Do.	.032	9	38.08	6.72	6.14	31.22	.68	.65
21-H, Hastings, Nebr.	.006	13	22.83	3.40	3.69	19.16	.03	.41
3-H, Do.	.006	14	23.25	5.22	5.31	18.03	.01	.66
3-H, Do.	.006	9	23.45	4.75	5.41	17.98	.01	.68
P-1, Watkinsville, Ga.	.010	3	47.54	8.72	8.30	33.03	6.95	.46
P-2, Do.	.005	2	44.26	5.94	6.46	29.15	9.30	.53
104, Coshocton, Ohio	.002	8	38.04	.35	.61	32.39	4.75	.39
104, Do.	.002	4	35.40	.88	1.14	28.67	4.73	.92
129, Do.	.004	34	35.73	.83	.84	29.81	5.15	.33
130, Do.	.003	33	35.50	.95	.84	30.58	4.18	.45
132, Do.	.001	21	35.32	2.08	2.18	28.70	4.57	.51
115, Do.	.003	30	37.07	1.93	2.33	31.22	3.53	.56
110, Do.	.002	29	35.38	1.70	1.78	30.81	2.76	.43
118, Do.	.003	33	36.53	2.01	2.23	30.95	3.35	.53
106, Do.	.002	31	34.60	2.06	1.73	30.21	2.63	.33
192, Do.	.012	28	34.71	2.61	1.88	30.19	2.87	.48
R-5, Chickasha, Okla.	.037	8	30.14	1.76	1.95	27.31	.70	.73
R-7, Do.	.030	8	30.14	5.98	5.30	24.19	.41	.86
C-4, Do.	.047	9	32.21	3.45	2.79	29.33	.10	.59
C-5, Do.	.020	9	27.46	2.02	1.92	25.32	.09	.35
W-6, Cherokee, Okla.	.003	19	23.74	3.33	3.58	20.05	.17	.45
W-7, Do.	.003	19	23.74	3.59	3.59	20.03	.17	.53
W-13, Do.	.003	7	21.79	1.66	2.11	19.56	.02	.59
W-2, Guthrie, Okla.	.005	10	28.00	4.18	3.74	23.34	1.17	.85
W-1, Do.	.004	7	27.41	.67	.89	25.54	1.57	.46
W-2, Vega, Tex.	.150	5	18.54	.97	.80	17.91	.00	.27
W-1, Spur, Tex.	.018	19	20.07	1.93	2.05	18.03	.00	.67
W-2, Do.	.015	19	20.07	2.68	2.56	17.51	.00	.70
W-3, Do.	.018	18	20.10	1.55	1.72	18.35	.00	.72
63105, Lucky Hills, Ariz.	.001	10	11.15	1.14	1.00	10.32	.00	.84
01, Ft. Stanton, New Mex.	.038	10	14.40	.02	.05	14.96	.00	.24
02, Do.	.050	10	14.64	.00	.06	14.94	.00	.001
66001, Moorefield, W. Va.	.013	9	30.16	2.90	2.77	25.95	.93	.51
62014, Holly Spr., Miss.	.002	3	46.46	15.08	15.98	27.95	1.82	.80
62015, Do.	.002	3	33.73	9.48	9.65	24.22	1.19	.65
22003, Guthrie Ctr., Iowa	.019	4	24.31	1.16	1.10	22.59	.28	.74
Z, Tifton, Ga.	.001	6	50.65	2.96	3.04	41.25	7.17	.26
A, Sidney, Mont.	.003	3	14.50	1.70	1.32	13.65	.00	.72
W-3, Garland, Tex.	.016	8	41.02	9.14	8.92	30.66	.86	.84
W-1, Do.	.039	8	42.24	5.11	6.42	31.68	3.50	.86
W-3, Tyler, Tex.	.012	9	42.35	1.31	1.79	31.93	8.20	.36
W-5, Do.	.003	9	41.56	8.23	7.25	30.90	3.52	.58
W-4, Do.	.093	11	41.03	7.63	6.90	30.32	3.52	.60

Table I-4.—Runoff model test results (events)

Watershed location	R^2	Runoff volume Standard deviation		Peak runoff rate Mean		Peak runoff rate Standard deviation	
		Measured	Predicted	Measured	Predicted	Measured	Predicted
SW-2, Riesel, Tex.	0.85	0.74	0.74	2.16	1.74	3.12	2.39
SW-12, Do.	.69	.74	.55	1.72	1.46	2.53	2.34
Y-6, Do.	.90	.68	.84	5.34	5.36	7.48	8.78
Y-8, Do.	.64	.64	.50	5.90	5.76	8.76	8.29
21-H, Hastings, Nebr.	.46	.42	.37	1.88	1.37	2.66	2.23
3-H, Do.	.65	.47	.41	3.06	1.51	4.05	2.45
3-H, Do.	.55	.55	.56	3.06	2.96	4.05	4.64
P-1, Watkinsville, Ga.	.60	.61	.48	4.47	3.01	6.72	4.94
P-2, Do.	.64	.45	.44	1.88	1.48	2.63	2.49
104, Coshocton, Ohio	.28	.10	.11	.48	.28	.75	.43
104, Do.	.88	.42	.36	.48	.45	.75	1.09
129, Do.	.24	.25	.17	.57	.67	1.05	1.30
130, Do.	.29	.26	.16	.33	.33	.74	.58
132, Do.	.46	.33	.24	.08	.09	.11	.16
115, Do.	.55	.32	.29	.73	.57	1.31	1.16
110, Do.	.37	.32	.23	.34	.45	.82	.86
118, Do.	.52	.29	.23	.59	.60	1.10	1.02
106, Do.	.31	.22	.19	.54	.49	1.15	.87
192, Do.	.41	.37	.25	1.05	1.24	2.88	2.41
R-5, Chickasha, Okla.	.72	.35	.32	5.69	5.39	9.99	8.15
R-7, Do.	.86	.45	.44	7.65	6.59	11.50	10.81
C-4, Do.	.64	.42	.36	3.75	1.69	3.74	2.75
C-5, Do.	.46	.32	.29	1.45	1.07	1.48	2.00
W-6, Cherokee, Okla.	.35	.44	.41	1.32	1.48	1.68	2.28
W-7, Do.	.42	.49	.42	1.48	1.35	1.87	2.09
W-13, Do.	.59	.31	.33	1.33	1.21	1.74	1.95
W-2, Guthrie, Okla.	.67	.38	.36	1.65	1.61	2.45	2.09
W-I, Do.	.15	.22	.15	.44	.75	.55	.89
63105, Lucky Hills, Ariz.	.64	.23	.17	.29	.13	.46	.20
01, Ft. Stanton, New Mex.	.003	.02	.12	1.43	1.84	1.03	2.23
02, Do.	.10	.00	.13	1.00	1.49	.36	2.49
66001, Moorefield, W. Va.	.71	.70	1.54	.48	.64	.44	1.11
62014, Holly Spr., Miss.	.82	.74	.64	.89	1.39	1.21	1.65
62015, Do.	.62	.57	.50	.89	1.00	1.21	1.39
22003, Guthrie Ctr., Iowa	.44	.16	.17	.63	.95	.28	1.28
Z, Tifton, Ga.	.08	.23	.23	.48	.61	.54	.77
A, Sidney, Mont.	.68	.34	.28	.48	.38	.88	.59

Table I-5.—Percolation model results at Tifton, Ga., watershed Z

Year	Annual percolation (in)		Month	Average monthly percolation (in)	
	Measured	Predicted		Measured	Predicted
1970	17.90	19.74	1	2.15	2.67
1971	9.23	15.77	2	2.95	2.67
1972	11.72	12.13	3	1.52	1.58
1973	17.42	12.78	4	2.54	1.89
1974	8.41	10.60	5	.78	1.03
1975	14.83	9.81	6	.57	.71
Mean	13.25	13.47	7	.40	.13
Std. deviation	4.09	3.70	8	.98	.72
			9	.71	.24
			10	.00	.42
			11	.00	.25
			12	.72	1.16
			Mean	1.11	1.12
			Std. deviation	.97	.90

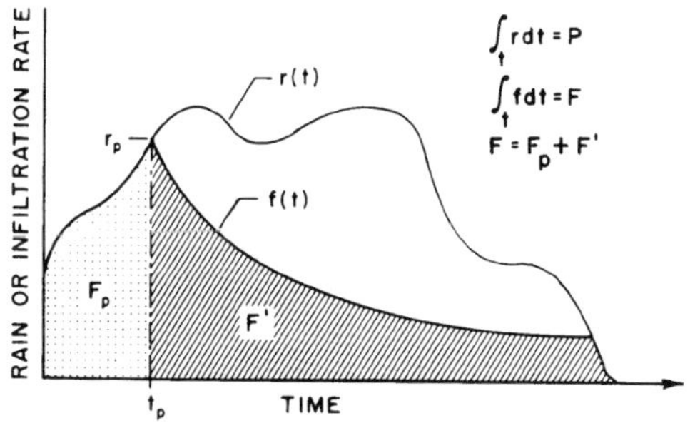

Figure I-3.—Definition diagram for infiltration model.

Table I-6.—Percolation and evapotranspiration model results at Coshocton, Ohio, watershed 115

		Annual				Average monthly				
		Percolation (in)		ET (in)			Percolation (in)		ET (in)	
Year	Crop	Measured[1/]	Predicted	Measured[1/]	Predicted	Month	Measured[1/]	Predicted	Measured[1/]	Predicted
1944	Meadow	5.06	2.85	30.37	28.80	1	1.11	1.39	0.77	0.76
1945	Corn	11.00	11.58	33.49	32.69	2	1.22	1.27	1.20	.96
1946	Wheat	4.99	4.28	38.30	36.63	3	1.80	1.75	2.21	1.72
1947	Meadow	9.97	6.42	33.79	35.99	4	1.34	.91	3.17	3.40
1948	Meadow	6.30	6.26	35.62	31.94	5	.41	.33	5.45	5.19
1949	Corn	5.54	8.46	34.95	32.87	6	.24	.31	5.40	4.96
1950	Wheat	12.88	10.42	36.46	37.58	7	.04	.16	5.52	4.84
1951	Meadow	13.70	11.96	34.85	31.55	8	.03	.02	4.61	4.27
1952	Meadow	9.14	10.17	34.93	31.02	9	.04	.02	2.93	2.39
1953	Corn	2.50	3.23	30.05	27.99	10	.03	.06	1.89	1.99
1954	Wheat	.64	.73	30.70	30.31	11	.05	.10	.97	1.31
1955	Meadow	5.27	8.01	35.36	29.77	12	.31	.48	.84	.93
1956	Meadow	6.37	7.77	39.57	40.60	Mean	.55	.57	2.91	2.73
1957	Corn	7.14	9.65	33.58	31.12	Std. dev.	.64	.61	1.90	1.71
1958	Wheat	3.70	1.33	37.02	39.12					
1959	Meadow	5.11	9.69	38.78	33.97					
1960	Meadow	3.86	6.80	38.61	35.96					
1961	Corn	7.23	5.56	33.84	31.58					
1962	Wheat	5.23	4.18	33.89	31.75					
Mean		6.61	6.81	34.96	33.22					
Std. dev.		3.40	3.38	2.77	3.51					

1/ Measured percolation is from a nearby lysimeter (Y103A) with land use same as watershed 115 each year. Period of record is 1944-1962.

called time of ponding, t_p. The dept21of rain which enters the soil prior to t_p is analogous to the SCS CN parameter I_a, and is here called F_p. Unlike I_a, F_p is a function of rainfall rate, r.

Rainfall infiltration models must describe both the occurrence of t_p and the shape of the subsequent infiltration curve, f(t), ($t > t_p$). Clearly, some storms will have short total time so that $t_r \leq t_p$.

The rainfall model based on soil water flow theory employed here is related to an infiltration curve describing infiltration from an instantaneous ponding condition. It is important to understand that the sudden ponding condition is distinct from the condition where water arrives at the soil surface at a certain rate, as for rainfall, yet the infiltration curves are mathematically related (5).

Others could be chosen, but this model employs the Green and Ampt (1) infiltration relation

$$K_s t = F - \Phi H_c(S_o - S_i)\ln\left[1 + \frac{F}{\Phi H_c(S_o - S_i)}\right] \qquad [I\text{-}9]$$

in which K_s = effective saturated conductivity, $[LT^{-1}]$
t = time from start of ponding, [T]
H_c = effective capillary tension, a soil parameter, [L], and
F = cumulative depth of infiltration, [L].

Derivation of this expression may be found elsewhere (1, 5). This relation will be used in different forms below to obtain expressions for ponding time and the inter-storm infiltration-rate curve.

Ponding Time

The key to predicting ponding time is a relation of infiltrated depth, infiltration rate, and time. The relationship used is derived from experiments and theory which indicate that

$$F_p = \int_0^{t_p} r(t)dt = \int_0^{t(r_p)} f_o(t)dt \qquad [I\text{-}10]$$

in which r is rainfall rate, r_p is rainfall rate at t_p. In this equation, f_o is the infiltration rate curve for the instantaneous ponding condition ($r = \infty$), which equals dF/dt, with F(t) defined by equation [I-9] above. In other words, for the rainfall rate, r_p, the depth of water, F_p, infiltrated at t_p is equal to the depth of water infiltrated from t = 0 to the time at which $f_o = r_p$ for the case of infiltration from sudden ponding.

The Green and Ampt (1) model comes originally from an assumption of a piston-type movement of soil water downward in the soil, as in figure I-4. This model can be used with equation [I-10] to derive an expression for ponding time. Clearly, by continuity,

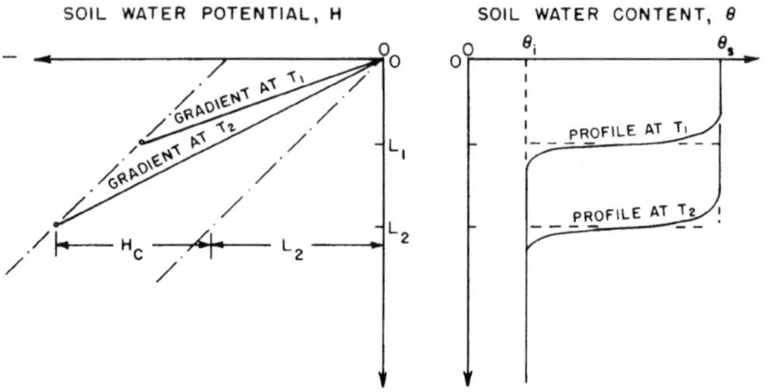

Figure I-4.—Conceptual assumptions in the Green-Ampt (1) infiltration model.

$$F = L\Phi(S_o - S_i) \qquad [I\text{-}11]$$

where L is the depth of wetting. Total head across the wetting front is $-(H_c + L)$, and by Darcy's law, infiltration rate, f, is

$$f = K_s\left(\frac{H_c + L}{L}\right) \quad . \qquad [I\text{-}12]$$

Solving equation [I-11] for L, and substituting into equation [I-12]

$$f = K_s\left[\frac{H_c\Phi(S_o - S_i) + F}{F}\right] \qquad [I\text{-}13]$$

This can be solved for F(f), as

$$F = \frac{H_c\Phi(S_o - S_i)K_s}{(f - K_s)} \quad . \qquad [I\text{-}14]$$

For a ponding time expression, equation [I-10] can be written as

$$F_p = \int_0^{t(r_p)} f dt = F(r_p)$$

so that ponding depth is estimated as

$$F_p = \frac{H_c \Phi(S_o - S_i)K_s}{(r_p - K_s)} = \frac{GDK_s}{r-K_s} \quad [I\text{-}15]$$

in which G is used for H_c, and D represents $\Phi(S_o - S_i)$.

This expression is used to find when ponding is expected to occur in a histogram of rainfall rate pulses of variable height (as with breakpoint rainfall data). If ponding occurs within a pulse $[t_i < t_p < t_{i+1}]$, interpolation is used.

Infiltration Curve

To obtain an expression for infiltration within a breakpoint interval where $t > t_p$, we start with equation [I-9], and as above for simplicity let

$$G = H_c,$$

and

$$D = (S_o - S_i)\Phi,$$

so that

$$K_s t = F - GD \ln\left(1 + \frac{F}{GD}\right) . \quad [I\text{-}16]$$

This expression may be derived from equation [I-13] by setting $f = dF/dt$ and integrating. We assume a finite difference perturbation of equation [I-16] to be equally correct, so that

$$K_s(t + \Delta t) = F + \Delta F - GD \ln\left(1 + \frac{F + \Delta F}{GD}\right) . \quad [I\text{-}17]$$

Subtracting equation [I-16] from equation [I-17] and rearranging the logarithmic expression, we form the difference expression

$$K_s \Delta t = \Delta F - GD \ln\left(1 + \frac{\Delta F}{GD + F}\right) . \quad [I\text{-}18]$$

Then using the first term of the following series approximation for the natural logarithm,

$$\ln(1 + x) = 2\left[\frac{x}{2 + x} + \frac{1}{3}\left(\frac{x}{2 + x}\right)^3 + \ldots\right],$$

we can solve for ΔF as

$$\Delta F = \sqrt{2K_s\Delta t(GD + F) + (F - \frac{K_s}{2}\Delta t)^2} - (F - \frac{K_s}{2}\Delta t) \quad [I\text{-}19]$$

with an approximation error of approximately 8% (3).

Let $A = K_s \Delta t/2$, so that

$$\Delta F = \sqrt{4A[GD + F] + (F - A)^2} + A - F \; . \qquad [I-20]$$

Mean infiltration rate can be calculated for time interval Δt_i as

$$\bar{f}_i = \frac{\Delta F_i}{\Delta t_i} \; . \qquad [I-21]$$

Runoff during interval i is $q_i = r_i - \bar{f}_i$. Calculation proceeds through the storm, with F being updated at each interval i as

$$F_{i+1} = F_i + \Delta F_i \qquad [I-22]$$

in any interval where $t > t_p$. Where $r_i \Delta t < \Delta t \, \Delta F_i$ from equation [I-18], then

$$F_{i+1} = F_i + r_i \Delta t \; . \qquad [I-23]$$

Total runoff for a storm having n intervals is simply

$$Q = \sum_{i=1}^{n} q_i \Delta t_i \, . \qquad [I-24]$$

Adjustments for Hourly Data

On the basis of rainfall record analysis, it has been found (2) that storm intensity changes significantly during intervals shorter than 60 minutes, and that peak intensity will be significantly biased (reduced) by hourly data. Therefore, employment of hourly data such as are commonly available through the National Weather Service (formerly U.S. Weather Bureau) suggests an adjustment of the infiltration procedure. Sufficient research has not been completed to know an optimum adjustment, nor how the adjustment should be changed to reflect various climatic zones. In this model, the procedure adopted in the interim is to base predicted ponding times for hourly data on 133% of the hourly intensity for the early storm periods, still using equation [I-15]. Storm EI (see volume III, chapter 1 for definition) is calculated using a 30-minute maximum intensity which is assumed to be twice the maximum hourly intensity.

Multiple Storms

More than one storm is assumed to occur on a day when a rainfall hietus of t_e = 180 minutes is found. In this case, $D_2 = S_o - S_i$ is estimated for the subquent storm as

$$D_2 = \Phi[0.9 - (t_e/180)0.05]. \qquad [I-25]$$

This estimates a rather wet initial condition for the next storm, and Q_2 for the second storm is added to Q_1 for the first to get the daily Q. The same process can be used if more than two storms occur.

Estimating Runoff Peak Rates

The breakpoint rainfall infiltration simulation produces a histogram of excess rainfall rates which are sequences of time intervals and associated rainfall excess rates. These rates are typically much larger than that seen as the output runoff rate from a field or small watershed. To estimate peak runoff rates from field areas, we can use kinematic surface water flow equations. The description here is brief, and a more detailed explanation may be obtained by referring to Woolhiser (10).

Runoff begins when free surface water is generated from the excess of rainfall rate above the infiltration rate. A certain "lag" period must occur, however, in which rate of rainfall excess far exceeds the runoff rate at the catchment outlet. If rainfall excess rate is uniform and lasts long enough, "equilibrium" will eventually occur when the two rates are equal.

Since rainfall excess usually varies rather abruptly during a storm and rarely lasts long enough, equilibrium flow is practically a hypothetical concept. Shallow water flow hydraulics allows estimation, nevertheless, of peak runoff rates.

Figure I-5 illustrates some of the aspects of the flow regimes occurring during surface water flow response to rainfall. This figure presents, graphically, the movement of the characteristic "waves" which originate from the

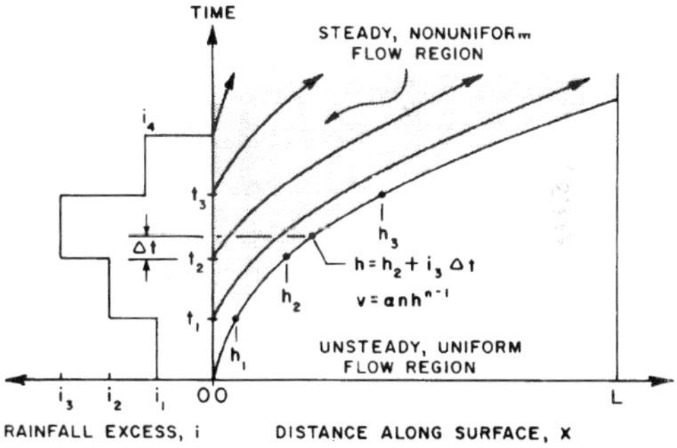

Figure I-5.—Illustration of flow regimes occurring during serface-flow response to rainfall.

upstream point (x = 0), and illustrates the usual case where the peak rainfall excess occurs prior to the "equilibrium" time. A hypothetical pattern of rainfall excess rates is illustrated on the left of the ordinate.

The curve from x = 0, t = 0 is called the upstream characteristic. To the right of this line, flow is unsteady (rising), but uniform. To the left of this characteristic, flow is steady but nonuniform, varying with location along the surface.

Along any characteristic, between points (2) and (3) for example, the depth is

$$h_j = h_{j-1} + i_j \Delta t_j \qquad [I\text{-}26]$$

where i_j is rainfall excess rate during interval j, Δt is time from start of interval j, and h_{j-1} is depth on the characteristic at time t_{j-1}. The velocity of the characteristics is

$$v = m\alpha h^{m-1} = dx/dt \qquad [I\text{-}27]$$

where v = wave celerity (not flow velocity)
m = uniform flow exponent (1.5 for the Chezy roughness law)
α = uniform flow coefficient (= $C\sqrt{S}$ for the Chezy roughness law)
S = plane slope, and C = Chezy roughness coefficient.

Combining equations [I-26] and [I-27] and integrating for the characteristic starting at t_{j-1}, we have

$$x_j = m\alpha \int_0^{\Delta t} (h_{j-1} + i_j st)^{m-1} ds + x_{j-i}$$

$$= \frac{\alpha}{i_j} (h_{j-1} + i_j \Delta t)^m + x_{j-1} \quad . \qquad [I\text{-}28]$$

This gives us the position of the "wave" from the upstream edge at any time. This equation may be applied successively with all increments of i(t) to obtain the distance the peak (or any other disturbance) moves. In any case, the peak runoff rate may be estimated from the greatest depth h reached at x = L according to equation [I-26] for the fastest characteristic since at all points for kinematic flow,

$$q = \alpha\, h^m \quad . \qquad [I\text{-}29]$$

This estimation procedure is based on the peak flow occurring with monotonic rise of depth, since if flow recession occurs prior to a second rainfall burst, recession calculations are necessary. In a very few cases, this could cause underestimation of the flow peak.

To estimate the peak outflow from a complex rainfall pattern, we must choose the characteristic path along which (in time) the largest rates of rainfall excess occur, and thus, the largest depth h at the downstream edge of the surface or watershed outlet. Obviously, this characteristic would usually include the time in which the largest rainfall excess occurs, plus the intervals with largest i before and after, necessary for the characteristic to traverse the distance L (figure I-5). One may estimate the peak, therefore, by looking at the characteristic incremental depth h as given in equation [I-26] for the peak excess interval, and adding adjacent intervals (of largest positive excess rate) to each side of the peak, thus choosing the fastest characteristic for which x from equation [I-28] equals or exceeds the length L.

In estimating peak runoff rates using kinematic surface water flow equations, complex slopes can be represented by hydraulically equivalent uniform slopes. If the length is broken into N regions of different slope and roughness, as illustrated in figure I-6, the equivalent single plane values can be determined. For each segment or sub-plane j, where j = 1, N, there is an α_j as in equation [I-27]

$$\alpha_j = C_j\sqrt{S_j} \quad . \qquad [I-30]$$

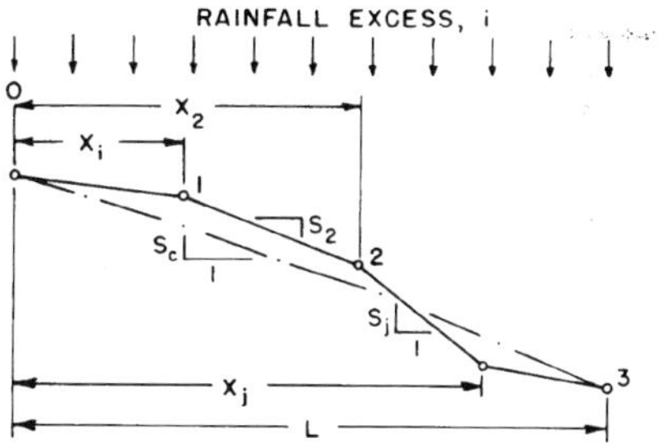

Figure I-6.—Representation of a complex slope in terms of a single equivalent plane.

For the Manning roughness law, C is the same as 1.49/n where n is Manning's roughness coefficient. The objective is to determine the α_c for a single plane which best represents the composite hydraulic response of the set of N planes. From equation [I-30], if S_c is the overall slope as in figure I-6, then composite α_c will specify a C_c and vice versa.

Research by Wu (11) indicates that the best hydraulically equivalent single plane is the one that gives equilibrium surface detention storage equal to that of the set of different planes. Detention storage, A, (10) is

$$A = \int_0^L h dx \qquad [I-31]$$

where h is local surface water depth. Thus equation [I-31] is the equivalence criteria. Equilibrium discharge for any plane of length x is, by continuity, ix, where i is rainfall excess. From equation [I-29], then

$$ix = \alpha\, h(x)^m \qquad [I\text{-}32]$$

and from which equilibrium depth is

$$h = \left(\frac{ix}{\alpha}\right)^{1/m} . \qquad [I\text{-}33]$$

Equating storage on the equivalent single plane to that on the set of planes, we have $A = \sum_{j=1}^{n} A_j$. This combined with equations [I-31] and [I-32], yields

$$\int_0^L \left(\frac{ix}{\alpha_c}\right)^{1/m} dx = \int_0^{x_1} \left(\frac{ix}{\alpha_1}\right)^{1/m} dx + \int_{x_1}^{x_2} \left(\frac{ix}{\alpha_2}\right)^{1/m} dx + \ldots \int_{x_{N-1}}^{x_N} \left(\frac{ix}{\alpha_N}\right)^{1/m} dx$$

or

$$\frac{1}{b}\left(\frac{i}{\alpha_c}\right)^{1/m} L^b = \left[\frac{1}{b}\left(\frac{i}{\alpha_1}\right)^{1/m} x_1^{\,b} + \left(\frac{i}{\alpha_2}\right)^{1/m}\left(x_2^{\,b} - x_1^{\,b}\right) + \ldots \left(\frac{i}{\alpha_N}\right)^{1/m}\left(L^b - x_{N-1}^{\,b}\right)\right] \quad [I\text{-}34]$$

where $b = (m + 1)/m$. Dividing both sides by $(\frac{1}{b})\ (i)^{1/m}\ (L)^b$ gives

$$\left(\frac{1}{\alpha_c}\right)^{1/m} = \sum_{j=1}^{N} \left(\frac{1}{\alpha_j}\right)^{1/m} \frac{\left(x_j^{\,b} - x_{j-1}^{\,b}\right)}{L^b} \qquad [I\text{-}35]$$

and rearranging,

$$C_c\sqrt{S_c} = \alpha_c = \left[\frac{L^b}{\sum_{j=1}^{N} \frac{(x_j^{\,b} - x_{j-1}^{\,b})}{\alpha_j^{1/m}}}\right]^m \qquad [I\text{-}36]$$

where $x_0 = 0$ and $x_N = L$.

Evaluation Results

Table I-7 presents results of model testing performed on those watersheds where breakpoint rainfall and runoff records were obtained, including two watersheds where lysimeter data provided estimates of amounts of seepage below

Table I-7.—Comparison of observed and simulated runoff and percolation for selected watersheds using hydrology option 2

Watershed location and number	Area	Record length	Observed values: Average annual precip.	Observed values: Average annual runoff	Observed values: Peak discharge Mean	Observed values: Peak discharge Std. dev.	Observed values: Average annual percolation	Simulated values: Average annual runoff	Simulated values: Peak dishcarge Mean	Simulated values: Peak dishcarge Std. dev.	Simulated values: Average annual percolation	Annual mean runoff event r^2
	(acre)	(yr)	(in)	(in)	(in/hr)	(in/hr)	(in)	(in)	(in/hr)	(in/hr)	(in)	
Watkinsville, Ga.												
P-1	6.7	3 1/2	46.7	7.03	0.25	0.71	NA 1/	0.82	0.46	0.96	6.40	0.735
P-2	3.2	2 1/2	43.2	6.29	.26	.62	NA	5.54	.28	1.05	7.86	.898
Chickasha, Okla.												
R-5	23.7	9	28.2	1.40	.035	.19	NA	.99	.051	.33	2.65	.644
R-7	19.2	9	28.2	4.77	.19	.48	NA	3.69	.24	.66	1.38	.733
Cherokee, Okla.												
W-7	2.0	19	23.2	3.53	.23	.64	NA	3.51	.27	.67	.31	.665
Walnut Gulch, Ariz.												
LH-5	.6	11	10.7	1.04	.33	.77	0	.90	.22	.68	.0	.81
Tifton, Ga.												
Z	.85	5	47.7	2.71	.19	.43	14.5	2.18	.19	.50	12.1	.68
Coshocton, Ohio												
129	2.71	19	2/ 34.3	0.74	.026	.16	2/ 8.5	.60	.038	.23	5.7	* 3/
Hastings, Nebr.												
3-H	3.77	26	22.2	3.12	.346	.94	NA	3.10	.308	.79	.28	.63

1/ Seepage data not available for most watersheds; NA = not available.
2/ Seepage data from lysimeters which indicate 41 in annual precipitation, as compared with 34.3 in. in the data for watershed 129.
3/ Correlation for most years invalid since only 1 or 2 events measured per year.

the root zone. Complicating the ability to accurately predict runoff on agricultural watersheds, as mentioned above, is the occurrence of often unrecorded cultivation practices which severely modify the soil's infiltration properties. This is demonstrated in the relative predictive accuracy for the cultivated watershed at Tifton, Ga., and the rangeland at Lucky Hills, Ariz. In addition, the data include many instances of errors in time, such as major storms recorded by only two breakpoints, or records of runoff attributed to days where the major portion of the rainfall was a day earlier. Other common data errors include blank periods where major storms occur with no recorded runoff, and runoff peaks with greater rates than the associated rainfall rate. Curiously, most of the examples in this table show correlation coefficients for daily runoff in the 0.80 to 0.90 range, yet with occasional years having contrastingly low r^2 of 0.1 to 0.2.

EVAPOTRANSPIRATION AND SOIL WATER ROUTING

From either infiltration submodel, water that enters the soil, F, becomes either evapotranspiration, storage, or seepage below the root zone. A daily time interval is used between storm events, and the components of the water balance equation are evaluated. In equation form,

$$SM_i = SM_{i-1} + F_i - ET_i - O_i + M_i \qquad [I\text{-}37]$$

where
- F_i = infiltration on day i
- ET_i = plant and soil evapotranspiration on day i
- O_i = seepage below the root zone on day i
- M_i = snowmelt amount on day i
- SM = soil water storage in the root zone.

Snowmelt

A simple snow accumulation and snowmelt equation is used by the model taken from Stewart and others (6). For all those days where precipitation occurs when the temperature is less than 0°C, that precipitation is stored in the form of snow. When snow storage exists and the temperature, T, is above 0° C, snowmelt occurs, and input to the soil at the surface is calculated by

$$M_i = 0.18T \qquad [I\text{-}38]$$

unless M is greater than the amount of surface snow. Although this model is quite simplistic, it does help account for spring melt input, and would be difficult to improve without detailed daily temperature and radiation information.

Evapotranspiration

As illustrated in figure I-7, the soil water balance model considers both soil and plant evaporation losses, and treats the growth of plant leaf area and depth of root extraction explicitly. The evapotranspiration (ET) component of the runoff model is taken from Ritchie (4). To compute potential evaporation, the model uses the equation

$$E_0 = \frac{1.28\ \Delta\ H_0}{\Delta + \gamma} \qquad [I\text{-}39]$$

where E_0 is the potential evaporation; Δ is the slope of the saturation vapor pressure curve at the mean air temperature; H_0 is the net solar radiation; and γ is a psychrometric constant. Δ is computed with the equation

$$\Delta = \frac{5304}{T^2}\ e^{(21.255\ -\ 5304/T)}$$

where T is the daily temperature in degrees kelvin. H_0 is calculated with the equation

$$H_0 = \frac{(1 - \lambda)(R)}{58.3} \qquad [I\text{-}41]$$

where R is the daily solar radiation in langleys and λ is the albedo for solar radiation.

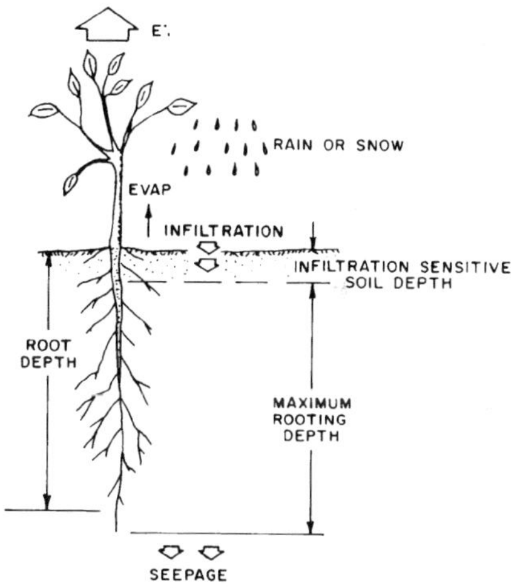

Figure I-7.—Schematic representation of the water-balance model.

Soil Evaporation

The model computes soil and plant evaporation separately. Potential daily soil evaporation is predicted with the equation

$$E_{so} = E_o\ e^{-0.4\ LAI} \qquad [I-42]$$

where E_{so} is the potential evaporation at the soil surface and LAI is the leaf area index defined as the area of plant leaves relative to the soil surface area. Actual soil evaporation is computed in two stages. In the first stage, soil evaporation is limited only by the energy available at the surface, and thus is equal to the potential soil evaporation. When the accumulated soil evaporation exceeds the stage one upper limit, the stage two evaporative process begins. Here the stage one upper limit is estimated with the equation

$$U = 9\ (\alpha_s - 3)^{0.42} \qquad [I-43]$$

where U is the stage one upper limit in mm and α_s is a soil evaporation parameter dependent on soil water transmission characteristics (ranges from about 3.3 to 5.5 mm/d$^{1/2}$). Ritchie (4) suggests a value of 4.5 for loamy soils, 3.5 for clays, and 3.3 for sands.

Stage-two daily soil evaporation is predicted with the equation

$$E_s = \alpha_s\ [t^{1/2} - (t - 1)^{1/2}] \qquad [I-44]$$

where E_s is the soil evaporation for day t, and t is the number of days since stage two evaporation began.

Plant Transpiration

Plant evaporation is computed with the equations

$$E_p = \frac{(E_o)(LAI)}{3},\ 0 \leq LAI \leq 3 \qquad [I-45]$$

$$E_p = E_o - E_s\ ,\ LAI > 3\ . \qquad [I-46]$$

If soil moisture is limited, plant evaporation is reduced with the equation

$$E_{pL} = \frac{(E_p)(SM)}{0.25FC}\ ,\ SM \leq\ 0.25FC \qquad [I-47]$$

where Ep is normal plant evaporation: E_{pL} is plant evaporation reduced by limited SM, and FC is the field capacity of the soil. Evapotranspiration, the sum of plant and soil evaporation, cannot exceed E_o.

Drought

When soil moisture falls below 15 bar amount (estimated), plant growth is stopped by holding leaf area index constant until water becomes available.

This allows an interaction between rainfall data and leaf area index description, to account in an approximate manner for drought conditions.

PERCOLATION

The model uses a soil storage routing technique to predict flow through the root zone (8). When the SCS (7) curve number method is used, the root zone is divided into seven layers or storages for routing. Root-zone depth is usually estimated to be three feet, although it may vary with various crops and soils. The routing equation is

$$O = \sigma\left(F + \frac{ST}{\Delta t}\right), \left(F + \frac{ST}{\Delta t}\right) > FC \qquad [I-48]$$

where F is the infiltration or inflow rate; ST is the storage volume; σ is the storage coefficient; and Δt is the routing interval (1 day). If inflow plus storage does not exceed field capacity, FC, percolation is not predicted to occur. The storage coefficient is a function of the travel time through the storage expressed by the equation

$$\sigma = \frac{2\Delta t}{2t + \Delta t} \qquad [I-49]$$

where t is the travel time through a storage. Travel time is estimated with the equation

$$t = \frac{SM - FC}{r_c} \qquad [I-50]$$

where SM is soil water storage, and r_c is the saturated conductivity of the soil.

Besides percolation losses, each soil storage is subject to ET losses. Therefore, the daily predicted ET must be distributed properly through the storages. A model for simulating root growth is used for this purpose. The water-use rate as a function of root depth is expressed by the equation

$$u = u_o\ e^{-4.16RD} \qquad [I-51]$$

where u is the water-use rate by the crop at depth, RD, and u_o is the rate at the surface. The total water use within any depth can be computed by integrating equation [I-51] to obtain

$$ET = \frac{u_o}{4.16}\ (1 - e^{-4.16RD}) \quad . \qquad [I-52]$$

The value of u_o is determined for the root depth each day, and the water use in each storage is computed with the equation

$$uw_i = \frac{v_o}{4.16}\left(e^{-4.16RD_{i-1}} - e^{-4.16RD_i}\right) \qquad [I-53]$$

where uw_i is the water use in storage, i, and RD_{i-1} and RD_i are the depths at the top and bottom of storage, i.

When the breakpoint infiltration model is used for runoff calculations, the soil water movement and percolation calculation involves only two storage elements, a surface soil zone, and a root soil zone. The surface soil zone is subject to soil evaporation from the evapotranspiration model, plus a portion of the plant root extraction. It is the region of the soil which determines initial conditions to which the infiltration model is sensitive. The lower zone is subject to root extraction during the growing season. A root growth model is used in this option which simulates relative root depth proportional to relative leaf area index.

Water moves from the upper soil zone to the root zone as a function of the positive difference in saturation between the two zones as:

$$q_s = C_s S_s^3 (S_s - S_p) \Phi D_s, \quad (S_s > S_p), \qquad [I-54]$$

in which q_s = daily water movement from surface to root zone
C_s = coefficient (normally 0.1)
S_s = saturation by volume in surface zone
S_p = saturation by volume in root zone
Φ = porosity
D_s = depth of surface zone (2 to 5 cm)

This is designed as a crude analogy to Darcy's law, with $C_s S_s^3$ approximating the relation between conductivity and saturation.

Seepage from the root zone is predicted to occur when S_p exceeds field capacity, and is estimated as the daily excess of S_s over field capacity. Root extraction occurs from both surface and root zones in proportion to the relative root depth, which varies with leaf area index up to the maximum depth. Thus, if root depth = $2D_s$, evapotranspiration water is taken equally from D_s and root zone, D. Total soil water storage UL is estimated as porosity times surface depth, D_p, plus field capacity in the root zone. Field capacity is a ratio, F_c, of porosity, so that

$$UL = \Phi D_s + f_c \cdot D_p . \qquad [I-55]$$

REFERENCES

(1) Green, W. A., and G. A. Ampt.
1911. Studies on soil physics, I. The flow of air and water thru soils. Journal of Agricultural Science, 4:1-24.

(2) Hershfield, D. M.
1961. Rainfall frequency atlas of the United States. Weather Bureau Technical Paper No. 40, 115 pp. (Weather Bureau is now the National Weather Service.)

(3) Li, R. M., M. A. Stevens, and D. B. Simons.
1976. Solutions to Green-Ampt infiltration equation. Journal of Irrigation and Drainage Division, American Society of Civil Engineers, 102 (IR2):239-248.

(4) Ritchie, J. T.
1972. A model for predicting evaporation from a row crop with incomplete cover. Water Resources Research 8(5):1204-1213.

(5) Smith, R. E., and J. Y. Parlange.
1978. A parameter-efficient hydrologic infiltration model. Water Resources Research 14(3):533-538.

(6) Stewart, B. A., and others.
1975. Control of water pollution from cropland. Vol. 1, A manual for guideline development. U. S. Department of Agriculture, Agricultural Research Service, Headquarters, ARS-H-5-1. (Series discontinued; Agricultural Research Service now Science and Education Administration-Agricultural Research.)

(7) U.S. Department of Agriculture, Soil Conservation Service.
1972. National engineering handbook, Hydrology, Section 4, 1972. 548 pp.

(8) Williams, J. R., and R. W. Hann.
1978. Optimal operation of large agricultural watersheds with water quality constraints. Texas A&M University, Texas Water Resources Institute, TR-96. 152 pp.

(9) ------, and W. V. LaSeur.
1976. Water yield model using SCS curve numbers. Journal of the Hydraulics Division, American Society of Civil Engineers 102(HY9): 1241-1253.

(10) Woolhiser, D. A.
1974. Unsteady free-surface flow problems. Ch. 12, Proceeding of the Institute on Unsteady Flow in Open Channels, Colorado State University, July 17-28, Water Resources Publications, Ft. Collins, Colo.

(11) Wu, Y. H.
1978. Effects of roughness and its spatial variability on runoff hydrographs. PhD Dissertation, Colorado State University, Civil Engineering Department Report No. CED77-78 YHW7, Fort Collins, Colo., 174 pp.

Chapter 3. A MODEL TO ESTIMATE SEDIMENT YIELD FROM FIELD-SIZED AREAS: DEVELOPMENT OF MODEL

G. R. Foster, L. J. Lane, J. D. Nowlin, J. M. Laflen, and R. A. Young[1/]

INTRODUCTION

Estimates of erosion and sediment yield on field-sized areas are needed to wisely select best management practices to control erosion for maintenance of soil productivity and to control sediment yield for prevention of excessive degradation of water quality. A field is a typical management unit for farmers. The selection of a management practice is usually based on site-specific conditions. Soil conservationists have used the Universal Soil Loss Equation (USLE) (31) for several years to select practices specifically tailored to a given farmer's situation. Assuming that sediment yield tolerance for maintenance of water quality will be established for given local areas, best management practices can then be selected based on a given farmer's needs and the tolerable water loading for fields in his area using a model such as the one described herein (8).

Sediment yield is a function of detachment of soil particles and the subsequent transport of these particles (sediment). On a given field, either detachment or sediment transport capacity may limit sediment yield depending on topography, soil characteristics, cover, and rainfall/runoff rates and amounts. Control of sediment yield by detachment or transport can change from season to season, from storm to storm, and even within a storm. The relationship for detachment is different from the one for transport so that they cannot be lumped together into a single equation. Since detachment and transport for each storm are best considered separately, lumped equations such as the USLE (an erosion equation), or Williams' (29) modified USLE (a flow transport, sediment yield equation) cannot give the best results over a broad range of conditions on field-sized areas. Furthermore, the interrelation between detachment and transport is nonlinear and interactive for each storm, which prevents using separate equations to linearly accumulate amount of detached sediment or sediment transport capacity over several storms. Therefore, to simulate erosion and sediment yield on an individual storm basis and to satisfy the need for a continuous simulation model, a rather fundamental approach was selected where separate equations are used for soil detachment and sediment transport.

A number of fundamentally based models (1, 20) compute detachment and transport at various times during the runoff event. While these models are powerful, their excessive use of computer time practically prohibits simulating 20 to 30 years of record. The model described herein uses characteristic

1/ Hydraulic engineer, USDA-SEA-AR, Lafayette, Ind.; hydrologist, USDA-SEA-AR, Tucson, Ariz.; computer programmer, Agricultural Engineering Department, Purdue University, Lafayette, Ind.; agricultural engineer, USDA-SEA-AR, Ames, Iowa; and agricultural engineer, USDA-SEA-AR, Morris, Minn.

rainfall and runoff factors for a storm to compute detachment and sediment transport for that storm. In terms of computational time, this amounts to a single time step for models which simulate over the entire runoff event.

The model is intended to be useful without calibration or collection of research data to determine parameter values. Therefore, established relationships such as the USLE were modified and used in the model.

OVERVIEW OF THE MODEL

Every model is a representation and a simplification of the prototype. Various techniques, including planes and channels (20), square grids (1), converging sections (28), and stream tubes (24) have been used. Most erosion/sediment yield models have adequate degrees of freedom to fit observed data. Some models, depending on their representation scheme, distort parameter values more than do others. Distortion of parameter values greatly reduces the transferability of parameter values from one area to another (18). An objective in this model development was to represent the field in a way that minimizes parameter distortion. Hydrologic input to the erosion/sediment yield component consists of rainfall amount, rainfall erosivity (EI), runoff volume, and peak rate of runoff. These terms drive soil detachment and subsequent transport in overland and open channel flow.

Overland flow, channel flow, and impoundment (pond) elements are used to represent major features of a field. The user selects the best combination of elements and enters the appropriate sequence number according to table I-8. The model (computer program) calls the elements in the proper sequence. Typical systems that the model can represent are illustrated in figure I-8.

Table I-8.—Possible elements and their calling sequence used to represent field-sized area

Sequence number	Elements and their sequence
1	Overland
2	Overland-Pond
3	Overland-Channel
4	Overland-Channel-Channel
5	Overland-Channel-Pond
6	Overland-Channel-Channel-Pond

Computations begin in the uppermost element, which is always the overland flow element, and proceed downstream. Sediment concentration (for each particle type) is the output from each element which becomes the input to the next element in the sequence.

BASIC CONCEPTS

Basic Equations

Sediment load is assumed to be limited by either the amount of sediment made available by detachment or by transport capacity (11). Also, quasisteady state is assumed so that a rainfall and a runoff rate characteristic of each storm can be used in the computations. Sediment movement downslope obeys continuity of mass expressed by:

$$\frac{dq_s}{dx} = D_L + D_F \qquad [I\text{-}56]$$

where q_s = sediment load per unit width per unit time, x = distance, D_L = lateral inflow of sediment (mass/unit area/unit time), and D_F = detachment or deposition by flow (mass/unit area/unit time). The assumption of quasisteady state allows deletion of time terms from equation [I-56]. The major sequence of computations is illustrated in figure I-9.

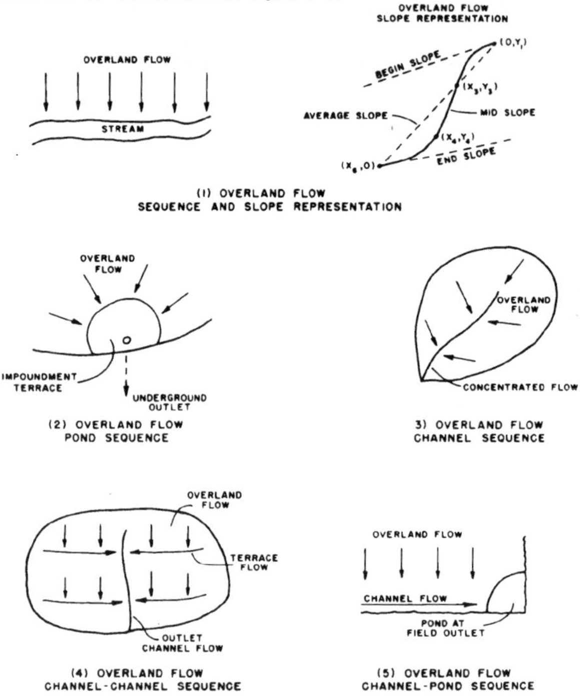

Figure I-8.—Schematic representation of typical field systems in the field-scale erosion/sediment yield model.

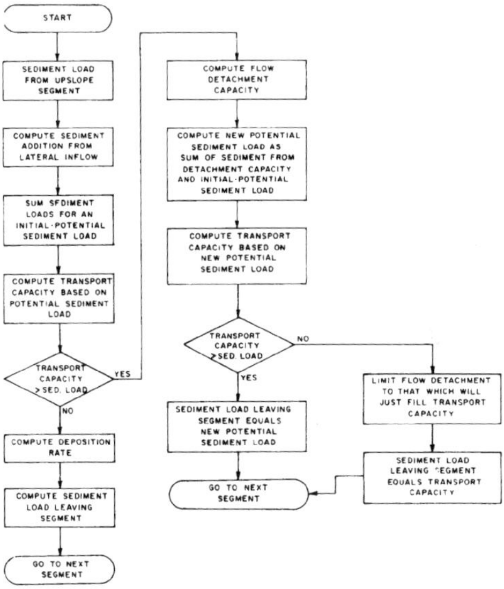

Figure I-9.—Flow chart for detachment-transport-deposition computations within a segment of overland flow or concentrated flow elements.

Lateral sediment inflow is from interrill erosion on overland flow elements, or it is from overland flow (or a channel, if two channel segments are in the sequence) for the channel elements. Flow in rills on overland flow areas or in channels transports the sediment load downstream. Lateral sediment inflow is assumed regardless of whether the flow is detaching or depositing.

For each segment, either on the overland flow element or in a channel, the model computes an initial potential sediment load which is the sum of the sediment load from the immediate upslope segment plus that added by lateral inflow within the segment. If this potential load is less than the flow's transport capacity, detachment occurs at the lesser of the detachment capacity rate or the rate which will just fill transport capacity. When detachment by flow occurs, it adds particles, having the particle-size distribution for detached sediment given as input. No sorting is allowed during detachment.

If the initial potential sediment load is greater than the transport capacity, deposition is assumed to occur at the rate of:

$$D = \alpha\ (T_c - q_s) \qquad [I\text{-}57]$$

where D = deposition rate (mass/unit area/unit time), α = a first order reaction coefficient ($length^{-1}$), and T_c = transport capacity (mass/unit width/unit time). The coefficient α is estimated from:

$$\alpha = \varepsilon \frac{V_s}{q_L x} \qquad [I\text{-}58]$$

where ε = 0.5 for overland flow (5), and 1.0 for channel flow (7), V_s= particle fall velocity, and $q_L x = q_w$ = discharge per unit width (volume/unit width/unit time). Fall velocity is estimated assuming standard drag relationships for a sphere of a given diameter and density falling in still water.

Detachment-Deposition Limiting Cases

Four possible cases may exist for a segment: (1) Deposition may occur over the entire segment; (2) detachment by flow in the upper end and deposition in the lower end may (but not necessarily) occur when transport capacity decreases in a segment; (3) deposition on the upper end and detachment by flow in the lower end may (but not necessarily) occur when transport capacity increases within the segment; (4) detachment by flow may occur all along the segment.

Case 1 occurs when $T_c < q_s$ all along the segment. Where deposition occurs over the entire segment length, deposition rate is:

$$D = [\phi/(1+\phi)](dT_c/dx - D_L)\left[1 - (x_u/x)^{1+\phi}\right] + D_u(x_u/x)^{1+\phi} \qquad [I\text{-}59]$$

where

$$\phi = \varepsilon V_s/q_L \qquad [I\text{-}60]$$

where dT_c/dx is assumed constant over the segment and D_u = deposition rate at x_u. The deposition rate D_u may be estimated from:

$$D_u = \alpha\ (T_{cu} - q_{su}) \qquad [I\text{-}61]$$

where T_{cu} and q_{su}= respectively, the transport capacity and sediment load at x_u. Sediment load at x is:

$$q_s = T_c - D/\alpha \qquad [I\text{-}62]$$

Case 2 occurs when $T_{cu} > q_{su}$, $dT_c/dx < 0$, and T_c becomes less than q_s within the segment. If $dT_c/dx < 0$ for a segment where $T_{cu} > q_{su}$, T_c may decrease below q_s within the segment. The point where $q_s = T_c$ is determined as x_{db}. This becomes x_u in equation [I-60], with D_u = 0. Deposition and sediment load are computed from equations [I-59], [I-60], and [I-62].

Case 3 occurs when $T_{cu} < q_{su}$, $dT_c/dx > 0$, and T_c becomes greater than q_s within the segment. In situations like a grass buffer strip, the transport capacity at the upper edge may drop abruptly to a level below the sediment load. Within the upper end of the strip, the sediment load decreases due to deposition while the transport capacity increases from the point of the abrupt decrease. Somewhere upslope from the lower edge of the strip, the sediment load equals the transport capacity. At this point, x_{de}, deposition ends, that is, $D_u = 0$, and $T_c = q_s$. Downslope, detachment by flow occurs. The point where deposition ends is given by:

$$x_{de} = x_u \left\{1 - [(1+\phi)/\phi][D_u/(dT_c/dx - D_L)]\right\}^{1/(1+\phi)} \qquad [I-63]$$

where:

$$D_u = \alpha\,(T_{cu} - q_{su}) \qquad [I-64]$$

and T_{cu} = transport capacity after the abrupt decrease at x_u and q_{su} = sediment load at x_u. Continuity of sediment load is maintained, but D may be discontinuous at segment ends.

Downslope from x_{de}, where detachment by flow occurs, the sediment load is given by:

$$q_s = (D_{Fu} + D_{Lu} + D_{FL} + D_{LL})\Delta x/2 + g_{su} \qquad [I-65]$$

where the second subscript u or L indicates upper or lower, and Δx = length of the segment where detachment by flow is occurring. In this case, Δx is from x_{de} to the lower end of the segment; q_{su} is at x_{de}, which is T_c at x_{de}, $D_{Fu} = 0$ at x_{de}, and D_{FL} is either detachment capacity at x or that which will just fill the transport capacity.

Case 4 occurs when $T_c > q_{su}$ over the entire segment. Sediment load is computed with equation [I-65].

The equation for sediment transport capacity, T_c, shifts total transport capacity among the various particle types. If transport capacity exceeds availability for one particle type while it is less for another, transport capacity is shifted from the particle type having the excess to the one having the deficit. Furthermore, logic checks within the model prevent simultaneous deposition and detachment of particles by flow.

Eroded sediment is a mixture of particles having various sizes and densities. The distribution is broken into classes, with each class represented by a particle diameter and density. Equations [I-58] to [I-65] are solved for each particle type within the constraints noted above.

SEDIMENT CHARACTERISTICS

Sediment eroded on field-sized areas is a mixture of primary particles and aggregates (conglomerates of primary particles). The distribution of these particles as they are detached is a function of soil properties, management, and rainfall and runoff characteristics. If deposition occurs, usually the coarse and dense particles are deposited first, leaving a finer sediment mixture. The input to the model is the distribution of the sediment as it is detached; the model calculates a new distribution if it calculates that deposition occurs.

Based on our survey of existing data, values given in table I-9 are an example of input for many midwestern silt loam soils.

Table I-9.—Sediment characteristics assumed for detached sediment before deposition; assumed typical of many midwestern silt loam soils

Particle type	Diameter	Specific gravity	Fraction of total amount (mass basis)
	(mm)	(g/cm^3)	
Primary clay	0.002	2.60	0.05
Primary silt	.010	2.65	.08
Small aggregate	.030	1.80	.50
Large aggregate	.500	1.60	.31
Primary sand	.200	2.65	.06

If the particle distribution is not known, the model assumes five particle types, and estimates the distribution from the primary particle-size distribution of the soil mass from the following equations:

$$PSA = (1.0 - ORCL)^{2.49}\ ORSA \qquad [I-66]$$

$$PSI = 0.13\ ORSI \qquad [I-67]$$

$$PCL = 0.2\ ORCL \qquad [I-68]$$

$$SAG = \begin{cases} 2\ ORCL & ORCL < 0.25 \qquad [I-69] \\ 0.28(ORCL - 0.25) + 0.5 & 0.25 \leq ORCL \leq 0.50 \qquad [I-70] \\ 0.57 & 0.5 < ORCL \qquad [I-71] \end{cases}$$

$$LAG = 1.0 - PSA - PSI - PCL - SAG \qquad [I-72]$$

if LAG < 0.0, multiply all others by the same ratio to make LAG = 0.0 where ORCL, ORSI, and ORSA are, respectively, fractions for primary clay, silt, and sand in the original soil mass, and PCL, PSI, PSA, SAG, and LAG are, respectively, fractions for primary clay, silt, sand, and small and large aggregates in the detached sediment.

The diameters for the particles are given by:

$$DPCL = 0.002\ mm \qquad [I-73]$$

$$DPSI = 0.010\ mm \qquad [I-74]$$

$$DPSA = 0.20\ mm \qquad [I-75]$$

$$
DSAG = \begin{cases} 0.03 \text{ mm} & ORCL < 0.25 \quad [I\text{-}76] \\ 0.20(ORCL - 0.25) + 0.03 \text{ mm} & 0.25 \leq ORCL \leq 0.60 \quad [I\text{-}77] \\ 0.1 \text{ mm} & 0.60 < ORCL \quad [I\text{-}78] \end{cases}
$$

$$DLAG = 2(ORCL) \text{ mm} \qquad [I\text{-}79]$$

where DPCL, DPSI, DPSA, DSAG, and DLAG are, respectively, the diameters of the primary clay, silt, and sand, and the small and large aggregates in sediment. The assumed specific gravities are shown in table I-9. The primary particle composition of the sediment load is estimated from:

Small aggregates:

$$CLSAG = SAG \cdot ORCL/(ORCL + ORSI) \qquad [I\text{-}80]$$

$$SISAG = SAG \cdot ORSI/(ORCL + ORSI) \qquad [I\text{-}81]$$

$$SASAG = 0.0 \qquad [I\text{-}82]$$

Large aggregates:

$$CLLAG = ORCL - PCL - CLSAG \qquad [I\text{-}83]$$

$$SILAG = ORSI - PSI - SISAG \qquad [I\text{-}84]$$

$$SALAG = ORSA - PSA \qquad [I\text{-}85]$$

where CLSAG, SISAG, and SASAG = fractions of the total for the sediment of, respectively, primary clay, silt, and sand in the small aggregates in the sediment load, and CLLAG, SILAG, and SALAG are corresponding fractions for the large aggregates.

If the clay in the large aggregate expressed as a fraction for that particle alone is less than 0.5 times ORCL, the distribution of the particle types is recomputed so that this constraint can be met. A sum, SUMPRI, is computed whereby:

$$SUMPRI = PCL + PSI + PSA. \qquad [I\text{-}86]$$

The fractions PSA, PSI, and PCL are not changed. The new SAG is:

$$SAG = (0.3 + 0.5\ SUMPRI)(ORCL + ORSI)/[1 - 0.5\ (ORCL + ORSI)]. \qquad [I\text{-}87]$$

Equation [I-87] is derived given previously determined values for PCL, PSI, and PSA; the sum of primary clay fractions for the total sediment equals the clay fraction in the original soil, and the assumption that the fraction of primary clay in LAG equals half of the primary clay in the original soil.

The model also computes an enrichment ratio using specific surface areas for organic matter, clay, silt, and sand. Organic matter is distributed among the particle types based on the proportion of primary clay in each type. Enrichment ratio is the ratio of the total specific surface area for the sediment to that for the original soil.

Although these relationships are approximations to the data found in the literature (33), they represent the general trends.

OVERLAND FLOW ELEMENT

Detachment Equation

Detachment on interrill and rill areas and transport and deposition by rill flow are the erosion-transport processes on the overland flow element. Detachment is described by a modified USLE (12) written as:

$$D_{Li} = 0.210\ EI\ (s + 0.014)\ KCP\ (\sigma_p/V_u) \qquad [I\text{-}88]$$

and

$$D_{Fr} = 37983\ mV_u\ \sigma_p^{1/3}\ (x/72.6)^{m-1}\ s^2\ KCP\ (\sigma_p/V_u) \qquad [I\text{-}89]$$

where D_{Li} = interrill detachment rate (lb/ft^2/s), D_{Fr} = rill detachment capacity rate (lb/ft^2/s), EI = Wischmeier's rainfall erosivity (energy times 30-minute intensity) [100(ft-tons/acre)(in/hr)], x = distance downslope (ft), s = sine of slope angle, m = slope length exponent, K = USLE soil erodibility factor [(tons/acre)(acre/100 ft-tons)(hr/in)], C = soil loss ratio of the USLE cover-management factor, P = USLE contouring factor, V_u = runoff volume [volume/unit area (ft)], and σ_p = peak runoff rate [volume/unit area/unit time (ft//s)]. Note that for P only the contouring factor of the USLE is used. The model is structured to directly account for other USLE P-factor effects such as strip cropping and deposition in terrace channels. These P factors are highly variable, and the model can account for a number of them.

Storm Erosivity

The hydrologic processes of rainfall and runoff drive the erosion-transport processes. Storm EI (storm energy times maximum 30-minute intensity), volume of runoff, and peak discharge are the variables used to characterize hydrologic inputs. Values for these factors are generated by the hydrology component of CREAMS. When daily rainfall amounts are used, storm EI is estimated from (21):

$$EI = 8.0\ V_R^{1.51} \qquad [I\text{-}90]$$

where EI = storm EI [(100 ft-tons/acre)(in/hr)] and V_R = volume of rainfall (in). This equation is very approximate. It was developed by regression analysis from about 2,700 data points used in the development of the USLE and has a coefficient of determination (R^2) of 0.56. When breakpoint rainfall is used, storm EI is computed using standard USLE procedures (31). Storm energy per unit of rainfall is given by:

$$e = 916 + 331\ \log_{10} i \qquad [I\text{-}91]$$

where e = rainfall energy per unit of rainfall (ft-tons/acre-in) and i = rainfall intensity (in/hr). The energy for each segment of the rainfall hyetograph is the product of e and the rainfall amount for the segment. Total energy for the storm is the sum of these incremental energies.

Slope Length Exponent, m

For slopes less than 150 ft, m is set to 2.0, but for slopes longer than 150 ft, m is limited by:

$$m = 1.0 + 5.011/\ln (x). \qquad [I-92]$$

This limit avoids excessive erosion for very long slopes (12). Equation [I-92] limits the effective slope length exponent for the total of rill and interrill erosion to 1.67 so far as it is a function of length. The effective exponent is a function of slope (smaller for flatter slopes), runoff erosivity relative to rainfall erosivity (greater for relatively greater runoff erosivity), and slope length (greater with longer length, except with the above restriction).

The Yalin sediment transport equation (32) is used to describe sediment transport capacity. It gave reasonable results when compared with experimental data for deposition of sand and coal by overland flow in a laboratory study (5, 9) and on field plots (vol. III, ch. 10). The Yalin equation was modified to distribute transport capacity among the various particle types. The discussion of the method given below is abstracted from Foster and Meyer (10), Davis (5), and Khaleel and others (14).

The Yalin equation is given by:

$$\frac{W_s}{(S_g) g \rho_w \, dV_*} = 0.635\, \delta \left[1 - \frac{1}{\sigma} \ln (1 + \sigma)\right] = P_s \qquad [I-93]$$

where:

$$\sigma = A \cdot \delta \qquad [I-94]$$

$$\delta = \frac{Y}{Y_{cr}} - 1 \text{ (when } Y < Y_{cr},\ \delta = 0) \qquad [I-95]$$

$$A = 2.45 (S_g)^{-0.4} (Y_{cr})^{1/2} \qquad [I-96]$$

$$Y = \frac{V_*^2}{(S_g - 1.0) g d} \qquad [I-97]$$

$$V_* = (g R S_f)^{1/2} \qquad [I-98]$$

where V_* = shear velocity $=(\tau/\rho_w)^{1/2}$, τ = shear stress, g = acceleration of gravity, ρ_w = mass density of the fluid, R = hydraulic radius, S_f = slope of the energy gradeline, S_g = particle specific gravity, d = particle diameter, Y_{cr} = critical lift force given by the Shields' diagram extended to low particle Reynolds numbers (22), and W_s = transport capacity (mass/unit width/unit time). The constant 0.635 and Shields' diagram were empirically derived.

The sediment load may have fewer46articles of a given type than the flow's transport capacity for that type. At the same time, the sediment load of other particle types may exceed the flow's transport capacity for those types. The excess transport capacity for the deficit types is assumed to be available to increase the transport capacity for the types where available sediment exceeds transport capacity.

The Yalin equation was modified to shift excess transport capacity. For large sediment loads (sediment loads for each particle type clearly in excess of the respective transport capacity for each particle type), or for small loads (sediment loads for each particle type clearly less than the respective transport capacity for each particle type), the flow's transport capacity is distributed among the available particle types based on particle size and density and flow hydraulics (10).

Yalin assumed that the number of particles in transport is proportional to δ. For a mixture, the number of particles of a given type i is assumed to be proportional to δ_i. Values of ${}_i$ for each particle type in a mixture are calculated and summed to give a total:

$$T = \sum_{i=1}^{n_s} \delta_i \qquad [I\text{-}99]$$

where n_s = number of particle types. The number of transported particles of type i in a mixture is given as:

$$(N_e)_i = N_i\ (\delta_i/T) \qquad [I\text{-}100]$$

where N_i = number of particles transported in sediment of uniform type i for a δ_i.

As derived by Yalin, the nondimensional transport, P_s, of equation [I-93] is proportional to the number of particles in transport. Then

$$(P_e)_i = \frac{P_i \delta_i}{T} \qquad [I\text{-}101]$$

where $(P_e)_i$ = the effective P_s for particle type i in a mixture, and P_i is the P_s calculated for uniform material of type i. The transport capacity W_{si} of each particle type in a mixture is then expressed by:

$$W_{si} = (P_e)_i\ (S_g)_i \rho_w g d_i V_*. \qquad [I\text{-}102]$$

This is the transport capacity assuming that the supply of all particle types is either greater than or less than their respective W_{si}. When availability of some types is greater than their W_{si} and others are less than their W_{si}, transport capacity shifts from those types where supply is less than capacity so that all of the total transport capacity is used.

The steps given below are followed to redistribute the transport capacity when excesses and deficits occur.

1. For those particles where $W_{si} \geq q_{si}$ (q_{si} = sediment load for particle type i), compute the actual required P_{ireq} from equation [I-93], that is,

$$P_{ireq} = q_{si}/(S_g)_i g \rho_w d_i V_* . \qquad [I\text{-}103]$$

2. For those particle types where $W_{si} \geq q_{si}$, the sum:

$$SPT = \sum_{i-1}^{n_s} (P_{ireq}/P_i) \qquad [I\text{-}104]$$

   .s computed where $k_i = 1$ if $W_{si} \geq q_{si}$ and $k_i = 0$ if $W_{si} < q_{si}$. The sum SPT represents the fraction of the total transport capacity used by those particle types where sediment availability is less than transport capacity.

3. The excess (expressed as a fraction of the total) to be distributed is:

$$E_{xc} = 1 - SPT . \qquad [I\text{-}105]$$

4. For those particle types where $W_{si} < q_{si}$, sum δ_i as:

$$SDLT = \sum_{i=1}^{n_s} \delta_i l_i \qquad [I\text{-}106]$$

   where $l_i = 0$ if $W_{si} \geq q_{si}$ and $l_i = 1$ if $W_{si} < q_{si}$.

5. The excess is distributed according to the distribution of δ_i among the particle types, that is,

$$T_{ci} = (\delta_i/SDLT)(E_{xc})(P_i)(S_g)_i g \rho_w d_i V_* l_i , \qquad [I\text{-}107]$$

   and

$$T_{ci} = q_{si} k_i . \qquad [I\text{-}108]$$

6. Repeat steps 1-5 until either all $T_{ci} \leq q_{si}$ or all $T_{ci} \geq q_{si}$. When the former occurs, the proper T_{ci}'s have been found. If the latter occurs, one particle type will have all of the excess transport capacity. The excess for this one type should be equally distributed among all of the types. This is done by:

$$SMUS = \sum_{i=1}^{n_s} (P_{ireq}/P_i) \qquad [I\text{-}109]$$

$$T_{ci} = (1.0/SMUS) q_{si} . \qquad [I\text{-}110]$$

Conversion from Storm to Rate Basis

Without the (σ_p/V_u) term, equations [I-88] and [I-89], as originally developed (12), were on a storm basis, while the transport equation is on an instantaneous rate basis. The two are combined by assuming that the computed sediment concentration represents an average for the runoff event, and that the peak discharge represents a characteristic discharge that can be used to compute the average concentration.

Since most field-sized areas are relatively small, time of concentration is usually small and is assumed to be less than rainfall duration. Thus, for a given storm, discharge at a location is assumed to be directly proportional to upstream drainage area.

Shear Stress

The transport equation requires an estimate of shear stress. The sediment transport concept, where shear is divided between form roughness and grain roughness, is used to estimate the shear stress acting on the soil, the portion assumed responsible for sediment transport (13). Mulch or vegetation reduces this stress. The shear stress acting on the soil, τ_{soil}, is estimated by:

$$\tau_{soil} = \gamma y s (n_{bov}/n_{cov})^{0.9} \qquad [I\text{-}111]$$

where γ = weight density of water, y = flow depth for bare, smooth soil, s = sine of slope angle, n_{bov} = Manning's n for bare soil (0.01 assumed), and n_{cov} = total Manning's n for rough surfaces or soil covered by mulch or vegetation. Flow depth is estimated by the Manning equation as:

$$y = [q_w n_{bov}/s^{1/2}]^{0.6} \qquad [I\text{-}112]$$

where q_w = rate of discharge per unit width. Although the Darcy-Weisbach equation with a varying friction factor for laminar flow might be more accurate for y in some cases, most users are better acquainted with estimating Manning's n. The error in estimating a value for the roughness factor is probably greater than the error in using the Manning equation for laminar flow.

Slope Segments

Computations begin at the upper end of the slope. Sediment is routed downslope much the same as it is in most erosion models. The output is the sediment concentration for each particle type. Concentration multiplied by the runoff volume and overland flow area represented by the overland flow profile gives the sediment yield for the storm on the overland flow area of the field.

The overland flow area is represented by a typical land profile selected from several possible overland flow paths. Its shape may be uniform, convex, concave, or a combination of these shapes. Inputs are total slope length, average steepness, steepness at the upper end, steepness at the lower end, and location of the end points of a mid-uniform section.

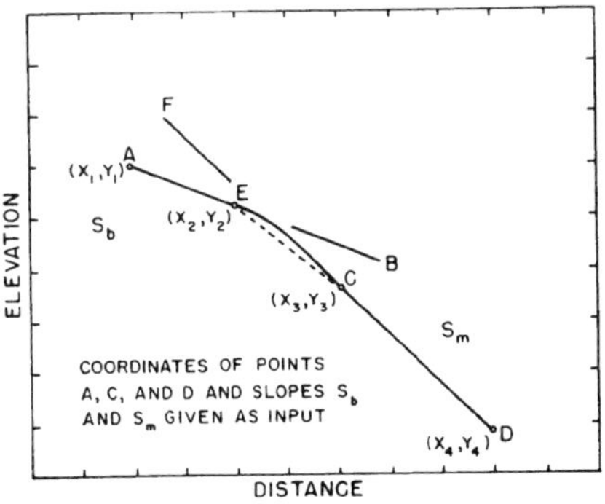

Figure I-10.—Representation of convex slope profile for overland flow.

Given this minimum of information, the model establishes segments along the profile. The procedure is illustrated by the convex shape shown in figure I-10. Coordinates of points A, C, and D are given, as are slopes S_b and S_m. A quadratic curve will pass through point C tangent to line CD and through point E tangent to line AB. The location of point E is the intersection of a line having a slope equal to the average of S_b and S_m with line AB. If X_2 is less than X_1, X_3 is shifted downslope so that $X_1 = X_2$.

Each uniform section is one segment. In figure I-10, AE and CD are segments. Convex sections like EC are divided into only three segments, because detachment and transport computations are not especially sensitive to the number of segments on convex slopes. Concave segments are divided into 10 segments because deposition computations on concave slopes are especially sensitive to the number of segments. Furthermore, several segments are required to accurately determine where deposition begins.

Additional segment ends are designated where K, C, P, or n change. Given locations where these changes occur as input, the model computes the coordinates for all the segments for the overland flow slope.

Selection of Parameter Values

Slope length is, perhaps, the most difficult of the overland flow parameters to estimate. Williams and Berndt's (30) contour method is a possible technique to use. Another is to sketch flow lines from the watershed boundary to concentrated flow. Topography in most fields causes overland flow to converge into concentrated flow within about 300 ft. Certainly a grass waterway or a terrace channel is the end of overland slope length.

Values for the parameters K, C, and P (contouring) are selected from Wischmeier and Smith (31) according to crop stage. Values for Manning's n_{cov} may be selected from Lane and others (18) or from vol. II, ch. 2.

CHANNEL ELEMENT

The channel element is used to represent flow in terrace channels, diversions, major flow concentrations where topography has caused overland flow to converge, grass waterways, row middles or graded rows, tail ditches, and other similar channels. The channel element does not describe gully or large stream channel erosion.

With the exception that shear stress and detachment by flow are estimated differently, the same concepts and equations are used in both the channel and overland flow elements. Discharge along the channel is assumed to vary directly with upstream drainage area. A discharge greater than zero is permitted at the upper end to account for upland contributing areas. As with the overland flow element, changes in the controlling variables along the channel are allowed. Thus, a channel with a decreasing slope or a change in cover can be analyzed.

Spatially Varied Flow Equations

Flow in most channels in fields is spatially varied, especially for outlets restricted by ridges and heavy vegetation, and for very flat terrace channels. Also, discharge generally increases along the channel. The model approximates the energy gradeline along the channel using a set of normalized curves and assuming steady flow at peak discharge. As an alternative, the model will set the friction slope equal to the channel slope.

The equation for spatially varied flow (3) with increasing discharge in a triangular channel may be normalized as:

$$\frac{dy_*}{dX_*} = [S_* - C_2\, X_*^2/y_*^{16/3} - C_3\, X_*/y_*^4]/[1 - C_3\, X_*^2/y_*^5] \qquad \text{[I-113]}$$

where $y_* = y/y_e$, y = flow depth, y_e = flow depth at the end of the channel, S_* = $S \cdot L_{eff}/y_e$, X = distance along channel, $X_* = X/L_{eff}$, and L_{eff} = effective channel length (that is, the length of the channel if it is extended upslope to where discharge would be zero with the given lateral inflow rate). Constants C_1, C_2, and C_3 are given by:

$$C_1 = [z^{5/2}/2(z^2 + 1)^{1/2}]^{2/3} \qquad \text{[I-114]}$$

$$C_2 = [Q_e\, n\, L_{eff}^{1/2}/C_1 y_e^{19/6}]^2 \qquad \text{[I-115]}$$

$$C_3 = 2\, \beta\, Q_e^2/g\, z^2\, y_e^5 \qquad \text{[I-116]}$$

where n = Manning's n, z = side slope of channel, Q_e = discharge at end of channel, β = energy coefficient [1.56 used from McCool and others (23)], and g = acceleration due to gravity. Equation [I-113] was solved for a range of typical values of C_1, C_2, and C_3. The curves given by equations [I-117] to [I-126] were fitted by regression to the solutions.

Range of C_3: $C_3 > 0.3$

where $0.0 \leq S_* \leq 1.2$ and $X_* \leq 0.9$,

$$SSF = 0.2777 - 3.3110\, X_* + 9.1683\, X_*^2 - 8.9551 X_*^3 \qquad \text{[I-117]}$$

where $1.2 \leq S_* \leq 4.8$ and $X_* \leq 0.9$, [I-117]

$$SSF = 2.6002 - 8.0678X_* + 15.6502X_*^2 - 11.7998X_*^3 \quad [I\text{-}118]$$

where $4.8 < S_* \leq 20.0$ and $X_* \leq 0.9$,

$$SSF = 3.8532 - 12.9501X_* + 21.1788X_*^2 - 12.1143X_*^3, \quad [I\text{-}119]$$

and where $20.0 < S_*$ and $X_* \leq 0.9$

$$SSF = 0. \quad [I\text{-}120]$$

Range of C_3: $0.3 \geq C_3 \geq 0.03$

Where $S_* > 0$ and $X_* \leq 0.8$,

$$SSF = 2.0553 - 6.9875X_* + 11.418X_*^2 - 6.4588X_*^3, \quad [I\text{-}121]$$

and where $S_* = 0$ and $X_* \leq 0.9$,

$$SSF = 0.0392 - 0.4774X_* + 1.0775X_*^2 - 1.3694X_*^3. \quad [I\text{-}122]$$

Range of C_3: $0.03 > C_3 \geq 0.007$

Where $S_* > 0$ and $X_* \leq 0.8$,

$$SSF = 1.5386 - 5.2042X_* + 8.4477X_*^2 - 4.740X_*^3, \quad [I\text{-}123]$$

and where $S_* = 0.0$ and $X_* \leq 0.9$,

$$SSF = 0.0014 - 0.0162X_* - 0.0926X_*^2 - 0.0377X_*^3. \quad [I\text{-}124]$$

Range of C_3: $0.007 > C_3$

Where $S_* > 0$ and $X_* < 0.7$,

$$SSF = 1.2742 - 4.7020X_* + 8.4755X_*^2 - 5.3332X_*^3, \quad [I\text{-}125]$$

and where $S_* = 0$ and $X_* \leq 0.9$,

$$SSF = - 0.0363X_*^2. \quad [I\text{-}126]$$

With these values of SSF, the friction slope is:

$$S_f = (S_* - SSF)\, y_e/L_{eff}. \quad [I\text{-}127]$$

Flow depth y_e at the end of the channel is estimated by assuming at the user's option, either critical depth, depth of uniform flow in an outlet control channel, or depth from a rating curve.

A triangular channel section, a reasonable approximation to most field channels, was used to develop the friction slope curves because the equations are simpler. In the model, a triangular channel must be used to estimate the slope of the energy gradeline, but the user may select a triangular, rectangular, or "naturally eroded" section in other computational components of the channel element.

Concentrated Flow Detachment

In the spring after planting, concentrated flow from intense rains on a freshly prepared seedbed often erodes through the finely tilled layer to the depth of secondary tillage or, perhaps, primary tillage. Once the channel erodes to the nonerodible layer, it widens at a decreasing rate.

Data from observed rill erosion rates (vol. III, ch. 11) suggests that detachment capacity (lb/ft^2/s) by flow over a loosely tilled seed-seedbed may be described by:

$$D = K_{ch}(1.35\ \bar{\tau} - \tau_{cr})^{1.05} \qquad [I\text{-}128]$$

where K_{ch} = an erodibility factor $[(lb/ft^2/s)(ft^2/lb)^{1.05}]$, $\bar{\tau}$ = average shear (lb/ft^2) of the flow in the channel, and τ_{cr}= a critical shear stress (lb/ft^2) below which erosion is negligible. Critical shear stress seems to increase greatly over the year as the soil consolidates (13).

Shear stress is assumed to be triangularly distributed in time during the runoff event in order to estimate the time that shear stress is greater than the critical shear stress. For the time that shear stress is greater than critical shear stress, shear stress is assumed constant and equal to peak shear stress for the storm.

Until the channel reaches the nonerodible layer, an active channel is assumed that is rectangular with the width obtained from figures I-11 and I-12 and equations [I-130] and [I-131]. The solution requires finding a value of x_c. Given the discharge Q, Manning's n, friction slope S_f, a value $g(x_c)$ is calculated from:

$$g\ (x_c) = \left[\frac{Q\ n}{1.49\ S_f^{1/2}}\right]^{3/8} \frac{\gamma S_f}{\tau_{cr}}\ . \qquad [I\text{-}129]$$

Given a particular value $g(x_c)$, a value of x_c is obtained from figure I-11. Having determined x_c, a value for R_* = hydraulic radius/wetted perimeter and W_* = width/wetted perimeter is read from figure I-12. The width of the channel is then calculated from:

$$W_{ac} = \left[\frac{Q\ n}{1.49\ S_f^{1/2}}\right]^{3/8} \frac{W_*}{R_*^{5/8}}\ . \qquad [I\text{-}130]$$

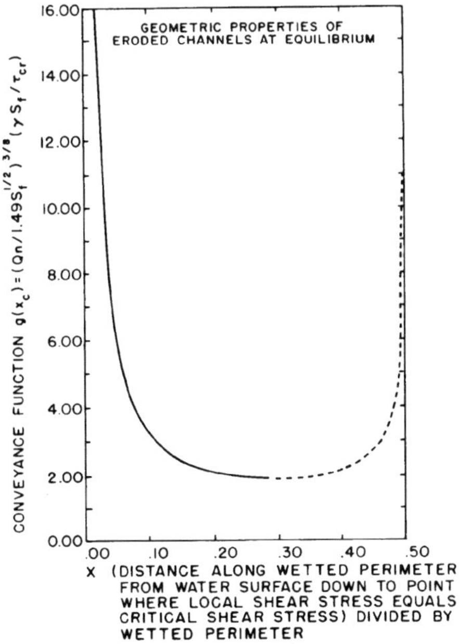

Figure I-11.—Function $g(x_c)$ for equilibrium eroded channel

The functions shown in figures I-11 and I-12 are stored piecewise in the model.

The channel moves downward at the rate d_{ch}:

$$d_{ch} = e_m/\rho_{soil} = K_{ch}\ (1.35\ \bar{\tau} - \tau_{cr})^{1.05}/\rho_{soil} \qquad [I\text{-}131]$$

where e_m = erosion rate calculated using the maximum shear stress (mass/unit area/unit time), and ρ_{soil} = mass density of the soil in place. The erosion rate in the channel is:

$$E_{ch} = W_{ac}\ K_{ch}(1.35\ \bar{\tau} - \tau_{cr})^{1.05} \qquad [I\text{-}132]$$

where E_{ch} is the soil loss per unit channel length (mass/unit channel length/unit time).

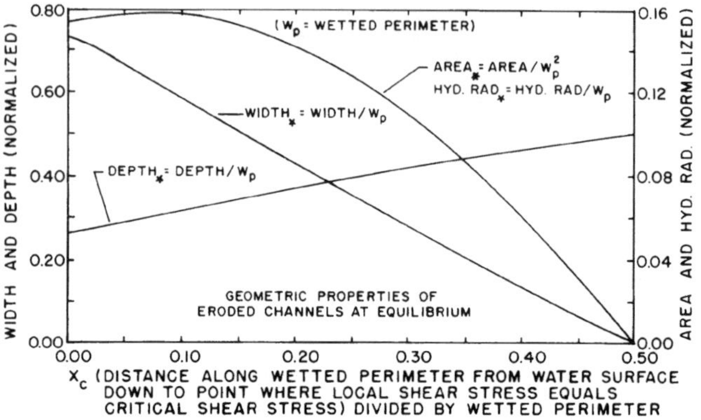

Figure I-12.—Equilibrium eroded channel geometric properties

Once the channel hits the nonerodible boundary, the erosion rate begins to decrease with time. The width W of the channel at any time after the channel has eroded to the nonerodible layer is estimated by:

$$\omega = 1 - \exp(-t_*) \qquad [I\text{-}133]$$

where ω is the nondimensional channel width given by:

$$\omega = (W - W_i)/(W_f - W_i) \qquad [I\text{-}134]$$

$$t_* = (t - t_i)(dw/dt)_i/(W_f - W_i) \qquad [I\text{-}135]$$

where W_i = width at $t = t_i$, W = width at t, W_f = final eroded width for $t \rightarrow \infty$, and the given Q, t = time, and $(dw/dt)_i$ = rate that channel widens at $t = t_i$. The initial widening rate is given by:

$$(dW/dt)_i = 2\ K_{ch}\ (\tau_b - \tau_{cr})^{1.05}/\rho_{soil} \qquad [I\text{-}136]$$

where τ_b is given by:

$$f(x_b) = \tau_b/\bar{\tau} = \tau_m[(8\ x_b)^{1/2} - 2\ x_b] \qquad [I\text{-}137]$$

and:

$$\tau_m = \tau_{max}/\bar{\tau} = 1.35 \qquad [I\text{-}138]$$

where x_b = flow depth/wetted perimeter, and τ_{max} = maximum shear stress at center of channel

The final width w_f is determined by finding the x_{cf} that gives:

$$\left[\frac{Q\ n}{1.49 S_f^{1/2}}\right]^{3/8} \frac{\gamma S_f}{\tau_{cr}} \qquad \frac{1}{x_{cf}\ (1 - 2x_{cf})^{3/8} f(x_{cf})} \qquad [I-139]$$

where $f(x_{cf})$ is the function given by equations I-137 and I-138 and evaluated at x_{cf}. The final width is:

$$W_f = \left[\frac{Q\ n}{1.49 S_f^{1/2}}\right]^{3/8} \left[\frac{1-2x_{cf}}{x_{cf}^{5/3}}\right]^{3/8} . \qquad [I-140]$$

Sediment Transport and Partitioning of Shear Stress

Sediment transport capacity for the channel is described using the Yalin equation in exactly the same form as it was used in the overland flow element.

The shear stress acting on the soil is the shear stress used to compute detachment and transport. Grass and mulch reduce this stress. Total shear is divided into that acting on the vegetation or mulch and that acting on the soil using sediment transport theory (13).

First, velocity is estimated using n_t, the total Manning's n [See (3), (26) for estimates of n_t]. The hydraulic radius due to the soil is:

$$R_{soil} = (V\ n_{bch}/S_f^{1/2})^{3/2} \qquad [I-141]$$

where n_{bch} = Manning's n for a bare channel and S_f = friction slope. Shear stress acting on the soil is:

$$\tau_{soil} = \gamma R_{soil}\ S_f \qquad [I-142]$$

$$\tau_{cov} = \gamma\ S_f[V(n_t - n_{bch})/1.49\ S_f^{1/2}]^{3/2} . \qquad [I-143]$$

If τ_{cov} exceeds the shear stress at which the cover starts to move, the cover fails, thereby increasing the flow's shear stress on the soil.

Variations in parameters such as Manning's n and slope along the channel can be considered. In addition, the model breaks the channel into segments of length $L_{eff}/10$. Calculations begin at the upper end of the channel and proceed downstream.

IMPOUNDMENT (POND) ELEMENT

The impoundment (pond) element describes deposition behind impoundments (including parallel tile outlet) that drain following each storm.

Deposition is the main sedimentation process occurring in impoundments.

Since transport capacity in the impoundments is essentially nonexistent, the amount of sediment trapped in an impoundment is basically a function of time available for sediment to settle to the bottom of the impoundment before flow leaves the impoundment. The equations for the pond element were developed from regression analyses that fit relationships to output from a more complex model (15, 16) which had been previously validated with field data.

The fraction of particles of a specific size and density that passes through the impoundment is:

$$f_{pi} = A_1 \exp(B_1 \cdot d_{eqi}) \qquad [I-144]$$

where f_{pi} = fraction passing through pond for particle type i, A_1 and B_1 = coefficients given below, and d_{eqi} = the equivalent sand diameter of particle type i (microns). The particle types in the model represent classes rather than specific particles. Therefore, equation [I-144] was integrated over the class range and divided by the class width to obtain average for the class as:

$$F_{pi} = A_1 [\exp(B_1 d_u) - \exp(B_1 d_1)]/(B_1 \cdot \Delta d) \qquad [I-145]$$

where F_{pi} = fraction passed for particle class i. The equivalent sand diameters are arranged in ascending order, and d_u is the d_{eqi} for the class and d_1 is the next smallest d_{eqi}. The diameters d_u and d_1 are not centered around d_{eqi} because d_{eqi} is assumed to represent the maximum diameter in the class. The class width $\Delta d = d_u - d_1$. Values of F_{pi} are limited to a maximum of 1.0. The coefficients A_1 and B_1 are given by:

$$A_1 = 1.136 \exp(Z_s) \qquad [I-146]$$

$$B_1 = -0.152 \exp(Y_s) \qquad [I-147]$$

with Z_s and Y_s in turn given by:

$$Z_s = (-6.68 \times 10^{-6})f - 0.0903B + (1.19 \times 10^{-4})C_{or} - (3.42 \times 10^{-6})V_r - 20400I \qquad [I-148]$$

$$Y_s = (3.28 \times 10^{-5})f + 0.123B - (2.4 \times 10^{-4})C_{or} + (8.10 \times 10^{-6})V_r - 11880I \qquad [I-149]$$

where f and B = coefficient and exponent in a power equation relating surface area to depth $S_a = f y_p{}^B$, y_p = depth in pond (ft), S_a = surface area (ft^2), V_r = volume of runoff (ft^3), and I = infiltration rate in the pond (ft/s). The coefficient C_{or} is related to the orifice in the pipe outlet by:

$$C_{or} = 13968\, d_{or}{}^2 \qquad [I-150]$$

where d_{or} = diameter of the orifice (ft). Also, the coefficient C_{or} is related to discharge and the depth above the outlet point by:

$$C_{or} = 3600\ Q_p/y_d^{1/2} \qquad [I\text{-}151]$$

where Q = discharge (ft^3/s) and y_d = depth (ft).

All of the water which enters the pond will not leave. The volume leaving is estimated by:

$$V_{out} = 0.95\ V_{in}\ \exp(Z_r) \qquad [I\text{-}152]$$

where V_{out} = volume of runoff discharged, V_{in} = volume of runoff reaching the pond, and Z_r is given by:

$$Z_r = -\ (9.29 \times 10^{-6})\ f + 0.282B + (1.25 \times 10^{-4})C_{or} - (3.08 \times 10^{-6})\ V_r - 33359I \qquad [I\text{-}153]$$

$$\text{If } I = 0.0,\ V_{out} = V_{in} \qquad [I\text{-}154]$$

$$\text{If } V_{out} > V_{in},\ V_{out} = V_{in} \qquad [I\text{-}155]$$

are additional constraints on V_{out} for equation [I-152].

VALIDITY OF THE MODEL

Comparison with Other Models

The validity of the model can be partially assessed by comparing it with other models that might be used in this application. The detachment relationships used in the overland flow element gave good results for a watershed at Treynor, Iowa. Estimates were considerably better than those from the USLE using storm EI (12) and better than those obtained from a procedure using runoff volume and peak discharge alone as an erosivity factor (25). Both rainfall and runoff appear to be important for estimating detachment on overland flow areas. More comprehensive models like ARM (6) or ANSWERS (1) use modifications of the USLE and/or require data for calibration. The CREAMS erosion/sediment yield component preserves the USLE form when erosion is simulated for a range of storms and slope lengths and steepnesses. On long-term simulation, the model produces results comparable with those of the USLE. Information to select overland flow erosion parameters is as readily available for CREAMS as it is for the USLE.

Comparison of Output from Model with Observed Data

The validity of the model has been partially assessed by comparing output from the model with measured sediment yield from concave field plots under simulated rainfall, single terrace watersheds, small watersheds with impoundment terraces, and a small watershed with conservation tillage. The simulations were made using measured rainfall and runoff values. Parameter values were selected from volume II, chapter 2 without calibration, except as noted.

Concave Plots

Three concave plots 35 ft long were carefully shaped in a soil where soil properties were uniform within the depth of shaping. Slope along the plots continuously decreased from 18% at the upper end to 0% at the lower end. Simulated rainfall at 2.5 in/hr was used to detach and provide runoff to transport sediment (vol. III, ch. 10). The measured particle distribution of the sediment reaching the deposition area was used as input to the model. The soil erodibility factor and Manning's n were adjusted in the model to give observed soil loss entering the deposition area at the lower end of the plots. The estimated sediment yield for the 29-ft plot was 0.0026 lb/ft/s compared with 0.0017 lb/ft/s observed. For the 35-ft plot, the estimated and observed values were 0.0014 and 0.00094 lb/ft/s, respectively.

Single Terrace Watersheds

Soil loss was simulated for 8 yr of data from small, single terrace watersheds at Guthrie, Okla. (4). The simulations were made without calibration. Table I-10 gives computed and measured results.

Table I-10.—Comparison of simulated sediment yield from single terrace watersheds with observed values

Terrace	Grade	Sediment yield	
		Observed	Simulated
		(tons/acre)	(tons/acre)
2B	Variable, 0.0033 at outlet to 0.0 at upper end.	54	28
3B	Variable, 0.005 at outlet to 0.0 at upper end.	62	53
3C	Constant, 0.005.	54	47
5C	Constant, 0.0017.	21	20

Impoundment Terraces

Soil loss was simulated for selected storms representing a range of rainfall and runoff characteristics for three locations in Iowa; Eldora, Charles City, and Guthrie Center, from an impoundment terrace study (17). The model was run without calibration. The results are given in table I-11.

Small Watershed

Simulations were run without calibration for approximately 2-1/2 years of data from the P2 watershed at Watkinsville, Ga. in a conservation tillage system for corn (27). Deposition in the backwater from the flume at the watershed outlet was modeled. Deposition measured in the flume backwater was about equal to the measured sediment yield on a similar nearby watershed (19). The computed total sediment yield for the period of record was 6.6 tons/acre, while the measured value was 8.3 tons/acre.

Table I-11.—Summary of observed and simulated sediment yield from impoundment terraces in Iowa

Watershed	Area	Julian date	Sediment yield	
			Observed	Simulated
	(acres)		(lb)	(lb)
Charles City	4.6	70147	1,197	52
		70152	72	14
		70244	4	160
		70323	58	5
		71151	280	294
		71157	209	160
Eldora	1.8	68198	283	150
		68220	58	55
		69187	1,057	554
		69232	124	227
		71163	335	139
Guthrie Center	1.4	69207	256	273
		69249	23	89
		70144	122	63
		70162	198	123
		70167	21	28
		70229	10	52

Overland Flow Sediment Transport

Estimates for sediment transport capacity of overland flow may be in error by a factor of two (vol. III, Ch. 10). However, the sediment transport equations used by other models have not been tested against field data where deposition was known with certainty to be limiting sediment load. Overland flow conditions are outside the range of most sediment transport equations developed for streamflow, and consequently, many give results greatly in error for overland flow (vol. III, ch. 10). Given the present state of the art, the transport relationship used in this model is believed to be as adequate as any available, especially when the equation is not calibrated.

Channel Erosion

The channel erosion relationships are the ones most likely to be in error even though they fit data from a rill erosion study very well (vol. III, ch. 11). Data from the rills (12 in wide) may not scale up to channel size (that is, 10 ft wide). However, computed final channel width agreed well with observed widths for a wide range of streams. While the channel erosion rate for a single storm may be in error, the upper limit for annual channel erosion should be a reasonable for soils having a nonerodible layer beneath the soil surface.

Proven parameter values for the channel soil erodibility and critical shear stress are not available. CREAMS considers the decay in erosion with time due to previous erosion; most models do not, with the exception of Bruce's and others (2). This component of CREAMS require calibration.

Backwater

Most erosion models as applied to fields use a kinematic runoff simulation model to generate values for hydraulic variables. That is, friction slope is set equal to the channel slope. This prevents modeling deposition in a backwater area at the field outlet. Such deposition occurs often and is important in estimating chemical yields associated with enrichment of fine sediment during deposition. The solutions to the spatially varied flow equations discussed earlier account for these outlet controls, and thus can be used to simulate sediment deposition.

SIMULATION COSTS

Comprehensive models that simulate erosion over space and over time through a runoff event are potentially more powerful than this model. However, detailed downslope spatial variability (slope, cover, and so forth) can be analyzed with this model. The expected slight increase in improved estimates with a more comprehensive model probably does not offset the additional costs for computing, and moreover, many of these models require lumped parameter estimates which prevents their consideration of slope shape and buffer strips, for example, that can be analyzed with this model.

While computer costs vary from site to site and change often, rough estimates are, nontheless, important for qualitative comparisons. Using the CDC 6500 Computer at Purdue University, simulation costs for the erosion/sediment yield component of CREAMS were about $0.10 per storm event. Therefore, the erosion/sediment yield component can simulate individual storm events for a cost of about $1 to $3 per year. Although the model is quite comprehensive, the programming is efficient and simulation costs are not prohibitive.

SUMMARY

An erosion/sediment yield model for field-sized areas was developed for use on a storm-by-storm basis. The overall objective was to develop a model, incorporating fundamental erosion/sediment transport relationships, to evaluate best management practices. Although the procedure does not consider changes in parameter values within individual storms, it does allow these parameters to change from storm to storm throughout the season. Moreover, parameters of the model allow for distribution of field characteristics along overland flow slopes and along waterways. Many of the model parameters are selected using tested methods developed for the well-known Universal Soil Loss Equation. For this reason, we feel that the model has immediate applications without the need for extensive calibration.

Limited testing has shown that the procedures developed herein give improved estimates over the USLE and modified USLE procedures. Specific components of the model were tested using experimental data from overland flow, erodible channel, and impoundment studies. Sensitivity analyses are described in chapter 6. Application of the model is demonstrated in volume II, chapter 2. Initial results suggest that the model produces reasonable results and is a useful tool for analyzing the influence of alternate management practices.

REFERENCES

(1) Beasley, D. B., E. J. Monke, and L. F. Huggins.
1977. The ANSWERS model: A planning tool for watershed research. American Society of Agricultural Engineers, Paper No. 77-2532. St. Joseph, Mich.

(2) Bruce, R. R., L. A. Harper, R. A. Leonard, W. M. Snyder, and others.
1975. A model for runoff of pesticides from small upland watersheds. Journal of Environmental Quality 4(4):541-548.

(3) Chow, V. T.
1959. Open-Channel Hydraulics. McGraw-Hill Book Company, Inc., New York, NY. 680 pp.

(4) Daniel, H. A., H. M. Elwell, and M. B. Cox.
1943. Investigation in erosion control and reclamation of eroded land at the Red Plains Conservation Experiment Station, Guthrie, Oklahoma, 1930-40. U.S. Department of Agriculture, Technical Bulletin No. 837.

(5) Davis, S. S.
1978. Deposition of nonuniform sediment by overland flow on concave slopes. M.S. Thesis, Purdue University, West Lafayette, Ind.

(6) Donigian, A. S., Jr., and N. H. Crawford.
1976. Modeling nonpoint source pollution from the land surface. U.S. Environmental Protection Agency, EPA-600/376-083. 279 pp.

(7) Einstein, H. A.
1968. Deposition of suspended particles in a gravel bed. Journal of the Hydraulics Division, Proceedings of the American Society of Civil Engineers 94(HY5):1197-1205.

(8) Foster, G. R.
1979. Sediment yield from farm fields: The Universal Soil Loss Equation and onfarm 208 plan implementation. In Universal Soil Loss Equation: Past, Present, and Future, Chapter 3. Soil Science Society of America. Madison, Wisc.

(9) ------, and L. F. Huggins.
1977. Deposition of sediment by overland flow on concave slopes. In Soil Erosion Prediction and Control. Soil Conservation Society of America Special Publication No. 21, pp. 167-182. Ankeny, Iowa.

10) Foster, G. R., and L. D. Meyer.
1972. Transport of soil particles by shallow flow. Transactions of the American Society of Agricultural Engineers 15(1):99-102.

11) ------, and L. D. Meyer.
1975. Mathematical simulation of upland erosion by fundamental erosion mechanics. In Present and Prospective Technology for Predicting Sediment Yields and Sources. U.S. Department of Agriculture, Agricultural Research Service, Southern Region, ARS-S-40, pp. 190-207. (Service discontinued; Agricultural Research Service is now Science and Education Administration-Agricultural Research.)

12) ------, L. D. Meyer, and C. A. Onstad.
1977. A runoff erosivity factor and variable slope length exponents for soil loss estimates. Transactions of the American Society of Agricultural Engineers 20(4):683-687.

13) Graf, W. H.
1971. Hydraulics of Sediment Transport. McGraw-Hill Book Co., New York, NY. 544 pp.

14) Khaleel, R., G. R. Foster, K. R. Reddy, M. R. Overcash, and others.
1980. A nonpoint source model for land areas receiving animal wastes: III. A conceptual model for sediment and manure transport. Transactions of the American Society of Agricultural Engineers. (In press.)

15) Laflen, J. M., and H. P. Johnson.
1976. Soil and water loss from impoundment terrace systems. In Proceedings of the Third Federal Inter-Agency Sedimentation Conference, Water Resources Council, Chapter 2, pp. 303-41, Washington, D.C.

16) ------, H. P. Johnson, and R. O. Hartwig.
1978. Sedimentation modeling of impoundment terraces. Transactions of the Amererican Society of Agricultural Engineers, 21(6):1131-1135.

17) ------, H. P. Johnson, and R. C. Reeve.
1972. Soil loss from tile outlet terraces. Journal of Soil and Water Conservation 27(2):74-77.

18) Lane, L. J., D. A. Woolhiser, and V. Yevjevich.
1975. Influence of simplification in watershed geometry in simulation of surface runoff. Colorado State University, Hydrology Paper No. 81, Fort Collins, Colorado. 50 pp.

19) Langdale, G. W., A. P. Barnett, R. A. Leonard, and W. G. Fleming.
1979. Reduction of soil erosion by no-till systems in the Southern Piedmont. Transactions of the American Society of Agricultural Engineers 22(1):82-86, 92.

20) Li, R. M.
1977. Water and Sediment routing from watersheds. Proceedings of River Mechanics Institute, Colorado State University, Fort Collins, Colorado. Chapter 9.

(21) Lombardi, F.
1979. Universal Soil Loss Equation (USLE), runoff erosivity factor, slope length exponent, and slope steepness exponent for individual storms. PhD Thesis, Purdue University, W. Lafayette, Ind.

(22) Mantz, P. A.
1977. Incipient transport of fine grains and flakes of fluids - extended Shields diagram. Journal of Hydraulics Division, Proceedings of the American Society of Civil Engineers 103(HY6):601-615.

(23) McCool, D. K., W. R. Gwinn, W. O. Ree, and J. E. Garton.
1966. Spatially varied steady flow in a vegetated channel. Transactions of the American Society of Agricultural Engineers 9(3):440-444.

(24) Onstad, C. A., and G. R. Foster.
1975. Erosion modeling on a watershed. Transactions of the American Society of Agricultural Engineers 18(2):288-292.

(25) ------, R. F. Piest, and K. E. Saxton.
1976. Watershed erosion model validation for Southwest Iowa. In Proceedings of the Third Federal Inter-Agency Sedimentation Conference, Water Resources Council, Chapter 1, pp. 22-24. Washington, D.C.

(26) Ree, W. O., and F. R. Crow.
1977. Friction factors for vegetated waterways of small slopes. U.S. Department of Agriculture, Agricultural Research Service, Southern Region, ARS-S-151, 56 pp. (Series discontinued; Agricultural Research Service is now Science and Education Administration-Agricultural Research.)

(27) Smith, C. N., R. A. Leonard, G. W. Langdale, and G. W. Bailey.
1978. Transport of agricultural chemicals from upland Piedmont watersheds. U.S. Environmental Protection Agency, EPA-600/3-78-056.

(28) Smith, R. E.
1977. Field test of a distributed watershed erosion/sedimentation model. In Soil Erosion: Prediction and Control. Soil Conservation Society of America, Special Publication No. 21, pp. 201-209, Ankeny, Iowa.

(29) Williams, J. R.
1975. Sediment-yield prediction with universal equation using runoff energy factor. In Present and Prospective Technology for Predicting Sediment Yields and Sources, U.S. Department of Agriculture, Agricultural Research Service, Southern Region, ARS-S-40, pp. 244-252. (Series discontinued; Agricultural Research Service is now Science and Education Administration-Agricultural Research.)

(30) ------, and H. D. Berndt.
1977. Determining the Universal Soil Loss Equation's length-slope factor for watersheds. In Soil Erosion: Prediction and Control. Soil Conservation Society of America, Special Publication No. 21, pp. 217-225, Ankeny, Iowa.

(31) Wischmeier, W. H., and D. D. Smith.
1978. Predicting rainfall erosion losses. U.S. Department of Agriculture, Agriculture Handbook No. 537, 58 pp.

(32) Yalin, Y. S.
1963. An expression for bedload transportation. Journal of the Hydraulics Division, Proceedings of the American Society of Civil Engineers 89(HY3):221-250.

(33) Young, R. A.
1978. Review of eroded sediment particle size and density data. Personal Correspondence. U.S. Department of Agriculture, Science and Education Administration, Morris, Minn.

Chapter 4. THE NUTRIENT SUBMODEL

M. H. Frere, J. D. Ross, and L. J. Lane[1/]

INTRODUCTION

Nutrients are naturally occurring chemicals essential for plant growth. A total of 16 chemical elements are necessary for the growth and reproduction of most plants, although the most significant are nitrogen (N), phosphorus (P), and potassium (K). Most soils are deficient in N, P, and K for optimum plant production, and thus commercially-available fertilizers contain these nutrients essential to maintain the current level of agricultural production. The other nutrient elements may be added as impurities in the fertilizer or applied to treat specific nutritional problems. Present evidence indicates that nitrogen and phosphorus are the principal nutrient pollutants and, therefore, only these nutrients are considered in this model.

A major source of nutrients reaching water bodies in this country is sewage, both from municipal treatment plants and nonsewered residences. These represent point sources of pollution and extensive efforts are underway to limit their contributions. Runoff from rural land is another major source, but unlike point sources, runoff integrates the contribution from a diffuse, dynamic source. It must be recognized that some nutrients leave the system even when fertilizer is not applied and while we cannot eliminate all nutrient losses, it is desirable to minimize them.

It must be emphasized at the beginning that the dynamic system under consideration is complex. The wide variety of climates and landscapes provides such a wide range of results that there is no typical case. Complications are introduced by difficulties in chemical analysis for nutrients in water samples. Numerous procedures have been followed for chemical analysis, but changes in nutrient form can occur between the times of sampling and sample analysis. Some nutrient data have been reported as the soluble form when actually they could have been associated with colloidal material not removed from the sample. The practical significance of these complications is unknown, but they are noted at this time to alert the reader of the limitation of the data associated with nutrient pollution.

THE PROBLEMS

Two problems are associated with nutrients in the aquatic environment: (1) Water may be toxic to humans, animals, or fish when the concentration of

1/ Soil scientist, USDA-SEA-AR, Southern Region Office, New Orleans, La.; mathematician, USDA-SEA-AR, Durant, Okla.; and hydrologist, USDA-SEA-AR, Tucson, Ariz., respectively.

certain nutrient forms exceeds a critical level, and (2) eutrophication may be accelerated. The nitrite form of nitrogen, which is the most toxic, interacts with components in the blood to interfere with oxygen transport. Methemoglobinema, the technical name of this illness, is often called "the blue baby syndrome" because infants are very susceptible. Most of the problems with drinking water have been associated with farm wells with faulty well casings, located close to manure concentrations such as barnyards.

Nitrate is 5 to 10 times less toxic than nitrite. Children convert some nitrate to nitrite in their stomachs and can develop methemoglobinemia. The U. S. Drinking Water Standards set the limit for nitrate at 10 mg nitrogen/l and recommendations for livestock are 10 times higher. Dissolved ammonia is another form of nitrogen that can occur at levels toxic to fish. Micro-organisms can generate free ammonia from organic matter in lake bottoms during summer stagnation periods. Trout are sensitive to 1 to 2 ppm ammonia while goldfish appear to be less sensitive.

Eutrophication is the enrichment of waters by nutrients and the ensuing luxuriant growth of plants. Rapid growth of algae is the greatest and most widespread eutrophication problem in most states. Algae can create obnoxious conditions in ponded waters, increase water treatment costs by clogging screens and requiring more chemicals, and cause serious taste and odor problems. When a large mass of algae dies and begins to decay, the oxygen dissolved in the water decreases and certain toxins are produced, both of which may kill fish.

Aquatic plants require a number of nutrients for growth, but nitrogen and phosphorus appear to be the ones accounting for most of the excessive growth. Eutrophication appears to become a problem when the concentration of inorganic nitrogen exceeds about 0.3 ppm and inorganic phosphorus exceeds about 0.015 ppm. These concentrations of inorganic forms of nutrients are maintained by microbial conversion of organic forms so the total input of nitrogen and phosphorus per unit area of the lake (loading rate) is important. Current international quidelines for eutrophication control are 2.0 to 5.0 kg of P and 50 to 100 kg of nitrogen per surface hectare of lake per year (1.8 - 4.5 lb P/acre and 45 - 90 lb N/acre).

CONTROL OF NUTRIENT POLLUTION

Nutrients as related to water quality are transported from the watershed by three processes: runoff, erosion, and leaching. Soluble forms of nitrogen and phosphorus are transported in the runoff. Insoluble forms and forms adsorbed to sediment particles are moved by erosion. Nitrate is the principal nutrient form leached to groundwater or base flow by percolating water.

To reduce the concentration of nutrients in receiving water (amount of nutrient per unit volume of water) or the total amount (load), either the amount of nutrient available for transport or the transport process must be reduced. The amount of nutrient available for transport can be reduced by practices such as applying fertilizers, manures, and wastes when the runoff, erosion, and leaching processes are at a minimum, or by incorporating the nutrients into the soil so that they are not accessible to runoff water. Conservation practices such as contour farming, conservation tillage, terracing, and grassed waterways can reduce the amount of runoff or erosion or leaching.

Unfortunately, each practice has its limitations and a combination of practices are often needed. In addition, a practice that controls one problem may induce another. As an example, practices reducing runoff may create leaching problems. Finally, we must recognize that the effect of practices cannot be evaluated by a single storm event. Few, if any, storms produce the same results. Consequently, a practice or combination of practices must be evaluated for different types and sizes of storms occurring at different stages of growth or times of the year.

NUTRIENT MODEL INPUT AND OUTPUT

Weather, soils, topography, and land use all effect the performance of a conservation or pollution control practice, and a comprehensive mathematical model is useful in the evaluation. As shown in figure I-13, the model requires information about hydrology, erosion, and particular nutrient characteristics of the field to predict the nitrogen and phosphorus moving in runoff, with erosion, and by leaching. The hydrology model provides estimates of the volume of runoff, percolation, soil water and temperature, and plant growth and water-use, while the erosion model provides estimates of sediment loss on a field scale. The model predicts the average concentration of soluble N and P in the runoff. Multiplying the average concentration by the volume of runoff estimates the total amount or load produced by the storm. The model provides an estimate of the amount of nitrate leached and its average concentration. The estimates of N and P associated with sediment from erosion corresponds to the total N and total P often reported in water quality studies.

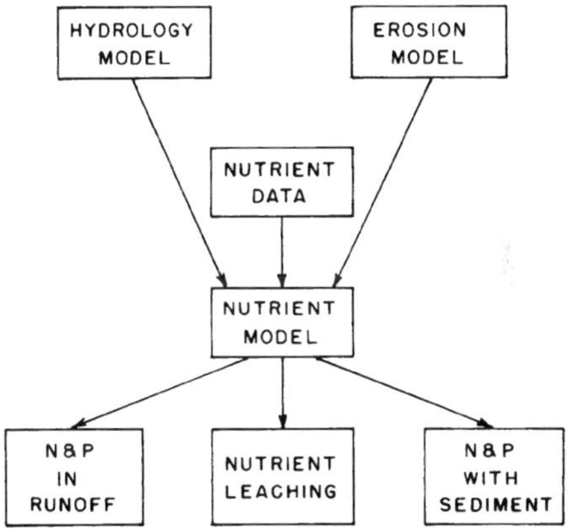

Figure I-13.—Flow diagram of input and output for the nutrient model.

SEDIMENT TRANSPORT OF NUTRIENTS

The loss of total N and total P from cropland ranges from about 1 to 50 or 100 kg/ha per year. Sediment can be a major transport vehicle for phosphorus and organic nitrogen. Raindrop splash and flowing water detach soil particles and organic matter containing nitrogen and phosphorus. The transport capacity of the flowing water depends primarily on the volume and velocity of water flow. Whenever the velocity is reduced, such as by a flatter slope, the transport capacity is reduced and any sediment in excess of the reduced capacity settles out. Since larger and heavier particles settle out first, the remaining sediment contains a larger percentage of finer particles which have a higher capacity per unit of sediment to absorb phosphate and organic nitrogen. Also organic matter is lighter and tends to be associated with the fine particles. Thus, the transported sediment is richer in phosphorus and nitrogen than the original soil. Figure I-14(a) is a flow diagram for this process in the nutrient model.

The sediment of concern in this model is limited to that from surface rill and interrill erosion. Sediment from gully, channel, or other sources of erosion is not included in the calculation because the nutrient content of the soil changes significantly with soil depth. In fact, some subsoils high in clay content may absorb phosphate and can deplete the solution concentration of phosphate from the soil.

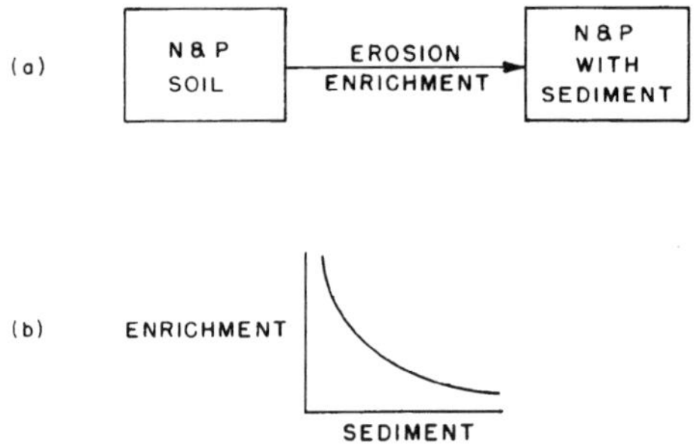

Figure I-14.—(a) Diagram for estimating nitrogen and phosphorus losses with sediment; (b) relation of enrichment to amount of sediment.

Algorithm

The kilograms/hectare of nitrogen or phosphorus transported by sediment (SEDN or SEDP) is predicted in this model by the following equation:

SED- = SOIL- * SED * ER- [I-156]

and

ER- = A- * SED ** B- [I-157]

where SOIL- is the N (SOILN) or P (SOILP) content (kg/kg soil) in the field, SED is the kg/ha of sediment predicted by the erosion model. ER- is the enrichment ratio for N or P, A- is a coefficient for N or P, and B- is an exponent for N or P.

Soils typically contain 0.05 to 0.3 percent nitrogen and 0.01 to 0.13 percent phosphorus (see vol. III, ch. 13, table 1). This sixfold to tenfold range indicates that a measurement of the specific field involved is highly desirable. Applications of fertilizers, manures, wastes, and crop residues increase the N and P content above natural levels while intensive cropping without nutrient additions reduces the N and P content (vol. III, ch. 15). Many samples of soils are analyzed each year by State and commercial soil testing laboratories. Adequate data for a specific field might already be available from this source or could be measured in a short period of time. If no other information is available, approximate values (vol. III, ch. 14 and 15) could be used, recognizing the error that can be made.

The enrichment of sediment, as described above, occurs because of selective erosion and deposition processes. An evaluation of all available data indicates that the logarithmic relation, figure I-14(b), between the enrichment ratio and the amount of sediment holds for wide ranges of soil and vegetative conditions (vol. III, ch. 12). It is suspected that changing soils, crops, or management practices should result in different coefficients and exponents in the equation. However, the amount of data available at the present time does not permit a statistically significant distinction in these parameters. Considerable research work is presently being conducted that should be useful in improving this relation. Using a value of 7.4 for A- and -0.2 for B-, the enrichment ratio for both nitrogen and phosphorus can be predicted within a factor of two for an annual average and a factor of five for individual storm events. The implications of these relations for sediment transport of nutrients is that reducing sediment transport will not reduce nutrient transport by an equal amount. Conservation practices that reduce erosion will reduce nutrient transport but to a smaller degree.

SOLUBLE NUTRIENTS IN RUNOFF WATERS

Runoff waters contain soluble forms of nitrogen and phosphorus ranging from 0.01 to 1 ppm P and 0.1 to 10 ppm N with loads from less than 0.1 to as much as 10 kg/ha/yr or lb/acre/yr (figures I-15 and I-16). This is one of the most difficult areas to model because of the variety of nutrient sources and processes of extraction. While considerable data have been reported on the

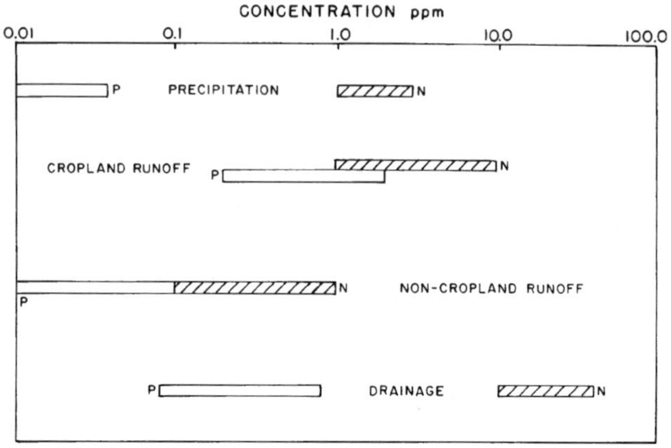

Figure I-15.—Range of nitrogen and phosphorus concentrations in different waters.

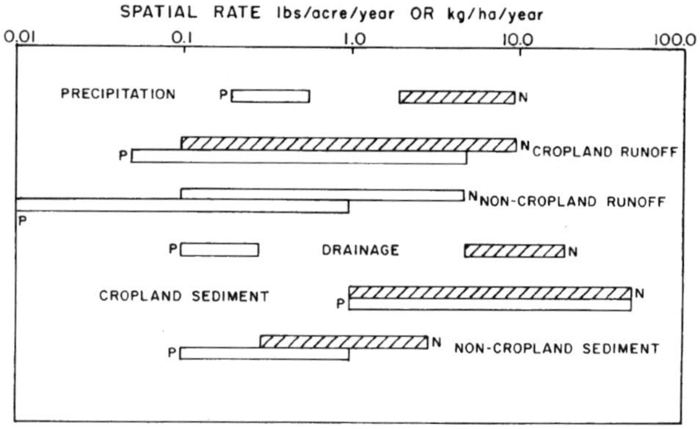

Figure I-16.—Range of spatial rates of nitrogen and phosphorus in waters and sediments.

integrated or gross effects, very little research has been reported on the individual processes (vol. III, ch. 14 and 15). Figure I-17 illustrates the concepts used in this model for predicting the soluble forms of N and P in runoff waters.

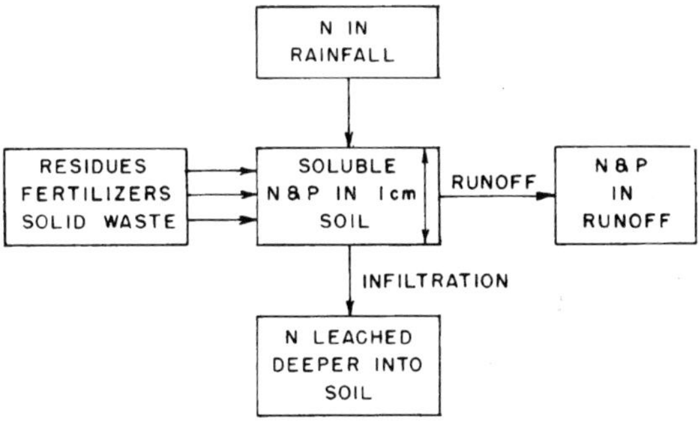

Figure I-17.—Diagram for estimating nutrient losses in runoff.

Rainfall

Rainfall is the driving force for the system and it also contains nutrients. The concentrations of nutrients in rainfall not only vary across the country (figure I-18) but within short distances and during a storm. Most of the phosphate is associated with dust and is generally neglected as an input. Nitrate and ammonium are the principal nitrogen forms occurring in precipitation and their sum averages from 1 kg of N/hectare/yr in the West to over 3 kg of N/hectare/yr in the Great Lakes area (0.9 - 2.7 lb N/acre/yr). This level is not agronomically significant for cropland but could be for unfertilized range and forest areas. Seasonal charts indicate the highest concentrations occur in the spring and summer. Nitrogen concentrations are often found to be slightly higher during the first part of a storm.

The concentration of nitrogen in rain ranges from a little less than 1 ppm to a little over 1 ppm. This concentration range corresponds to the lower end of the concentration range for runoff from cropland and the upper end for runoff from noncropland (figure I-15).

Applied Nutrients

Other sources of nutrients are the fertilizers, manure, and plant residues placed on the surface. With the exception of slow-release fertilizers,

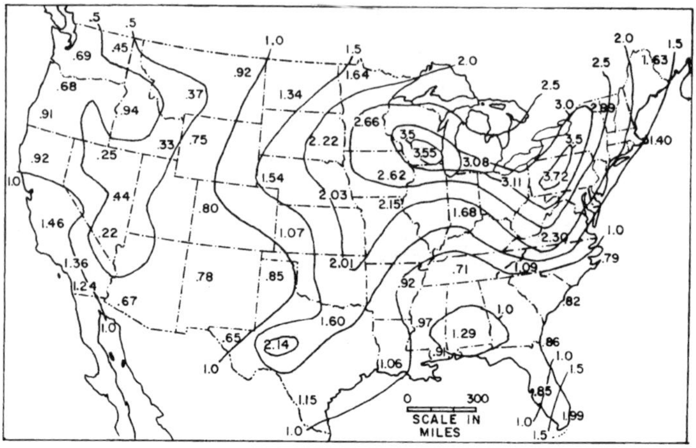

Figure I-18.—Nitrogen contributions (NO_3-N and NH_3-N), kilograms per hectare per year, from rainfall throughout the United states [from (1)].

nitrogen fertilizers are quite water soluble and phosphate fertilizers are moderately soluble. Consequently, water from the soil and light rains dissolves the granules. Most fertilizer and manure applied to cropland is mixed into the soil by plowing or disking leaving only a proportionate fraction near the surface.

Plant materials are not usually "applied" to a field, rather they are the living crop canopy or the residues left after the harvest of the crop. They include the stems and leaves of such crops as small grains, corn, sorghum, soybeans, and cotton. On rangeland, pasture, hay meadows, and cropland with sod crops, there is a significant amount of litter and mulch on the soil surface besides the living crops. This plant material contains organic forms of nitrogen and phosphorus, some of which can be leached from the residue during storms. Plant material changes the nitrogen content of runoff on the order of $\pm$ 10 percent but increases the phosphorus content as much as fourfold (vol. III, ch. 15). Table I-12 gives some estimates for amounts of residues and their nutrient content. These values can vary by a factor of two across the country and the fraction of the nutrient content that can be leached out is probably 50% $\pm$ 20%. Concentrations on the order of 0.1 ppm P have been observed in washoff from mature cotton plants.

Table I-12. Approximate yield and nutrient content of selected crops (values can vary by a factor of two across the country)

Crop		Yield		Nitrogen		Phosphorus[1/]	
		(kg/ha)	(units/acre)	(kg/ha)	(lb/acre)	(kg/ha)	(lb/acre)
Alfalfa[2/]		8,960	(4 tons)	224	(200)	20	(18)
Barley	grain	2,150	(40 bu)	39	(35)	7	(6)
	straw	2,240	(1 ton)	17	(15)	2	(2)
Beans[2/]	(dry)	1,950	(30 bu)	84	(75)	11	(10)
Bermudagrass		17,920	(8 tons)	224	(200)	34	(30)
Bluegrass		4,480	(2 tons)	67	(60)	9	(8)
Cabbage		44,800	(20 tons)	168	(150)	18	(16)
Clover[2/]	red	4,480	(2 tons)	90	(80)	11	(10)
	white	4,480	(2 tons)	146	(130)	11	(10)
Corn	grain	9,400	(150 bu)	151	(135)	27	(24)
	stover	10,080	(4.5 tons)	112	(100)	18	(16)
	silage	56,000	(25 tons)	224	(200)	34	(30)
Cotton	lint & seed	2,240	(1 ton)	67	(60)	13	(12)
	stalks	2,240	(1 ton)	50	(45)	7	(6)
Cowpea hay[2/]		4,480	(2 tons)	134	(120)	11	(10)
Lettuce		44,800	(20 tons)	100	(90)	13	(12)
Lespedeza[2/]		4,480	(2 tons)	95	(85)	9	(8)
Oats	grain	3,200	(90 bu)	62	(55)	17	(10)
	straw	4,480	(2 tons)	28	(25)	9	(8)
Onions		16,800	(7.5 tons)	50	(45)	9	(8)
Oranges		62,720	(28 tons)	95	(85)	13	(12)
Peanuts[2/]	nuts	3,360	(1.5 tons)	123	(110)	7	(6)
Potatoes	tubers	44,800	(400 cwt)	106	(95)	13	(12)
	vines	2,240	(1 ton)	100	(90)	9	(8)
Rice	grain	4,540	(90 bu)	62	(55)	13	(12)
	straw	5,600	(2.5 tons)	34	(30)	4	(4)
Rye	grain	1,880	(30 bu)	39	(35)	4	(4)
	straw	3,360	(1.5 tons)	17	(15)	4	(4)
Sorghum	grain	3,360	(60 bu)	56	(50)	11	(10)
	stover	6,720	(3 tons)	73	(65)	9	(8)
Soybean[2/]	grain	3,020	(45 bu)	179	(160)	18	(16)
	straw	2,240	(1 ton)	28	(25)	4	(4)
Sugarbeets	roots	44,800	(20 tons)	95	(85)	16	(14)
	tops	26,880	(12 tons)	123	(110)	11	(10)
Sugarcane	stalks	67,200	(30 tons)	112	(100)	22	(20)
	tops	29,120	(13 tons)	56	(50)	11	(10)
Timothy		5,600	(2.5 tons)	67	(60)	11	(10)
Tobacco		3,360	(1.5 tons)	129	(115)	11	(10)
Tomatoes	fruit	56,000	(25 tons)	162	(145)	22	(20)
	vines	3,360	(1.5 tons)	78	(70)	11	(10)
Wheat	grain	3,360	(50 bu)	73	(65)	16	(14)
	straw	3,360	(1.5 tons)	22	(20)	2	(2)

1/ Pounds P = 0.436 lbs P_2O_5.
2/ Legumes that do not require fertilizer nitrogen.

Animal manures may be applied to crop, pasture, and even rangelands if the manure is available. The manure is available and even a disposal problem when animal production is part of the total agricultural unit. Beef production poses different problems than dairy, swine, and poultry operations. Most beef animals spend part of their lives in an open grazing situation and the rest of the time in a confined feedlot where the manure can be collected and used. The nutrient content of manure varies with the animal and the type of feed used. An average content is given in table I-13. About 50% of the N is lost during handling, storage, and application. In addition, only about 50% of the applied organic N is mineralized and available to plants in the first cropping season.

Table I-13.—Nutrient content of manures[1/]

Animals	N	P
	(%)	(%)
Beef	2.5	0.8
Dairy	2.0	.6
Swine	2.8	1.0
Laying hens	4.3	1.3
Broilers	3.8	1.3

1/ On a dry weight basis after losses during handling, storage, and application.

The P content remains relatively stable during handling, storage, and application. The fraction of the nutrient content in manure that can be leached out is probably about the same as for plant residues 50% ± 20%.

Soluble Nitrogen and Phosphorus in the Surface Soil

The surface layer of soil contains a certain amount of soluble N and P. Estimating this value along with a runoff extracting or efficiency factor is the weakest part of the nutrient model. Fortunately, the concentration and load of soluble nutrients in runoff are not usually the dominant factor. However, conservation practices that reduce sediment transport of nutrients, like notill, can increase their concentration in runoff water and the relative importance of this pathway.

At the present time with our limited understanding of the system, we assume that a 1 cm layer of soil interacts with the rain. All the soluble N and P are expected to exist in the water of the pores. Only a fraction of the nutrients are extracted into the flowing water. Analysis of pesticide and nutrient data suggest that this extraction coefficient ranges from 0.01 to 0.4. The loss of freshly applied fertilizer is also seldom more than 1 to 5% (vol. III, ch. 15). Based on the range of extraction coefficients and the observed range of P concentration in runoff, the concentration of soluble P in the surface layer would range from an upper level of 5 ppm to less than 2 ppm.

The porosity of soils is usually in the order of 40% ± 10% and a hectare would contain 4×10^4 kg of water in the upper centimeter layer. At 5 ppm this is equivalent to 0.20 kg/ha. For the lack of better information, we assume similar concentrations for nitrogen.

Leaching from the Surface Layer

Only part of the rainfall leaves the field as runoff. Frequently there will be no runoff from a storm. The part of the rainfall that does not runoff fills the surface layer and leaches soluble nutrients deeper into the soil. In this model the amount leached is proportional to the fraction of the rainfall that does not run off. It is subtracted before runoff to account for non-runoff producing rains. Soluble nitrogen compounds leached into the soil are assumed to be nitrate or quickly converted to nitrate and are added to the nitrate pool in the soil. Soluble phosphate compounds are leached out of the surface layer but do not move on through the soil because of its large buffering capacity. The buffering capacity of the soil also prevents the phosphate concentration from dropping below a level characteristic of the soil.

Algorithm

The basic assumption is that the rate of change in concentration of soluble nitrogen in the water in the surface (1 cm) of soil is proportional to the difference between the existing concentration and the concentration of nitrogen in the rainfall. That is, we assume

$$\frac{dc}{dt} = K_1 \; f(t)(C_r - C) \qquad [I-158]$$

where K_1 is a rate constant for downward movement, f(t) is infiltration rate, C_r is concentration in the rainfall, and C is concentration in the soil surface. The mean concentration during infiltration is

$$\overline{C}_1 = ((C_o - C_r)/K_1F)(1-\exp(-K_1F)) + C_r \qquad [I-159]$$

where C_o is initial concentration and F is total infiltration for the storm. The concentration at the end of infiltration and the start of runoff is

$$C_1 = (C_o - C_r)\exp(-K_1F) + C_r \quad . \qquad [I-160]$$

The final concentration after runoff is

$$C_2 = (C_1 - C_r)\exp(-K_2Q) + C_r \qquad [I-161]$$

and the mean concentration during runoff is

$$\overline{C}_2 = ((C_1 - C_r)/K_2Q)(1-\exp(-K_2Q)) + C_r \qquad [I-162]$$

where K_2 is a rate constant for movement into runoff and Q is total runoff.

The extraction coefficients are

$$EXKN_1 = d \; POR \; K_1 \qquad [I-163]$$

for downward movement and

$$EXKN_2 = d\ POR\ K_2 \qquad [I-164]$$

for movement in runoff where d = 10 mm is the depth of the surface layer and POR is the porosity.

The equations for soluble phosphorus are similar except the rainfall concentration C_r is assumed zero but it is replaced by a base concentration C_B due to the buffering action described earlier.

Downward movement of nitrogen is calculated as

$$DWN = \bar{C}_1 * EXKN_1 * FI * 0.01 \qquad [I-165]$$

where FI is the total infiltration minus an initial abstraction assumed equal to the volume of pore space in the surface soil (d POR). Similarly, the amount of soluble nitrogen in runoff is

$$RON = \bar{C}_2 * EXKN_2 * Q * 0.01 \quad . \qquad [I-166]$$

The amount of souble phosphorus is calculated as

$$ROP = \bar{C} * EXKP_2 * Q * 0.01 \qquad [I-167]$$

where $EXKP_2$ is the phosphorus extraction coefficient for movement into runoff.

The concentration in the soil solution is 10 times (depth of active surface layer = 10 mm) the kilograms/hectare of soluble N or P, SOL-, divided by the porosity, POR. The amount of soluble N or P is increased by the amount in rainfall and additions of fertilizers, manure, and plant residues, F-, on several application dates, DF. These additions are assumed instantaneous inputs or impulse additions to the soil solution while the rainfall nitrogen is distributed throughout the storm.

The application factor, FA, is 1 for surface-applied nutrients and equal to the fraction of the application remaining in the 1 cm surface layer if the nutrients are incorporated. The initial value of SOL- reflects the soil contribution only. All fertilizers, wastes, and residues are input via additions. If the addition occurred before the first simulation date, the date of application can be set as zero.

In the simplified model described above, soluble nutrients are moved out of the surface layer with infiltration. Soil evaporation may contribute to the nitrogen concentration in the surface layer if nitrogen is transported upward with the water flux due to evaporation. This contribution may be estimated as the product of the NO_3 concentration in the root zone and the water flux due to evaporation.

Exact values of the nitrogen and phosphorus extraction coefficients are unknown. For the downward movement coefficients, values of 1.0 imply complete efficiency for infiltrating water while values equal to the extraction coefficients for movement in surface runoff imply the same efficiency for downward movement as for the movement in runoff. As a first approximation, and in the absence of experimental data, we assumed the downward movement extraction coefficients were less than 1.0 but greater than the extraction coefficients for movement in surface runoff. Simulation runs were made with the downward movement extraction coefficients varying from 0.1 to 1.0. Values near 1.0 result in very rapid decreases in concentration due to infiltration. Values in the range 0.2 to 0.5 resulted in somewhat higher soil concentrations and more reasonable concentrations in surface runoff. Therefore, we arbitrarily fixed the downward movement coefficients at 0.25. Recognizing the interaction between the extraction coefficients, this assumption forces all the variability into the extraction coefficients for movement in runoff. However, since the extraction coefficient is a number reflecting the complex movement of soluble nutrients, and as such is a simplification, we fixed two of the coefficients and let the two remaining coefficients represent extraction efficiency. The constraint is that the extraction coefficients for movement in runoff must be less than the downward movement extraction coefficients.

NITROGEN CYCLING AND NITRATE LEACHING

One of the important pathways of nitrogen loss is the leaching of nitrate from the root zone to ground water, tile drains, or base flow. Figure I-15 illustrates that the concentration of nitrate in drainage can exceed 10 ppm. In order to predict this source of pollution it is necessary to maintain a budget of nitrate and water in the root zone. The water budget and movement are calculated in the hydrology model. Figure I-19 shows the inputs and withdrawals for the nitrogen system. The nitrogen cycle in the soil is extremely

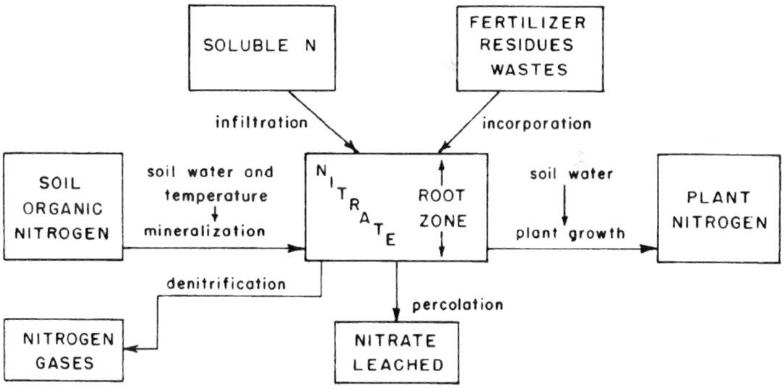

Figure I-19.—Diagram for estimating nitrate leaching

complex and very complex models have been developed to describe the system. Unfortunately, complex models require numerous parameters which are not usually available. Therefore, we have chosen to include the minimum number of relations and parameters that will provide an acceptable estimate of the system behavior.

The active root zone of a crop depends upon the crop, the soil, and the stage of growth. The root zone is important because it determines the depth of soil from which the plant is removing nitrate. In this model we neglect the fact that early in the season the root zone is shallower than late in the season. We use only the final root zone depth because the nitrate is not lost to potential uptake until it has moved below this zone. The density of roots and extraction of water and nutrients are concentrated in the upper part of the root zone. Therefore, while a few roots may grow very deep, the effective root zone is much shallower. Table I-14 gives depths of active root zones for some crops. Soil conditions, like hard pans, sand or gravel layers, or acid subsoil can often limit the depth of the root zones. Therefore, it is important to check the characteristics of the field under study.

Table I-14.—Root zones of various crops

Crop	Active root zone	
	(in)	(mm)
Corn Sorghum Alfalfa Tomatoes Wheat	48	1,200
Sugarbeets Soybeans Field beans Potatoes	36	900
Pasture	24	600
Onions Bluegrass lawns	18	450

Soluble N from the Surface Layer

As described in a previous section, soluble nitrogen is leached from the surface layer by infiltrated water into the root zone. All of the soluble N is nitrate or forms that are quickly converted to nitrate.

Applied Nutrients

Fertilizers, wastes, and plant residues are often injected, plowed under, disked in, or otherwise incorporated into the soil. Not all of the nitrogen in these materials is in the nitrate form or forms rapidly converted to nitrate. However, for simplicity we have assumed that only nitrate or forms easily converted to nitrate are included in the application amount. Other forms of N would add to the organic nitrogen pool which is already very large. In this version of the model the potentially mineralizable nitrogen in the soil is not increased by the applied nutrients.

Soil Organic Nitrogen

Micro-organisms in the soil convert organic forms of nitrogen to nitrate. This process, mineralization, is sensitive to temperature and moisture conditions in the soil. Only a part of the organic nitrogen is readily converted and a chemical test of the sample of soil from the field in question is the best method for determining the potentially mineralizable nitrogen. In the event this is not possible, a less exact estimate can be made from the total N and organic carbon contents for various soil orders (vol. III, ch. 13, table 1). Increasing temperature increases the rate of mineralization in an exponential manner up to a peak at 35°C (308° kelvin, 95°F). The optimum moisture content for mineralization is "field capacity" (when grapitational water has drained away, about 1/3 bar tension). Mineralization decreases linearly with a decrease in water content below this optimum.

Algorithm

The following equations are used to calculate the kilograms/hectare of mineralized nitrogen, MN, during a period, DAYS, between storm events:

$$TK = EXP(15.807 - 6350/TA) \qquad [I\text{-}168]$$

$$WK = AWC/FC \qquad [I\text{-}169]$$

$$MN = POTM*WK*(1-EXP(-TK*DAYS) \qquad [I\text{-}170]$$

where TK, the temperature coefficient, is calculated from the average temperature during the period, TA, in degrees kelvin; WK, the water coefficient, is calculated from the average volumetric water content during the period, AWC, and the field capacity, FC, both in fractions (mm^3/mm^3); and POTM is the kilograms/hectare of potentially mineralizable nitrogen in the soil. The average temperature and the average soil water content are supplied by the hydrology model. While the temperature of the soil in the root zone is preferred, air temperature can be used because the error introduced is small compared with other sources of error.

The potential mineralizable nitrogen, POTM, should reflect agricultural practices such as the amount of plant residue. However, for a single cultural practice, POTM can be initialized each year (for example, at the date of crop emergence) to the original measured value for the soil.

Plant Nitrogen

Crop growth is the most important process that removes nitrate from the root zone. Under good weather conditions and with good management practices, most of the nitrate in the root zone is taken up by plants. In this model two options are provided for calculating plant uptake of nitrogen.

Algorithms

Option I simulates plant growth as a function of plant water use and N uptake as a function of plant N content. Accumulated dry matter production (grain + stover + roots) is calculated with the equation

$$DM_i = \frac{\Sigma WU_i}{PWU}(YP)(K) \qquad [I\text{-}171]$$

where DM is accumulated dry matter production on day i, WU is daily plant water use, PWU is total plant water use for the growing season, YP is the crop yield potential, and K is the ratio of dry matter to crop yield at maturity.

The N poncentration in plants is a function of plant maturity as expressed by the equation

$$c_i = \text{MINIMUM} \begin{matrix} b_1(DM_i/TDM)^{b_2} \\ b_3(DM_i/TDM)^{b_4} \end{matrix} \quad [I-172]$$

where c_i is the N concentration in the plant on day i, TDM is the total dry matter production for the growing season, and b_1, b_2, b_3, and b_4 are parameters defined by Smith and others, table 3 (vol. III, ch. 13).

The accumulated N uptake can be computed for any day during the growing season with the equation

$$UN_i = c_i \ DM_i \quad [I-173]$$

where UN is the accumulated N uptake for day i. The N uptake during any period of the growing season is simply the difference between beginning and ending accumulated amounts.

Option II assumes nitrogen uptake follows a normal probability (S shaped) curve which is reduced for moisture stress. The equations are:

$$PUN = 1 - 1/2 \ (S)^{-4} \quad [I-174]$$

$$S = 1.0 + 0.196854X + 0.115194X^2 + 0.000344X^3 + 0.01957X^4 \quad [I-175]$$

$$X = (T-M)/SD \quad [I-176]$$

where PUN is the fraction of the potential annual plant nitrogen taken up by T days of growth. M is the days of growth required to take up 50% of the annual amount. SD is the number of days between 50% and 84% uptake and corresponds to 1 standard deviation. When X is negative, PUN = 1-PUN, the lower part of a symetrical curve. The amount of nitrogen uptake between storms, UN, kilograms/hectare, is:

$$UN = (PUN-PPUN)*PU*TR \quad [I-177]$$

where PPUN is the previous level of uptake at the last storm; PU is the potential annual nitrogen uptake for the crop in kilograms/hectare; and TR is the ratio of actual to potential transpiration during the period.

This option should be most useful if concentration parameters used in equation [I-172] are not available.

Nitrate Leaching

The amount of nitrate leached below the root zone is the fraction of the water in the root zone that percolates out of the root zone times the amount of nitrate in the root zone. The amount of percolation between storms is calculated by the hydrology model. The amount of water remaining in the root zone after percolation is the field capacity times the depth of the root zone.

Most nitrate leaching occurs in the winter and spring when plants are not extracting much water. October 1 is usually the date when the soil profile is driest and that date (Julian day 274) is used in this model as the initial day for accumulating the annual amounts of water percolated and nitrate leached. Thus, DRAIN is the accumulated amount of percolated water from each storm event and TOTNL is the accumulated amount of nitrate leached.

Algorithms

One estimate of the leaching fraction, Method A, assuming complete mixing, is:

$$FL = PERC/(PERC + RZC) \quad [I\text{-}178]$$

where PERC is the mm of percolation and RZC is the mm of water remaining in the root zone. The kilograms/hectare of nitrate leached after each storm, NL, is FL * NO3 where NO3 is the kilograms/hectare of nitrate in the root zone at the time of the storm.

There is a second way, (Method B in vol. III, ch. 13), to estimate the annual amount of nitrate leached. This method calculates the leaching fraction, FL, as

$$FL = (DRAIN/(DRAIN + 10FC))^{**}EX \quad [I\text{-}179]$$

where

$$EX = (RZMAX - 300)/10 \quad [I\text{-}180]$$

where DRAIN is the annual depth of percolation in millimeters, RZMAX is the root zone depth in millimeters, and FC is the water fraction of the soil at field capacity. The constant, 300, assumes that the nitrate was uniformally distributed in the top 600 mm (2 ft) of the soil.

Denitrification

Under anaerobic conditions in the soil, nitrate can be reduced to nitrogen gases. The process is considered to be a first-order reaction sensitive to organic carbon, temperature, and moisture. In this simple model, the rate constant at 35°C is calculated from the amount of organic matter in the soil and adjusted for temperature assuming a twofold reduction for each ten degrees decrease in temperature. The effect of moisture is accounted for by permitting denitrification only for the number of days of drainage in excess of a half day; that is, when the moisture content of the soil exceeds field capacity.

Algorithm

The amount of soil carbon is calculated from the amount of organic matter:

$$SC = OM/0.1724 \quad [I\text{-}181]$$

where SC is milligrams carbon/gram of soil and OM is percent organic matter. The rate constant, DK, at 35°C is calculated according to:

$$DK = 24*(0.011 * SC + 0.0025) \quad [I\text{-}182]$$

where DK has units of day^{-1}. The temperature adjusted rate constant is:

$$DKT = \exp(0.0693 * ATP + DB) \quad [I\text{-}183]$$

with

$$DB = \ln DK - 2.4255 \quad [I\text{-}184]$$

and ATP is equal to the average temperature, degrees celsius, used in the mineralization process. The amount of denitrification, DNI, in kilograms/hectare, between storms is:

$$DNI = NO3 * (1.-\exp(-DKT*(DT-.5))) \quad [I\text{-}185]$$

where NO3 is the kilograms/hectare of nitrate in the root zone and DT is the number of days of drainage since the last storm. The half day subtraction is justified on the basis that many short drainage periods have only short periods of anaerobic conditions.

TESTING THE NUTRIENT SUBMODEL

A limited test of the submodel was made using data from watershed P2 at Watkinsville, Ga. Nitrogen losses in runoff, with sediment, and in percolation below the root zone and phosphorus losses in runoff and with sediment were simulated for 1974. The simulated losses in runoff and with sediment were compared to observations (3). Fertilizer (N-P-K) was applied at the rates of 38-33-127 kg/ha and incorporated to an average depth of 10 cm. Corn was planted immediately after fertilization. Forty-three days after planting 101 kg/ha of N was applied to the surface soil by spray.

Model input values for the simulation are given in table I-15. The potential N uptake for Option 2 was estimated from the sampled grain and stover yield. Values for the parameters in table I-15 were taken from Smith and others (3).

The climatic and hydrologic data given in table I-16 are, except for average soil water content, measured values. The transpiration ratio was set at 75%.

Measured and simulated runoff and sediment losses of nitrogen and phosphorus are compared in table I-17 and by relating observed and computed yields for individual storms. The relation between computed and observed soluble nitrogen in runoff was

$$\hat{N}_Q = -0.04 + 0.89\ N_Q \quad [I\text{-}186]$$

$$R^2 = 0.94$$

where $\hat{N}_Q$ is computed nitrogen yield and N_Q is observed nitrogen yield. For soluble phosphorus the regression equation was

Table I-15.—Model input values used for simulation, watershed P2 at Watkinsville, Ga., 1974

Input parameters	Parameter description	Parameter values
Initial conditions		
SOLN	Soluble nitrogen in the surface cm (kg/ha)	0.2
SOLP	Soluble phosphorus in the surface cm (kg/ha)	.2
NO3	Nitrate in the root zone (kg/ha)	21
SOILN	Soil nitrogen associated with sediment (kg/kg)	.00035
SOILP	Soil phosphorus associated with sediment (kg/kg)	.00018
Nutrient parameters		
EXKN	Extraction coefficient of nitrogen into runoff	.075
EXKP	Extraction coefficient of phosphorus into runoff	.075
AN	Enrichment coefficient of nitrogen in sediment	16.8
BN	Enrichment exponent of nitrogen in sediment	- .16
AP	Enrichment coefficient of phosphorus in sediment	11.2
BP	Enrichment exponent of phosphorus in sediment	- .146
POTM	Potentially mineralizable nitrogen (kg/ha)	47.0
RCN	Nitrogen concentration in rain (ppm)	.8
Nutrient additions		
NF	Number of additions	2
DF	Date of application	74119 74174
FN	Amount of nitrogen in application (kg/ha)	38 102
FP	Amount of phosphorus in application (kg/ha)	33 0
FA	Fraction of application in top cm	.1 1.0
Nitrogen uptake		
OPT	Plant growth simulation	1
YP	Potential yield of crop (kg/ha)	5700
C1	Coefficients for the nitrogen concentration	.0209
C2	in the plant.	- .157
C3		.0128
C4		- .415
Common to hydrology		
DEMERG	Date of crop emergence	74125
DHRVST	Date of crop harvest	74303
RZMAX	Depth of potential root zone (mm)	450
POR	Porosity (mm^3/mm^3)	.45
FC	Water content at field capacity (mm/mm)	.20
OM	Organic matter (%)	.65

Table I-16.—Climatic and hydrology data for Watershed P2, Watkinsville, Ga.

	Date	Rain-fall	Run-off	Perco-lation	Average temperature between events	Average soil water content	Sediment
Model parameter	DATE	RAIN	RUNOFF	PERC	ATP	AWC	SED
Month, day	Year and Julian day						
		(mm)	(mm)	(mm)	(°C)	(mm/mm)	(kg/ha)
April 4	74094	33.0	3.0	29.11	10	0.200	9.6
April 12	74102	1.0	.0	.0	14.83	.195	0
April 13	74103	24.0	4.0	13.13	15.83	.192	14.5
April 22	74112	8.0	.0	3.26	16.67	.197	0
May 2	74122	2.0	.0	.0	18.17	.195	0
May 4	74124	9.0	.0	1.55	19.28	.192	0
May 5	74125	19.0	1.0	17.56	19.56	.200	10.1
May 11	74131	3.0	.0	.0	20.11	.198	0
May 12	74132	13.0	.0	12.02	20.67	.200	0
May 15	74135	3.0	.0	.66	21.00	.199	0
May 23	74143	70.0	7.0	58.25	21.89	.197	92.0
May 26	74146	7.0	.0	4.54	22.94	.199	0
May 31	74151	13.0	.0	9.89	23.61	.199	0
June 8	74159	8.0	.0	3.84	24.39	.197	0
June 10	74161	6.0	.0	4.42	24.89	.200	0
June 20	74171	12.0	1.0	4.21	25.39	.196	1.4
June 27	74178	108.0	42.9	57.82	25.83	.196	966.5
July 17	74198	3.0	.0	.0	26.33	.189	0
July 23	74204	3.0	.0	.0	26.56	.182	0
July 24	74205	15.0	1.0	2.08	26.56	.184	23.4
July 26	74207	13.0	.0	12.32	26.56	.200	0
July 27	74208	72.0	45.6	26.30	26.56	.200	661.3
Aug 5	74217	1.0	.0	.0	26.44	.193	0
Aug 7	74219	27.0	.0	18.06	26.38	.189	0
Aug 10	74222	28.0	2.0	25.03	26.33	.200	22.6
Aug 14	74226	8.0	.0	5.95	26.28	.199	0
Aug 16	74228	51.0	8.0	39.47	26.17	.199	70.7
Aug 17	74229	15.0	1.0	13.51	26.06	.200	7.3
Aug 29	74241	17.0	1.0	2.58	25.67	.189	3.8
Sept 1	74244	11.0	1.0	7.82	25.06	.199	.5
Sept 3	74246	8.0	.0	6.86	24.89	.199	0
Sept 6	74249	23.0	.0	21.38	24.56	.199	0
Sept 25	74268	4.0	.0	.0	22.78	.193	0
Oct 16	74289	9.0	.0	.0	18.61	.188	0

Table I-17.—Observed and simulated nutrient losses for watershed P2 near Watkinsville, Ga., 1974 growing season

Date	Total nitrogen				Total phosphorus			
	Runoff		Sediment		Runoff		Sediment	
	Observed	Simulated	Observed	Simulated	Observed	Simulated	Observed	Simulated
	(kg/ha)							
May 5	0.0086	0.0104	0.0091	0.0448	0.0002	0.0035	0.0061	0.0159
May 23	.2152	.1253	.0733	.2622	.0061	.0289	.1460	.0958
June 20	.0534	.1809	---	---	.0004	.0005	.0021	.0026
June 27	1.8586	1.5284	1.7922	1.9026	.3301	.1394	.9275	.7183
July 24	.0861	.2175	.0811	.0868	.0019	.0049	.0158	.0311
July 27	1.0204	1.2950	2.0722	1.3846	.0353	.1518	.3464	.5200
Aug 10	.0603	.0480	.0671	.0801	.0022	.0073	.0122	.0287
Aug 16	.1669	.1117	.1565	.2129	.0074	.0276	.0406	.0775
Aug 17	.0248	.0216	.0245	.0291	.0016	.0043	.0050	.0103
Aug 29	.0169	.0192	.0128	.0207	.0015	.0017	.0015	.0073
Sept 1	.0102	.0234	.0123	.0030	.0126	.0019	.0016	.0010
Total	3.521	3.581	4.301	4.027	.399	.372	1.505	1.508

$$\hat{P}_Q = 0.02 + 0.40\ P_Q \qquad [I\text{-}187]$$

$$R^2 = 0.48$$

where $\hat{P}_Q$ is computed phosphorus yield in runoff and P_Q is the corresponding observed value.

Nutrients transported with sediment were predicted for individual storms and related to observed values. Predicted and observed nitrogen with sediment were related as

$$\hat{N}_s = 0.05 + 0.81\ N_s \qquad [I\text{-}188]$$

$$R^2 = 0.92$$

where $\hat{N}_s$ is predicted and N_s is observed nitrogen yield with sediment. The corresponding equation for phosphorus was

$$\hat{P}_s = 0.02 + 0.82\ P_s \qquad [I\text{-}189]$$

$$R^2 = 0.91$$

where $\hat{P}_s$ is predicted and P_s is observed phosphorus yield with sediment.

With the exception of soluble phosphorus (equation [I-187]) the model predicted total yields and reproduced trends in the observed data. In this case the regression coefficient of 0.40 suggests a bias in the predicted phosphorus yields with runoff. However, the extraction coefficients for movement of nitrogen and phosphorus in runoff were obtained by calibration with the observed data while the remainder of the parameters were estimated prior to the simulation runs.

The computed nitrogen balance for 1974 on watershed P2 at Watkinsville, Ga. is summarized in table I-18. Soil N refers to the soluble nitrogen content in the surface 1 cm of soil and soil NO_3 refers to the NO_3 content in the root zone. The nitrogen balance is maintained since the change in storage equals the difference between the input and output. The data shown in table I-18 do not include nitrogen losses with soil loss. As shown in table I-17, these losses amount to over 4 kg/ha.

SUMMARY

A simplified nutrient model has been developed for field-sized areas. The simplified model predicts nutrient losses in runoff, in leached water, and with soil loss. These predictions are not as accurate as desired but they may be within the limitations of our capability at this time. Additional testing and evaluation are described in chapter 6.

Table I-18.—Nitrogen balance for watershed P2 near Watkinsville, Ga., 1974

Source or process	Input	Output	Change in storage
	(kg/ha)	(kg/ha)	(kg/ha)
Rainfall	8.18		
Fertilizer	140		
Mineralization	43.13		
Soil N			+ 0.30
Soil NO_3			- 9.23
Runoff		3.76	
Leaching		33.98	
Dentrification		37.83	
Uptake		124.66	
Total	191.3	200.2	- 8.9

REFERENCES

(1) Chapin, J. D., and P. D. Uttormark.
1973. Atmospheric contribution of nitrogen and phosphorus. The University of Wisconsin Water Resources Center, Hydraulic and Sanitary Laboratory, Technical Report WIS-WRC 73-2. Madison, Wis.

(2) Chichester, F. W. and S. J. Smith.
1978. Disposition of ^{15}N-Labeled fertilizer nitrate applied during corn culture in field lysimeters. Journal of Environmental Quality 7(2): 227-233.

(3) Smith, C. N., G. W. Bailey, R. A. Leonard, and G. W. Langdale.
1978. Transport of agricultural chemicals from small upland Piedmont watersheds. U.S. Environmental Protection Agency, EPA-600/3-78-056.

Chapter 5. THE PESTICIDE SUBMODEL

R. A. Leonard and R. D. Wauchope[1/]

The pesticide submodel was developed on simplified concepts of processes and designed to be responsive to different management options. Foliar- and soil- applied pesticides are separately described so that different decay rates can be used for each source of the same chemical if necessary. Usually pesticide residing on foliage dissipates more rapidly than that from soil. Also decay rates can be made site-specific if information is available. Movement of pesticides from the soil surface as a result of infiltrating water is estimated using differences of rainfall and runoff for the storm and pesticide mobility parameters. Pesticide in runoff is partitioned between the solution or water phase and the sediment phase. This aspect is particularly important when examining management options that limit sediment yield. Comprehensive discussions are provided in volume III, Supporting Documentation, to aid the user in assigning appropriate parameter values.

THE RUNOFF SYSTEM

A simple conceptualization of the runoff system is shown in figure I-20. The primary source of pesticide available to enter the runoff stream is visualized as a surface layer of soil defined arbitrarily as having a depth of 1 cm. This definition is based on observations by Leonard and others (10) that runoff concentrations of both dissolved and adsorbed pesticides were strongly correlated with pesticide concentrations in this layer. Actually the thickness of this layer depends on many factors. Pesticide extraction by raindrop splash and interrill soil movement may occur in a very shallow layer, whereas extraction from rills may extend several centimeters deep. In models by Bruce and others (2) and Frere and others (5), rill and interrill extractions were described separately, but here the process was conceptually combined for simplicity. Others have defined this effective thickness to be about 0.5 cm (3) and 2.5 cm (12).

Washoff of pesticide applied to foliage is another source that may enter the runoff stream. In this model, the fraction of applied pesticide intercepted by foliage is specified initially. Dislodgeable residue remaining on the foliage at the time of rainfall is estimated from information given in volume III, chapter 18. The fraction of this dislodgeable residue removed by rainfall is then added to the soil surface 0 to 1 cm zone and a new concentration for this zone is computed for the runoff event.

1/ Soil scientist, USDA-SEA-AR, Southeast Watershed Research Program, Athens, Ga., and chemist, USDA-SEA-AR, U.S. Delta States Agricultural Research Center, Stoneville, Miss., respectively.

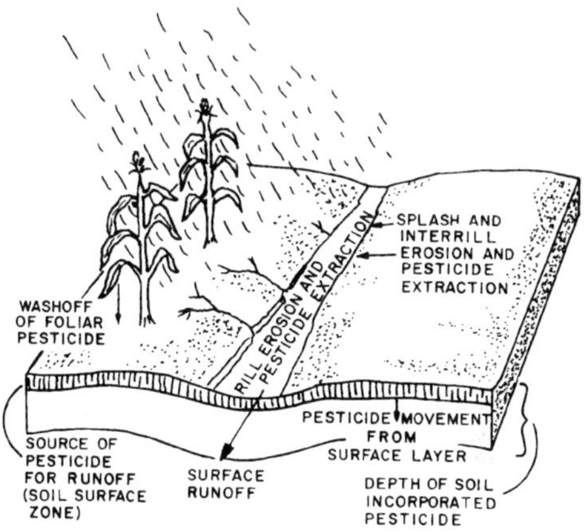

Figure I-20.—Schematic representation of the conceptualized runoff process.

Pesticide dissipates from the surface zone primarily by degradation and volatilization processes. During rainfall events, pesticide may move below the surface zone in the infiltrating water and across the surface in runoff. In the model, initial concentrations of unincorporated pesticides are computed as if they were uniformly incorporated into the 0 to 1 cm depth. Concentrations of incorporated pesticides are computed based on their incorporation depth and efficiency of incorporation. A simplified schematic of the pesticide submodel is shown in figure I-21.

DESCRIPTION OF THE PESTICIDE SOURCE

As stated previously, the source zone for extraction into runoff was arbitrarily defined as the 0 to 1 cm depth increment of the soil surface. Concentrations are computed in units of micrograms/gram or parts per million. For pesticides applied directly to the soil surface the concentration resulting from the application, C_1, is:

$$C_1 = R \times \frac{10}{BD} \qquad \text{[I-190]}$$

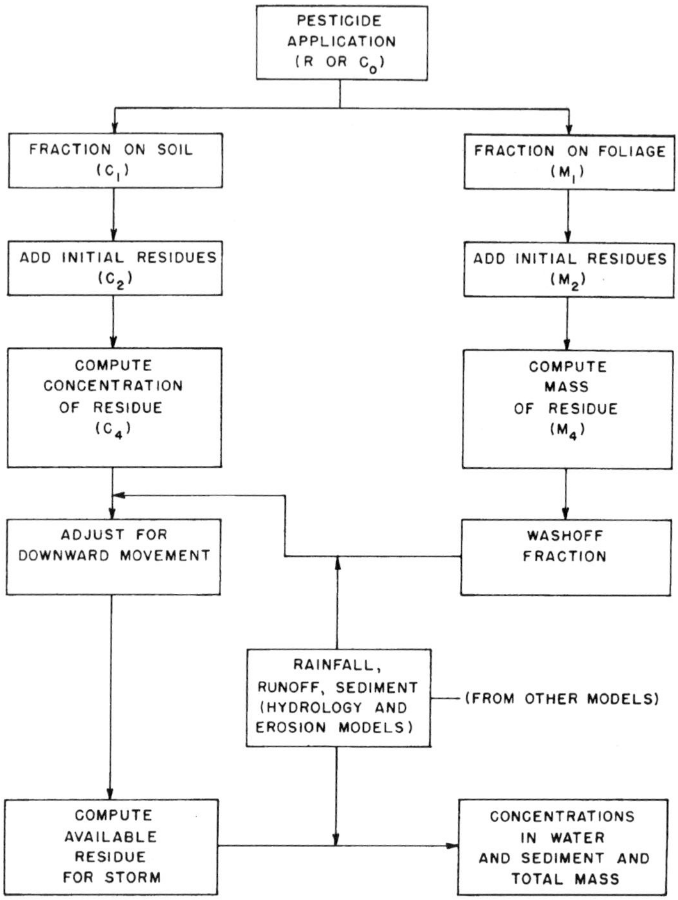

Figure I-21.—Simplified schematic representation of the pesticide model.

where BD is the bulk density of the surface soil layer, and R is the application rate in units of kilograms/hectares. Assuming an average BD of 1.5,

$$C_1 = R \times 6.7. \qquad [I-191]$$

For soil incorporated pesticides,

$$C_1 = 6.7\ R \times EF/ID \qquad [I-192]$$

where EF is a unitless factor to compensate for nonuniform incorporation and ID is the incorporation depth. If uniform mixing is assumed, EF = 1; however, experience has shown that uniform mixing is rarely achieved (11). Concentrations in the surface 0 to 1 cm layer are usually higher than computed assuming uniformity so that EF probably ranges from 1 to 3. In situations where pesticide is injected or banded below the soil surface, EF may be less than 1. A range of 0.5 to 1 is suggested. Normally EF would be assigned a value of 1 unless information is available on the incorporation pattern in a specific situation of interest. If some pesticide residue, C_2, was initially present in the soil at the time of application, the total or net concentration would be $C_3 = C_2 + C_1$.

When pesticides are applied to foliage, the areal concentration expressed in units of milligrams/square meters is

$$M_1 = R \times FF \times 100 \qquad [I-193]$$

where FF is the fraction of the application intercepted by the foliage. M is not concentration on the leaf surface, but a concentration based on the projected ground area. Unless the canopy is dense with complete closure, a fraction of the application, SF, will also be intercepted by the soil surface. Soil concentration resulting from this application is computed as above as $C_1 = R \times SF$. When aerial applications are made, losses by drift and volatilization may occur so that FF + SF will not equal 1. Information on foliar interception is found in volume III, chapter 18.

Residues of the same pesticide from previous applications, if present in either the soil or foliage compartment, are added to that resulting from the new application for the total residue level. At the beginning of the model application period, any initial residues present are specified. When pesticide residues are redistributed in the soil by major tillage, a new model application period should be begun, with the resultant surface concentration input as initial residue for this period. The surface concentrations at the beginning of the period may be estimated from the residue remaining and the tillage depth.

DISSIPATION OF PESTICIDE FROM THE RUNOFF-ACTIVE ZONE

A simple exponential dissipation rate is assumed for both soil and foliar residues throughout the model application period. For soil residue, C_4, the concentration remaining at time, t, in days after application of the pesticide or in days after specifying the concentration of initial residue is:

$$C_4 = C_3 e^{-k_s t}. \qquad [I-194]$$

Likewise, mass remaining on foliage, M_4, at time, t, is:

$$M_4 = M_3 e^{-k_f t}, \qquad [I-195]$$

or

$$M_4 = M_3 e^{-\left(\frac{0.693\ t}{C_{1/2}}\right)} \qquad [I-196]$$

where $C_{1/2}$ is the "half-life" or half-concentration time of the foliar residue in days. In the model, the mass of foliar pesticide of concern is that "dislodgeable" or potentially removed by rainfall. Rate constants, k_s, for dissipation from soil are tabulated in volume III, chapter 17 for selected pesticides based on reviews of published data, along with discussions on their use and limitations. Half-lives of foliar residues are given in volume III, chapter 18.

FOLIAR WASHOFF -- CONTRIBUTIONS TO RUNOFF

Little information is available in patterns of pesticide removal by rainfall. In the model, the assumption is made that once rainfall exceeds a threshold value corresponding to the amount that can be retained as droplets on the canopy, the fraction potentially dislodgeable is removed during the event (see vol. III, ch. 18, for estimates of percent dislodgeable for selected pesticides). This amount is then added to the soil pesticide residue present at the time of the event. For computation of concentrations consistent with the conceptual thickness of the soil surface, this mass is distributed evenly in the 0 to 1 cm zone. In reality, washoff may occur during the storm such that foliar contributions may fall directly into the runoff stream and be transported off the field. Also, spatial patterns of washoff are likely not uniform and washoff may fall into rills under the plant formed by previous rainfall. Therefore, the assumptions made may tend to underestimate the foliar contribution to runoff.

Some pesticides, particularly dust formulations, may reach the soil surface by dry-fall between runoff events. Also, drip losses from heavy dew may remove pesticide from foliage.

VERTICAL MOVEMENT OF PESTICIDE DURING RAINFALL

Runoff potential of mobile pesticides is reduced as infiltrating water moves some of the pesticide below the soil surface (1, 4). Pesticide mobility in soil has been studied extensively using thin layer chromatography technique (8, 9). With this technique, mobility is expressed relative to the movement of water (R_f values). In volume III, chapter 19 of this publication, R_f values are related to K_d, a coefficient describing distribution of pesticide between the solution phase and the soil phase, defined as a constant for a simple linear adsorpiton isotherm as:

$$K_d = \frac{C_s}{C_w} \qquad [I-197]$$

where at equilibrium C_s is the concentration, micrograms/gram, in the soil or solid phase and C_w is the concentration in solution, micrograms/milliliter. Other procedures for estimating K_d for a number of common pesticides in soil, along with limitations and possible inherent errors in its use, are also discussed in chapter 19, volume III.

The following algorithm was developed to estimate vertical movement of pesticide from the soil surface.

The rate of change of pesticide mass, Z, in the soil surface is

$$- dZ = C_w \cdot f \cdot dt \qquad [I-198]$$

where C_w is the pesticide concentration in water or mobile phase and f is the water flux. At saturation,

$$Z = C_w \cdot p + C_s \ (1 - p) \ D \qquad [I-199]$$

where p is the soil porosity, C_s is the concentration of the adsorbed or immobile phase, and D is the particle density. Introducing $C_s = K_d C_w$ and rearranging the equation above becomes:

$$C_w = \frac{Z}{p + DK_d(1-p)}. \qquad [I-200]$$

The rate equation can now be written:

$$- dZ = \frac{Z \cdot f}{p + DK_d(1-p)} \ dt \qquad [I-201]$$

and integrated between limits of Z_0, t_0 and Z, t to yield

$$Z = Z_0 e^{-\left(\frac{f \cdot t}{DK_d(1-p) + p}\right)} \qquad [I-202]$$

where Z_0 is the mass of pesticide present per unit volume of soil surface at the beginning of the storm. The water flux through the surface during a storm is

$$f = \frac{RF - RO - S}{t} \qquad [I-203]$$

where RF is the amount of rainfall, RO is runoff, S is the surface storage or initial abstraction to reach saturation, and t is the storm duration. Making the substitution for f, t can be eliminated so that

$$Z = Z_o e^{-\left(\frac{RF - RO - S}{DK_d(1-p) + p}\right)} \quad . \qquad [I-204]$$

The value of S is estimated from porosity and the average soil water content plus canopy stored water. In the model, $C_4 \times BD = Z_0$ and $Z = C_{AV} \times BD$ where C_{AV} is the runoff available pesticide concentration and C_4 is as previously described.

Where pesticide is foliar applied, the amount assumed to reach the soil by washoff is added to the surface pesticide residue before estimation of vertical translocation. This method provides only a crude approximation of the process compared to other methods (6, 9). However, it is developed for use where only total storm rainfall and runoff amounts are available. Its primary function is to reduce surface concentrations of those compounds with high soil mobility. Since the amount of vertical translocation will be small in a single storm for relatively insoluble compounds, this calculation is by passed in the model if the compound solubility is < 1 µg/g.

PESTICIDE EXTRACTION INTO RUNOFF

The source zone supplying pesticide to runoff was defined as the surface (0 to 1 cm) depth increment. At the time of runoff, this increment of soil contains a pesticide residue specified in the model as the concentration of "available residue." This is the concentration computed using the appropriate decay functions, adding any foliar washoff, and allowing for vertical translocation. The concentration units are expressed in micrograms/gram of dry soil as is the convention when a soil sample is removed and analyzed for its pesticide content.

Pesticide is extracted by water flowing over the surface and by dispersion and mixing of the soil material by the flow and by raindrop impact. Instantaneous pressure gradients at the surface caused by raindrop impact on a water-saturated soil could also contribute to exchange of pesticide between the soil water and the flowing water. At the interface between the soil matrix and the runoff stream, some mass of soil is "extracted" or is effective in supplying pesticide to some volume of runoff. The mass of pesticide in this mass of soil is:

$$Y = B \cdot C_{AV} \qquad [I-205]$$

where B is the soil mass per unit volume and C_{AV} is the concentration of available residue. As this soil mass mixes or "equilibrates" with the runoff stream

$$Y = (C_w \cdot V) + (C_s \cdot B) \qquad [I-206]$$

where C_w is the pesticide concentration in the water, V is the volume of water per unit volume of runoff interface, and C_s is the pesticide concentration remaining in the soil or solid phase. Ignoring the volume occupied by the soil mass compared to the larger volume of water; that is, V = total unit volume of runoff interface = 1,

$$1 \cdot C_w + C_s \cdot B = B \cdot C_{AV}. \qquad [I-207]$$

By assuming that the distribution between the solution and the soil is approximated by the equilibrium expression:

$$K_d = \frac{C_s}{C_w}, \qquad [I-208]$$

$$C_w = \frac{B \cdot C_{AV}}{1 + B\, K_d}, \qquad [I-209]$$

or

$$C_s = \frac{B\, C_{AV}\, K_d}{1 + B\, K_d}. \qquad [I-210]$$

In these expressions it can be seen that when $K_d = 0$, then $C_w = B\ C_{AV}$; e.g., if 100 g of soil containing 1 µg/g of pesticide that partitions completely to the solution phase is extracted by or is dispersed in 1 liter of water, C_w = 100 µg/l. Also, as K_d becomes large, $C_s = C_{AV}$. The numerical value of the parameter, B, in the above equation cannot be obtained by direct measurement, and probably is dependent on runoff conditions. However, it will be shown later that the value ranges from 0.05 to 0.2, with 0.1 giving adequate fit in most situations.

As material flows from the field, it is assumed that the pesticide concentration in the runoff solution is equal to C_w. However, not all the affected soil material will become sediment at the field edge. The coarser soil material will be deposited or left in place. As a result, the transported soil will have a higher per unit mass adsorptive capacity and adsorbed pesticide concentration than that of the whole soil. Therefore, an enrichment factor is required and is provided by computations in the erosion submodel (see vol. I, ch. 3).

Total storm loads are computed as: mass in solution phase = C_w · storm runoff volume, and mass in sediment phase = C_s · enrichment factor · sediment yield.

The approach taken by these procedures differs from other models in that that the runoff stream is not forced to equilibrate at the soil/water ratio determined by the composition of the saturated soil matrix (5) nor at a ratio determined by the concentration of the transported sediment, assuming sediment has the same adsorptive capacity as the soil (3). The weakest assumption, probably, is that associated with using K_d to partition between the solution and the soil phase. In addition to the limitations discussed in volume III, chapter 19, the runoff proccess is dynamic, and true equilibration is probably never reached. Also, pesticide apparently partitions differently depending on time of contact with the soil (11); that is, the "apparent K_d" based on observed partitioning in runoff from experimental watersheds differs from the laboratory determined values and increases throughout the observation period.

For this reason, K_d may be best used to differentiate between behavior of pesticide classes, with K_d ranges differing perhaps by orders of magnitude, that is, 1 to 10, 10 to 100, and so forth.

TESTING AND EVALUATION

The submodel was tested using data from several experiments conducted under widely different conditions and on pesticides with different properties. Observed rainfall, runoff, and sediment yield were used as available. No attempt was made to calibrate the submodel or adjust parameters for best fits. Parameters were estimated either from site-specific information reported or from information provided in volume III, chapter 16 through chapter 19. Assignments based on subjective judgment or experience are indicated with explanations. Usually where only a limited number of storms were examined after a single pesticide application, the computations were made with a desk calculator.

Table I-19 lists the parameters estimated for a simulated runoff experiment with lindane and dieldrin under simulated rainfall conducted at Watkinsville, Ga., on a Cecil sandy loam soil of about 6% slope (A. W. White, 1970, unpublished). A total of 6.35 cm of rainfall was applied in 1 hr by procedures described in White and others (15). Rainfall was applied on 1, 8, and 28 days after pesticide application. Runoff was about 50% of applied rainfall. An enrichment ratio of 1.5 was used based on observations of sand, silt, and clay in the eroded soil material. Experience has shown that an extraction ratio of 0.2 is required for extreme conditions as prevalent in this experiment. The pesticides decay constant were computed from analysis of soil samples taken during the experiment. The K_d values were estimated from information in volume III, chapter 19.

Results of the simulation are given in table I-20. In general, predicted compared to observed were reasonably close. Although these compounds are not used to a significant extent presently, these data illustrate the difference in behavior in runoff as a result of differences in pesticide adsorption (K_d), solubility, and mode of application which were adequately described.

White and others (15) conducted a similar exeriment with atrazine at 1 hr and 96 hr after pesticide application. Samples were taken throughout the runoff event so that discharge-weighted mean concentrations can be computed for different portions of the runoff event. These data are useful in illustrating how the model responds to different sized storms. In the model, no upper limit for single storm loss is used. Since atrazine is somewhat mobile (see vol. III, ch. 19), however, storms of increasing size reduced the computed concentrations of runoff-available pesticide, and thereby reduced the predicted runoff concentrations (table I-21). The model reduces the concentration of runoff-available pesticide by an amount proportional to that lost in runoff only at the end of each event. Therefore, for very large storms where runoff losses become a significant vehicle for surface depletion, the model may overpredict total losses, as apparently happened in the prediction for the 6.35-cm storm (table I-21).

Table I-19.—Inputs and parameters for lindane and dieldrin simulation, Cecil sandy loam soil, Watkinsville, Ga.

Input	Lindane		Dieldrin	
	Surface	Incorporated	Surface	Incorporated
Application rate, kg/ha.	11.4	11.4	11.4	11.4
Incorporation depth, cm.	1	7.5	1	7.5
Incorporation efficiency.	1	1	1	1
Fraction on foliage.	0	0	0	0
Fraction on soil.	1	1	1	1
Initial foliar residue, mg/m^2.	0	0	0	0
Initial soil residue, μg/g	0	0	0	0
Foliar washoff threshold, cm.	0	0	0	0
Washoff fraction.	0	0	0	0
Water solubility, ppm.	10	10	.12	.12
Foliar residue half-life, days.	0	0	0	0
Enrichment ratio.	1.5	1.5	1.5	1.5
Extraction ratio.	.2	.2	.2	.2
Decay constant, k_s.	.046	.015	.02	.01
Distribution coefficient, K_d.	30	30	500	500

Atrazine simulations were compared with observations by Hall (7) on atrazine runoff from small plots in Pennsylvania. Values for the decay constant, k_s, and K_d were estimated from information in volume III, chapters 17 and 19. An enrichment factor of 2 was arbitrarily chosen. Simulations gave reasonable predictions; however, the model underestimated solution concentration in the first event (table I-22). Predicted concentrations in sediment were close to observed except for the last two events. This problem apparently is associated with using K_d as a constant throughout the season. Other data discussed later show an even greater discrepancy associated with partitioning pesticide between water and sediment.

In table I-23 comparisons are made of predicted vs observed concentrations of 2,4-D from a treated rangeland watershed (L. J. Lane, 1978, unpublished data). Acceptable predictions were obtained throughout the observation period.

Table I-20.—Lindane and dieldrin in runoff; comparison of observed[1] vs. predicted values

Pesticide	Time after application	Concentration in 0-1 cm soil		Concentration in runoff		Concentration in sediment	
		Observed	Predicted	Observed	Predicted	Observed	Predicted
	(days)	(μg/g)	(μg/g)	(mg/l)	(mg/l)	(μg/g)	(μg/g)
Lindane, Surface applied.	1	84	69	2.6	2.0	88	89
	7	49	45	2.0	1.3	43	58
	28	19	15	.8	.42	29	19
Lindane, Incorporated.	1	10.4	9.5	.24	.27	6.5	12.2
	7	9.1	7.5	.23	.21	8.6	9.6
	28	6.8	4.7	.16	.13	5.7	6.1
Dieldrin, Surface applied.	1	100	74	.14	.12	189	109
	7	70	62	.12	.12	176	92
	28	51	38	.14	.08	126	60
Dieldrin, Incorporated.	1	10.2	9.9	.07	.02	15	15
	7	9.4	8.8	.07	.02	21	13
	28	8.3	6.8	.07	.01	13	10

1/ Unpublished data from A. W. White, 1970, Watkinsville, Ga.

Table I-21.—Atrazine in runoff from simulated rainfall; comparison of observed[1] vs. predicted

Size of Storm	Concentration in runoff		Concentration in sediment		Percentage of application	
	Observed[2]	Predicted	Observed[2]	Predicted	Observed	Predicted
(cm)	(mg/l)	(mg/l)	(μg/g)	(μg/g)	(%)	(%)
1 hr after application						
1.25	7.2	2.2	24	13	4.3	1.4
3.18	2.3	1.9	9.4	11	12.0	10
6.35	1.3	1.7	4.6	10	17.0	24
96 hr after application						
1.25	3.3	1.3	11	7.8	2.0	.85
3.18	1.0	1.1	4.2	6.6	5.3	6.0
6.35	.55	1.0	2.0	6.0	7.3	14

Enrichment ratio = 1.5.
Extraction ratio = 0.2.
Decay constant = 0.14.
Distribution coefficient, K_d = 4.

1/ White and others (15). Simulated rainfall, Cecil sandy loam soil.
2/ Estimated based on reported losses in water and sediment.

Table I-22.—Atrazine in runoff; comparison of observed[1/] vs. predicted; Hagerstown silty clay loam, Pa.

Time after application	Concentration in runoff		Concentration in sediment		Percentage of application	
	Observed	Predicted	Observed	Predicted	Observed	Predicted
(days)	(mg/l)	(mg/l)	(mg/kg)	(mg/kg)	(%)	(%)
5	4.6	2.3	10	9.2	1.1	0.55
14	2.2	1.1	5.1	4.4	.70	.38
19	.92	.74	4.1	3.0	1.1	.90
27	.75	.34	3.3	1.4	.42	.19
28	.20	.24	1.8	1.0	1.0	1.1
37	.18	.08	1.7	.3	.13	.10
43	.20	.05	2.2	.2	.41	.08
TOTAL					4.9	3.3

Enrichment ratio = 2. Extraction ratio = 0.2. Decay constant = 0.05. Distribution coefficient, K_d = 2 (based on reported organic matter content of 1.3%).

1/ Hall, J. K. (7), observed values estimated from data reported graphically.

Table I-23.—2,4-D in runoff from a semiarid desert rangeland; comparison of observed[1/] vs. predicted

Time after application	Concentration in runoff		Percentage of application	
	Observed	Predicted	Observed	Predicted
(days)	(mg/l)	(mg/l)	(%)	(%)
7	0.240	0.26	0.424	0.454
17	.017	.064	.009	.035
17	.013	.054	.019	.074
25	.010	.018	.018	.032
26	.010	.012	.014	.017
31	.0067	.0064	.002	.002
40	.0029	.0015	.018	.009
41	.0042	.0011	.004	.001
TOTAL			.508	.624

Enrichment ratio = (sediment concentration not computed). Extraction ratio = 0.1. Decay constant = 0.1. Distribution coefficient, K_d = 1.

1/ Lane, L. J., 1978, unpublished data.

Another set of comparisons on 2,4-D were obtained with the data of White and others (14) (table I-24). Here the soil was a loamy sand with high infiltration rates. Slope was about 3% and soil loss from the plots was relatively low. However, the sediment phase was probably high in organic matter and clay compared to the original soil, and consequently, the enrichment ratio would be high. A value of 10 was arbitrarily assigned. Other parameters were estimated from information in this report and observed 2,4-D persistence in the Cowarts soil. The model gave good estimates of concentrations in the first storm, but underestimated concentrations in the later storm. The greatest discrepancy occurs in the comparison of concentrations in sediment. A K_d value of 1 was assigned to 2,4-D based on information in volume III, chapter 19. However, even with the assumption of a larger enrichment ratio, the observed partitioning would indicate an effective K_d of about 10 or greater which increased with time. However, the objective of this evaluation is not to adjust parameters by rationalizations to obtain fit to a particular data set. Since the 2,4-D formulation used (alkanolamine salt) is very sparingly soluble in water, the K_d value of 1 used was probably inappropriate. However, using a higher K_d value would not have greatly affected the predicted concentrations in the water phase. Because of the large storms applied, predicted downward movement of the pesticide significantly lowered the predicted runoff-available pesticide. Use of a larger K_d would have prevented this, but in turn the portion entering the water phase would be lowered by the higher K_d, and thus the predicted concentrations in solution would be similar. This illustrates how model predictions can give good fit with observed total mass lost in runoff but not adequately describe processes or "prove" that processes have been represented correctly.

The data on pesticide runoff reported by Smith and others (11) provide an opportunity to test the model in several ways. Tables I-25 through I-27 give results for atrazine on a small watershed for three consequtive years. Tables I-28 through I-30 show results for three pesticides of differing properties on another watershed in the same year, 1973. Many comparisons can be made; however, only a few points will be discussed. Predicted sediment concentrations in the first storm agree reasonably well with observed. The agreement is very poor for the later storms, indicating that assumptions regarding a constant K_d are inadequate. It has been generally observed that pesticide desorption from soil is nonlinear and appears to become more difficult with time. However, sufficient research information on pesticide behavior in soil is not available for development of some model algorithm to account for this behavior in a direct manner. Again, the objective here is not to calibrate some function to give close fit.

The data in tables I-28 through I-30 show that the model can give reasonable predictions for pesticides of widely different properties and behavior. However, the prediction for trifluralin is excessive. Since this compound is quite volatile, the immediate soil surface subjected to runoff may have become more depleted of pesticide than predicted in the model. In this particular experiment, the data indicated that major transport was in the water phase. Others (13) have indicated that sediment transport is the major route for this compound.

Data in tables I-20 through I-30 are summarized in figures I-22 through I-25 for visual comparison. Observed vs predicted values are plotted in exponential scale in figures I-22 through I-24 because of the range of values

Table I-24.—2,4-D in runoff obtained from a Cowarts loamy sand with simulated rainfall of 8.25 cm in 30 min; comparison of observed[1/] vs. predicted

Time after application	Concentration in runoff		Concentration in sediment		Percentage of application	
	Observed	Predicted	Observed	Predicted	Observed	Predicted
(days)	(mg/l)	(mg/l)	(mg/kg)	(mg/kg)	(%)	(%)
1	0.022	0.025	2.8	0.25	1.53	1.54
8	.004	.002	1.5	.01	.37	.10
35	.0003	< .0001	.6	< .001	.06	< .01
TOTAL					1.96	1.65

Enrichment ratio = 10. Extraction ratio = 0.1. Decay constant = 0.4
Distribution coefficient, K_d = 1.

1/ White, A. W. and others (14).

Table I-25.—Atrazine in runoff; comparison of observed[1/] vs. predicted watershed P2, 1973, Watkinsville, Ga.

Time after application	Concentration in water		Concentration in sediment		Percentage of application	
	Observed	Predicted	Observed	Predicted	Observed	Predicted
(days)	(mg/l)	(mg/l)	(mg/kg)	(mg/kg)	(%)	(%)
8	0.200	0.460	3.232	3.686	0.002	0.004
12	.179	.224	.856	1.793	.324	.421
17	.064	.094	.892	.752	.880	1.21
17	.044	.075	.785	.606	.528	.827
26	.014	.019	.879	.152	.118	.127
29	.017	.009	.749	.072	.008	.004
30	.036	.008	.900	.064	.006	.001
33	.007	.005	.425	.040	.021	.011
41	.009	.002	.587	.016	.015	.005
58	.002	< .001	.200	< .008	.015	< .005
TOTAL					1.92	1.61

Enrichment ratio = 2. Extraction ratio = 0.1. Decay constant = 0.14.
Distribution coefficient, K_d = 4.

1/ Smith and others (11).

Table I-26.—Atrazine in runoff; comparison of observed[1/] vs. predicted, watershed P2, Watkinsville, Ga., 1974

Time after application	Concentration in runoff		Concentration in sediment		Percentage of application	
	Observed	Predicted	Observed	Predicted	Observed	Predicted
(days)	(mg/l)	(mg/l)	(mg/kg)	(mg/kg)	(%)	(%)
6	1.90	0.630	4.10	5.04	0.080	0.027
24	.022	.018	2.56	.144	.049	.035
52	.020	.0003	.910	.002	.008	< .001
59	.003	< .0001	.488	< .001	.012	< .001
59	.003	< .0001	.483	< .001	.032	< .001
TOTAL					.18	.06

Enrichment ratio = 2. Extraction ratio = 0.1. Decay constant = 0.14. Distribution coefficient, K_d = 4.

1/ Smith and others (11).

Table. I-27.—Atrazine in runoff; comparison of observed[1/] vs. predicted, watershed P2, Watkinsville, Ga., 1975

Time after application	Concentration in runoff		Concentration in sediment		Percentage of application	
	Observed	Predicted	Observed	Predicted	Observed	Predicted
(days)	(mg/l)	(mg/l)	(mg/kg)	(mg/kg)	(%)	(%)
10	0.101	0.122	1.53	0.979	0.284	0.329
21	.017	.020	.987	.160	.098	.098
21	.010	.017	.299	.136	.295	.402
29	.012	.005	.039	.040	.008	.003
53	.001	.0002	.038	.001	.006	.001
TOTAL					.69	.83

Enrichment Ratio = 2. Extraction Ratio = 0.1. Decay constant = 0.14. Distribution coefficient, K_d = 4.

1/ Smith and others (11).

Table I-28.—Trifluralin in runoff; comparison of observed[1/] vs. predicted

Time after application	Concentration in runoff		Concentration in sediment		Percentage of application	
	Observed	Predicted	Observed	Predicted	Observed	Predicted
(days)	(mg/l)	(mg/l)	(mg/kg)	(mg/kg)	(%)	(%)
0	0.013	0.037	0.0304	1.11	0.18	1.03
8	.0057	.012	.0321	.36	.024	.073
15	.0020	.0046	.0100	.138	.0011	.0036
25	.0045	.0011	.0206	.033	.021	.0063
34	.0050	.0003	.0574	.009	.0045	.00053
47	.0021	.00005	.0197	.0015	.0027	.0031
TOTAL					.233	1.12

Enrichment ratio = 2. Extraction ratio = 0.1. Decay constant = 0.14. Distribution coefficient, K_d = 15.

1/ Smith and others (11).

Table I-29.—Paraquat in runoff; comparison of observed[1/] vs. predicted

Time after application	Concentration in runoff		Concentration in sediment		Percentage of application	
	Observed	Predicted	Observed	Predicted	Observed	Predicted
(days)	(mg/l)	(mg/l)	(mg/kg)	(mg/kg)	(%)	(%)
0	0	1.5×10^{-4}	36.8	30.4	9.70	8.02
8	0	1.4×10^{-4}	35.6	28.7	1.36	1.09
15	0	1.4×10^{-4}	61.5	27.2	.26	.12
25	0	1.3×10^{-4}	29.7	25.5	.65	.56
34	0	1.2×10^{-4}	38.0	24.0	.08	.05
47	0	1.1×10^{-4}	27.6	21.9	1.75	1.38
TOTAL					13.80	11.22

Enrichment ratio = 2. Extraction ratio = 0.1. Decay constant = 0.007. Distribution coefficient, K_d = 10^5.

1/ Smith and others (11).

Table I-30.—Diphenamid in runoff; comparison of observed[1/] vs. predicted

Time after application	Concentration in runoff		Concentration in sediment		Percentage of application	
	Observed	Predicted	Observed	Predicted	Observed	Predicted
(days)	(mg/l)	(mg/l)	(mg/kg)	(mg/kg)	(%)	(%)
0	1.65	1.90	0.64	3.80	6.82	8.42
8	.25	.21	.67	.54	.32	.35
15	.065	.077	.20	.15	.012	.014
25	.013	.008	.16	.016	.022	.012
34	.010	.001	.35	.002	.003	.0003
47	.002	.0001	.12	.0002	.01	.0004
TOTAL					7.19	8.80

Enrichment ratio = 2. Extraction ratio = 0.1. Decay constant = 0.18. Distribution coefficient, K_d = 1.

1/ Smith and others (11).

obtained. The broken lines in each plot indicate 1:1 correspondence. Coefficients, r, shown are from linear correlation. Predictions of sediment concentrations in the first events were good except for trifluralin and diphenamid on watershed P1 (fig. I-22). Acceptable predictions of solution concentrations were obtained throughout most of the study periods (fig. I-23 and I-24). In figure I-24, all concentrations in excess of 1 ppb (0.001 ppm in tables) were used in the comparison. Concentrations were somewhat underpredicted in late events where they were very low. Use of different decay rates for different times after application, as suggested as a possibility in volume III, chapter 17, could rectify this discrepancy. However, except for persistent pesticides, runoff losses except during a short time period immediately following application may be insignificant. Total losses for the growing season are compared in figure I-25.

In summary, the comparisons presented here demonstrate the potential of the model and some of its shortcomings. Additional or other data sets probably would have given different results, but most likely they would have fallen within the range of accuracy indicated by these data.

Rather than use a comprehensive data set in an attempt to evaluate the model in situations of multiple foliar applications of insecticide, hypothetical cases were set up that should closely resemble toxaphene application to cotton as reported by Willis and others (16). It was assumed that six applications of 2.2 kg/ha were applied on days 2, 9, 16, 23, 30, and 37. Six rainfall events of 2.0 cm each were imposed in two sequences: on days 3, 10, 17, 24,

31, and 38, and on days 8, 15, 22, 29, 36, and 43. The first sequence gave rainfall 1 day after each application; the second sequence, 6 days after application. Additionally, in each sequence above, it was assumed in one case that 10% of the application was intercepted by the soil with 50% intercepted by foliage, and in another, none was intercepted by soil and 50% was intercepted by foliage. The remaining pesticide was assumed off-target losses. An initial toxaphene residue in the soil of 2 μg/g was also assumed. Parameter values and inputs are summarized in table I-31. Values of parameters were estimated based on information provided by Willis and others (16) and in volume III, chapter 18. Predicted toxaphene concentrations in water, sediment, and soil are shown in figure I-26 for three of the four scenarios. Concentrations increased in all phases in response to application. However, when the insecticide application was intercepted by foliage only, predicted concentrations

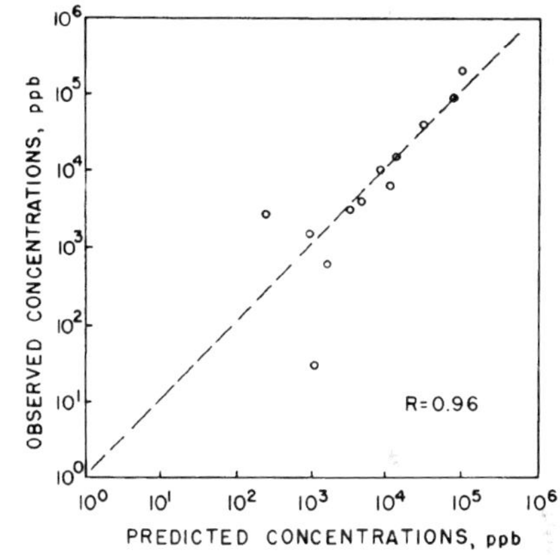

Figure I-22.—Comparison of predicted and observed concentrations of pesticide in sediment (first storm events after application).

increased little since only 10% washoff was allowed and the foliar half-life was only 7 days. The 10% washoff may be greater than actually observed for toxaphenes (see vol. III, ch. 18). The effects of delayed rainfall after application are also apparent.

In the study by Willis and others (16), 10 kg/ha of toxaphene was applied in six applications to cotton on a watershed containing about 2 ppm toxaphene soil residue. Average toxaphene concentration observed in sediment during the application period was 12.9 ppm. The maximum toxaphene sediment concentration predicted here was 22 ppm after application of 13.4 kg/ha. Although no direct comparisons can be made, model predictions appear to be reasonable compared with observations. Total predicted losses in the six hypothetical storms are summarized in table I-32. These data illustrate the utility of the model in examining relative effects of rainfall probabilities and pesticide application efficiency. If the same simulations were performed with an insecticide of shorter foliar half-life such as methyl parathion, the results would have been affected more by timing than is reflected here. Since toxaphene with an estimated K_d of 3500 is transported primarily by sediment, total losses also would be sensitive to factors that reduce sediment yield.

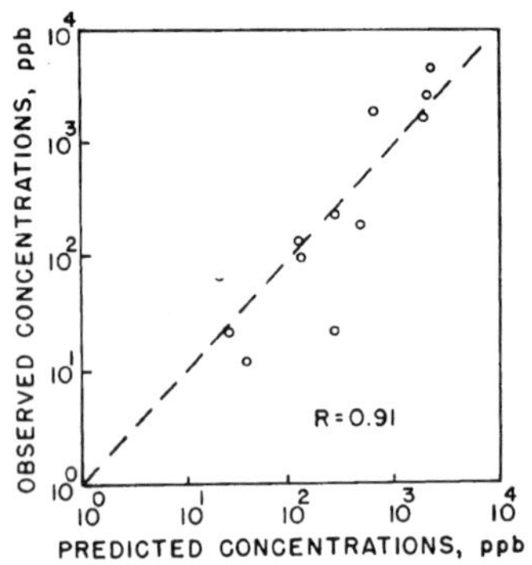

Figure I-23.—Comparison of predicted and observed concentrations of pesticides in solution phase of runoff for first events after application.

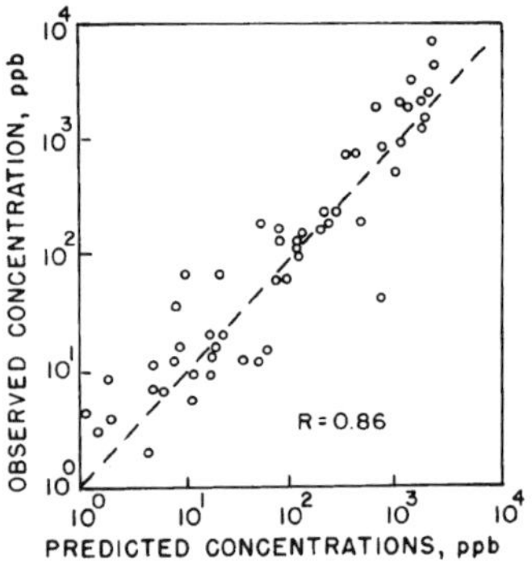

Figure 1-24.—Comparison of observed and predicted concentrations of pesticide in solution phase of runoff for all events with concentrations greater than 1.0 ppb.

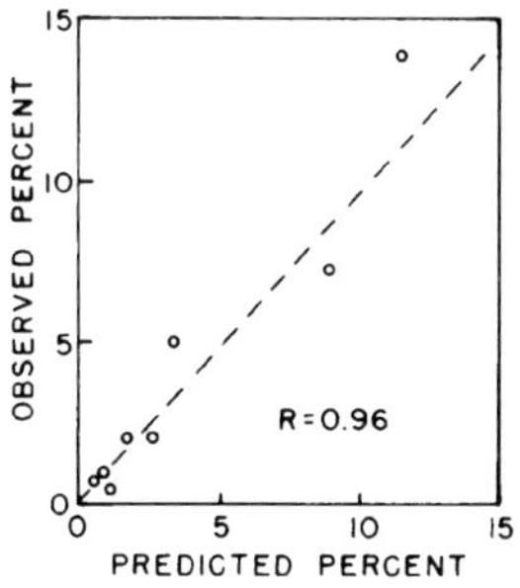

Figure I-25.—Comparison of predicted and observed seasonal losses of pesticide in runoff, percent of total applied or present during the year.

Table I-31.—Parameters and inputs for simulation of hypothetical cases of toxaphene applied to cotton

Application dates	Days 2, 9, 16, 23, 30, 37
Application rate	2.2 kg/ha
Rainfall dates	3, 10, 17, 24, 31, 38 8, 15, 22, 29, 36, 43
Rainfall amount	2.0 cm each event
Runoff amount	1.0 cm each event
Sediment yield	200 kg/ha each event
Depth of incorporation	1.0 cm
Incorporation efficiency	1
Fraction on foliage	0.50
Fraction on soil	0.0 0.10
Foliar washoff rainfal threshold.	0.10 cm
Initial foliar residue	0
Initial soil residue	2.00 μg/g
Foliar washoff fraction	0.10
Water solubility	0.40 ppm
Foliar residue half-life	7 days
Enrichment ratio	1.50
Extraction ratio	0.10
Decay constant for soil residue.	0.005
Distribution coefficient, K_d.	3,500

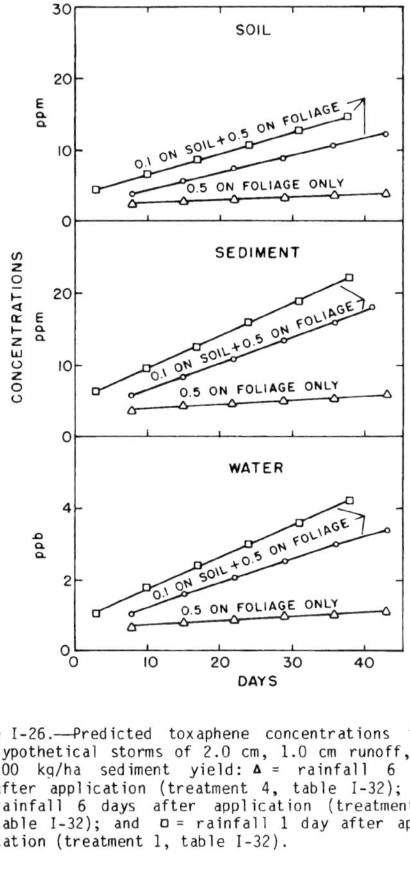

Figure I-26.—Predicted toxaphene concentrations in 6 hypothetical storms of 2.0 cm, 1.0 cm runoff, and 200 kg/ha sediment yield: Δ = rainfall 6 days after application (treatment 4, table I-32); o = rainfall 6 days after application (treatment 2, table I-32); and □ = rainfall 1 day after application (treatment 1, table I-32).

Table I-32.—Predicted losses of toxaphene in six hypothetical storms of 2.0 cm each with 1.0 cm runoff and 200 kg/ha sediment yield

Treatment[1/]	Mass in water	Mass in sediment	Total mass
	(g/ha)	(g/ha)	(g/ha)
1	1.62	17.20	18.82
2	1.37	14.37	15.74
3	.74	7.74	8.48
4	.54	5.68	6.22

1/ (1) Soil residue + 6 applications at 2.2 kg/ha; 0.5 of application on foliage; 0.1 on soil. Rainfall 1 day after applications (refers to □-symbols on figure I-26); (2) Soil residue + 6 applications at 2.2 kg/ha; 0.5 of application on foliage; 0.1 on soil. Rainfall 6 days after applications (refers to o-symbols on figure I-26); (3) Soil residue + 6 applications at 2.2 kg/ha; 0.5 of application on foliage; 0 on soil. Rainfall 1 day after applications; (4) Soil residue + 6 applications at 2.2 kg/ha; 0.5 of application on foliage; 0 on soil. Rainfall 6 days after applications (refers to Δ- symbol on figure I-26).

REFERENCES

(1) Baldwin, F. L., P. W. Santelmann, and J. M. Davidson. 1975. Movement of fluometuron across and through the soil. Journal of Environmental Quality 4:191-194.

(2) Bruce, R. R., L. A. Harper, R. A. Leonard, W. M. Snyder, and A. W. Thomas. 1975. A model for runoff of pesticides from small upland watersheds. Journal of Environmental Quality 4:541-548.

(3) Crawford, N. H., and A. S. Donigian, Jr. 1974. Pesticide transport and runoff model for agricultural lands. Hydrocomp, Inc., Palo Alto, Calif., prepared for U.S. Environmental Protection Agency, Athens, Ga., Publication No. EPA-600/2-74-013. 211 pp.

(4) Davidson, J. M., G. H. Brusewitz, D. R. Baker, and A. L. Wood. 1975. Use of soil parameters for describing pesticide movement through soils. U.S. Environmental Protection Agency, Publication No. USEPA-660/2-75-009. 149 pp.

(5) Frere, M. H., C. A. Onstad, and H. N. Holtan.
1975. ACTMO, an agricultural chemical transport model. U.S. Department of Agriculture, Agricultural Research Service, Headquarters, ARS-H-3, 54 pp. (Series discontinued; Agricultural Research Service now Science and Education Administration-Agricultural Research.)

(6) Genuchten, M. Th. van, J. M. Davidson, and P. J. Wierengen.
1974. An evaluation of kinetic and equilibrium equations for the prediction of pesticide movement through porous media. Soil Science Society of America Proceedings 38:29-35.

(7) Hall, J. K.
1974. Erosional losses of s-triazine herbicides. Journal of Environmental Quality 3:174-180.

(8) Helling, C. S.
1971. Pesticide mobility in soils. II. Application of soil thin-layer chromatography. Soil Science Society of America Proceedings 35:737-743.

(9) ------.
1971. Pesticide mobility in soils. III. Influence of soil properties. Soil Science Society of America Proceedings 35:743-748.

(10) Leonard, R. A., G. W. Langdale, and W. G. Fleming.
1979. Herbicide runoff from upland Piedmont watersheds - Data and implications for modeling pesticide transport. Journal of Environmental Quality 8:223-229.

(11) Smith, C. N., R. A. Leonard, G. W. Langdale, and G. W. Bailey.
1978. Transport of agricultural chemicals from small upland Piedmont watersheds. U.S. Environmental Protection Agency, Athens, Ga. and U. S. Department of Agriculture, Watkinsville, Ga. Final Report on Interagency Agreement No. D6-0381. Publication No. EPA 600/3-78-056. 363 pp.

(12) Steenhuis, T. S., and M. F. Walter.
1978. Closed form solution for pesticide loss in runoff water. American Society of Agricultural Engineers Technical Paper No. 78-2031, presented at the 1978 summer meeting of the American Society of Agricultural Engineers, Logan, Utah, June 27-30.

(13) Wauchope, R. D.
1978. The pesticide content of surface water draining from agricultural fields - A review. Journal of Environmental Quality 7:459-472.

(14) White, A. W., L. E. Asmussen, E. W. Hauser, and J. W. Turnbull.
1976. Loss of 2,4-D in runoff for plots receiving rainfall and from a small agricultural watershed. Journal of Environmental Quality 4:487-490.

(15) White, A. W., A. P. Barnett, B. G. Wright, and J. H. Holladay. 1967. Atrazine losses from fallow land caused by runoff and erosion. Environmental Science Technology 1:740-744.

(16) Willis, G. H., L. L. McDowell, J. F. Parr, and C. E. Murphree. 1976. Pesticide concentrations and yields in runoff and sediment from a Mississippi Delta watershed. Proceeding of the 3d Federal Interagency Sedimentation Conference, Denver, Colo.

Chapter 6. SENSITIVITY ANALYSIS

L. J. Lane and V. A. Ferreira[1/]

INTRODUCTION

Sensitivity analysis is a technique for assessing the relative change in a model response or output resulting from a change in inputs or in model parameters. For simple, explicit models, it is possible to take derivatives of the output with respect to input or parameters, and express the sensitivity as explicit functions. However, as the models become more complex, sensitivity is more easily expressed in the form of differentials, relative changes, graphs, and tables, rather than as functions. This is the approach used for the field-scale model.

Based on derived parameter values and representative values of the input variables, base values are selected. For a given set of base parameter values, computations are performed, and then the input variables are varied over a range of values and the computations repeated. For given values of the input variables, the procedure is repeated with the parameters varying about their base values. The resulting computations show how the model outputs vary with changes in the input and parameters. This shows how the model functions and how important each of the parameters is in determining the output. Such analyses also aid in parameter estimation.

The main shortcomings of this procedure are (1) the parameters are varied individually so that complex interactions are difficult to determine, and (2) the number of simulation runs increases rapidly with the number of parameters and inputs and with the number of points selected to vary about the base values. For example, nm + 1 simulation runs are required for a model with n parameters and input variables, and with simulation runs for the base values and m points around the base value of each parameter and input variable. In some cases, it may be necessary to limit the sensitivity analysis to a subset of the model parameters. Finally, the sensitivity analyses given in this chapter are for a complex watershed including detachment, transport, and deposition processes in overland flow and in concentrated flow. Sensitivity for other conditions may be much different. Users should determine model sensitivity for the particular application.

FIELD-SCALE MODEL: HYDROLOGIC COMPONENTS

The hydrologic components consist of two versions. The first, option 1, uses daily rainfall to predict runoff volume and peak discharge rates. The

1/ Hydrologist and hydrologic technician, respectively, USDA-SEA-AR, Southwest Rangeland Watershed Research Center, Tucson, Ariz.

second, option 2, uses breakpoint precipitation data for individual events, and also produces runoff volumes and peak discharge rates as output. Both options also predict daily plant transpiration, potential transpiration, average soil moisture, and percolation.

Option 1, Daily Rainfall Model

This model is essentially a modified Soil Conservation Service (SCS) runoff curve number water balance model. As in the analysis for the other components, data from watershed P2 at Watkinsville, Ga., were used to determine model sensitivity. This watershed and cultural practices on it are described in detail by Smith and others (2), and are summarized in table I-33.

Table I-33.— Summary of watershed characteristics and cultural practices for watershed P2, Watkinsville, Ga., 1973-75

Item	Description	
Area	3.19 acres	
Soil	60% Cecil sandy loam 30% Cecil sandy clay loam 10% Loam	
Cover	Corn, rows nearly on the contour	
Cultural practices.	April 18, 1973	Tilled 20 cm deep with moldboard.1/
	May 11, 1973	Corn planted, 50,000 plants/ha, 90 cm rows.
	November 2, 1973	Harvest.
	November 5, 1973	Stalks shredded, 3,100 kg/ha stover. Estimated near 30% residue cover.
	April 23, 1974	Disked.
	April 25, 1974	Chisel plowed, 20 cm deep.
	April 29, 1974	Corn planted, 50,000 plants/ha, 90 cm rows.
	October 29, 1974	Harvest, 6,300 kg/ha stover. Estimated near 50% residue cover.
	May 15, 1975	Disked.
	May 21, 1975	Corn planted, 54,000 plants/ha, 90 cm rows.
	October 3, 1975	Harvest, 6,800 kg/ha stover.

1/ See Smith and others (2) for additional practices, including fertilizer and pesticide application rates and summary of hydrologic data.

Initial estimates of parameter values used in the sensitivity analysis were made by J. R. Williams[2/] as summarized in table I-34. These parameter values are denoted "base values," and the parameters were then varied about the base values to determine model sensitivity.

Table I-34.—Summary of model parameters for prediction of runoff volume and peak discharge using the option 1 hydrologic model for watershed P2, Watkinsville, Ga., 1974-75

Parameter	Base value	Comments
FUL	0.75	Portion of plant available water storage filled at field capacity. Maximum value 1.0.
CONA	3.50	Soil evaporation parameter.
CN2	81	Runoff curve number for antecedent moisture condition II. Selected using SCS National Engineering Handbook.
CHS	0.022	Channel slope determined from topographic map.
UL(I)	7 Values	Plant available soil water storage in 7 layers of the soil profile.
X(I)	9 Values	Leaf area index for corn throughout the growing season.
WLW	2.1	Watershed length-width ratio determined from topographic map.
TEMP(I)	12 Values	Mean monthly air temperature.
RADI(I)	12 Values	Mean monthly radiation.

Note: Additional parameters not listed in this table were selected from watershed characteristics using procedures outlined in volume II, chapter 1.

For 1974-75, 138 precipitation events were analyzed. Observed and predicted runoff volume and peak discharge for 48 runoff-producing events were related as follows:

$$\hat{Q} = 0.08 + 0.72\ Q \qquad [I-211]$$

$$R^2 = 0.53$$

and

$$\hat{Q}_p = 0.39 + 0.77\ Q_p \qquad [I-212]$$

$$R^2 = 0.30$$

2/ Hydraulic engineer, USDA-SEA-AR, Temple, Tex., personal communication.

where:

- Q = observed runoff volume, in,
- $\hat{Q}$ = predicted runoff volume using base values, in,
- Q_p = observed peak discharge, in/hr,
- $\hat{Q}_p$ = predicted peak discharge using base values, in/hr, and
- R^2 = coefficient of determination with 100 R^2 as the percent variance explained by the model.

Therefore, the model generally overpredicted runoff volumes for observed volumes less than 0.30 in and underpredicted for larger values. The coefficient of determination for equation [I-211] is 0.53, meaning that the model explains 53% of the variance in runoff volume. For peak discharge, equation [I-212], the model overpredicted for values of peak discharge less than 1.67 in/hr, or virtually all but the very largest events. Also, the model only explained 30% of the variance in peak discharge (R^2 = 0.30 in equation [I-212]). From these results, it appears that the model predicts runoff volume better than peak discharge. However, these results are for a specific watershed, and may not be typical, especially for larger watersheds where daily rainfall may be a better predictor for peak discharge. Finally, the results summarized by equations [I-211] and [I-212] represent predictions rather than fitting or optimization results.

Based on these analyses, the base values of the parameters were judged as adequate values to use in determining model sensitivity.

Sensitivity Analyses for Mean Runoff Volume and Peak Discharge

Mean values of predicted runoff volume and peak discharge were computed for the 138 precipitation events for each run. As each parameter was varied about the base value, the mean values were compared to the means from the "base value" predictions as a measure of the sensitivity. The simulation data are summarized in table I-35. Column 2 of table I-35 shows which parameters were varied, and by how much. Columns 3 and 5 show the mean predicted runoff volume and peak discharge, and columns 4 and 6 show the mean values divided by the corresponding mean values predicted using the base values of all parameters. Values of 1.0 in these two columns represent no change in runoff volume and peak; values less than 1.0 represent decreases. For example, a 50% decrease in the portion of plant-available water storage in the soil filled at field capacity, FUL, resulted in a 54% decrease in mean runoff volume and peak rate. A 33% increase in FUL resulted in over 100% increase in mean runoff volume, and nearly 100% increase in the mean peak discharge. The results for FUL are summarized in rows 2 to 5 of table I-35.

As expected, there was an inverse relation between the evaporation parameter, CONA, and runoff as summarized in rows 6 to 9 of table I-35. However, runoff is more sensitive to decreases in CONA than to increases. Runoff volumes and peaks were more sensitive to the runoff curve number, CN2, than to any other parameter. This is significant in that this parameter best reflects the influence of management practices and crop cover. Moreover, estimation techniques for this parameter are well developed in the SCS National Engineering Handbook (3).

Table I-35.— Hydrologic model, option 1, sensitivity analyses, watershed P2, Watkinsville, Ga., 1974-75 data; effect of parameter variation on runoff volume and peak

Run No.	Variation	Parameter	Mean runoff volume $\overline{Q}$	$\overline{Q}$/Base	Mean peak discharge $\overline{Q}_p$	$\overline{Q}_p$/Base
(1)	(2)		(3)	(4)	(5)	(6)
	(%)		(in)		(in/hr)	
1	Base		0.092	1.00	0.284	1.00
2	-50	FUL	.042	.457	.132	.465
3	-25		.059	.641	.185	.651
4	+25		.159	1.728	.474	1.669
5	1/ +33		.192	2.087	.565	1.989
6	-50	CONA	.143	1.554	.431	1.518
7	-25		.130	1.413	.395	1.391
8	+25		.079	.859	.245	.863
9	+50		.073	.793	.228	.803
10	2/ -10	CN2	.048	.522	.152	.535
11	-05		.067	.728	.210	.739
12	+05		.125	1.359	.379	1.335
13	+10		.172	1.870	.512	1.803
14	-50	CHS	.092	1.00	.254	.894
15	-25		.092	1.00	.271	.954
16	+25		.092	1.00	.294	1.035
17	+50		.092	1.00	.302	1.063
18	-50	UL(I)	.076	.826	.237	.835
19	-25		.084	.913	.261	.919
20	+25		.097	1.054	.299	1.053
21	+50		.101	1.098	.308	1.085
22	-50	X(I)	.099	1.076	.304	1.070
23	-25		.096	1.043	.297	1.046
24	+25		.091	.989	.283	.996
25	+50		.090	.978	.279	.982
26	-50	WLW	.092	1.00	.323	1.137
27	-25		.092	1.00	.299	1.053
28	+25		.092	1.00	.272	.958
29	+50		.092	1.00	.263	.926
30	-50	TEMP(I)	.104	1.130	.321	1.130
31	-25		.095	1.033	.294	1.035
32	+25		.090	.978	.278	.979
33	+50		.089	.967	.274	.965
34	-50	RADI(I)	.112	1.217	.341	1.201
35	-25		.099	1.076	.305	1.074
36	+25		.088	.957	.273	.961
37	+50		.087	.946	.269	.947

1/ Upper limit for FUL is 1.0, a 33% increase over base value.
2/ Limits for CN2 are 0 to 100.

Fortunately, the model is less sensitive to plant-available soil water storage, UL(I), leaf area index, X(I), average daily temperature, TEMP(I), and average daily radiation, RADI(I). In terms of parameter estimation, this means that rather coarse estimates of these parameters can be used with correspondingly smaller errors in computed runoff. For example, monthly averages can be used to interpolate daily values or point records can be used to represent large geographic areas. This is important in applying the model to ungaged watersheds.

Additional sensitivity analyses were conducted with respect to average soil water content and mean volume of percolation. Table I-36 shows the sensitivity of soil water and percolation to leaf area index, temperature, and radiation. In general, percolation was more sensitive than soil moisture with more relative change in percolation with changes in temperature and radiation. However, a 50% error in temperature or radiation would represent very gross errors. In a negative sense, these results suggest that the model is not very sensitive to changes in crop growth and canopy, as reflected in the leaf area index. A possible exception would be interactions, and thus changes in runoff curve number with changes in crop canopy. Additional research may be needed to determine interactions between runoff curve number and crop canopy development.

Table I-36.—Effect of parameter variation on average soil water and percolation

Run No.	Variation	Parameter	Mean $\overline{SW}$	$\overline{SW}$/Base	Mean percolation	Perc/Base
	(%)		(in)		(in)	
1	Base		0.353	1.0	0.137	1.0
2	-50	X(I)	.361	1.02	.150	1.09
3	-25	X(I)	.358	1.01	.141	1.03
4	+25	X(I)	.352	.99	.135	.99
5	+50	X(I)	.352	.99	.134	.98
6	-50	TEMP(I)	.364	1.03	.184	1.34
7	-25	TEMP(I)	.357	1.01	.156	1.14
8	+25	TEMP(I)	.351	.99	.127	.93
9	+50	TEMP(I)	.350	.99	.121	.88
10	-50	RADI(I)	.368	1.04	.209	1.53
11	-25	RADI(I)	.360	1.02	.164	1.20
12	+25	RADI(I)	.350	.99	.123	.90
13	+50	RADI(I)	.349	.99	.112	.82

Model sensitivity for the hydrologic model, option 1, daily rainfall model, is summarized in table I-37. As described at the bottom of table I-37, a parameter's sensitivity is judged as "significant" when changes in runoff due to a parameter change exceed the absolute magnitude of the parameter change. This criterion identifies areas where errors are "magnified" by the mode. For this model and the particular data set analyzed, CN2, CONA, and FUL are the most sensitive parameters, with the runoff curve number, CN2, as the single most important parameter. Therefore, the user must exercise good judgment and particular care in selecting the runoff curve number.

Table I-37.— Summary of sensitivity of daily runoff model for watershed P2, Watkinsville, Ga.[1/]

Parameter	Mean volume	Mean peak	Comments
CHS	None	Moderate	Not considered in volume calculation, peak coefficient is directly related to CHS.
CN2	Significant	Significant	Critical parameter; small variation causes gross change in runoff.
CONA	Significant	Significant	Strongly affects ET; increasing CONA produces higher SW, inversely affects runoff.
FUL	Significant	Significant	Portion of plant-available water storage filled at field capacity; related to soil properties.
POROS	None	None	Parameter in percolation calculation; no effect on runoff.
RADI(I)	Moderate	Moderate	Monthly average radiation.
RC	None	None	Parameter in percolation calculation no effect on runoff.
TEMP(I)	Moderate	Moderate	Monthly average temperatures.
UL(I)	Moderate	Moderate	Plant-available soil water storage in up to 7 soil layers.
WLW	None	Moderate	Not considered in volume calculation; influences peak discharge.
X(I)	Slight	Slight	Monthly leaf area index; measure of crop canopy.

1/ A ± 50% change in parameters produces a change in mean runoff volume or peak of: Slight < 10%; moderate 10-50%; significant > 50%. These sensitivity analyses are for a particular watershed and are thus site specific.

Option 2, Breakpoint Rainfall Model

This model uses breakpoint precipitation data as input to compute runoff volume and peak rate, as well as soil moisture and percolation. As in the analysis described above, data from watershed P2, Watkinsville, Ga., were used to determine model sensitivity.

Initial estimates of parameter values used in the sensitivity analysis were made by R. E. Smith[3/] as summarized in table I-38. These parameter values are denoted "base values," and then were varied about the base values to determine model sensitivity.

Table I-38.—Summary of model parameters for prediction of runoff volume and peak discharge using the option 2 hydrologic model for watershed P2, Watkinsville, Ga., 1974-75

Parameter	Base value	Comments
FUL	0.75	Same as in option 1.
CONA	3.5	Same as in option 1.
DS	2.0	Depth of surface soil layer (in).
DP	26.0	Depth of maximum root growth layer (in).
GA	13.0	Effective capillary tension of soil (in).
RMN	.08	Manning's n for overland flow.
SLOPE	.025	Slope of plane of watershed (ft/ft).
XLP	350.	Length of plane (ft).
TEMP (I)	12 Values	Same as in option 1.
RADI (I)	12 Values	Same as in option 1.
X (I)	7 Values	Same as in option 1.
POROS	.41	Same as in option 1.
RC	.15	Same as in option 1.

For the period 1974-75, 138 precipitation events were analyzed. Observed and predicted runoff volume and peak discharge for 58 runoff-producing events were related as follows:

$$\hat{Q} = 0.104 + 1.033\, Q \qquad [I\text{-}213]$$

$$R^2 = 0.76$$

and

$$\hat{Q}_p = -0.060 + 2.084\, Q_p \qquad [I\text{-}214]$$

$$R^2 = .75$$

where: Q = observed runoff volume, in,
$\hat{Q}$ = predicted runoff volume using the base values, in,
Q_p = observed peak discharge, in/hr,
$\hat{Q}_p$ = predicted peak discharge using base values, in/hr, and
R^2 = coefficient of determination.

3/ Hydraulic engineer, USDA-SEA-AR, Fort Collins, Colo., personal communication.

Thus, the model generally overpredicted runoff volume and explained 76% of the variance in runoff volume for the 3.2-acre test watershed. Runoff peak was also generally overpredicted; the coefficent of variation is 0.75. As in the option 1 sensitivity test, base parameter values were chosen as they would be chosen by the user, using available measurements and handbook values. No fitting or optimization techniques were employed.

Sensitivity Analyses for Runoff Volume and Peak, Average Soil Moisture, and Percolation

The effects of parameter variation on runoff volume and peak, average soil moisture, and percolation were studied. As in the tests of option 1, the mean of each predicted value was calculated (for 138 precipitation events) and compared to the mean values obtained in the base run. The simulation data are summarized in table I-39. Columns 2, 4, 6, and 8 contain the mean predicted values of each variable for each parameter variation. Columns 3, 5, 7, and 9 contain the ratios of the calculated means to their respective base values to indicate the relative effect of each parameter variation. This relative effect is compiled in table I-40, which corresponds to table I-37 of the option 1 sensitivity study. Again, the results are for a particular watershed and relative sensitivity may be different depending upon site specific conditions.

Model use objectives should be carefully determined before parameter values are chosen. For instance, if runoff volume prediction is the user's sole interest, eight parameters affect this variable slightly, four moderately, and only one, RC, significantly. Peak flow is not significantly affected by changes in any single parameter, but eight parameters have moderate influence; their cumulative influence warrants some judicious choosing. Average soil moisture is affected much like volume: slightly sensitive to seven parameters, moderately sensitive to four, and significantly sensitive to two. Percolation has been shown to be an extremely sensitive variable, responding significantly to seven parameters, moderately to three, and slightly to three. Therefore, if the model is to be used primarily for percolation estimates, seven of the parameters must be very carefully determined.

The parameters defining temperature, radiation, and leaf area index are used in the calculation of evapotranspiration, and therefore, have greatest effect on percolation and soil moisture, and a resultant effect on runoff. Temperature and radiation variation resulted in significant changes in percolation and moderate changes in soil moisture and runoff volume. Because the parameters are both directly related to ET, their effect on runoff, percolation, and soil moisture is inverse, that is, increasing these parameters increases ET, and thus decreases soil moisture, percolation, and runoff. The other ET parameter, CONA, is similarly related.

In option 2, a surface control layer, DS, is used to calculate runoff volume. As the plant canopy develops (increased LAI), soil evaporation is reduced proportional to exp(-0.4 LAI). At the same time, transpiration from the entire soil profile may increase. The result of these interactions is that moisture content in the surface control layer is not reduced as much as moisture content in the entire soil profile. For specific storm sequences, the runoff volume may be slightly increased (because of higher soil moisture in the surface control layer) while the overall soil moisture in the entire soil pro-

Table I-39.—Hydrologic model, option 2 sensitivity analysis, watershed P2, Watkinsville, Ga., 1974-75; effect of parameter variation on runoff volume and peak, average soil moisture, and percolation

Variation (1)	Parameter	Mean runoff volume Q (2)	Q/Base (3)	Mean peak disch. QP (4)	QP/Base (5)	Mean avg. soil moist. SW (6)	SW/Base (7)	Mean perc (8)	Perc/Base (9)
(%)		(in)		(in/hr)		(in)		(in)	
	Base	0.129	1.000	0.319	1.000	0.198	1.000	0.085	1.000
-50	RC	.180	1.395	.398	1.245	.188	.946	.050	.582
-25		.149	1.155	.354	1.109	.194	.980	.071	.829
+25		.112	.866	.290	.908	.201	1.015	.098	1.149
+50		.101	.787	.276	.863	.204	1.026	.105	1.232
-50	FUL	.125	.973	.320	1.000	.110	.554	.129	1.511
-25		.126	.978	.311	.990	.155	.780	.107	1.258
+25		.131	1.021	.324	1.015	.246	1.241	.067	.784
+33		.132	1.026	.325	1.018	.263	1.325	.063	.735
-50	POROS	.167	1.297	.388	1.213	.102	.517	.121	1.420
-25		.143	1.112	.349	1.092	.151	.762	.100	1.177
+25		.119	.928	.305	.954	.254	1.278	.073	.854
+50		.111	.863	.289	.905	.315	1.588	.070	.826
-50	DS	.120	.942	.310	.971	.207	1.042	.114	1.337
-25		.120	.959	.314	.983	.201	1.014	.095	1.118
+25		.130	1.018	.323	1.011	.197	.995	.078	.911
+50		.130	1.033	.326	1.022	.197	.995	.073	.857
-50	DP	.129	1.001	.321	1.004	.208	1.046	.123	1.444
-25		.129	.999	.320	1.003	.202	1.018	.104	1.217
+25		.129	1.004	.321	1.004	.200	1.006	.069	.813
+50		.129	1.005	.322	1.007	.201	1.015	.060	.707
-50	GA	.167	1.300	.384	1.203	.191	.960	.059	.695
-25		.143	1.113	.348	1.089	.196	.986	.075	.880
+25		.115	.895	.294	.922	.201	1.012	.095	1.111
+50		.107	.833	.282	.883	.203	1.022	.100	1.172
-50	SLOPE	.128	.991	.258	.806	.199	1.001	.086	1.013
-25		.128	.997	.295	.922	.198	1.000	.086	1.005
+25		.129	1.000	.331	1.035	.198	1.000	.085	1.000
+50		.129	1.000	.341	1.067	.198	1.000	.085	1.000
-50	CONA	.171	1.329	.376	1.178	.237	1.192	.213	2.490
-25		.160	1.244	.360	1.128	.231	1.162	.195	2.288
+25		.121	.940	.309	.966	.191	.963	.068	.795
+50		.120	.929	.307	.962	.190	.960	.065	.763
-50	RMN	.130	1.010	.383	1.199	.198	.999	.084	.987
-25		.129	1.002	.351	1.100	.198	1.00	.085	.998
+25		.128	.994	.280	.878	.199	1.001	.086	1.009
+50		.127	.990	.249	.781	.199	1.001	.086	1.015
-50	XLP	.130	1.010	.383	1.199	.198	.999	.084	.987
-25		.129	1.002	.351	1.100	.198	1.000	.085	.998
+25		.128	.994	.280	.878	.199	1.001	.086	1.009
+50		.127	.990	.249	.781	.199	1.001	.086	1.015
-50	X(I)	.127	.988	.318	.997	.245	1.237	.113	1.329
-25		.125	.970	.313	.979	.215	1.086	.095	1.112
+25		.130	1.006	.321	1.006	.193	.974	.082	.967
+50		.130	1.008	.323	1.012	.191	.962	.080	.938
-50	TEMP(I)	.155	1.206	.362	1.132	.262	1.320	.321	3.769
-25		.131	1.019	.324	1.015	.217	1.096	.105	1.232
+25		.127	.988	.317	.993	.188	.950	.074	.865
+50		.126	.981	.315	.987	.182	.920	.066	.773
-50	RADI(I)	.141	1.094	.336	1.052	.270	1.359	.168	1.967
-25		.133	1.034	.327	1.024	.231	1.163	.117	1.379
+25		.126	.981	.315	.987	.182	.919	.068	.797
+50		.125	.969	.314	.982	.172	.869	.055	.645

Table I-40.—Summary of sensitivity of breakpoint runoff model for watershed P2, Watkinsville, Ga.[1/]

Parameter	Mean volume	Mean peak	Mean SW	Mean perc	Comments
FUL	Slight	Slight	Significant	Significant	Maximum value 1.0; important parameter in soil moisture and drainage.
CONA	Moderate	Moderate	Moderate	Significant	Soil evaporation and plant ET parameter.
DS	Slight	Slight	Slight	Moderate	Portion of soil profile which defines initial soil saturation.
DP	Slight	Slight	Slight	Significant	Maximum root growth layer (depth below DS).
GA	Moderate	Moderate	Slight	Moderate	Green and Ampt effective capillary tension; used in infiltration determination.
RMN	Slight	Moderate	Slight	Slight	Manning's n for overland flow; affects peak discharge.
SLOPE	Slight	Moderate	Slight	Slight	Used in peak discharge calculation.
XLP	Slight	Moderate	Slight	Slight	Used in peak discharge calculation.
TEMP(I)	Moderate	Moderate	Moderate	Significant	Used in calculation of ET; important to use reasonable (measured if possible) values.
RADI(I)	Moderate	Slight	Moderate	Significant	Used in calculation of ET; important to use reasonable (measured if possible) values.
X(I)	Slight	Slight	Moderate	Moderate	Used in calculation of ET.
POROS	Moderate	Moderate	Significant	Significant	Extremely important parameter; effects all variables considerably.
RC	Significant	Moderate	Slight	Significant	Used in infiltration determination.

1/ A ± 50% change in parameters produces a change in predicted variable of: Slight: < 10%; moderate: 10-50%; significant: > 50%.

file is decreased. Research is required for a better accounting of moisture balance in the surface control layer. Without this improvement, slight increases in runoff volume can occur with increases in leaf area index.

Though strongly influenced by the ET parameters, infiltration and drainage are driven by the parameters DS, DP, GA, RC, POROS, and FUL, which define and control the motion of water in the soil. DP and DS, soil depth parameters, have minimal effect on runoff and soil moisture, but considerable influence on percolation. GA and RC describe soil properties that govern the motion of water into and in the soil. Their effect on average soil moisture is minimal, but both have considerable effect on percolation (see table I-39) and runoff. Note that runoff is inversely related to both of these parameters. POROS and FUL define soil capacities. Soil porosity significantly affects soil moisture and percolation: greater porosity allowing greater quantities of water to be stored, and thus less to be percolated. FUL limits the amount of soil moisture that is available for plant use. It significantly affects percolation and soil moisture, and has a slight resultant effect on runoff.

Runoff is affected by parameters that primarily drive ET and infiltration. Both runoff volume and peak are directly affected by parameters that represent watershed geomorphology and physical characteristics. Peak flow is a function of flow volume and watershed characteristics. Because runoff peak is moderately sensitive to eight parameters, it is to be considered a sensitive variable. If peak estimation is the primary purpose for model use, these eight parameters must be carefully chosen. Fortunately, several are easily measurable or otherwise determinable (for instance, good temperature and radiation data are available for many locations; watershed slope is often known, and leaf area index can be reasonably estimated if crop type is known). The effective slope length, XLP, and Manning's n, RMN, have identical moderate influence on predicted peak discharge and slight resultant effect on volume of runoff, soil moisture, and percolation.

FIELD-SCALE MODEL: EROSION/SEDIMENT YIELD COMPONENT

The main processes in the erosion/sediment yield component are overland flow, concentrated flow, and impoundments (ponds). The overland flow component uses a modified form of the Universal Soil Loss Equation (USLE) to compute sediment detachment and the Yalin equation to compute sediment transport capacity. A first-order equation is used to compute sediment deposition. The concentrated flow component computes sediment transport and subsequent detachment or deposition, depending upon flow conditions, using the Yalin sediment transport equation, a flow detachment equation, and a deposition equation. The pond component estimates how much of the sediment settles to the bottom of a pond before the flow passes through the impoundment.

Overland Flow Component

Selection of Parameters

The overland flow component has three input variables: EI = storm erosivity, Q = runoff volume, and σ_p = peak discharge rate. The USLE factors KCP all occur in a linear form so that varying K shows the sensitivity to the other two factors as well. The Manning's n value for surface-cover conditions, n_{cov}, expresses hydraulic roughness in overland flow and C_{yal}, the coefficient in the Yalin sediment transport equation, represents the transport capacity. Therefore, we chose to vary three input variables (EI, Q, σ_p) and three parameters (K, n_{cov}, C_{yal}) to evaluate model sensitivity for overland flow.

Selection of Data and Base Values

Data from watershed P2 at Watkinsville, Ga., were also used to study model sensitivity for the overland and concentrated flow components. These data are summarized in table I-33.

Base values for many of the parameters are summarized in table I-41. Observed values of EI, Q, and $_p$ for 32 storm events were used to simulate sediment yield for the 1973-74 period. Estimated sediment yield from the watershed

Table I-41.— Summary of model parameters selected for prediction of sediment yield from watershed P2, Watkinsville, Ga., 1973-75

Parameter	Base value	Comments
K	0.23	Soil erodibility factor; does not consider seasonal variability (see vol. II, ch. 2.)
C	0.11-0.68	Soil loss ratio; reflects cover and cultural practices (see vol. II, ch. 2.)
P	1.0	Contouring factor; up and downhill tillage assumed (see vol. II, ch. 2.)
n_{cov}	0.010-0.035	Manning's n for overland flow; reflects effect of cover and cultural practices on flow's hydraulic resistance.
C_{yal}	0.635	Empirical constant in Yalin sediment transport equation; published value used (see vol. II, ch. 2.)
K_{rch}	0.135	Soil erodibility factor for erosion by concentrated flow; value empirically derived from rill erosion studies.
τ_{cr}	0.15-0.50	Critical shear stress for erosion by concentrated flow; reflects effect of cover and cultural practices on resistivity of soil to detachment by concentrated flow.
n_{ch}	0.03-0.12	Manning's n for concentrated flow; reflects effect of cover and cultural practice on flow's hydraulic resistance.
n_{bch}	0.03	Manning's n for concentrated flow over bare, tilled agricultural soil; value empirically determined.

Note: Parameters, including those not listed in this table, were selected using procedures outlined in the user Manual, vol. II, ch. 2.

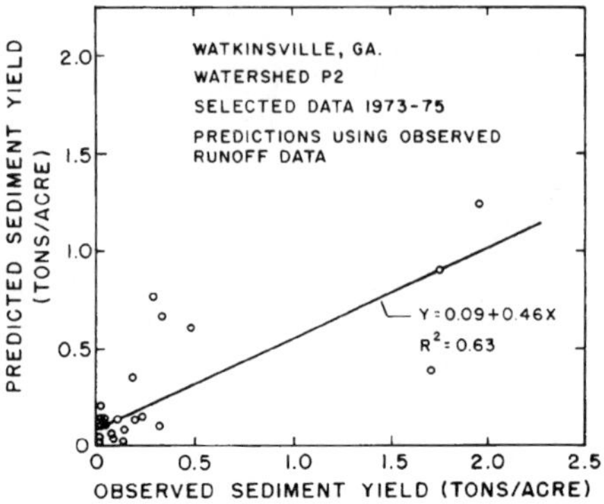

Figure I-27.—Relation between observed and predicted sediment yield for selected storms on watershed P2, Watkinsville, Ga.

was used as a model output, and these yields were then summed for the 3 years to produce a total sediment yield in tons per acre. The results of these initial predictions or base value runs are shown in figure I-27. The model underestimated sediment yield for three largest events. The model explained over 60% of the variance in observed sediment yield. The total observed sediment yield was 8.26 tons/acre, and the computed sediment yield was 6.56 tons/acre, with a ratio of estimated-to-observed of 0.79 representing a 20% error in total sediment yield. However, the 95% confidence limits for the mean observed sediment yield per event are 0.258 ± 0.188 or 0.07 to 0.45 tons/acre. The mean predicted sediment yield was 0.205 tons per/acre, well within the confidence limits.

Sensitivity Analysis for Overland Flow Component

Computed sediment yields for the sensitivity analysis are summarized in table I-42. Column 2 shows which item was varied and that each was varied over ± 25 and ± 50% of the base value. Column 3 shows the computed sediment yield from overland flow, and column 5 shows the ratio of this yield to the yield for the base values. Column 7 shows the ratio of overland to total watershed sediment yield and is similar to a delivery ratio. Except for run no. 21, all simulations indicate net deposition in the channel system, so that sediment yields from the watershed were less than sediment yields in overland flow. However, for 5 of the 32 events the model predicted net erosion in the channel system producing a delivery ratio greater than one. For the other 27 events, the model predicted net deposition in the channel system producing an overall effect of less sediment yield from the watershed than from overland flow.

Sensitivity of the model output to changes in EI, Q, and σ_p for the 32

Table I-42.—Overland flow sensitivity analysis, watershed P2, Watkinsville, Ga., selected data, 1973-75

Run no. (1)	Variation (2)		Overland sed. yield QSO (3)	Watershed sed. yield QSW (4)	$QSO/Base_o$ (5)	$QSW/Base_w$ (6)	Ratio of watershed to overland sed. yield QSW/QSO (7)
	(%)		(tons/acre)	(tons/acre)			
1	Base		8.700	6.560	1.00	1.00	0.754
2	-50	EI	6.685	5.186	.768	.791	.776
3	-25	EI	7.677	5.823	.882	.888	.758
4	+25	EI	9.481	7.004	1.131	1.068	.739
5	+50	EI	10.155	7.605	1.167	1.159	.749
6	-50	Q	5.603	4.178	.644	.637	.746
7	-25	Q	7.252	5.375	.834	.819	.741
8	+25	Q	9.810	7.420	1.128	1.131	.756
9	+50	Q	11.004	8.432	1.265	1.285	.766
10	-50	σ_p	6.136	4.773	.705	.728	.778
11	-25	σ_p	7.616	5.749	.875	.875	.755
12	+25	σ_p	9.376	6.953	1.078	1.060	.742
13	+50	σ_p	9.930	7.373	1.141	1.124	.742
14	-50	K	5.365	4.391	.617	.669	.818
15	-25	K	7.281	5.600	.829	.854	.769
16	+25	K	9.905	7.352	1.139	1.121	.742
17	+50	K	10.634	7.840	1.222	1.195	.737
18	-50	n_{cov}	13.784	7.650	1.584	1.166	.555
19	-25	n_{cov}	11.161	7.154	1.283	1.091	.641
20	+25	n_{cov}	6.505	5.672	.748	.865	.872
21	+50	n_{cov}	4.978	5.043	.572	.769	1.013
22	-50	C_{yal}	5.977	4.666	.687	.711	.781
23	-25	C_{yal}	7.623	5.821	.876	.887	.764
24	+25	C_{yal}	9.427	7.057	1.084	1.076	.749
25	+50	C_{yal}	10.029	7.527	1.153	1.147	.751

storms is shown in figure I-28. Overall, the percent change in sediment yield was approximately half of the percentage change in input variables. The model was more sensitive to decreases in the input variables than to increases. Except for a large increase in EI, changes in sediment yield were nearly linear with changes in input variables, and the model appears to be most sensitive to volume of runoff, Q.

Overall, the percent change in sediment yield was approximately half the percent change in K and C_{yal}, the soil erodibility and Yalin transport equation coefficient. Again, the relative changes were greater for decreases in the parameters than for increases. For the hydraulic roughness, Manning's n_{cov}, the changes in sediment yield were larger than the relative changes in the parame-

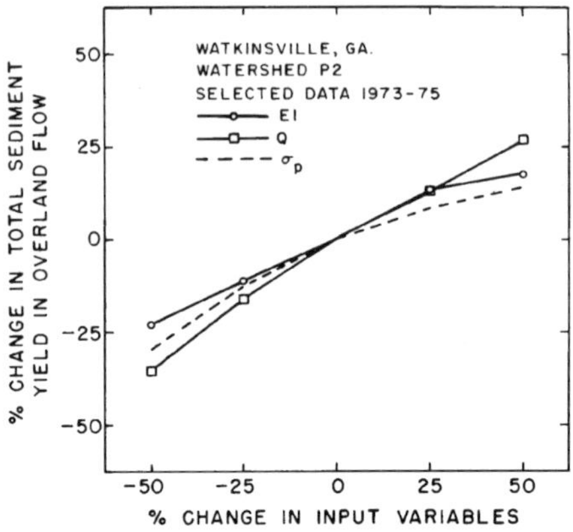

Figure I-28.—Sensitivity of sediment yield in overland flow to input variables.

ter. This is significant in that this roughness parameter has a direct control on transport capacity, and thus on deposition in overland flow. The change in overland flow sediment yield appeared to vary nearly linearly with changes in the roughness parameter over the range that the parameter was varied. For the observed data on this particular watershed, the computed overland flow sediment yield was most sensitive to this parameter.

Interpretations and Summary

Typical slope values for watershed P2 varied from 2 to 6% for overland flow and 1.5 to 3.5% for the channel or concentrated flow section. Changes in sediment yield with changes in roughness indicates deposition in overland flow. As discussed earlier, the channel component produced a "delivery ratio" of less than one. Sediment yield was limited by transport capacity for many of the 32 storms, and was primarily controlled by deposition. These sensitivity analyses represent results obtained for a watershed situation where transport capacity was moderately limiting for most storms. These results may be typical for cultivated fields with low to moderate concave slopes and with relatively high values of hydraulic roughness due to tillage and residue cover conditions. In this situation, the hydraulic roughness factor for overland flow should be carefully chosen.

Sensitivity Analysis for Concentrated Flow Component

Sensitivity analyses for the concentrated flow (channel) component can be separated into two sections: (1) inputs and parameters affecting overland flow sediment delivery to the channel system, and (2) parameters directly affecting detachment, transport, and deposition in the concentrated flow.

Overland Flow Factors Affecting Sediment Delivery to the Channel

Column 4 in table I-42 shows total sediment yield from the entire watershed, and the difference between these values and corresponding values listed in column 3 show the net effect of the channel on sediment yield.

In all cases except run no. 21, the net effect of the channel was a sink rather than a source for the total sediment yield from the 32 storms. Some of the larger events showed net channel erosion, but the overall effect was to reduce sediment yield. Also, erosion was often predicted near the upper end of the channel, with deposition predicted near the outlet where water was ponded by a runoff measuring flume.

The influence of changes in EI, Q, and σ_p on sediment yield from the watershed is shown in table I-42. The results for EI and σ_p are very similar to those in overland flow only (figure I-28), but the influence of runoff volume, Q, is more pronounced, suggesting that this may be a most sensitive input. Influence of the "overland" parameters (K and n_{cov}) are shown in table I-42. Note that the influence of overland flow hydraulic roughness is damped out by the channel system so that its sensitivity is comparable to the other parameters.

Sediment routing was extended to the entire watershed system with the result that the influence of runoff volume was accentuated and the influence of overland flow parameters was dampened.

Parameters Directly Affecting Sediment Yield from Concentrated Flow

Channel routing assumptions--The general case for concentrated flow in a field situation is a channel of length L with an upstream inflow rate, Q_i, and a la-a lateral inflow rate, q_*, along the channel reach. This configuration is illustrated in figure I-29, with the runoff rates corresponding to the peak discharge at steady state, that is, steady-state spatially varied flow with increasing discharge. The effective channel length, L_{eff}, is the length of channel required to produce the outflow discharge, Q_e given the lateral inflow rate. The procedure used here is to solve the spatially varied flow equations for a channel of length L_{eff} to produce depth, velocity, and shear stress along the channel reach, and then apply the transport and detachment capacity equations along the original length of channel, L, to compute sediment yield for the channel.

Assuming a wide range of channel and flow conditions, the spatially varied flow equations were solved, and polynomial equations were fitted by regression to the solutions (vol. I, ch. 3). As part of these sensitivity analyses, the regression equations were reviewed and found to be quite accurate over a wide range of conditions representing subcritical flow in channels with triangular cross-sections. Although the model has user options for rectangular and naturally eroded channels, the regression equations for spatially varied flow have not been checked under these conditions. Therefore, results of the sensitivity analysis presented here are for triangular channels only. A second user option is to assume the friction slope in the routing procedure to be equal to the channel slope. This assumption has been tested under a limited number of

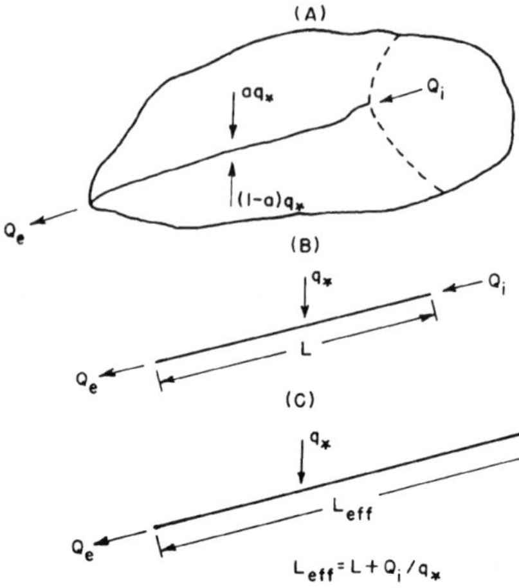

Figure I-29.—Illustration of general case for concentrated flow in a field-sized channel: (A) schematic of watershed channel system, (B) channel reach with upstream inflow and uniform lateral inflow, and (C) effective channel reach for spatially varied flow computations.

conditions and has been found to be a poor approximation except under special circumstances. The conditions under which this assumption is appropriate are (1) no outlet controls producing backwater; (2) channel slopes relatively steep; and (3) lateral inflow rate per unit length of channel is small compared with the outlet discharge.

In summary, the model assumes a triangular shaped channel to estimate friction slopes. The same friction slope estimates are used for rectangular channels. Preliminary analyses suggested that this approximation can be quite accurate. However, additional tests under a variety of conditions are required before the general applicability of the procedure can be determined.

Channel erosion/sediment transport parameters--The parameters selected for analysis here are summarized in table I-41. In addition, selected user options were used to determine model sensitivity to these inputs. Simulation runs for the concentrated flow sensitivity analysis are summarized in table I-43. Most items were varied over ± 25 and ± 50% of their base values except for runs 34 to 41, where Manning's $\bar{n}$ for the cover-practice was restricted to be equal to or larger than Manning's n for bare soil in the channel. Column 4 shows the computed sediment yield from the entire watershed; column 5 shows the ratio of this yield to the yield using the base values. Column 6 shows the ratio of overland to total watershed sediment yield and is similar to a delivery ratio for the simulation results. In all runs except 44 and 45, simulations suggest that the channel system was a "sink," so that sediment yields from the watershed were less than sediment yields from overland flow. For a few events, net channel erosion was estimated, but the overall effect was net deposition in the channel.

Sensitivity of the model output to changes in K_{rch}, τ_{cr}, and n_{ch} for the 32 storms is shown in table I-43. For the erodibility factor, K_{rch}, and critical shear stress, τ_{cr}, the model was more sensitive to decreases than to increases in the parameters. For hydraulic roughness, n_{ch}, the situation was reversed in part due to the constraint that $n_{ch} \geq n_{bch}$. However, total sediment yield was more sensitive to the critical shear stress than to the erodibility or roughness parameters.

Changes in sediment yield were nearly linear with the side slope, but were nonlinear and very large with changes in the channel slope. Although a ± 50% change in channel slope is an extreme error in user input, nonetheless, the simulated sediment yield is very sensitive to channel slope.

The influence of assuming a rectangular or naturally eroded channel cross-section was a 30% increase in estimated sediment yields over similar predictions using a triangular channel. The most significant increase in estimated sediment yield (+88%) was due to assuming that the friction slope was equal to the channel slope. The reason for this is that the H-flume measuring structure used on this watershed caused significant backwater, which is ignored when using the channel slope as an approximation to the friction slope. This is a most serious error, yet a common one in runoff and sediment routing, and requires a good deal of judgment by the user. The fact that the simulated sediment yield was more sensitive to the "channel slope" assumption than to ± 50% errors in input variables or parameters suggests that the user exercise caution in specifying the outlet control. Site specific conditions such as grass around field edges or other outlet controls may pond the water and induce significant deposition.

Interpretations and Summary

Model simulations and limited field observations indicate that sediment yield from watershed P2 is primarily transport limited, and thus is primarily controlled by deposition. However, for certain parameter values and for the largest storms, the model predicted significant channel erosion. From these results we may conclude that the concept of a constant delivery ratio is, at best, a gross approximation.

Table I-43.— Concentrated flow sensitivity analysis, watershed P2, Watkinsville, Ga., selected data 1973-75

Run no. (1)	Variation (2)	Parameter (3)	Watershed sed. yield QSW (4)	$QSW/Base_W$ (5)	Ratio of watershed to overland sed. yield QSW/QSO (6)
	(%)		(tons/acre)		
26	-50	K_{rch}	6.261	0.954	0.720
27	-25		6.414	.978	.737
28	+25		6.696	1.021	.770
29	+50		6.828	1.041	.785
30	-50	τ_{cr}	7.911	1.206	.909
31	-25		7.184	1.095	.826
32	+25		6.204	.946	.713
33	+50		6.016	.917	.691
34	-50	n_{ch}	1/6.753	1.029	.776
35	-25		6.635	1.011	.763
36	+25		6.064	.924	.697
37	+50		5.676	.865	.652
38	-50	n_{bch}	2/5.546	.996	.637
39	-25		5.558	.998	.639
40	+25		5.574	1.001	.641
41	+50		5.601	1.006	.644
42	-50	S	3.520	.537	.405
43	-25	Slope of	4.138	.631	.476
44	+25	Channel	9.269	1.413	1.065
45	+50		11.788	1.797	1.355
46	-50	Z	7.697	1.173	.885
47	-25	Side	6.994	1.066	.804
48	+25	Slope of	6.262	.955	.720
49	+50	Channel	6.048	.922	.695

1/ Lowest values of n_{ch} set to n_{bch} = 0.03 to insure $n_{ch} \geq n_{bch}$.
2/ Lowest values of n_{ch} set to 0.045 to insure that $n_{ch} \geq n_{bch}$. Therefore, the base value for these 4 runs was QSW = 5.568 tons/acre.

The common "normal flow" or "kinematic assumption" of friction slope equal to the bed slope can lead to serious errors in cases where an outlet control causes significant backwater effects. Therefore, in applying the model, site specific conditions causing possible backwater effects should be carefully evaluated to determine the proper outlet control (discharge-depth relationship).

Model sensitivity for the overland and concentrated flow components is summarized in table I-44. In this situation, overall watershed sediment yield was less than overall overland flow sediment yield, suggesting a transport limiting or depositional channel. For this reason, sediment yield was more sensitive to "transport-deposition" inputs and less sensitive to "erosion-detachment" inputs. In steep watersheds with actively eroding channels without outlet controls, this situation can be reversed. In either case, site-specific conditions such as outlet controls on channel depth-discharge relationships can be significant in determining sediment yield.

Impoundment (Pond) Component

The impoundment or pond component was tested using data from three small watersheds in Iowa with impoundment terraces (1). The data are from watersheds at Charles City, Eldora, and Guthrie Center, Iowa, with drainage areas of 4.6, 1.8, and 1.4 acres, respectively. Sediment yield data were measured at the outlets with overland flow runoff volumes and peak rates used to estimate the overland flow sediment input to the ponds. The impoundment component is based on the rate particles reach the bottom of the impoundment versus the rate that they leave the impoundment with the outflow. With this simplified model, the fraction of particles of a specific size and density that passes through the pond follows an exponential distribution.

Observed and computed sediment yield data from the three impoundment terraces are shown in table I-45. Relations between observed and computed sediment yields is shown in figure I-30. In general, sediment yields were underestimated a, as shown by the regression equations in figure I-30. However, the simulation results, especially those for Guthrie Center, were judged adequate to use as base values in determining model sensitivity.

Sensitivity of the model to the infiltration rate into the pond bottom is summarized in table I-46. In this analysis the volume of water discharged from the pond was held constant. Therefore, as infiltration in the pond was increased from 0 to 0.5 in/hr the volume of runoff reaching the pond had to be increased to maintain the same outflow. The estimated sediment yield in overland flow increased two to three times with the increase in volume of runoff. Simulated sediment yield from the ponds increased only 1 to 40%. Only 3 to 11% of the input sediment passed through the ponds. Although the 1 to 40% changes in sediment yield at the impoundment outlets are significant, they are an order of magnitude, or more, less than the corresponding changes in overland flow sediment yield. However, as expected, there was significant clay/silt enrichment with a much higher proportion of the sediment leaving the impoundment in the clay-size range.

Increasing the soil erodibility factor K by 50% resulted in sediment yield increases from the impoundments of 45, 45, and 18% for the three watersheds, respectively. These changes are of the same order of magnitude as the increased sediment yield in overland flow. This suggests that the impoundment component is sensitive to the sediment load entering the impoundment and its particle-size distribution.

Table I-44.—Summary of sensitivity of erosion sediment yield model for watershed P2, Watkinsville, Ga.[1]

Input	Type	Relative sensitivity of total sediment yield	Comments
EI	Variable	Moderate[2]	Measure of detachment capacity of rainfall.
Q	Variable	Moderate	Measure of detachment and transport capacity of overland and concentrated flow.
σ_p	Variable	Moderate	Measure of detachment and transport capacity of overland and concentrated flow.
KCP	Parameter	Moderate	Overland flow detachment parameters.
C_{yal}	Parameter	Moderate	Transport parameter.
n_{cov}	Parameter	Significant[3]	Overland flow hydraulic roughness. Less sensitivity for watershed sediment yield with depositional channels.
K_{rch}	Parameter	Moderate	Not sensitive for depositional channel.
τ_{cr}	Parameter	Moderate	Not sensitive for depositional channel.
n_{ch}	Parameter	Moderate	Sensitive for depositional channel.
n_{bch}	Parameter	Slight	Sensitive for depositional channel.
S	Parameter	Significant	Channel slope should be carefully determined.
Z	Parameter	Moderate	Triangular cross-section side slope.
Channel shape.	User option	Moderate	Site specific.
Friction slope.	User option	Significant	Most critical user option. Outlet control should be carefully determined.

1/ Watershed with "depositional" channel on the average.

2/ A ± 50% change in input produces a change in sediment yield of: Slight - 0 to 10%; moderate - 10 to 50%; significant - greater than 50%.

3/ These results are for a particular watershed and a given sequence of storms. Relative sensitivity of the model to input parameters can be different under different conditions.

Table I-45.—Summary of observed and simulated sediment yield from impoundment terraces in Iowa

Watershed	Julian date	Observed sediment yield	Computed sediment yield
		(lb)	(lb)
Charles City	70147	1,197	52
	70152	72	14
	70244	4	160
	70323	58	5
	71151	280	294
	71157	209	160
Eldora	68198	283	150
	68220	58	55
	69187	1,057	554
	69232	124	227
	71163	335	139
Guthrie Center	69207	256	273
	69249	23	89
	70144	122	63
	70162	198	123
	70167	21	28
	70229	10	52

The assumed particle size distribution for the eroded soil is shown in table I-47. Two alternate distributions with smaller clay and smaller clay and silt particles are also summarized in table I-47. Smaller particles should have a lower fall velocity, and thus more of them should pass through an impoundment. These simulation results are summarized in table I-48. Decreasing the diameter of clay particles from 0.002 to 0.001 mm resulted in an average 18% increase in sediment yield from the impoundments. With the same reduction in clay size and reducing the diameter of silt particles from 0.010 to 0.005 mm resulted in an average 56% increase in sediment yield from the impoundments (table I-48). This means that the model predicts significant clay and silt enrichment, and that the impoundment model is quite sensitive to the assumed particle size distribution. Therefore, for depositional systems such as impoundments, accurate particle-size data are critical.

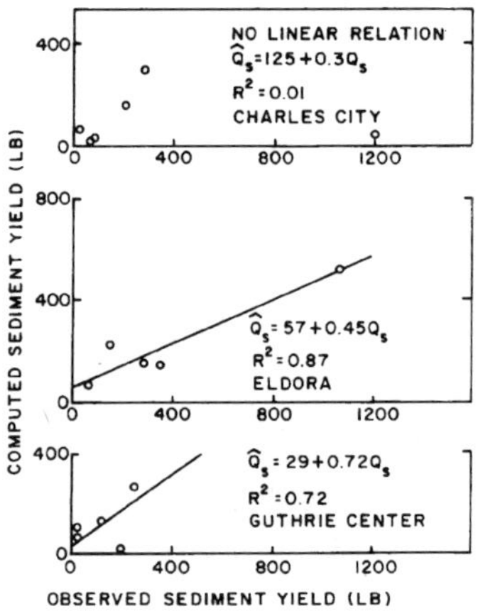

Figure I-30.— Relation between observed and computed sediment yield from impoundment terrace outlets in Iowa.

Table I-46.—Summary of simulated sediment yield from impoundment terraces in Iowa with the infiltration rate in the pond assumed to be 0, 0.2, and 0.5 in/hr

Watershed, Julian date	Sediment yields in overland flow from pond outlets					
	I = 0.0 in/hr		I = 0.2 in/hr		I = 0.5 in/hr	
	Overland	Pond	Overland	Pond	Overland	Pond
Charles City						
70147	1,257	60	1,361	58	1,824	52
70152	150	14	219	15	315	14
70244	346	151	420	155	585	160
70323	110	5	72	5	72	5
71151	2,775	326	6,005	323	11,571	294
71157	1,824	125	3,580	121	6,176	160
TOTAL	6,462	681	11,657	677	20,543	685
DR[1/]	--	0.11	--	0.06	--	0.03
Eldora						
68198	1,714	134	2,559	140	4,612	150
68220	646	41	1393	50	1,577	55
69187	8,091	490	13,495	413	27,122	554
69232	2,250	178	4,982	189	7,172	227
71163	1,349	90	2,296	87	4,222	139
TOTAL	14,050	933	24,725	879	44,660	1,125
DR	--	0.07	--	0.04	--	0.03
Guthrie Center						
69207	3,849	192	4,999	241	5,935	273
69249	825	61	1,104	61	1,844	89
70144	356	20	413	21	1,301	63
70162	578	112	1,149	120	1,704	123
70167	121	25	195	26	390	28
70299	113	9	258	49	451	52
TOTAL	5,842	449	8,118	518	11,625	628
DR	--	0.08	--	0.06	--	0.05

1/ DR is the delivery ratio; sediment yield from the pond outlet divided by the sediment yield in overland flow.

Table I-47.—Assumed particle-size distributions for eroded sediment at 3 small watersheds in Iowa

Particle type1/	Assumed distribution			Alternate distribution 1			Alternate distribution 2		
	Diameter	Specific gravity	Fraction	Diameter	Specific gravity	Fraction	Diameter	Specific gravity	Fracti
	(mm)	(g/cm^3)		(mm)	(g/cm^3)		(mm)	(g/cm^3)	
1	0.002	2.60	0.03	2/0.001	2.60	0.03	2/0.001	2.60	0.03
2	.010	2.65	.03	.010	2.65	.03	2/ .005	2.65	.03
3	.030	1.80	.28	.030	1.80	.28	.030	1.80	.28
4	.280	1.60	.21	.280	1.60	.21	.280	1.60	.21
5	.200	2.65	.45	.200	2.65	.45	.200	2.65	.45

1/ Particle types: 1 = clay, 2 = silt, 3 = small aggregates, 4 = large aggregates, and 5 = sand.
2/ Values which changed from the assumed distribution.

Table I-48.—Influence of particle-size distribution on sediment yield from impoundment terrace outlets

Watershed	Total sediment yield from impoundment outlet		
	Assumed distribution	Alternate distribution 1	Alternate distribution 2
	(lb)	(lb)	(lb)
Charles City	677	908	1,085
Eldora	879	961	1,517
Guthrie Center	518	581	702

SUMMARY

We emphasize that the results of this sensitivity analysis are very site specific and also specific to the observed storm sequences. Therefore, the results are indicative of a particular application and do not necessarily apply in general. The model user should conduct sensitivity analysis for his specific conditions.

The Watkinsville watershed is a mixed, complex watershed. By that we mean the control of sediment yield was mixed between detachment and deposition, between overland flow and channel flow, between storm sizes and sequences, and between particle size classes. While primarily deposition controlled, detachment had a significant effect. For some storms, little deposition occurred either in overland flow or in channel flow. Often the upper ends of the overland and channel flow areas eroded while sediment was deposited in the lower ends. Furthermore, for some storms the net effect for the coarse particles was deposition while the net effect for fine particles was erosion. Consequently, sediment yield was sensitive to both detachment and transport parameters.

In more simple situations such as uniform overland flow slopes or uniform grade channels, many of these complex interactions would not occur. This would be true also for analysis of single storm events. For example, in overland flow where transport capacity exceeds incoming sediment load, the sediment yield may be controlled by detachment parameters. As a result, sediment yield may not be sensitive to Manning's n. However, with increasing roughness as the transport capacity becomes controlling, the sediment yield would be more sensitive to changes in Manning's n. Again, whether sediment yield is detachment or transport limited, will depend upon site specific conditions and the rates and amounts of runoff. Similar analysis can be made for a channel. In any event, it is necessary to consider simple flow systems to isolate and study the behavior of model components with changes in parameter values. For this reason, we recommend that these sensitivity analyses be considered as examples of a specific application and that the model user conduct similar analyses for his particular application.

FIELD-SCALE MODEL: CHEMISTRY COMPONENT

The chemistry component of the model consists of a nutrient submodel to account for plant nutrients and pesticide submodel to account for pesticides. The hydrologic component provides input to the erosion/ sediment yield component which in turn provides input to the chemistry component. The input to the nutrient submodel consists of rainfall, runoff, and sediment yield as well as climatic variables necessary to simulate a water balance including runoff, evapotranspiration, soil moisture, and percolation. Since the water balance calculations are not as critical in the pesticide processes, primary inputs to the pesticide submodel consist of rainfall, runoff, sediment yield, and an enrichment factor. In these analyses, we used observed rainfall, runoff, and sediment yield as input to the chemistry component.

Nutrient Submodel

The nutrient submodel is an accounting and transport model to estimate nitrogen and phosphorus losses from fields. Nutrients are added to the system as fertilizer; in addition, nitrogen is added by rainfall and by mineralization of organic matter from crop residue. Chemical transport is in the solution and sediment phases. Nitrate leaching, plant uptake, and dentrification are calculated to complete the mass balance. Analyses in this section are limited to option 1, wherein the amount of dry matter is estimated from the yield potential and the ratio of actual transpiration to potential transpiration. Using this procedure, the fraction of the total plant growth expected is calculated and the amount of nitrogen currently in the plant material is then the product of the dry matter and average concentration in the dry matter. Incremental plant uptake of nitrogen is then the difference between the current and previous value of nitrogen in the plants.

Output from the nutrient submodel used in sensitivity analysis are the total yield (loss) of nitrogen and phosphorus in runoff and with sediment, total nitrate leached, total plant nitrogen uptake, and total denitrification during 1974 on watershed P2 at Watkinsville, Ga.

Initial estimates of parameter values used in the sensitivity analysis were made using the procedures outlined in volume II, chapter 3. These initial estimates are the base values; the parameters are varied about the base values to determine model sensitivity. The hydrology and erosion input variables (from the erosion component) are also varied about the observed values to determine model sensitivity to errors in this input. Base values for the parameters are summarized in table I-49.

Comparison of observed and computed nitrogen yield in runoff for 11 storm events in 1974 produced

$$\hat{N}_Q = 0.039 + 0.895\ N_Q \qquad [I\text{-}215]$$

$$R^2 = 0.94$$

Table I-49.—Summary of input variables and parameters varied in nutrient submodel sensitivity analyses, watershed P2, Watkinsville, Ga., 1974

Variable or parameter	Base value	Comments
P	Measured data	Daily precipitation.
Q	Measured data	Daily runoff volume.
SED	Measured data	Daily sediment yield.
PERC	HYDONE predictions	Percolation.
$\overline{SM}$	HYDONE predictions	Average soil moisture since last precipitation event.
$\overline{TEMP}$	HYDONE predictions	Average daily air temperature since last precipitation event.
AWU	HYDONE predictions	Actual water use since last precipitation event.
SOLPOR	0.45	Soil porosity.
FC	0.20	Field capacity.
OM	0.65	Organic matter in root zone.
SOLN	0.20	Soluble nitrogen in surface layer.
SOLP	0.20	Soluble phosphorus in surface layer.
NO3	21.0	Nitrate in root zone.
SOILN	0.00035	Soil nitrogen.
SOILP	0.00018	Soil phosphorus.
EXKN	0.075	Nitrogen extraction coefficient.
EXKP	0.075	Phosphorus extraction coefficient
AN	16.8	Nitrogen enrichment coefficient.
BN	-0.160	Nitrogen enrichment exponent.
AP	11.2	Phosphorus enrichment coefficient
BP	-0.146	Phosphorus enrichment exponent.
POTM	47.0	Potential mineralizable nitrogen.
RCN	0.80	Concentration of nitrogen in rainfall.
YP	5700	Potential grain yield.
TOTAL AWU	225	Total actual water use, from hydrology option 1, must be < PWU.
TOTAL PWU	329	Total potential water use, from hydrology option 1.
DMY	2.5	Dry matter yield ratio to convert YP to total dry matter.

where $\widehat{N_Q}$ is predicted nitrogen yield (kg/ha) in runoff and N_Q is the corresponding observed value. The total yields of nitrogen in runoff for the year were 3.52 kg/ha observed and 3.58 kg/ha predicted. Therefore, the model followed trends in the observed data and explained 94% of the variance. The corresponding regression equation for nitrogen yield with sediment, N_S was

$$\widehat{N_S} = 0.054 + 0.810\ N_S \qquad [I\text{-}216]$$

$$R^2 = 0.92$$

with total observed yield for the year 4.3 kg/ha and computed total yield of 3.95 kg/ha. In this case, the total yields are comparable and the bias in predictions was similar to the corresponding predictions for soluble nitrogen.

The regression equation for yield of phosphorus in runoff, P_Q, was

$$\widehat{P_Q} = 0.019 + 0.399\ P_Q \qquad [I\text{-}217]$$

$$R^2 = 0.48.$$

In this case the reqression slope of 0.399 suggests a significant bias from underprediction, but the relatively large intercept of 0.019 means that on the average the predictions were not nearly as biased as the slope suggests. The total observed yield was 0.40 kg/ha while the corresponding predicted value was 0.37 kg/ha. Although the totals are comparable, the low values for R^2 and the slope indicate that the model does not explain the trend in the data and only explains 48% of the variance. The regression equation for phosphorus with sediment was

$$\widehat{P_S} = 0.024 + 0.825\ P_S \qquad [I\text{-}218]$$

$$R^2 = 0.91$$

with observed total yield of 1.505 kg/ha and computed value of 1.509 kg/ha.

From this analysis using the base values of the parameters, we conclude that the selected parameter values produce reasonable predictions. However, the regression results summarized above were dominated by a few large storms.

Sensitivity Analysis for Nutrient Yields in Runoff and with Sediment

The results of the surface transport sensitivity analysis are tabulated in table I-50, which contains information about the amount of variation and its resulting total yield of nitrogen and phosphorus in runoff and with sediment. For comparison, each yield column is followed by a column containing the ratio of yield to base-value yield.

Table I-50.—Nutrient submodel sensitivity analysis: chemical transport in runoff and sediment; watershed P2, Watkinsville, Ga., 1974

Variation	Parameter or variable	Nitrogen				Phosphorus			
		N_Q	$\frac{N_Q}{Base}$	N_S	$\frac{N_S}{Base}$	P_Q	$\frac{P_Q}{Base}$	P_S	$\frac{P_S}{Base}$
(%)		(kg/ha)		(kg/ha)		(kg/ha)		(kg/ha)	
	Base	3.764	1.000	4.134	1.000	0.401	1.000	1.544	1.000
-50	P	74.969	19.915	4.134	1.000	0.847	2.113	1.544	1.000
-25		13.044	3.465	4.134	1.000	.425	1.061	1.544	1.000
+25		1.873	.498	4.134	1.000	.396	.989	1.544	1.000
+50		1.400	.372	4.134	1.000	.395	.985	1.544	1.000
-50	Q	1.185	.315	4.134	1.000	.200	.499	1.544	1.000
-25		2.166	.575	4.134	1.000	.300	.749	1.544	1.000
+25		6.577	1.747	4.134	1.000	.502	1.252	1.544	1.000
+50		11.745	3.120	4.134	1.000	.603	1.505	1.544	1.000
-50	SED	3.764	1.000	2.310	.559	.401	1.000	.854	.553
-25		3.764	1.000	3.247	.785	.401	1.000	1.207	.782
+25		3.764	1.000	4.987	1.206	.401	1.000	1.868	1.210
+50		3.764	1.000	5.812	1.406	.401	1.000	2.182	1.414
-50	PERC	3.967	1.054	4.134	1.000	.401	1.000	1.544	1.000
-25		3.855	1.024	4.134	1.000	.401	1.000	1.544	1.000
+25		3.690	.980	4.134	1.000	.401	1.000	1.544	1.000
+50		3.627	.963	4.134	1.000	.401	1.000	1.544	1.000
-50	$\overline{TEMP}$	3.858	1.025	4.134	1.000	.401	1.000	1.544	1.000
-25		3.830	1.018	4.134	1.000	.401	1.000	1.544	1.000
+25		3.607	.958	4.134	1.000	.401	1.000	1.544	1.000
+50		3.340	.887	4.134	1.000	.401	1.000	1.544	1.000
-50	$\overline{SW}$	4.897	1.301	4.134	1.000	.401	1.000	1.544	1.000
-25		4.147	1.102	4.134	1.000	.401	1.000	1.544	1.000
+25		3.530	.938	4.134	1.000	.401	1.000	1.544	1.000
+50		3.369	.895	4.134	1.000	.401	1.000	1.544	1.000
-25	AWU	4.981	1.323	4.134	1.000	.401	1.000	1.544	1.000
-50		6.401	1.700	4.134	1.000	.401	1.000	1.544	1.000
-75		8.148	2.165	4.134	1.000	.401	1.000	1.544	1.000
-90		9.527	2.531	4.134	1.000	.401	1.000	1.544	1.000
-50	SOLPOR	1.420	.377	4.134	1.000	.788	1.966	1.544	1.000
-25		2.226	.591	4.134	1.000	.528	1.316	1.544	1.000
+25		5.844	1.552	4.134	1.000	.331	.825	1.544	1.000
+50		8.209	2.181	4.134	1.000	.293	.730	1.544	1.000
-50	FC	3.683	.978	4.134	1.000	.401	1.000	1.544	1.000
-25		3.738	.993	4.134	1.000	.401	1.000	1.544	1.000
+25		3.780	1.004	4.134	1.000	.401	1.000	1.544	1.000
+50		3.790	1.007	4.134	1.000	.401	1.000	1.544	1.000

Table I-50.—Nutrient submodel sensitivity analysis: chemical transport in runoff and sediment; watershed P2, Watkinsville, Ga., 1974 -- continued

Variation	Parameter or variable	Nitrogen N_Q	$\frac{N_Q}{\text{Base}}$	N_S	$\frac{N_S}{\text{Base}}$	Phosphorus P_Q	$\frac{P_Q}{\text{Base}}$	P_S	$\frac{P_S}{\text{Base}}$
(%)		(kg/ha)		(kg/ha)		(kg/ha)		(kg/ha)	
-50	OM	3.996	1.062	4.134	1.000	0.401	1.000	1.544	1.000
-25		3.871	1.028	4.134	1.000	.401	1.000	1.544	1.000
+25		3.674	.976	4.134	1.000	.401	1.000	1.544	1.000
+50		3.596	.955	4.134	1.000	.401	1.000	1.544	1.000
-50	SOLP	3.764	1.000	4.134	1.000	.204	.509	1.544	1.000
-25		3.764	1.000	4.134	1.000	.303	.755	1.544	1.000
+25		3.764	1.000	4.134	1.000	.499	1.245	1.544	1.000
+50		3.764	1.000	4.134	1.000	.598	1.491	1.544	1.000
-50	NO3	3.712	.986	4.134	1.000	.401	1.000	1.544	1.000
-25		3.738	.993	4.134	1.000	.401	1.000	1.544	1.000
+25		3.791	1.007	4.134	1.000	.401	1.000	1.544	1.000
+50		3.817	1.014	4.134	1.000	.401	1.000	1.544	1.000
-50	SOLIN	3.764	1.000	2.008	.486	.401	1.000	1.544	1.000
-25		3.764	1.000	3.071	.743	.401	1.000	1.544	1.000
+25		3.764	1.000	5.198	1.257	.401	1.000	1.544	1.000
+50		3.764	1.000	6.143	1.486	.401	1.000	1.544	1.000
-50	SOILP	3.764	1.000	4.134	1.000	.401	1.000	.772	.500
-25		3.764	1.000	4.134	1.000	.401	1.000	1.201	.778
+25		3.764	1.000	4.134	1.000	.401	1.000	1.887	1.272
+50		3.764	1.000	4.134	1.000	.401	1.000	2.316	1.500
-50	EXKN	2.619	.696	4.134	1.000	.401	1.000	1.544	1.000
-25		3.237	.860	4.134	1.000	.401	1.000	1.544	1.000
+25		4.222	1.122	4.134	1.000	.401	1.000	1.544	1.000
+50		4.619	1.227	4.134	1.000	.401	1.000	1.544	1.000
-50	EXKP	3.764	1.000	4.134	1.000	.201	.500	1.544	1.000
-25		3.764	1.000	4.134	1.000	.301	.751	1.544	1.000
+25		3.764	1.000	4.134	1.000	.501	1.250	1.544	1.000
+50		3.764	1.000	4.134	1.000	.601	1.499	1.544	1.000
-50	AN	3.764	1.000	2.067	.500	.401	1.000	1.544	1.000
-25		3.764	1.000	3.101	.750	.401	1.000	1.544	1.000
+25		3.764	1.000	5.168	1.250	.401	1.000	1.544	1.000
+50		3.764	1.000	6.202	1.500	.401	1.000	1.544	1.000
-50	BN	3.764	1.000	2.559	.619	.401	1.000	1.544	1.000
-25		3.764	1.000	3.247	.785	.401	1.000	1.544	1.000
+25		3.764	1.000	5.280	1.277	.401	1.000	1.544	1.000
+50		3.764	1.000	6.761	1.635	.401	1.000	1.544	1.000

Table I-50.—Nutrient submodel sensitivity analysis: chemical transport in runoff and sediment; watershed P2, Watkinsville, Ga., 1974 -- continued

Variation	Parameter or variable	Nitrogen				Phosphorus			
		N_Q	$\frac{N_Q}{Base}$	N_S	$\frac{N_S}{Base}$	P_Q	$\frac{P_Q}{Base}$	P_S	$\frac{P_S}{Base}$
(%)		(kg/ha)		(kg/ha)		(kg/ha)		(kg/ha)	
-50	AP	3.764	1.000	4.134	1.000	0.401	1.000	0.772	0.500
-25		3.764	1.000	4.134	1.000	.401	1.000	1.158	.750
+25		3.764	1.000	4.134	1.000	.401	1.000	1.930	1.250
+50		3.764	1.000	4.134	1.000	.401	1.000	2.316	1.500
-50	BP	3.764	1.000	4.134	1.000	.401	1.000	.994	.644
-25		3.764	1.000	4.134	1.000	.401	1.000	1.237	.801
+25		3.764	1.000	4.134	1.000	.401	1.000	1.931	1.251
+50		3.764	1.000	4.134	1.000	.401	1.000	2.421	1.568
-50	POTM	3.608	.958	4.134	1.000	.401	1.000	1.544	1.000
-25		3.686	.979	4.134	1.000	.401	1.000	1.544	1.000
+25		3.843	1.021	4.134	1.000	.401	1.000	1.544	1.000
+50		3.921	1.042	4.134	1.000	.401	1.000	1.544	1.000
-50	RCN	3.225	.857	4.134	1.000	.401	1.000	1.544	1.000
-25		3.495	.928	4.134	1.000	.401	1.000	1.544	1.000
+25		4.034	1.072	4.134	1.000	.401	1.000	1.544	1.000
+50		4.304	1.143	4.134	1.000	.401	1.000	1.544	1.000
-50	RZMAX	4.728	1.256	4.134	1.000	.401	1.000	1.544	1.000
-25		4.109	1.092	4.134	1.000	.401	1.000	1.544	1.000
+25		3.540	.940	4.134	1.000	.401	1.000	1.544	1.000
+50		3.380	.898	4.134	1.000	.401	1.000	1.544	1.000
-50	YP	4.252	1.130	4.134	1.000	.401	1.000	1.544	1.000
-25		4.008	1.065	4.134	1.000	.401	1.000	1.544	1.000
+25		3.539	.940	4.134	1.000	.401	1.000	1.544	1.000
+50		3.373	.896	4.134	1.000	.401	1.000	1.544	1.000
-25	TOTAL AWU	3.873	1.029	4.134	1.000	.401	1.000	1.544	1.000
-50		4.019	1.068	4.134	1.000	.401	1.000	1.544	1.000
-75		4.200	1.116	4.134	1.000	.401	1.000	1.544	1.000
-90		4.375	1.162	4.134	1.000	.401	1.000	1.544	1.000
+25	TOTAL PWU	3.960	1.052	4.134	1.000	.401	1.000	1.544	1.000
+50		4.090	1.086	4.134	1.000	.401	1.000	1.544	1.000
+75		4.183	1.111	4.134	1.000	.401	1.000	1.544	1.000
+100		4.252	1.130	4.134	1.000	.401	1.000	1.544	1.000
-50	DMY	4.252	1.130	4.134	1.000	.401	1.000	1.544	1.000
-25		4.008	1.065	4.134	1.000	.401	1.000	1.544	1.000
+25		3.539	.940	4.134	1.000	.401	1.000	1.544	1.000
+50		3.373	.896	4.134	1.000	.401	1.000	1.544	1.000

Table I-51 contains a summary of the results, with the same criteria for significance as in the hydrology and erosion sensitivity analyses.

Variations in precipitation and runoff both have significant effects on the runoff transport of nitrogen and phosphorus, with nitrogen being much more sensitive because of its greater downward movement. Phosphorus leaching is relatively insignificant because of buffering action. Phosphorus in runoff is also affected by soil porosity, soluble phosphorus, and extraction coefficient. Nitrogen shows similar sensitivity, except to soluble nitrogen, which caused little variation in yield. Both nutrients are sensitive to soil porosity because the nutrient accounting scheme used first calculates an initial abstraction, then allows downward movement by infiltration; the remaining nutrients are available for runoff transport.

Constant surface soil-root zone interactions take place for nitrogen but not for phosphorus. Therefore, nitrogen in runoff is influenced strongly by plant water use and soil evaporation, and is influenced moderately by potential yield, root zone depth, temperature, soil moisture, organic matter, and dry matter yield. Nitrogen in runoff is slightly sensitive to NO_3 in the root zone potential mineralization, rainfall nitrogen concentration, percolation, field capacity, and total plant water use.

Sediment transport of each nutrient is considered to be the product of nutrient content of the soil, sediment amount, and a coefficient times sediment amount to an exponent; the sensitivity of this result is therefore easily predictable, with nutrient yields significantly responsive to changes in sediment yield, nutrient quantities in the soil, and enrichment coefficients and exponents. It should be noted that sediment transport is a function of soil type and is not affected by surface nutrient application or by removal during leaching.

Sensitivity for Subsurface Nutrient Movement

Table I-52 presents the results of parameter and hydrology variation in subsurface nitrogen movement. Nitrogen uptake, denitrification, and nitrogen leaching are the subsurface nutrient variables which were monitored for sensitivity. These constitute the nitrogen mass balance variables in the root zone; they are important also because of the complex interactions between the surface and root zone nitrogen concentrations. When choosing parameter values, it is necessary to consider the results of errors in root zone processes. Table I-53 summarizes the significance of subsurface variable response to parameter changes.

As would be expected, nitrogen leached is most strongly affected by field capacity, root zone depth, and percolation. Note, however, that it is moderately sensitive to 11 other parameters and slightly sensitive to 4. Uptake is

Table I-51.—Results of nutrient submodel parameter and variable sensitivity analysis, watershed P2, Watkinsville, Ga., 1974, surface transport of chemicals[1/]

Parameter or variable	Nitrogen in runoff	Nitrogen in sediment	Phosphorus in runoff	Phosphorus in sediment	Comments
P	Significant	None	Significant	None	Leaching affects soluble nutrients.
Q	Significant	None	Significant	None	Runoff affects soluble transport.
SED	None	Significant	None	Significant	
PERC	Slight	None	None	None	
$\overline{TEMP}$	Moderate	None	None	None	
$\overline{SW}$	Moderate	None	None	None	
AWU	Significant	None	None	None	
SOLPOR	Significant	None	Significant	None	Soil porosity.
FC	Slight	None	None	None	
OM	Moderate	None	None	None	
SOLP	None	None	Significant	None	
NO3	Slight	None	None	None	
SOILN	None	Significant	None	None	
SOILP	None	None	None	Significant	
EXKN	Significant	None	None	None	
EXKP	None	None	Significant	None	
AN	None	Significant	None	None	Enrichment parameter.
BN	None	Significant	None	None	Enrichment parameter.
AP	None	None	None	Significant	Enrichment parameter.
BP	None	None	None	Significant	Enrichment pameter.
POTM	Slight	None	None	None	
RCN	Slight	None	None	None	
RZMAX	Moderate	None	None	None	
YP	Moderate	None	None	None	
TOTAL AWU	Moderate	None	None	None	
TOTAL PWU	Slight	None	None	None	
DMY	Moderate	None	None	None	

1/ A ± 50% change in input produces a change in chemical yield of: Slight 0.01 - 10%; moderate 10 - 50%; significant > 50%.

Table I-52.—Nutrient submodel sensitivity analysis: subsurface nitrogen movement, watershed P2, Watkinsville, Ga., 1974

Variation	Parameter	Nitrate leached	NL/Base	Nitrogen uptake	NU/Base	Denitri-fication	D/Base
(%)		(kg/ha)		(kg/ha)		(kg/ha)	
	Base	33.980	1.000	124.661	1.000	37.828	1.000
-50	P	23.170	.682	76.877	.617	27.325	.722
-25		29.430	.866	124.661	1.000	33.476	.885
+25		36.829	1.084	124.661	1.000	40.071	1.081
+50		39.046	1.149	124.661	1.000	43.330	1.146
-50	Q	34.414	1.013	124.661	1.000	38.138	1.008
-25		34.240	1.008	124.661	1.000	38.007	1.005
+25		33.552	.987	124.661	1.000	37.555	.993
+50		32.799	.965	124.661	1.000	37.100	.981
-50	PERC	21.215	.624	124.661	1.000	42.117	1.114
-25		28.334	.834	124.661	1.000	39.778	1.052
+25		38.536	1.134	124.661	1.000	36.174	.956
+50		42.266	1.244	124.661	1.000	34.761	.919
-50	$\overline{TEMP}$	33.073	.973	124.661	1.000	11.369	.301
-25		33.646	.990	124.661	1.000	20.906	.553
+25		32.775	.965	124.661	1.000	64.384	1.702
+50		30.741	.905	112.011	.899	98.944	2.616
-50	$\overline{SW}$	27.235	.801	124.661	1.000	31.608	.836
-25		30.972	.911	124.661	1.000	35.015	.926
+25		36.541	1.075	124.661	1.000	40.278	1.065
+50		38.780	1.141	124.661	1.000	42.460	1.123
-25	AWU	37.343	1.099	105.352	.845	40.164	1.062
-50		41.126	1.210	83.107	.667	42.755	1.130
-75		45.689	1.345	55.402	.444	45.817	1.211
-90		49.773	1.465	29.218	.234	48.442	1.281
-50	SOLPOR	41.153	1.211	119.883	.962	45.686	1.208
-25		36.858	1.085	124.661	1.000	41.054	1.085
+25		31.542	.928	124.661	1.000	35.230	.931
+50		29.382	.865	124.661	1.000	33.001	.873
-50	FC	58.569	1.724	124.661	1.000	38.791	1.026
-25		43.180	1.271	124.661	1.000	38.294	1.012
+25		27.941	.822	124.661	1.000	37.462	.990
+50		23.697	.697	124.661	1.000	37.177	.983
-50	OM	39.410	1.160	124.661	1.000	23.697	.627
-25		36.449	1.074	124.661	1.000	31.394	.830
+25		31.783	.935	124.661	1.000	43.250	1.144
+50		29.855	.879	124.661	1.000	47.852	1.265
-50	SOLN	33.934	.999	124.661	1.000	37.780	.999
-25		33.957	.999	124.661	1.000	37.804	1.000
+25		34.002	1.001	124.661	1.000	37.852	1.001
+50		34.025	1.001	124.661	1.000	37.876	1.001

Table I-52.—Nutrient submodel sensitivity analysis: subsurface nitrogen transport, watershed P2, Watkinsville, Ga., 1974 data -- continued

Variation	Parameter	Nitrate leached	NL/Base	Nitrogen uptake	NU/Base	Denitri-fication	D/Base
(%)		(kg/ha)		(kg/ha)		(kg/ha)	
-50		29.119	0.857	124.661	1.000	32.839	0.868
-25	NO3	31.549	.928	124.661	1.000	35.333	.934
+25		36.410	1.072	124.661	1.000	40.323	1.066
+50		38.840	1.143	124.661	1.000	42.817	1.132
-50		34.152	1.005	124.661	1.000	37.935	1.003
-25	EXKN	34.059	1.002	124.661	1.000	37.878	1.001
+25		33.909	.998	124.661	1.000	37.784	.999
+50		33.848	.996	124.661	1.000	37.744	.998
-50		28.346	.834	123.587	.991	32.479	.859
-25	POTM	31.089	.915	124.661	1.000	35.113	.928
+25		36.870	1.085	124.661	1.000	40.544	1.072
+50		39.760	1.170	124.661	1.000	43.259	1.144
-50		33.113	.975	124.661	1.000	37.009	.978
-25	RCN	33.546	.987	124.661	1.000	37.418	.989
+25		34.413	1.013	124.661	1.000	38.238	1.011
+50		34.846	1.025	124.661	1.000	38.648	1.022
-50		45.898	1.351	124.661	1.000	31.554	.834
-25	RZMAX	39.162	1.153	124.661	1.000	35.244	.932
+25		29.945	.881	124.661	1.000	39.736	1.051
+50		26.739	.787	124.661	1.000	41.199	1.089
-50		44.670	1.315	62.331	.500	44.913	1.187
-25	YP	39.325	1.157	93.496	.750	41.371	1.094
+25		31.117	.916	137.745	1.105	35.666	.943
+50		29.625	.872	140.806	1.130	34.266	.906
-25	TOTAL	36.087	1.062	110.632	.887	39.100	1.034
-50	AWU	38.736	1.140	93.498	.750	40.746	1.077
-75		42.801	1.260	70.126	.563	43.536	1.151
-90		46.649	1.373	47.944	.385	46.158	1.220
+25	TOTAL	38.256	1.126	99.729	.800	40.662	1.075
+50	PWU	41.107	1.210	83.107	.667	42.551	1.125
+75		43.143	1.270	71.235	.571	43.901	1.161
+100		44.670	1.315	62.331	.500	44.913	1.187
-50		44.670	1.315	62.331	.500	44.913	1.187
-25	DMY	39.325	1.157	93.496	.750	41.371	1.094
+25		31.117	.916	137.745	1.105	35.666	.943
+50		29.625	.872	140.806	1.130	24.266	.906

Table I-53.—Results of nutrient submodel parameter and variable sensitivity analyses, watershed P2, Watkinsville, Ga., 1974, subsurface transport of nitrogen[1/]

Parameter or variable	Nitrate leached	Nitrate uptake	Denitrification	Comments
P	Moderate	Significant	Moderate	Affects leaching, so forth
Q	Slight	None	Slight	
PERC	Significant	None	Moderate	
TEMP	Slight	Moderate	Significant	
SW	Moderate	None	Moderate	
AWU	Moderate	Significant	Moderate	
SOLPOR	Moderate	Slight	Moderate	
FC	Significant	None	Slight	
OM	Moderate	None	Significant	
SOLN	Slight	None	Slight	
NO3	Moderate	None	Moderate	
EXKN	Slight	None	Slight	Mass balance determined in part by the extraction coefficient.
POTM	Moderate	Slight	Moderate	
RCN	Slight	None	Slight	
RZMAX	Significant	None	Moderate	
YP	Moderate	Significant	Moderate	
TOTAL AWU	Moderate	Significant	Moderate	
TOTAL PWU	Moderate	Moderate	Slight	
DMY	Moderate	Significant	Moderate	

1/ A ± 50% variation in input produces a change in chemical yield of: Slight - 0.01 - 10%; moderate 10-50%; significant > 50%.

significantly sensitive to 4 and moderately sensitive to 2 parameters, while it shows slight or no response to changes in the other 12 parameters. Denitrification is primarily a function of temperature and organic matter, but responds moderately to moisture movement in the root zone.

Pesticide Submodel

This model is essentially a simplified accounting and transport model which keeps track of pesticide concentrations on plant foliage and in the active (1 cm) soil surface, partitions transport into the water soluble and adsorbed phases, and predicts losses for up to 10 noninteractive pesticides. That is, each pesticide is considered separately without interaction with the plant nutrients or other pesticides.

Initial estimates of parameter values used in the sensitivity analysis were made by R. A. Leonard[4/] as summarized in table I-54. These parameter values are denoted base values, and the parameters were then varied about the base values to determine model sensitivity. Simulations were made for a weakly adsorbed pesticide (atrazine) and a highly adsorbed pesticide (paraquat). While these chemicals are only two of many possible compounds in use, they represent a wide range of variation in transport mechanisms and properties and thus should be indicative of many chemicals.

Observed yields of paraquat in runoff (transport in solution phase) were insignificant (2) and the model predicted 0.012 g of paraquat in runoff. Therefore, other than to say observed and computed paraquat yields in runoff were insignificant and possibly below detection limits, no other comparisons were made. Atrazine yields in runoff were significant and the relation between observed yields and computed yields for 1974 and using the base values was

$$\hat{A}_Q = -0.03 + 0.35\ A_Q \qquad [I\text{-}219]$$

$$R^2 = 0.63$$

where $\hat{A}_Q$ is computed yield of atrazine in runoff and A_Q is the corresponding observed value. The computed values underpredicted observed yields but the model explained 63% of the variance as indicated by $R^2 = 0.63$. For 1974, the observed total yield of atrazine in runoff was 8.45 g and the corresponding predicted yield was 2.68 g. Similar predictions for years 1973 and 1975 resulted in overprediction of the total atrazine yield in runoff. However, to be consistent with analyses for the other model components, data from 1974 were selected for the sensitivity analysis. The 1974 observed total yield of atrazine with sediment was 1.18 g and the model predicted 0.22 g. The model seemed to significantly underestimate atrazine transport with sediment.

The relation between observed and predicted yield of paraquat with sediment for individual storms was

$$\hat{P}_s = -0.38 + 0.42\ P_s \qquad [I\text{-}220]$$

$$R^2 = 0.88$$

where $\hat{P}_s$ is the computed yield of paraquat with sediment and P_s is the corresponding observed yield. Although the model explained 88% of the variance, the total observed paraquat yield with sediment was 102.2 g and the model predicted a total yield of 40.0 g. Again, there is a significant underprediction for 1974 but the model does explain the trend in the data. The yield of paraquat with sediment is a function of the calculated enrichment factor. This

4/ Soil scientist, USDA-SEA-AR, Athens, Ga., personal communication.

Table I-54.—Summary of input variables and parameters varied in pesticide submodel sensitivity analyses, watershed P2, Watkinsville, Ga., 1974

Variable or parameter	Base value	Comments
P	Measured data.	Daily precipitation.
Q	Measured data.	Daily runoff volume.
SED	Measured data.	Daily sediment yield.
ENRICH	Predicted by erosion model.	Sediment enrichment ratio.
SOLPOR	0.45	Soil porosity.
DEPINC	1.00	Depth of incorporation.
EFFINC	1.00	Efficiency of incorporation.
SOLFRC	1.00	Fraction applied to soil.
SOLH2O	33.0 A1/ $5x10^5$ P	Water solubility, ppm.
EXTRCT	0.10	Extraction ratio.
DECAY	0.14 A 0.007 P	Decay constant.
KD	4.00 A 10^6 P	
APRATE	3.40 A 2.049 P	Rate of chemical application.

1/ A refers to atrazine and P refers to paraquat.

enrichment factor is calculated based on organic matter content and selective deposition of larger size sediment fractions. The relative accuracy of these calculations is unknown.

Analysis of model predictions using the base values of the parameters suggests that predictions of weakly adsorbed atrazine yields with runoff and of highly adsorbed paraquat yields with sediment explain trends in the observed data. In both cases, the model underpredicts total yields. However, the model did not produce reasonable estimates of atrazine yields with sediment or paraquat yields with runoff. Nevertheless, the results represented by equations [I-219] and [I-220] are precise enough to enable analyses to determine model sensitivity.

Sensitivity Analysis for Pesticide Yields in Runoff and With Sediment

Total yield of atrazine and paraquat in runoff and with sediment were calculated for the 1974 data. As each parameter (or input variable) was varied about the base value, the total yields were compared to the totals obtained using the base values as a measure of the sensitivity. Column 2 of table I-55 shows which parameters were varied and by how much. Columns 3, 5, 7, and 9 show the predicted yields and columns 4, 6, 8, and 10 show the predicted yields divided by the corresponding yields using the base values of all parameters.

As expected, yield of atrazine in runoff was sensitive to changes in rainfall and runoff and not sensitive to changes in sediment yield or enrichment (see table I-55). Notice that each input variable was varied independently, so that a decrease in rainfall increased the pesticide yields because the runoff remained unchanged and thus proportionally more rainfall became runoff. On the other hand, the yield of paraquat with sediment was sensitive to sediment yield and enrichment. The sensitivity of pesticide yield with sediment to rainfall varies with the distribution coefficient KD. For atrazine, the base value is KD = 4.0 and for paraquat KD = 10^6; atrazine yield with sediment was sensitive to total rainfall while paraquat was not.

Very significant parameters are: EFFINC, the efficiency of incorporation of the pesticide into the soil; DEPINC, depth of pesticide incorporation into the soil; and SOLFRC, fraction of the pesticide applied directly to the soil. This is fortunate in the sense that these parameters can be accurately determined but unfortunate in the sense that detailed knowledge of pesticide application and cultivation techniques must be known and these vary with soil conditions.

The distribution coefficient, KD, is a parameter representing the distribution of pesticides between solution and soil phases. A low value of KD represents relatively more of the pesticide in solution while high values of KD represent the opposite. Various values of KD should be compared in terms of their order of magnitude rather than small differences, especially for large values. For small values of KD < 10, small differences may be significant. This can be seen by comparing the relative sensitivity of atrazine and paraquat yields for KD variations, in table I-55.

Sensitivity of the pesticide submodel is summarized in table I-56. As described at the bottom of the table, the most sensitive input variables and parameters are indicated as "significant" in table I-56. For the input variables and parameters listed as significant, errors in estimating the input or parameter values are magnified by the model. With this criterion, 10 of the 15 parameters and input variables listed in table I-56 are significant with respect to error magnification. Also, if KD is varied by orders of magnitude (as it must be to represent a broad spectrum of pesticides) then it, too, is sensitive. Compared to the hydrologic and erosion/sediment yield components, this is a relatively large number of "significant" parameters. This has the positive aspect that if a model is to be useful in reflecting management practices, it should be sensitive to parameters representing the practices. However, when more parameters are required to a higher degree of precision, then more effort and information are required to provide input to the model.

Table I-55.—Pesticide submodel sensitivity analyses, watershed P2, Watkinsville, Ga., 1974

Variation	Parameter	Atrazine in runoff A_Q	A_Q/ base	Atrazine with sediment A_S	A_S/ base	Paraquat in runoff P_Q	P_Q/ base	Paraquat with sediment P_S	P_S/ base
(1)	(2)	(3)	(4)	(5)	(6)	(7)	(8)	(9)	(10)
(%)		(g)		(g)		(g)		(g)	
	Base	2.678	1.000	0.215	1.000	0.012	1.000	40.041	1.000
-50	P	4.718	1.762	.278	1.293	.012	1.000	40.042	1.000
-25		3.574	1.335	.247	1.149	.012	1.000	40.042	1.000
+25		2.030	.758	.188	.874	.012	1.000	40.041	1.000
+50		1.577	.589	.166	.772	.012	1.000	40.041	1.000
-50	Q	1.295	.484	.213	.911	.006	.500	40.041	1.000
-25		1.975	.737	.214	.995	.009	.750	40.041	1.000
+25		3.405	1.271	.216	1.005	.014	1.167	40.041	1.000
+50		4.153	1.551	.216	1.005	.017	1.417	40.041	1.000
-50	SED	2.678	1.000	.107	.498	.012	1.000	20.086	.502
-25		2.678	1.000	.163	.758	.012	1.000	30.084	.751
+25		2.678	1.000	.270	1.256	.012	1.000	49.983	1.248
+50		2.678	1.000	.322	1.498	.012	1.000	59.866	1.495
-50	ENRICH	2.678	1.000	.107	.498	.012	1.000	20.086	.502
-25		2.678	1.000	.161	.749	.012	1.000	30.092	.752
+25		2.678	1.000	.268	1.247	.012	1.000	49.969	1.248
+50		2.678	1.000	.322	1.498	.012	1.000	59.865	1.495
-50	SOLPOR	2.845	1.062	.215	1.000	.012	1.000	40.041	1.000
-25		2.771	1.035	.215	1.000	.012	1.000	40.041	1.000
+25		2.527	.944	.214	.995	.012	1.000	40.041	1.000
+50		2.296	.857	.212	.986	.012	1.000	40.041	1.000
-50	DEPINC	5.356	2.000	.429	1.995	.023	1.917	80.083	2.000
-25		3.571	1.333	.286	1.330	.015	1.250	53.389	1.333
+25		2.142	.800	.172	.800	.009	.750	32.033	.800
+50		1.785	.667	.143	.665	.008	.667	26.694	.667
-100	EFFINC, SOLFRC	.000	.000	.000	.000	.000	.000	.000	.000
-75		.670	.250	.054	.250	.003	.250	10.010	.250
-50		1.339	.500	.107	.500	.006	.500	20.021	.500
-25		2.009	.750	.161	.750	.009	.750	30.031	.750

Table I-55.—Pesticide submodel sensitivity analyses, watershed P2, Watkinsville, Ga., 1974 -- continued

Variation	Parameter	Atrazine in runoff A_Q	A_Q/ base	Atrazine with sediment A_S	A_S/ base	Paraquat in runoff P_Q	P_Q/ base	Paraquat with sediment P_S	P_S/ base
(1)	(2)	(3)	(4)	(5)	(6)	(7)	(8)	(9)	(10)
(%)		(g)		(g)		(g)		(g)	
$x10^{-2}$		5.280	1.972	.915	4.260	.012	1.000	40.041	1.000
$x10^{-1}$	SOLH20	2.678	1.000	.215	1.000	.012	1.000	40.041	1.000
x10		2.678	1.000	.215	1.000	.012	1.000	40.041	1.000
$X10^2$		2.678	1.000	.215	1.000	.012	1.000	40.041	1.000
-50		1.563	.584	.125	.582	.012	1.000	40.041	1.000
-25	EXTRCT	2.163	.808	.174	.810	.012	1.000	40.041	1.000
+25		3.124	1.167	.251	1.169	.012	1.000	40.041	1.000
+50		3.514	1.312	.282	1.313	.012	1.000	40.041	1.000
-50		12.232	4.568	.444	2.067	.015	1.250	50.380	1.258
-25	DECAY	5.143	1.920	.290	1.350	.013	1.083	44.745	1.117
+25		1.550	.579	.166	.773	.010	.833	35.667	.891
+50		.998	.373	.131	.610	.009	.750	32.033	.800
10^0		.983	.367	.266	1.237	110.873	$9.2x10^3$	2.346	.059
10^1	KD1/	2.964	1.107	.457	2.128	197.277	$1.6x10^4$	7.615	.190
10^2		.744	.278	.959	4.465	92.516	$7.7x10^3$	32.604	.814
10^3		.084	.031	1.061	4.939	11.258	$9.4x10^2$	39.203	.979
10^4		.009	.003	1.072	4.991	1.149	95.750	39.957	.998
-50		1.339	.500	.107	.498	.006	.500	20.030	.500
-25	APRATE	2.009	.750	.161	.750	.009	.750	30.036	.750
+25		3.348	1.250	.269	1.252	.014	1.167	50.047	1.250
+50		4.017	1.500	.322	1.499	.017	1.417	60.072	1.500

1/ Figures for KD do not represent relative changes but actual KD values used from 10^0 to 10^4.

The pesticide submodel represents extreme simplifications of complex natural processes. Although the number of "significant" parameters is relatively large, the prediction results summarized by equations [I-219] and [I-220] indicate that additional simplifications would result in an oversimplified model which would not represent the known processes as well as this model does. Therefore, it seems likely that a model of at least the complexity used herein will be required to predict pesticide losses with improved accuracy.

Table I-56.—Variables and parameters varied in pesticide sensitivity analysis, watershed P2, Watkinsville, Ga., 1974

Parameter or variable	Atrazine in runoff	Atrazine with sediment	Paraquat in runoff	Paraquat with sediment	Comments
P	Significant[1/]	Significant	None	None	Sensitivity depends upon the value of KD for hydrologic inputs.
Q	Significant	Slight	Significant	None	
SED	None	Significant	None	Significant	
ENRICH	None	Significant	None	Significant	Measure of silt-clay enrichment important for adsorbed chemicals.
SOLPOR	Moderate	None	None	None	More significant for soluble chemicals.
DEPINC	Significant	Significant	Significant	Significant	Highly dependent on soil conditions.
EFFINC	Significant	Significant	Significant	Significant	Highly dependend on soil conditions.
SOLFRC	Significant	Significant	Significant	Significant	Highly dependent on soil conditions.
SOLH2O	2/	2/	None	None	
EXTRCT	Significant	Significant	None	None	
DECAY	Significant	Significant	Moderate	Moderate	
KD	2/	2/	2/	2/	Distribution coefficient most sensitive for small values of KD.
APRATE	Significant	Significant	Significant	Significant	All yields sensitive to application rate.

1/ A ± 50% change in parameter value produces a change in chemical yield of: Slight < 10%; Moderate 10-50%; Significant > 50%.

2/ Parameter varied by orders of magnitude; sensitivity relationship nonlinear but significant for large changes.

SUMMARY

Sensitivity analyses were conducted for the field scale model using data from watershed P2 at Watkinsville, Ga. Sensitivity to input data and parameter values was determined by considering the hydrology, erosion/sediment yield, nutrient, and pesticides submodels separately. Because of the complexity, interactions, and number of simulation runs required, no attempt was made to determine sensitivity of the entire model involving the simultaneous operation of its components. However, insight for sensitivity of the entire model can be gained by considering sensitivity of its components and their linkage in the field scale model.

Output from the hydrology component provides input to the erosion/sediment yield component. Both these components in turn provide input to the nutrient and pesticide components. Observed rainfall data were used in determining sensitivity of the hydrologic components. Observed rainfall and runoff data were used to determine sensitivity of the erosion/sediment transport component. Observed rainfall, runoff, sediment, and climatic data were used to determine sensitivity of the nutrient and pesticide submodels. By using observed data where possible, we sought to minimize compound errors and interactions due to errors in predictions from the hydrology and erosion/sediment yield components. Also, it is important to stress that model simulations represent predictions using base values of the parameters and that no attempts were made to optimize or calibrate model components using the observed data. Observed data were used for comparison and to evaluate model predictions.

The quality or accuracy of the model predictions made using base values of the parameters represent the type and magnitude of errors which might be expected in applying the model to predict runoff, sediment yield, nutrient losses, and pesticide losses on a complex agricultural watershed. Moreover, conclusions regarding model sensitivity refer to the results for a specific watershed with initial parameter estimates derived using procedures outlined in the volume II, User Manual. Somewhat different results could result from application of the model under different conditions.

A qualitative assessment of the significance of and sensitivity to input variables and parameters for each component or submodel was made using the criterion that the sensitivity to a particular parameter is "significant" if errors in that parameter result in errors in the submodel output as large or larger than the parameter errors. In this case, the model was said to magnify the errors. Sensitivity assessments for the hydrology component are listed in tables I-37 and I-40, and for the erosion/sediment yield component in table I-44. Similar assessments for the nutrient submodel are given in tables I-51 and I53, and for the pesticide submodel in table I-56.

REFERENCES

(1) Laflen, J. M., H. P. Johnson, and R. O. Hartwig.
1978. Sedimentation modeling of impoundment terraces. Transactions of the American Society of Agricultural Engineers 21(6):1131-1135.

2) Smith, C. N., R. A. Leonard, G. W. Langdale, and G. W. Bailey. 1978. Transport of agricultural chemicals from small upland Piedmont watersheds. Environmental Protection Agency, EPA-600/3-78-056. U.S. Government Printing Office, Washington, D.C. 364 pp.

3) U.S. Department of Agriculture, Soil Conservation Service. 1972. National Engineering Handbook, Sec. 4, Hydrology. 548 pp.

CREAMS

A Field Scale Model for Chemicals, Runoff, and Erosion From Agricultural Management Systems

VOLUME II. USER MANUAL

CONTENTS

VOLUME II. USER MANUAL

W. G. Knisel and J. D. Nowlin[1/]

INTRODUCTION

CREAMS, volume II, is structured as an independent publication written for (1) the technical user to develop input and parameter information and (2) the computer programer to set up data files for running the model. Reference tables and figures in this section are for the user's convenience.

CREAMS is structured as three separate components: (1) hydrology, options one and two; (2) erosion/sedimentation; and (3) chemistry, plant nutrients, and pesticides. Although the influences of agricultural management systems on runoff, erosion and sediment transport, and chemical transport are complex and interactive, the user may wish to isolate these influences for specific components. A practice may have only a secondary influence on the hydrologic performance of a watershed, for example, while having a major influence on sediment yield or chemical transport. Rerunning the hydrologic component for every alternate practice considered for erosion control therefore is unnecessary. Rerunning the hydrology and erosion components to evaluate the effects of split applications or alternate pesticide applications also is unnecessary. Independent operation enables the user to consider management options and make comparisons while operating at a relatively low cost. The user also may want to run only the hydrology and erosion components rather than the chemistry component.

Running the model as three separate components places some restrictions on the user, who must record the files generated by a component to pass the correct file to the next component. Although this problem is not severe, the user should be aware of the many files that can be generated rather quickly for a specific problem.

Figure II-1 is a generalized chart of program flow that shows input and output files and pass files from one component to the next and the sequence of operations. As shown on the left side of figure II-1, the users also can supply the hydrology or sediment yield data, or both, if they so desire. If a historic record of runoff and sediment yield data is available for some location near the farm of interest, for example, the users might want to use observed runoff and sediment yield data and run the chemistry component rather than generate runoff and sediment yield with the model.

1/ Hydraulic engineer, USDA-SEA-AR, Southwest Rangeland Watershed Research Center, Tucson, Ariz., and computer programer, Agricultural Engineering Department, Purdue University, West Lafayette, Ind., respectively.

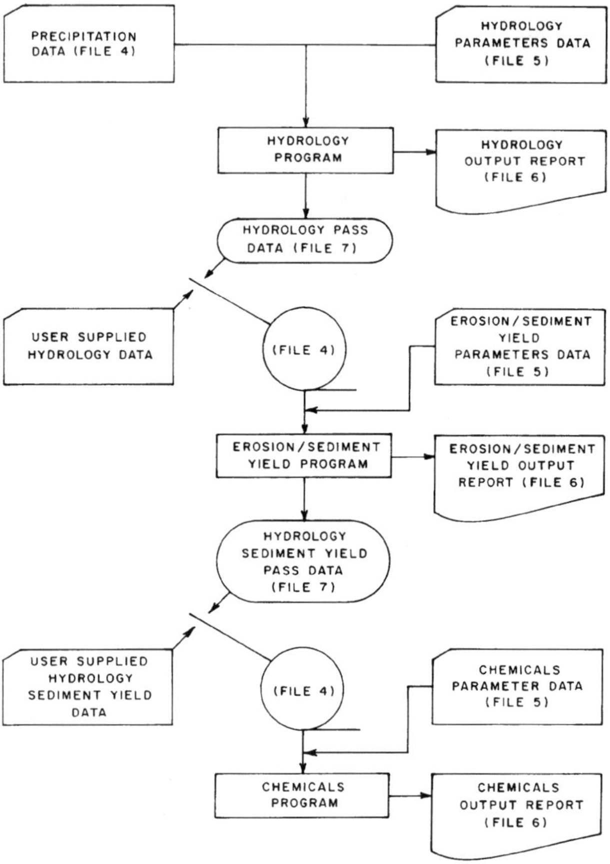

Figure II-1.—Schematic representation of flow of CREAMS programs

Each component of CREAMS and its input and output are treated separately in subsequent chapters. Sample data and parameter values are given in the respective chapters with schematic representations of card decks showing card order, variable name, and format on each schematic card image. Tables are included to show parameters, parameter definition, source of information, and relative accuracy of the estimation of parameters. The output information shows specified options to print the desired information. Samples of output are shown, and each output element is defined.

Chapter 5 gives examples of typical computer runs for three specific field situations: (1) Georgia Piedmont, (2) Mississippi Delta, and (3) western Tennessee. These situations provide a wide range of soils, topography, management practices, and climate. Specific parameter values are shown as well as some typical output. This information and description will help the user to understand and follow through the model operations.

COMPUTER REQUIREMENTS

The computer programs are written in standard FORTRAN IV. They were written and tested initially at Purdue University on a CDC-6500. They also were tested on a DEC-10 at the University of Arizona in Tucson and the IBM 370/158 at the U.S. Department of Agriculture, Washington Computer Center in Washington, D.C. Several other locations are testing experimental versions of individual programs in the model. These programs can adapt to hardware configurations with only a few minor modifications.

All three programs will compile, load, and execute in less than 55K words of memory on the CDC-6500. This figure varies according to the installation, but the programs can be run on a relatively small machine. Significant digits are hardware dependent but do not affect the numeric variable formats. Since alphanumeric variable formats can be affected, so they were kept small (A4) to facilitate the use of smaller computers.

The computer programs require the use of secondary input/output devices and two input files that can be accessed independently. The first and second programs in the series also require an extra output device to handle the pass files generated for the next program. Table II-1 estimates storage requirements and compilation, load, and execution time.

Table II-1.—Central processing unit (CPU) time in seconds for CREAMS on 3 computer systems

	Hydrology	Opt. 1	Opt. 2	Erosion	Chemistry	Nutrients	Pesticides
	(Seconds)						
Compile							
IBM-370/158	5.70			12.09	4.26		
CDC-6500	4.57			13.40	4.78		
DEC-10	3.00			16.90	5.45		
Link/Edit/Load							
IBM-370/158	2.30			4.51	1.79		
CDC-6500	1.10			1.46	.63		
DEC-10	.71			1.36	.58		
Total							
IBM-370/158	8.00			16.60	6.50		
CDC-6500	5.67			14.86	5.41		
DEC-10	3.71			18.26	6.03		
Run 1/							
IBM-370/158		1.63	4.16	5.35		1.54	2/ 1.30
CDC-6500		1.58	3.21	5.90		1.37	1.35
DEC-10		1.70	2.78	8.78		2.18	2/ 1.48

1/ Run time is for 1 year of simulation.
2/ Run time is for total of 7 pesticides with 10 applications for 2 pesticides.

Chapter 1. HYDROLOGY

J. R. Williams, R. E. Smith, J. D. Nowlin, and A. D. Nicks [1/]

INPUT DATA FILES

The information in this chapter will help the user assemble the precipitation data files, temperature and radiation files, and hydrology parameter files. As described in CREAMS, volume I, chapter 2 (Hydrology), two options are available to the user. Option 1 uses daily rainfall and requires 37 cards per year of precipitation data with 10 daily values per card, as described in table II-2. A schematic deck arrangement is shown in figure II-2. The format in figure II-2 and table II-2 (that is, 10X, 10F5.0) represents a read format for data, only. Sample data in table II-2 show the year, 74, in columns 4 and 5 and sequential card numbers within the year in columns 79 and 80. These are for identification of data only and not for use in the programs. Daily rainfall data are available from the climatological data of the National Weather Service and from several USDA-SEA-AR research locations.

Formats for breakpoint rainfall data are given in table II-3, and a sample deck arrangement is shown in figure II-3. Breakpoint rainfall data are available upon request from USDA-SEA-AR for several locations in the United States. As described in CREAMS, volume I, chapter 2, hourly rainfall data can be used as input for hydrology model option 2. Hourly data are available from the National Weather Service for many locations in the United States for the period since 1948. Hourly rainfall data would be input in the same format as that used for the breakpoint data. The hourly time entries would be in clock hour.

Users of the CREAMS model interested in running hydrology model option 2 who do not have breakpoint rainfall data should contact USDA-SEA-AR locations in their respective States for availability of breakpoint rainfall data. To evaluate management practices, rainfall data can be transferred some distance within climatic regions. In mountainous regions, orographic influences must be recognized. Since evaluation of management systems is relative for a given climatic record, the period of record is unimportant and records from 1930 to 1940 would be just as appropriate as data from 1968 to 1978. Since representativeness is important, the data should include years with above normal, near normal, and below normal annual rainfall so that results and interpretations are not biased unduly. Selection of record period may be critical for areas west of the 100th meridian. If the user has a suitable method of generating synthetic data, either daily or hourly, such data would be entirely satisfac-

---

1/ Hydraulic engineer, USDA-SEA-AR, Grassland-Soil and Water Research Laboratory, Temple, Tex.; hydraulic engineer, USDA-SEA-AR, Fort Collins, Colo.; computer programer, Agricultural Engineering Department, Purdue University, West Lafayette, Ind.; and hydraulic engineer, USDA-SEA-AR, Southern Great Plains Watershed Research Center, Chickasha, Okla.

Table II-2.—Daily precipitation data input files

Precipitation Data File

----- For Daily Rainfall model (option 1)

Card 1-37. R(1-365)

R() Daily rainfall (in) , e.g. 2.07

The rainfall data are on a separate file from the parameters. They're read in, 10 values per card, 37 cards per year, and are repeated for each year of simulation. The year along the left margin and the sequence number along the right margin are only to aid the user in putting together the data. The program doesn't read them in. The following sample is only a partial years data.

Format(10X,10F5.2)

74	2.07	.16	.37	.17	0	0	.34	0	0	0	1
74	.08	0	0	0	0	0	0	0	0	.87	2
74	0	0	0	.24	0	0	0	.10	.26	0	3
74	0	0	0	0	0	0	1.70	.20	0	0	4
74	0	0	0	0	.65	.90	0	0	0	.16	5

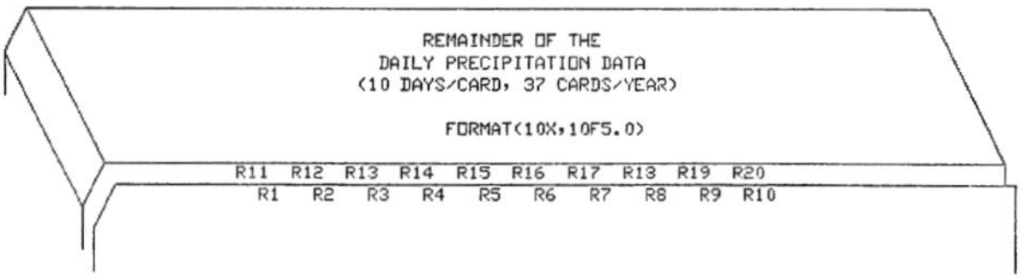

Figure II-2.—Sample deck arrangement for daily rainfall input for hydrology model option 1.

tory, and the generation scheme could be modified for output compatible with the input data formats described in tables II-2 and II-3.

Precipitation data are available to the user from different sources. Daily and hourly rainfall data are published by the National Weather Service (NWS) (7, 9). These data also are available on magnetic tape and can be purchased from the National Weather Data Center, Asheville, N. C.

Breakpoint rainfall data, required by hydrology model option 2, are available for several SEA-AR research watershed locations across the United States.

Table II-3.—Breakpoint rainfall input data files

-----	For Breakpoint Rainfall Model (option 2)		
Card 1.	JYR, JDAY, NP, MIDNI, PRE(JDAY)		
	JYR	Year of event (last 2 digits), e.g. 74	
	JDAY	Day of event (Julian day), e.g. 001	
	NP	Number of breakpoints in the event, e.g. 6	
	MIDNI	0 if event takes place during only one day 1 if event overlaps into two days	
	PRE()	Total rainfall for event (in), e.g. 2.07	
Card 2.	BP(1-NP), T(1-NP)		
	BP()	Accumulated rainfall at time T() (in), e.g. 1.96	
	T()	Time of measurement (min. from midnight), e.g. 38.0	

The rainfall data are on a separate file from the parameters. Card 2 is repeated for each breakpoint (NP, card 1) during the event. A card 1 and a series of card 2's are repeated for each event during the simulation. A partial sample, two events, follows.

Format(4I8,F8.0)
Format(2F8.0)

```
      74       1       6       0  2.0700
     0.0     1.0
  1.9600    38.0
  2.0300    47.0
  2.0400    54.0
  2.0500   154.0
  2.0700   180.0
      74       2       5       0   .1600
     0.0  1213.0
   .0500  1215.0
   .0800  1217.0
   .0900  1223.0
   .1600  1230.0
```

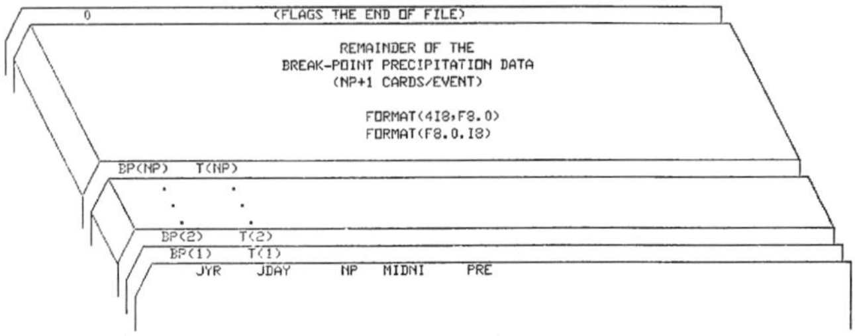

Figure II-3.—Sample deck arrangement for breakpoint rainfall input for hydrology model option 2.

A user might want to transfer these data for application within climatic regions. These data files are available in standard format on magnetic tape or cards at the Water Data Laboratory, USDA-SEA, Beltsville Agricultural Research Center, Beltsville, Md. 20705. The standard format of these data is different from the model input requirement and must be processed to conform to model input.

Several data sets were assembled for USDA-SEA research watersheds to test the CREAMS hydrology component. These data sets are available from the Water Data Laboratory and include runoff, air temperature, and solar radiation data. (table II-4). Data on soils, land use, and management have been published for these watersheds (4).

Hydrology Options

The hydrology submodel operates on a given rainfall data sequence plus a record of average monthly radiation and temperature, with information on crops, soil profile, and field shape to generate a sequence of information on runoff, evaporation, and seepage. This output information is used by the erosion, pesticide, and nutrient models in simulating chemical transport.

The hydrology model is designed to use physically related or easily estimable parameters as much as possible. It does not depend on extensive detail for soil or field topography. Plant growth patterns for crops grown are specified for a normal situation but are modified within the model for extreme stress (drought) conditions.

The simplifications used are dictated largely by data limitations rather than ignorance of the interrelations of the physical processes involved. Major limitations are:

(1) If only daily rainfall records are available, runoff is estimated by the SCS curve number (CN) procedure. Peak runoff rate

Table II-4. Watersheds selected for field-scale model testing and simulation

Water-shed ID.	Location	Watershed number	Area, acres	Land resource area	Soil and hydrologic group	Land use. managemer	Data available: Rainfall Daily	Rainfall Break-point	Runoff Daily	Runoff Break-point	Climatic Temp.	Humd.	Wind	Evap.	SR
26002	Coshocton OH	104	1.33	N-124	Muskng C	Pasture		MT3746		MT3746					
26003	Coshocton OH	129	2.71	N-124	Muskng C	Pasture		MT3875		MT3875					
26005	Coshocton OH	130	1.63	N-124	Muskng C	Meadow		MT3871		MT3871					
26008	Coshocton OH	132	0.59	N-124	Keene C	Woodland		MT4869		MT4869					
26011	Coshocton OH	115	1.61	N-124	Muskng C	Crop Rot.		MT3970		MT3970					
26015	Coshocton OH	110	1.27	N-124	Keene C	Crop Rot.		MT3970		MT3970					
26020	Coshocton OH	106	1.56	N-124	Muskng C	Crop Rot.		MT4072		MT4072					
26025	Coshocton OH	192	7.59	N-124	Muskng C	Crop Rot.		MT3970		MT3970					
34006	Cherokee OK	W-6	1.75	H-80	Grant B	Wheat		MT4067		MT4260	4067	4067	4067	4067	
34007	Cherokee OK	W-7	1.99	H-80	Grant B	Wheat		MT4067		MT4260	4067	4067	4067	4067	
34013	Cherokee OK	W-13	1.99	H-80	Grant B	Wheat SM		MT4067		CD6067	4067	4067	4067	4067	
35002	Guthrie OK	W-2	3.21	J-84	Sthnvil B	E. Cro Rot		MT3853		MT3948	3853	3853	3853	3853	
35006	Guthrie OK	W-I	2.5	J-84	Sthnvill B	Ex Range		MT3853		MT4248	3853	3853	3853	3853	
42014	Riesel TX	Y-6	16.3	J-86	Houston D	Tr Cont Cr		MT6876		MT6876	3276	3276	3276	3276	
42015	Riesel TX	Y-7	40.0	J-86	Houston D	Tr Cont Cr		MT6876		MT6876	3276	3276	3276	3276	
42016	Riesel TX	Y-8	20.8	J-86	Houston D	Tr Cont Cr		MT6876		MT6876	3276	3276	3276	3276	
42024	Riesel TX	SW-12	2.97	J-86	Houston D	Nat Past		MT6876		MT6876	3276	3276	3276	3276	
44007	Hastings NE	3-H	3.95	H-71	Hastings B	Cont Stp									
44025	Hastings NE	21-H	3.94	H-71	Hastings B	Cr. Rot. SM									
62014	Oxford MS	WC-2	1.45	P-134	Providnc C	Tr & No Til		MT5676		MT5676	5676				
62015	Oxford MS	WC-3	1.61	P-134	Providnc C	Tr & No Til		MT5676		MT5676	5676				
63105	Tombstone AZ	LH-5	.45	D-41	Laveen B	Range B		MT6375		MT6375					
66001	Moorefield WV	W-1	8.57	S-148	Litz C	Perm Past		MT5867		MT5867					
69033	Chickasha OK	C-4	29.9	H-80	McClain C	Cottonst.R	MT6774	MT6774	CD6774	CD6774	6674	6674	6674	6674	
69034	Chickasha OK	C-5	12.8	H-80	McClain C	Wheat	MT6774	MT6774	CD6774	CD6774	6674	6674	6674	6674	
69042	Chickasha OK	R-5	23.7	H-80	Renfrow D	Ex. Range	MT6774	MT6774	CD6774	CD6774	6676	6676	6676	6676	
69044	Chickasha OK	R-7	19.2	H-80	Renfrow D	Gul. Range	MT6774	MT6774	CD6774	CD6774	6676	6676	6676	6676	
73002	Ft. Stanton NM	W-2	32.2	D-39	C	Range UG		MT6676		MT6676					
00	Sidney MT			D-58		Range		PT7377		PT7377					
00	Laupahoehoe HA	1	1.52		Kaiwiki A	Sugar Cane		PT7277		PT7277					
00	Waialua Sugar HA	3	6.20		Pailoa B	Sugar Cane		PT7277		PT7277					
00	Kunia HA	6	7.07		Kolekole C	Pineapple		PT7277		PT7277					
00	Ames IA		20.0	M-108	Kenyon C	Corn to Tr		MT6972		MT6972					
00	Tifton GA	Z	.85	P-133	Cowarts B	Corn-Rot		MT6975		MT6975					
00	Watkinsville GA	P-1	6.7	P-136	Cecil B	Soy B		MT7275		MT7275					
00	Watkinsville GA	P-2	3.2	P-136	Cecil B	Corn		MT7375		MT7375					

and rainfall erosive energy index (EI) are predicted with regression equations based on runoff volume and watershed characteristics.

(2) Average daily values of net radiation and temperature are used for all years of a simulation. Since evapotranspiration (ET) depends strongly upon radiation and temperature, the standard deviation of ET is underestimated. The average radiation and temperature values give good estimates of long-term average ET, however.

(3) The soil profile is assumed to be constant in hydraulic properties throughout the growing season and constant (but different) in properties throughout the fallow period. This necessarily ignores the specific changes due to cultivations, rain splash crusting, and other time variations in soil properties. Model simulation methods to account for the effects of cultivation on infiltration and other properties of soil water movement await the results of current research.

(4) Soil water is assumed to move downward as a simple linear threshold model so that the elements of soil storage transfer water downward by gravity only when field capacity is exceeded. This simplification is necessary since the nonlinear differential equations for unsaturated water flow require input information and computational complexity far beyond the needs and resources of this management model.

Operation

Figures II-4 and II-5, flow charts for the hydrology simulation models, use the daily and breakpoint-infiltration options, respectively. These models operate on a time step of one day and use an alternate runoff model only on those days when rainfall occurs. The SCS-CN method or the infiltration method may be used for runoff simulation, depending on available rainfall data. Some hydrologic parameters that the user must specify will be different for these two models. Others will be common for either option. The hydrology model takes parameter input data from the parameter file and operates sequentially as precipitation information is read from the precipitation input file.

Input parameters necessary are listed in table II-5. This table refers to equations in the documentation section, volume I, chapter 2, where appropriate, and gives definitions and sources for parameter values.

Computer card format with variable names and definitions is shown in table II-6. The hydrology program reads from one file containing parameter values and another containing the actual precipitation data.

All numeric input formats consist of 8 column fields unless specifically stated otherwise in the card description. Integers are read with I8 formats, and real numbers are read with F8.0 formats. Integers must be right justified in columns 1 to 8, 9 to 16, 17 to 24, ... 73 to 80. Real numbers must be

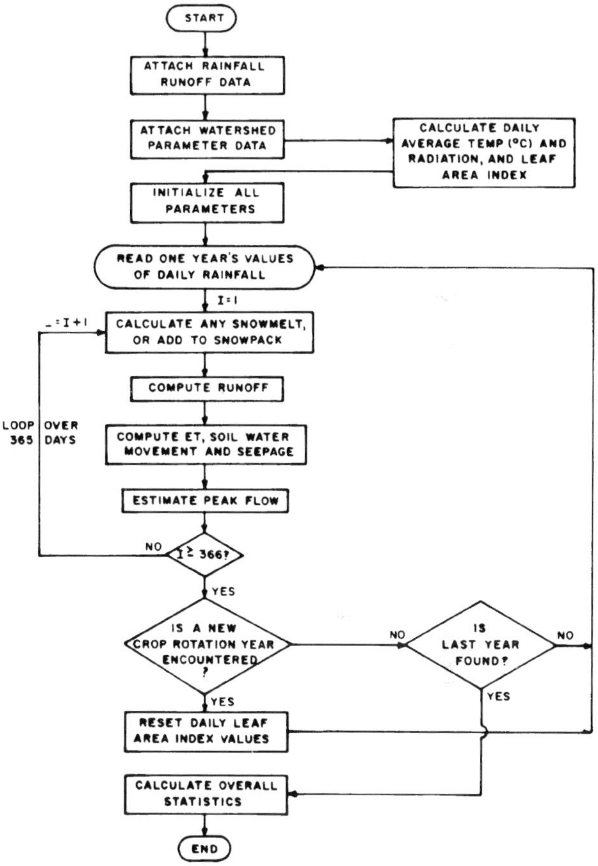

Figure II-4.—Generalized flow chart for HYDONE (hydrology option 1).

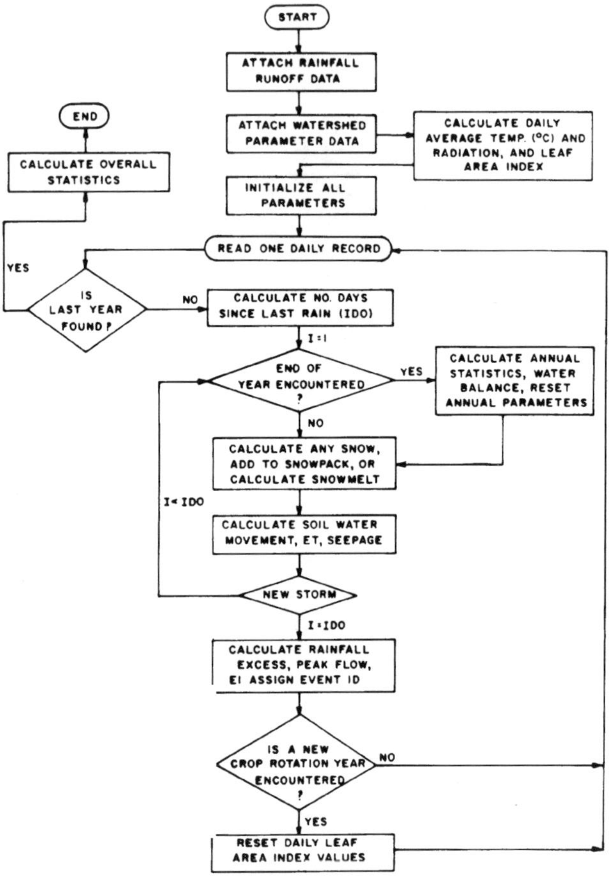

Figure II-5.—Generalized flow chart for HYDTWO (hydrology option 2).

Table II-5.—Hydrology model parameters

Parameter	Model option	Reference/definition	Source of estimate	Quality
DACRE	Both	Field area in acres.	Measurable	Good.
RC	Both	Saturated hydraulic conductivity, in/hr (K_s in equation I-9).	Estimate from SCS soil class; or measure, infiltrometer or in lab.	Poor to good; sensitive.
FUL	Both	Portion of plant-available water storage filled at field capacity.	Estimate or from reference.	Well defined quantity.
BST	Both	Portion of plant-available water storage filled when simulation begins.	Field measure or estimate.	Not sensitive.
CONA	Both	Soil evaporation parameter, α_s (eq. I-43).	Estimate from handbook.	Fair.
POROS	Both	ϕ, soil porosity (eq. I-8).	Estimate or measure.	Not sensitive.
BR15	Both	Immobile soil water content.	Estimate or measure, or from (1).	Not sensitive.
TEMP()	Both	Average monthly temperature (read values) °F.	Climatological data	Good, but only average.
RADI()	Both	Average monthly net radiation (read 12 values) langleys/day.	Climatological data	Good, but only average.
GR	Both	Winter cover factor (1 for crops, 0.5 for grass.)	Crop information	Rough.
X(I)	Both	Leaf area index, day I [must specify X(1) and X(366)].	Crop information handbook (table II-8).	Good.
SIA	1	Initial abstraction coefficient CN method (eq. I-2).	Use 0.2s in absence of measured value (5).	Fair.
CHS	1	Channel slope (CS in eq. I-7).	Field measurement	Good.
CN2	1	SCS curve number for AMC condition II.	Handbook; soils data.	Fair.
WLW	1	Watershed length/width ratio.	Watershed map	Good.
UL(1-7)	1	Plant-available water storage in 7 soil layers, in inches.	Difference between total soil porosity and 15 bar water content.	Fair to good.
DS	2	Depth of surface soil layer.	User estimate	Varies; subjective.
DP	2	Depth of root soil zone	Knowledge of soil; rooting depth.	Fair.
GA	2	G in equation I-16. Effective capillary tension.	Soil data; infiltrometer tests.	Fair to good.
RMN	2	Manning roughness for field surface (C_c in eq. I-30).	Handbooks; field observation.	Good; subjective.
SLOPE	2	Average field slope (S_c in eq. I-36).	Maps; field survey.	Good.
XLP	2	Length of flow plane (L in eq. I-36).	Maps; field survey.	Good.

Table II-6.—Hydrology parameters input file

----- Both Options

Card 1-3.	TITLE	
	TITLE	Three lines of 80 characters each for alphanumeric information to be printed at the beginning of the output. format (20A4)
Card 4.	BDATE, FLGOUT, FLGPAS, FLGOPT, FLGPRE	
	BDATE	The beginning date for simulation. It must be less than the first storm date (Julian date), e.g. 73138
	FLGOUT	0 for annual summary output 1 for storm by storm and annual summary output
	FLGPAS	0 if no hydrology file is to be created 1 if the program should create a hydrology file for use by the Erosion Program
	FLGOPT	1 for the daily rainfall model (option 1) 2 for the breakpoint or hourly rainfall model (option 2)
	FLGPRE	0 for breakpoint precipitation data 1 for hourly precipitation data (only used for hydrology model option 2)
Card 5.	DACRE, RC, FUL, BST, CONA, POROS, BR15	
	DACRE	Field area (acres), e.g. 3.2
	RC	Effective saturated conductivity of the soil (in/hr), e.g. 0.19
	FUL	Fraction of pore space filled at field capacity, e.g. 0.75
	BST	Fraction of plant-available water storage filled when simulation begins, e.g. 0.50
	CONA	Soil evaporation parameter, e.g. 3.75
	POROS	Soil porosity (cc/cc), e.g. 0.41
	BR15	Immobile soil water content at 15 bars tension (in/in), e.g. 0.17

Table II-6.—Hydrology parameters input file--Continued

-----	For Daily Rainfall Model (option 1)	
Card 6.	SIA, CN2, CHS, WLW, RD	
	SIA	Initial abstraction coefficient for SCS Curve Number method, e.g. 0.2
	CN2	Two condition SCS Curve Number, e.g. 80.0
	CHS	Channel slope, e.g. 0.022
	WLW	Watershed length/width ratio, e.g. 2.1
	RD	Maximum rooting depth (in), e.g. 24.0
Card 7.	UL(1-7)	
	UL()	Plant-available soil water storage for each of 7 soil storages (in), e.g. 0.16 (Top storage depth=1/36, 2nd storage depth=5/36, other storage depths=1/6 of rooting depth (RD, card 6))
-----	For Breakpoint Rainfall Model (option 2)	
Card 6.	DS, DP, GA, RMN, SLOPE, XLP	
	DS	Depth of surface soil layer (in), e.g. 2.0
	DP	Depth of maximum root growth layer (in), e.g. 22.0
	GA	Effective capillary tension of soil (in), e.g. 13.0
	RMN	Manning's n for overland flow, e.g. 0.03
	SLOPE	Effective hydrologic slope (ft/ft), e.g. 0.015
	XLP	Effective hydrologic slope length (ft), e.g. 350.0
-----	Both Options Continue	
Card 8,9.	TEMP(1-12)	
	TEMP()	Average monthly temperatures (degrees f.), e.g. 45.0
Card 10,11.	RADI(1-12)	
	RADI()	Average monthly solar radiation values (langleys/day), e.g. 218.0

Card 12. GR

GR Winter cover factor
1.0 for crops
0.5 for grass

Card 13. LDATE,AREA

LDATE Date (Julian day), e.g. 001

AREA Leaf area index for the crop grown the first year of simulation, e.g. 0.0

A card 13 is repeated as many times as is necessary to define the LAI curve. The first card 13 should always have the date 001. The last should always have the date 366.

- - - - - - - - - - -

Temperatures, solar radiation values, and leaf area index parameters can be updated at the end of each year. If they're to be updated, they will be read in the same sequence and format as the initial inputs. The winter cover factor (GR, card 12) will be read if the leaf area index is updated.

- - - - - - - - - - -

Card 14. NEWT, NEWR, NEWL

NEWT 0 use the temperatures from last year
1 read a new set of temperatures
[-] stop program execution

NEWR 0 use the solar radiation values from last year
1 read new solar radiation values

NEWL 0 use the leaf area index from last year
1 read a new set of leaf area index values and process them

A card 14 is read after each year of simulation. To stop execution of the program a negative value is read in NEWT. If any of the "NEW" parameters call for further input the appropriate data must be inserted after that card 14. That is cards 8 and 9 for NEWT, 10 and 11 for NEWR, and card 12 and a set of 13's for NEWL.

Table II-6.—Hydrology parameters input file--Continued

The following sample is a complete data set, good for a three year run, using the daily rainfall option.

CARD NO	HYDROLOGY PARAMETER DATA									
1	DAILY HYDROLOGY PARAMETERS - GEORGIA PIEDMONT									
2	MANAGEMENT PRACTICE ONE									
3	CONTINOUS CORN - CONVENTIONAL TILLAGE									
4	73138	0	1	1	0					
5	3.200	0.190	0.750	0.500	3.750	0.410	0.170			
6	0.200	80.000	0.022	2.100	24.000					
7	0.160	0.820	0.720	0.520	0.610	0.700	0.660			
8	45.0	47.0	52.0	61.0	70.0	77.0	79.0	78.0	73.0	63.0
9	51.0	44.0								
10	218.0	290.0	380.0	488.0	533.0	562.0	532.0	508.0	416.0	344.0
11	268.0	211.0								
12	1.000									
13	1	0.000								
13	122	0.000								
13	152	0.200								
13	166	0.200								
13	183	1.000								
13	192	2.500								
13	197	2.600								
13	202	2.700								
13	228	2.200								
13	255	0.000								
13	366	0.000								
14	0	0	0							
14	0	0	0							
14	-1	0	0							

contained within the same columns, and the decimal point must be entered in the number. If the example has a decimal in it, the parameter is real; otherwise, it is an integer. The alphanumeric input is read with A4 formats. Specific instructions are given whenever alphanumeric input is required. Sample deck representations are shown schematically in figures II-6 and II-7 for hydrology model options 1 and 2, respectively.

Climatic Data

Monthly mean air temperature and mean daily solar radiation data are required inputs used to calculate daily evapotranspiration. Daily values of temperature and radiation are calculated from the mean monthly values fitted to an annual curve by Fourier analysis (2). The user can use long-term averages or actual monthly data for the specific period of simulation. Temperature data are published regularly by the NWS (7). Current solar radiation data are not readily available. Daily and monthly data were published for selected locations from 1954 through 1973 (8). Publication was suspended in 1974, and only selected stations were published for the entire United States after that time.

The user can obtain monthly average daily radiation data from the Climatic Atlas of the United States (6). A summary of monthly radiation data is shown in table II-7 for the user's convenience.

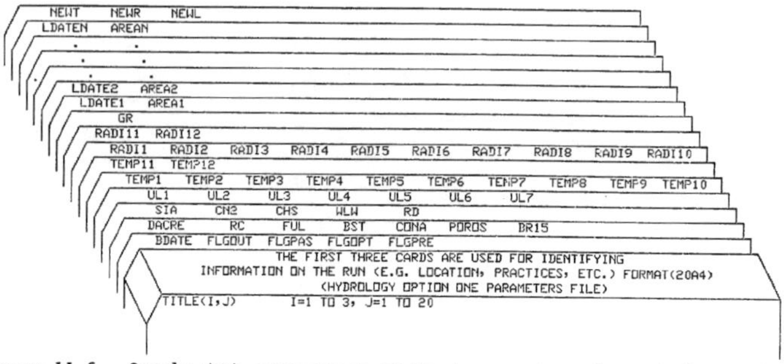

Figure II-6.—Sample deck arrangement of input parameters for hydrology model option 1.

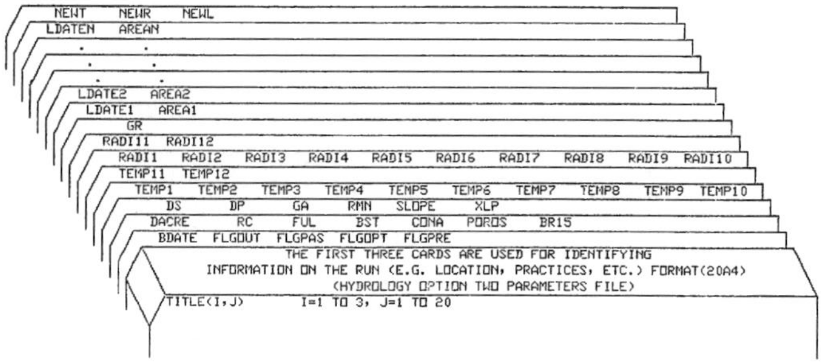

Figure II-7.—Sample deck arrangement of input parameters for hydrology model option 2.

Parameter Estimation

Daily Rainfall Model (Option 1)

Most parameters are easy to evaluate from existing data using the procedures outlined in table II-5. A slightly expanded explanation follows some procedures for estimation. The beginning soil water storage, BST, is generally unknown because most simulations begin several years in advance, and few locations measure soil water. Since yearly variations in soil water are usually small on January 1, BST can be estimated adequately for most areas. Since this estimate only affects model results before the first filling of soil storages, it is not very important except for short simulations (1 or 2 yr) or low-rainfall areas.

The winter cover factor, GR, reduces soil evaporation as a result of such ground cover as dormant pastures or heavy crop residue. The value of GR varies from 0.5 for excellent cover to 1.0 for bare soil.

The CN initial abstraction parameter, SIA, normally equals 0.2. The user may assign SIA any value grester than zero and less than one, however, for unusual applications. The 0.2 value generally is recommended unless justification is strong for another value.

Water storage, UL, is calculated for seven soil layers. Thickness of the layers normally is selected so that their sum equals maximum root depth. Thickness of the top layer is 1/36 of the maximum depth, the second layer is 5/36, and the remaining five layers are 1/6 each.

Leaf area index, X(I) is used in both model options. Table II-8 gives some typical leaf area index distributions for normalized times through a growing season for several crops. These values must be apportioned between actual local planting and harvesting dates. The distribution is specified as shown in figure II-8. Points for day = 1 and day = 366 are necessary.

The watershed length-width ratio is used in predicting peak runoff rates. Length is determined by measuring the distance from the watershed outlet along the main channel to the most distant point on the watershed boundary. The length-width ratio is computed by squaring length and dividing by the watershed area.

Breakpoint Infiltration Model (Option 2)

The breakpoint infiltration model (option 2) uses two parameters and a variable, as does the CN method, but can incorporate any additional data directly into its parameters. Any infiltrometer test can be used, for example, directly to yield values for parameters GA and RC.

Variable D is a straightforward estimate of porosity available in the soil surface at the beginning of the storm. Using S_w for relative saturation, S_w = 0 for air dry, and S_w = 1 for totally wet conditions, and ϕ for porosity,

$$D = \phi\ (1-S_w)\ . \qquad [II-1]$$

An AMC-III condition is analagous to a large value of D (D=0.3-0.4), AMC-I condition is similar to a small value of D (D=0.05-0.1). Water contents are assumed not to dry below the BR15 value. Water contents are assumed not to dry below the BR15 value. Values of this parameter may be estimated from information such as Holtan and others (1).

Table II-7.—Mean daily solar radiation (Langley's) and years of record used

States and stations	JAN	YRS	FEB	YRS	MAR	YRS	APR	YRS	MAY	YRS	JUNE	YRS	JULY	YRS	AUG	YRS	SEPT	YRS	OCT	YRS	NOV	YRS	DEC	YRS	ANNU
ALASKA																									
Annette- - - - - - -	63	6	115	6	236	7	364	7	437	6	438	6	438	6	341	6	258	7	122	7	59	7	41	7	243
Bethel - - - - - - -	38	9	108	10	282	9	444	10	457	10	454	10	376	10	252	10	202	10	115	10	44	9	22	9	233
Fairbanks- - - - - -	16	25	71	27	213	25	376	28	461	28	504	29	434	28	317	29	180	29	82	30	26	29	6	26	224
ARIZONA																									
Page - - - - - - - -	300	2	382	3	526	3	618	2	695	2	707	2	680	3	596	3	516	3	402	3	310	3	243	3	498
Phoenix- - - - - - -	301	11	409	11	526	11	638	11	724	11	739	11	658	11	613	11	566	11	449	11	344	11	281	11	520
Tucson - - - - - - -	315	5	391	5	540	4	655	5	729	5	699	5	626	6	588	6	570	6	442	6	356	6	305	6	518
ARKANSAS																									
Little Rock- - - - -	188	9	260	9	353	10	446	9	523	9	559	9	556	8	518	9	439	7	343	8	244	10	187	10	385
CALIFORNIA																									
Davis- - - - - - - -	174	18	257	17	390	18	528	18	625	18	694	18	682	18	612	18	493	18	347	19	222	19	148	19	431
Fresno - - - - - - -	184	31	289	31	427	31	552	31	647	31	702	32	682	32	621	31	510	31	376	32	250	31	161	32	450
Inyokern (China Lake)	306	11	412	11	562	11	683	11	772	11	819	11	772	11	729	10	635	8	467	9	363	11	300	12	568
La Jolla	244	19	302	18	397	19	457	20	506	19	487	21	497	22	464	22	389	22	320	21	277	20	221	20	380
Los Angeles WBAS - -	248	10	331	10	470	10	515	10	572	9	596	9	641	9	581	10	503	10	373	10	289	10	241	10	463
Riverside- - - - - -	275	8	367	8	478	9	541	9	623	9	680	9	673	9	618	9	535	9	407	9	319	9	270	9	483
Santa Maria- - - - -	263	11	346	11	482	11	552	10	635	11	694	11	680	11	613	11	524	11	419	11	313	11	252	11	481
Soda Springs - - - -	223	4	316	3	374	4	551	4	615	3	691	4	760	3	681	3	515	3	357	4	248	4	182	3	459
OLORADO																									
Boulder- - - - - - -	201	5	268	4	401	4	460	4	460	4	525	5	520	5	439	5	412	4	310	4	222	4	182	4	367
Grand Junction - - -	227	9	324	9	434	8	546	8	615	9	708	8	676	8	595	8	514	8	373	10	260	10	212	10	456
Grand Lake (Granby)-	212	6	313	7	423	7	512	8	552	8	632	8	600	8	505	7	476	6	361	7	234	6	184	7	417
American University-	158	39	231	39	322	39	398	39	467	39	510	39	496	39	440	38	364	38	278	38	192	39	141	39	333
LORIDA																									
Apalachicola - - - -	298	10	367	10	441	10	535	10	603	9	578	9	529	9	511	9	456	9	413	10	332	10	262	10	444
Belle Isle - - - - -	297	10	330	10	412	10	463	10	483	10	464	10	488	11	561	10	400	10	366	11	313	11	291	10	397
Gainsville - - - - -	267	11	343	10	427	12	517	12	579	12	521	10	488	10	483	8	418	9	347	8	300	10	233	10	410
Miami Airport- - - -	349	10	415	9	489	9	540	10	553	10	532	10	532	10	505	10	440	10	384	10	353	10	316	10	451
Tampa- - - - - - - -	327	8	391	8	474	8	539	8	596	8	574	9	534	9	494	9	452	9	400	9	356	9	300	9	453
GEORGIA																									
Atlanta- - - - - - -	218	11	290	11	380	11	488	11	533	11	562	11	532	10	508	10	416	10	344	11	268	11	211	11	396
Griffin- - - - - - -	234	9	295	9	385	10	522	11	570	11	577	11	556	11	522	11	435	11	368	11	283	11	201	11	413
HAWAII																									
Honolulu - - - - - -	363	4	422	4	516	4	559	5	617	5	615	5	615	5	612	5	573	5	507	5	426	5	371	5	516
Pearl Harbor - - - -	359	5	400	4	487	4	529	5	573	5	566	5	598	5	567	5	539	5	466	5	386	5	343	5	484
IDAHO																									
Boise- - - - - - - -	138	10	236	9	342	9	485	9	585	10	636	9	670	10	576	10	460	10	301	11	182	11	124	11	395
Twin Falls - - - - -	163	20	240	20	355	20	462	21	552	20	592	18	602	20	540	20	432	19	286	20	176	20	131	19	378
ILLINOIS																									
Chicago- - - - - - -	96	19	147	19	227	19	331	19	424	19	458	18	473	19	403	18	313	19	207	20	120	20	76	20	273
Lemont - - - - - - -	170	6	242	6	340	6	402	6	506	6	553	6	540	6	498	6	398	5	275	5	165	5	138	5	352
INDIANA																									
Indianapolis - - - -	144	10	213	10	316	10	396	10	488	9	543	11	541	10	490	11	405	11	293	11	177	11	132	11	345
IOWA																									
Ames - - - - - - - -	174	5	253	5	326	5	403	5	480	5	541	5	536	6	460	6	367	6	274	7	167	7	143	7	345
KANSAS																									
Dodge City - - - - -	255	7	316	7	418	7	528	7	568	7	650	7	642	8	592	9	493	9	380	9	285	10	234	10	447

Table II-7.—Mean daily solar radiation (Langley's) and years of record used--Continued

States and stations	JAN	YRS	FEB	YRS	MAR	YRS	APR	YRS	MAY	YRS	JUNE	YRS	JULY	YRS	AUG	YRS	SEPT	YRS	OCT	YRS	NOV	YRS	DEC	YRS	ANNU.
KANSAS																									
Manhattan- - - - - -	192	3	264	3	345	3	433	3527		4	551	4	531	4	526	4	410	4	492	4	227	4	156	4	371
KENTUCKY																									
Lexington- - - - - -	172	9	263	9	357	10	480	10	581	10	628	9	617	10	563	10	494	10	357	9	245	9	174	11	411
LOUISIANA																									
Lake Charles - - - -	245	11	306	11	397	11	481	11	555	11	591	11	526	11	511	11	449	11	402	11	300	10	250	10	418
New Orleans- - - - -	214	14	259	14	335	15	412	16	449	14	443	13	417	15	416	15	383	15	357	13	278	13	198	14	347
Shreveport - - - - -	232	3	292	3	384	3	446	4	558	4	557	4	578	4	528	4	414	4	354	4	254	4	205	4	400
MAINE																									
Caribou- - - - - - -	133	8	231	9	364	8	400	10	476	10	470	10	508	11	448	11	336	11	212	11	111	11	107	9	316
Portland - - - - - -	152	7	235	8	352	7	409	8	514	9	539	9	561	9	488	8	383	7	278	9	157	8	137	9	350
MASSACHUSETTS																									
Blue Hill- - - - - -	153	27	228	27	319	26	389	26	469	27	510	27	502	26	449	27	354	28	266	28	162	28	135	28	328
Boston - - - - - - -	129	16	194	17	290	17	350	17	445	16	483	16	486	16	411	16	334	17	235	16	136	16	115	15	301
East Wareham - - - -	140	13	218	13	305	12	385	14	452	14	508	14	495	14	436	14	365	13	258	14	163	14	140	13	322
MICHIGAN																									
East Lansing- - - - -	121	10	210	11	309	11	359	11	483	10	547	11	540	11	466	11	373	11	255	11	136	11	108	11	311
Sault Ste Marie - - -	130	10	225	9	356	10	416	10	523	10	557	11	573	11	472	10	322	10	216	9	105	9	96	9	333
MINNESOTA																									
St Cloud- - - - - - -	168	8	260	8	368	8	426	8	496	8	535	8	657	9	486	8	366	8	237	7	146	8	124	8	348
MISSOURI																									
Columbia- - - - - - -	173	10	251	10	340	11	434	11	530	11	574	12	574	10	522	10	453	10	322	10	225	10	158	9	380
MONTANA																									
Glasgow - - - - - - -	154	6	258	8	385	7	466	8	568	8	606	8	645	9	531	10	410	10	267	8	154	8	116	7	388
Great Falls - - - - -	140	8	232	9	366	9	434	8	528	8	583	8	639	9	532	9	407	10	264	10	154	10	112	10	366
Summit- - - - - - - -	122	3	162	2	268	3	414	3	462	3	493	3	560	2	510	2	354	2	216	2	102	2	76	2	312
NEBRASKA																									
Lincoln - - - - - - -	188	39	259	39	350	39	416	39	494	40	544	38	568	38	484	38	396	38	296	36	199	40	159	39	363
North Omaha - - - - -	193	3	299	3	365	3	463	3	516	3	546	4	668	4	419	4	410	4	298	4	204	4	170	4	379
NEVADA																									
Ely - - - - - - - - -	236	7	339	9	468	9	563	9	625	10	712	10	647	11	618	11	518	11	394	10	289	10	218	10	469
Las Vegas - - - - - -	277	11	384	11	519	11	621	11	702	11	748	10	675	11	627	11	551	11	429	11	318	11	258	11	509
NEW JERSEY																									
Seabrook- - - - - - -	157	8	227	8	318	8	403	8	482	9	527	8	509	8	455	9	385	9	278	7	192	8	140	8	339
NEW MEXICO																									
Albuquerque - - - - -	303	13	386	13	511	13	618	13	686	13	726	13	683	12	626	13	554	14	438	15	334	15	276	14	512
NEW YORK																									
Ithica- - - - - - - -	116	22	194	21	272	23	334	23	440	24	501	23	515	23	453	23	346	21	231	22	120	23	96	23	302
Central Park- - - - -	130	34	199	34	290	33	369	35	432	35	470	34	459	35	389	35	331	36	242	36	147	36	115	35	298
Sayville- - - - - - -	160	11	249	11	335	10	415	10	494	10	565	10	543	10	462	10	385	10	289	10	186	10	142	11	352
Schenectady - - - - -	130	8	200	9	273	9	338	9	413	9	448	8	441	8	397	8	299	8	218	8	128	8	104	8	282
Upton - - - - - - - -	155	8	232	8	339	8	428	8	502	8	573	8	543	7	475	7	391	7	293	6	182	7	143	7	355
NORTH CAROLINA																									
Greensboro- - - - - -	200	7	276	9	354	9	469	9	531	10	564	10	544	10	485	10	406	10	322	10	243	10	197	8	383
Hatteras- - - - - - -	238	10	317	9	426	8	569	9	635	10	652	10	625	10	562	11	471	11	358	11	282	11	214	11	443
NORTH DAKOTA																									
Bismarck- - - - - - -	157	7	250	8	356	6	447	8	550	8	590	9	617	10	516	11	390	11	272	11	161	10	124	10	369

Table II-7.—Mean daily solar radiation (Langley's) and years of record used--Continued

States and stations	JAN	YRS	FEB	YRS	MAR	YRS	APR	YRS	MAY	YRS	JUNE	YRS	JULY	YRS	AUG	YRS	SEPT	YRS	OCT	YRS	NOV	YRS	DEC	YRS	ANNUAL
OHIO																									
Cleveland	125	6	183	6	303	7	286	8	502	8	562	8	562	8	494	8	278	8	289	9	141	9	115	7	335
Columbus	128	7	200	7	297	7	391	7	471	6	562	4	542	5	477	4	422	4	286	4	176	4	129	5	340
Put-in-Bay	126	10	204	9	302	10	386	11	468	11	544	11	561	10	487	10	382	11	275	11	144	11	109	11	332
OKLAHOMA																									
Oklahoma City	251	10	319	10	409	9	494	10	536	10	615	7	610	8	593	8	487	9	377	10	291	9	240	9	436
Stillwater	205	8	289	8	390	9	454	9	504	9	600	10	596	10	545	10	455	11	354	10	269	9	209	8	405
OREGON																									
Astoria	90	7	162	8	270	8	375	8	492	8	469	8	539	8	461	7	354	7	209	8	111	8	79	8	301
Corvallis	89	2	*		287	3	406	3	517	3	570	3	676	4	558	4	397	4	235	4	114	4	80	4	---
Medford	116	11	215	11	336	11	482	11	592	11	652	11	698	10	605	11	447	11	279	11	149	11	93	11	389
PENNSYLVANIA																									
Pittsburgh	94	6	169	5	216	6	317	6	429	6	491	6	497	7	409	6	339	6	207	5	118	6	77	5	280
State College	133	19	201	19	295	20	380	20	456	20	518	20	511	20	444	20	358	20	256	20	149	20	118	20	318
RHODE ISLAND																									
Newport	155	23	232	22	334	23	405	23	477	23	527	24	513	24	455	24	377	24	271	24	176	24	139	24	338
SOUTH CAROLINA																									
Charleston	252	11	314	11	388	11	512	11	551	11	564	11	520	11	501	11	404	11	338	11	286	11	225	11	404
SOUTH DAKOTA																									
Rapid City	183	11	277	11	400	11	482	11	532	11	585	11	590	11	541	11	435	11	315	10	204	10	158	10	392
TENNESSEE																									
Nashville	149	18	228	19	322	19	432	19	503	18	551	18	530	17	473	17	403	17	308	19	208	18	150	19	355
Oak Ridge	161	11	239	11	331	11	450	11	518	11	551	11	526	11	478	11	416	11	318	11	213	10	163	11	364
TEXAS																									
Brownsville	297	10	341	10	402	10	456	11	564	10	610	9	627	8	568	11	475	11	411	11	296	11	263	10	442
El Paso	333	11	430	11	547	10	654	11	714	11	729	11	666	11	640	10	576	11	460	11	372	11	313	11	536
Ft. Worth	250	11	320	11	427	11	488	11	562	11	651	11	613	11	593	11	503	11	403	11	306	11	245	9	445
Midland	283	7	358	8	476	9	550	8	611	8	617	8	608	7	574	8	522	9	396	9	325	8	275	8	466
San Antonio	279	9	347	9	417	9	445	9	541	9	612	9	639	9	585	9	493	10	398	10	295	10	256	8	442
UTAH																									
Flaming Gorge	238	2	498	2	443	2	522	2	565	2	650	2	599	3	538	3	425	3	352	3	262	3	215	3	426
Salt Lake City	163	8	256	8	354	8	479	8	570	7	621	7	620	6	551	7	446	8	316	8	204	8	146	9	394
Mt. Weather	172	2	274	2	338	2	414	2	508	2	525	3	510	3	430	3	375	3	281	2	202	2	168	2	350
Friday Harbor	87	8	157	7	274	8	418	8	514	9	578	10	586	10	507	11	351	8	194	10	102	10	75	8	320
WASHINGTON																									
Prosser	117	4	222	4	351	4	521	5	616	4	680	4	707	4	604	4	458	4	274	4	136	4	100	4	399
University of Washington	67	9	126	9	245	10	364	9	445	10	461	10	496	11	435	10	299	8	170	9	93	9	59	9	272
Pullman	121	4	205	2	304	2	462	2	558	4	653	5	699	5	562	4	410	4	245	5	146	5	96	5	372
Seattle-Tacoma	75	9	139	9	265	9	403	9	503	9	511	9	566	9	452	10	324	10	188	10	104	9	64	10	300
WISCONSIN																									
Madison	148	46	220	46	313	45	394	47	466	47	514	47	531	47	452	47	348	47	241	47	145	44	115	46	324
WYOMING																									
Lander	226	8	324	9	452	9	548	11	587	11	678	11	651	11	586	10	472	8	354	9	239	9	196	9	443
Laramie	216	3	295	3	424	3	508	3	554	3	643	3	606	3	536	3	438	3	324	3	229	3	186	4	408
PUERTO RICO																									
San Juan	404	5	481	4	580	4	622	4	519	5	536	6	639	5	549	6	531	6	460	6	411	6	411	6	512

Table II-8.—Typical leaf area index distributions for crops

Portion of growing season	Leaf area index[1/]							
	Corn	Cotton	Sorghum	Oats	Wheat	Pasture[2/]	Barley	Soybeans
0.0	0.00	0.00	0.00	0.00	0.00	0.00	0.00	0.00
.1	.09	.13	.09	.42	.47	1.84	.44	.15
.2	.19	.28	.19	.84	.90	3.00	.88	.40
.3	.23	.50	.23	.90	.90	3.00	.90	2.18
.4	.49	2.14	.54	.90	.90	3.00	.90	2.97
.5	1.16	2.96	1.35	.98	.90	3.00	1.58	3.00
.6	2.97	3.00	2.98	2.62	1.62	3.00	3.00	2.96
.7	3.00	2.96	3.00	3.00	3.00	2.70	3.00	2.92
.8	2.72	2.92	2.72	3.00	3.00	1.96	3.00	2.30
.9	1.83	1.78	1.84	3.00	.96	2.14	1.15	2.00
1.0	.00	1.00	1.00	.00	.00	.50	.00	.50

1/ Good production assumed for all crops. LAI should be lowered for poor production.

2/ No grazing assumed. LAI must be lowered if grazed or mowed according to height of plants.

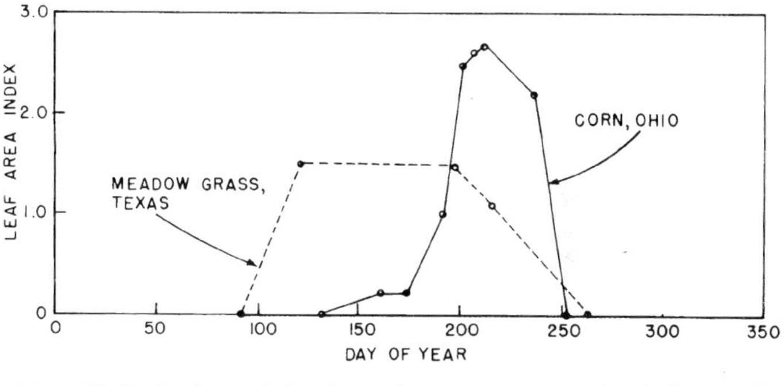

Figure II-8.—Leaf area index for meadow grass at a location in Texas and for corn at a location in Ohio.

GA is a parameter characteristic of the soil type and composition. Table II-9 shows how GA may be estimated in relation to the SCS hydrologic soil groups.

RC represents effective saturated hydraulic conductivity. Experiments and

Table II-9.—Parameter estimation of the infiltration model[1/]

		G (in)				RC (in/hr)			
		Hydrologic soil group				Hydrologic soil group			
		A	B	C	D	A	B	C	D
Expected range of values		3-6	7-11	12-17	18-22	0.3-3.0	0.1-1.0	0.05-0.5	0.01-0.2
Mean for estimation		4	9	15	20				
Land use[2/]	Hydrologic condition								
Row crops:									
Straight	Poor					[3/]0.4	0.18	0.06	0.03
	Good					.45	.21	.09	.04
Contoured	Poor					.4	.2	.1	.05
	Good					.5	.25	.12	.07
Fallow	--					.3	.12	.04	.02
Small grains and meadow:									
Straight	Poor					.05	.22	.09	.04
	Good					.6	.43	.12	.07
Contoured	Poor					.05	.25	.1	.07
	Good					.65	.3	.14	.09
Range/pasture:									
Straight	Poor					.5	.2	.08	.04
	Good					.8	.42	.2	.12
Contoured	Poor					.75	.35	.13	.04
	Good					2.0	.7	.24	.12
Meadow	Good					1.2	.46	.23	.13
Woods	Poor					.75	.36	.17	.09
	Good					1.5	.50	.24	.14

1/ Tentative.

2/ SCS presentation of CN for terraced systems is omitted since this affects routing, not condition of soil.

3/ Values given will reproduce CN estimates for a storm of 4.0 in covering 4-hr duration, with mean value of G.

theory suggest the approximation that variations of this parameter can relate changes to soil condition. Thus, table II-9 shows a proposed guide to use instead of table 9.1 of NEH4 (5). Although the variations shown for RC as a function of soil cover complex are tentative, they are useful in a relative sense. A better version of this table requires use of the data from which the SCS table was developed. Actual value for RC to reproduce the runoff predicted by the CN method depends on the depth and duration of the storm from which the tabulated value of CN was obtained. Should such data be available, as are much SEA-AR data for any model application, they can be used to improve parameter estimates. Notable exceptions to these RC values include a cracked clay, soil class D, which will exhibit apparent large RC early in a storm, confounding predictive accuracy.

The parameters RMN, SLOPE, and XLP are used to estimate peak rates of runoff. Effective length of flow in the field is determined by estimating the median flow path, including a fraction of concentrated flow path length. The fraction of concentrated flow path length to include will be large if the flow is through a rough or mildly sloping channel.

If the length of the watershed flow path is broken into N regions of different slope and roughness, equivalent single-plane values of RMN and SLOPE can be determined. For each segment or subplane, j, j = 1, N, there is an α_j as in equation [I-25], volume I, chapter 2, defined as

$$\alpha_j = C_j \sqrt{S_j} \qquad [II-2]$$

where C and S are roughness and slope, respectively. From volume I, chapter 2, equation [I-34] is

$$C_c \sqrt{S_c} = \alpha_c = \left[\frac{L^b}{\sum_{j=1}^{N} \frac{x_j{}^b - x_{j-1}{}^b}{(\alpha_j)^{1/m}}} \right]^m \qquad [II-3]$$

for $x_0 = 0$ to $X_N = L$ where C_c is composite roughness, S_c is composite slope, L is horizontal length of slope, and $b = (m + 1)/m$. If S_c is taken as the equivalent single slope (SLOPE),

$$C_c = \frac{\alpha_c}{\sqrt{S_c}} \qquad [II-4]$$

and the equivalent roughness, RMN, is $1.49/C_c$ for the Manning flow equation.

OUTPUT

Hydrology output is composed of input information and calculated values. Sample input information included in the output is shown in figures II-9 and II-10 for options 1 and 2, respectively. Daily and annual simulated output data are the same for both options. Sample output data are shown in figure II-11. These data are transmitted to the erosion model in the hydrology pass file. Figure II-12 is a sample of averages and statistics calculated for the period of simulation. Output for the simulation period also includes monthly totals and means of rainfall, runoff, ET, percolation, and average soil water (fig. II-13). Data include annual totals of each component.

HYDROLOGY OPTION ONE

(DAILY PRECIPITATION VALUES)

DAILY HYDROLOGY PARAMETERS - GEORGIA PIEDMONT
MANAGEMENT PRACTICE ONE
CONTINOUS CORN - CONVENTIONAL TILLAGE

MONTHLY MEAN TEMPERATURES, DEGREES FAHRENHEIT

43.43	45.64	52.15	61.21	70.39	77.23
79.90	77.69	71.18	62.12	52.94	46.10

MONTHLY MEAN RADIATION, LANGLEYS PER DAY

236.29	291.73	375.07	463.97	534.61	568.07
555.38	499.93	416.60	327.70	257.05	223.59

LEAF AREA INDEX TABLE

DATE	LAI
1	0.00
122	0.00
152	0.20
166	0.20
183	1.00
192	2.50
197	2.60
202	2.70
228	2.20
255	0.00
366	0.00

WINTER C FACTOR = 1.00
LAI-DAYS = 151.15

FIELD AREA	=	3.200	ACRES
ROOTING DEPTH	=	24.000	IN
RETENTION RATE	=	0.190	IN/HR
FIELD CAPACITY	=	0.750	
INITIAL STORAGE FRACTION	=	0.500	
INITIAL ABSTRACTION	=	0.200	
EVAPORATION COEFFICIENT	=	3.750	
SCS CURVE NUMBER	=	80.000	
CHANNEL SLOPE	=	0.022	
WATERSHED LEN/WIDTH RATIO	=	2.100	
PEAK FLOW RATE COEFFICIENT	=	9.087	
PEAK FLOW RATE EXPONENT	=	0.840	
UPPER LIMIT OF STORAGE	=	4.190	IN
IMMOBILE SOIL WATER CONTENT	=	0.235	IN/IN
INITIAL SOIL WATER STORAGE	=	2.095	IN

UPPER LIMIT OF STORAGES

0.160	0.820	0.720	0.520	0.610	0.700	0.660

INITIAL STORAGE

0.080	0.410	0.360	0.260	0.305	0.350	0.330

Figure II-9.—Sample output of input values for hydrology option 1.

HYDROLOGY OPTION TWO

(BREAKPOINT OR HOURLY PRECIPITATION VALUES)

BREAKPOINT HYDROLOGY PARAMETERS - GEORGIA PIEDMONT
MANAGEMENT PRACTICE ONE
CONTINOUS CORN - CONVENTIONAL TILLAGE

MONTHLY MEAN TEMPERATURES, DEGREES FAHRENHEIT

43.43	45.64	52.15	61.21	70.39	77.23
79.90	77.69	71.18	62.12	52.94	46.10

MONTHLY MEAN RADIATION, LANGLEYS PER DAY

236.29	291.73	375.07	463.97	534.61	568.07
555.38	499.93	416.60	327.70	257.05	223.59

LEAF AREA INDEX TABLE

DATE	LAI
1	0.00
122	0.00
152	0.20
166	0.20
183	1.00
192	2.50
197	2.60
202	2.70
228	2.20
255	0.00
366	0.00

WINTER C FACTOR = 1.00
LAI-DAYS = 151.15

EFFECTIVE HYDROLOGIC LENGTH	=	350.000	FT
EFFECTIVE HYDROLOGIC SLOPE	=	0.015	
EFFECTIVE MANNINGS N	=	0.030	
DEPTH OF SURFACE LAYER	=	2.000	IN
DEPTH OF REMAINING ROOT ZONE	=	22.000	IN
EFFECTIVE CAPILLARY TENSION	=	13.000	IN
EVAPORATION COEFFICIENT	=	3.750	
SAT. CONDUCTIVITY CULTIVATED	=	0.190	IN/HR
SAT. CONDUCTIVITY FALLOW	=	0.152	IN/HR
SOIL POROSITY	=	0.410	
IMMOBILE SOIL WATER CONTENT	=	0.170	IN/IN
UPPER LIMIT OF STORAGE	=	4.440	IN
INITIAL SURFACE STORAGE	=	0.240	IN
INITIAL REMAINING STORAGE	=	1.980	IN
TOTAL INITIAL STORAGE	=	2.220	IN

Figure II-10.—Sample output of input values for hydrology option 2.

DATE JULIAN	RAINFALL INCHES	RUNOFF INCHES	PERCOL. INCHES	AVERAGE TEMP. DEG. F.	AVERAGE SOIL W. IN./IN.	ACTUAL EP INCHES	POTENT. EP INCHES
74001	0.1100	0.0000	0.0000	44.0823	0.3242	0.0000	0.0000
74002	0.1600	0.0000	0.0000	44.0015	0.3278	0.0000	0.0000
74003	0.3700	0.0000	0.0000	43.9259	0.3401	0.0000	0.0000
74004	0.1700	0.0000	0.0000	43.8555	0.3441	0.0000	0.0000
74007	0.3400	0.0000	0.0000	43.7323	0.3426	0.0000	0.0000
74011	0.0800	0.0000	0.0000	43.5660	0.3419	0.0000	0.0000
74020	0.8700	0.0000	0.0000	43.4426	0.3320	0.0000	0.0000
74024	0.2400	0.0000	0.0000	43.5169	0.3500	0.0000	0.0000
74028	0.1000	0.0000	0.0000	43.6840	0.3453	0.0000	0.0000
74029	0.2600	0.0000	0.0000	43.8283	0.3515	0.0000	0.0000
74037	1.7000	0.2584	0.5470	44.1900	0.3426	0.0000	0.0000
74038	0.2000	0.0000	0.1085	44.6296	0.3664	0.0000	0.0000
74045	0.6500	0.0000	0.0823	45.1264	0.3558	0.0000	0.0000
74046	0.9000	0.1765	0.6243	45.6823	0.3664	0.0000	0.0000
74050	0.1600	0.0000	0.0000	46.0771	0.3582	0.0000	0.0000
74053	0.5000	0.0000	0.0182	46.6726	0.3573	0.0000	0.0000
74078	0.1500	0.0000	0.0000	49.6677	0.3397	0.0000	0.0000
74080	0.6500	0.0000	0.0000	53.0365	0.3402	0.0000	0.0000
74084	0.3500	0.0000	0.0000	53.8765	0.3487	0.0000	0.0000
74086	0.0800	0.0000	0.0000	54.7348	0.3513	0.0000	0.0000
74088	0.6900	0.0000	0.1064	55.3181	0.3577	0.0000	0.0000
74094	1.3000	0.1433	0.5355	56.5086	0.3547	0.0000	0.0000
74102	0.0500	0.0000	0.0000	58.6436	0.3501	0.0000	0.0000
74103	0.9500	0.0357	0.2315	60.0387	0.3664	0.0000	0.0000
74112	0.3000	0.0000	0.0000	61.6025	0.3503	0.0000	0.0000
74122	0.0900	0.0000	0.0000	64.5615	0.3425	0.0005	0.0005
74124	0.3500	0.0000	0.0000	66.4018	0.3413	0.0030	0.0030
74125	0.7400	0.0000	0.0000	66.8540	0.3650	0.0050	0.0050
74131	0.1000	0.0000	0.0000	67.8923	0.3515	0.0285	0.0285
74132	0.5000	0.0000	0.0000	68.9134	0.3596	0.0343	0.0343
74135	0.1000	0.0000	0.0000	68.4830	0.3519	0.0553	0.0553
74143	0.2600	0.0000	0.0000	70.9958	0.3428	0.1378	0.1378
74144	2.5000	0.6378	0.9712	72.1848	0.3664	0.1509	0.1509
74146	0.2800	0.0000	0.0000	72.5647	0.3595	0.1791	0.1791
74151	0.5000	0.0000	0.0000	73.4205	0.3523	0.2608	0.2608
74159	0.3000	0.0000	0.0000	74.8939	0.3433	0.4044	0.4044
74161	0.2500	0.0000	0.0000	75.5311	0.3376	0.4409	0.4409
74171	0.4800	0.0000	0.0000	77.0066	0.3279	0.7179	0.7179
74178	4.2600	1.2540	1.2569	78.2838	0.3221	1.1581	1.1581
74198	0.1100	0.0000	0.0000	79.4859	0.2880	3.5846	4.7379
74204	0.1100	0.0000	0.0000	79.8335	0.2358	3.6409	6.2028
74205	0.5800	0.0000	0.0000	79.8347	0.2520	3.8284	6.4397
74207	0.5100	0.0000	0.0000	79.7854	0.2506	4.1690	6.9062
74208	2.8400	0.5856	0.0000	79.7254	0.3385	4.3985	7.1358

ANNUAL TOTALS FOR 1974

PRECIPITATION	=	40.260
PREDICTED RUNOFF	=	3.516
DEEP PERCOLATION	=	4.916
TOTAL ET	=	30.929
BEGIN SOIL WATER	=	2.095
FINAL SOIL WATER	=	2.994
WATER BUDGET BAL.	=	0.000

Figure II-11.—Sample output of daily data and annual summary from the hydrology model. Full year of daily values not shown.

```
                    AVERAGE ANNUAL VALUES
                  PRECIPITATION      =  44.255
                  PREDICTED RUNOFF   =   5.511
                  DEEP PERCOLATION   =   9.372
                  TOTAL ET           =  29.765

               AVG. AVAL. STORAGE  =   2.270 IN
               FINAL AVAL. STORAGE =   2.881 IN

               FINAL STORAGE FOR EACH FRACTION
0.000     0.000     0.000     0.113     0.458     0.525     0.213

          MINIMUM TOTAL STORAGE WAS   0.000 ON  74200
          MAXIMUM TOTAL STORAGE WAS   3.143 ON  74144
```

Figure II-12.—Sample output of averages and statistics calculated for the period of simulation.

HYDROLOGY SUMMARY

DAILY HYDROLOGY PARAMETERS - GEORGIA PIEDMONT
MANAGEMENT PRACTICE ONE
CONTINOUS CORN - CONVENTIONAL TILLAGE

1974

MONTH	RAIN	RUNOFF	ET	PERC	AVG SW
JAN	2.700	0.008	2.157	0.000	2.514
FEB	4.110	0.436	2.270	1.380	2.830
MAR	1.920	0.008	1.618	0.106	2.544
APR	2.600	0.179	1.991	0.768	2.735
MAY	5.420	0.644	3.374	0.971	2.742
JUN	5.290	1.254	3.153	1.257	2.351
JUL	4.150	0.586	4.726	0.000	0.901
AUG	5.780	0.192	6.440	0.000	0.969
SEP	1.850	0.000	2.199	0.000	0.457
OCT	0.360	0.000	0.554	0.000	0.095
NOV	1.160	0.000	0.782	0.000	0.282
DEC	4.920	0.209	1.666	0.433	1.484
TOT	40.260	3.516	30.929	4.916	1.659

Figure II-13.— Sample output of monthly totals of rainfall, runoff, evapotranspiration, percolation, and average soil water, and averages for the period of simulation.

1975

MONTH	RAIN	RUNOFF	ET	PERC	AVG SW
JAN	5.020	0.490	2.321	2.522	2.977
FEB	7.170	1.035	2.460	3.611	2.920
MAR	9.780	2.961	2.930	3.682	2.864
APR	3.930	0.934	2.173	1.260	2.717
MAY	6.070	0.633	3.123	1.686	2.771
JUN	3.550	1.071	2.996	1.067	2.479
JUL	4.670	0.043	6.012	0.000	0.615
AUG	2.340	0.000	2.194	0.000	0.165
SEP	5.370	0.339	2.984	0.000	1.200
OCT	0.350	0.000	0.816	0.000	2.157
NOV	0.000	0.000	0.329	0.000	1.721
DEC	0.000	0.000	0.262	0.000	1.431
TOT	48.250	7.505	28.602	13.828	2.001

ANNUAL AVERAGES

MONTH	RAIN	RUNOFF	ET	PERC	AVG SW
JAN	3.860	0.249	2.239	1.261	2.745
FEB	5.640	0.735	2.365	2.496	2.875
MAR	5.850	1.485	2.274	1.894	2.704
APR	3.265	0.556	2.082	1.014	2.726
MAY	5.745	0.639	3.249	1.329	2.756
JUN	4.420	1.163	3.074	1.162	2.415
JUL	4.410	0.314	5.369	0.000	0.758
AUG	4.060	0.096	4.317	0.000	0.567
SEP	3.610	0.170	2.592	0.000	0.828
OCT	0.355	0.000	0.685	0.000	1.126
NOV	0.580	0.000	0.556	0.000	1.001
DEC	2.460	0.105	0.964	0.217	1.458
TOT	44.255	5.511	29.765	9.372	1.830

Figure II-13.—Sample output of monthly totals of rainfall, runoff, evapotranspiration, percolation, and average soil water, and averages for the period of simulation--Continued.

APPLICATION FOR EVALUATION OF MANAGEMENT OPTIONS

In hydrology option 2, changes in the parameters for infiltration and for surface flow can reflect rather directly the changes resulting from management options. Table II-10 indicates some general effects of management on infiltration parameter RC. A denser canopy or a denser stem count (grass in comparison to a row crop) generally shows a larger RC. The fallow season RC values are estimated as 0.8 of that for the growing season RC. No-till practices with good mulch cover should make year-round values for RC relatively stable. If mulch cover is poor for no-till practices, the RC value should be estimated lower because the soil surface is expected to crust.

Table II-10.—Effect of cultural practices on model parameters

Practices	Effective slope	Effective roughness, RMN[1/]	Effective length, XLP	Conductivity, RC
Grassed concentrated flow channel.	Same	Increase	Same	Same.
Standard terraces	Decrease	Same	Increase	Same.
Chisel-tillage	Same	Large increase	Same	Increase.
No-till	Same	Increase	Same	Decrease somewhat.
Contour plow	Decrease	Same	Increase	Same.

1/ High roughness implies low value of C_c.

Cultural practices that affect routing of the runoff water on the field surface also significantly affect runoff. Contour plowing, for example, can extend considerably the effective path of overland flow, XLP. This reduces the peak outflow, which dramatically affects the amount of erosion.

In evaluating the effective flow length, XLP, the actual mean overland flow path should be measured from a map, when available. A fraction of the mean channelized flow length also should be added. The fraction used should be small for steeper, smoother channels, and large (near 1.0) for rough (grassed or vegetated) and flat sloped channels. Terraces effectively increase storage volume along the flow path, which can be simulated for calculating runoff in this model by increasing length of the flow path and decreasing the effective slope.

Management practices affect both hydrology options through changes in leaf area index. More intense management (such as high fertilization rates) increases crop production and LAI. Increasing LAI causes greater water use which reduces soil water storage. Runoff is reduced when soil water storage is lowered. The LAI values in table II-8 are for high management levels that produce large plants, and they should be reduced for less intense management. The LAI values in table II-8 generally should be multiplied by 0.83, 0.67, and 0.5 for good, fair, and poor management, respectively.

Hydrology option 1 also reflects management changes through change of the SCS curve number for AMC-II condition. Tabulated values of CN2 are given as a function of soils, land use, and management level in the SCS Hydrology Handbook (5). Recent work on the effects of residue and tillage on the SCS curve number (3) provides more refined estimates of CN2 for such modern management practices as conservation tillage and no-till systems.

REFERENCES

(1) Holtan, H. N., C. B. England, G. P. Lawless, and G. A. Schumaker.
1968. Moisture-tension data for selected soils on experimental watersheds. U.S. Department of Agriculture, Agricultural Research Service, ARS 41-144. 609 pp. (Series discontinued; Agricultural Research Service now Science and Education Administration-Agricultural Research.)

(2) Kothandaraman, V., and R. L. Evans.
1972. Use of air-water relationships for predicting water temperature. Illinois State Water Survey, Report of Investigations No. 69, Urbana, Ill. 14 pp.

(3) Rawls, W. J., C. A. Onstad, and H. H. Richardson.
1979. Residue and tillage effects on SCS runoff curve numbers. (Accepted for publication in the Transactions of American Society of Agricultural Engineers.)

(4) U.S. Department of Agriculture.
1968. Hydrologic data from experimental agricultural watersheds. Miscellaneous Publication No. 1330.

(5) ------, Soil Conservation Service.
1972. National Engineering Handbook. Section 4. Hydrology. 548 pp

(6) U.S. Department of Commerce.
1968. Climatic atlas of the United States.

(7) ------
1979. Climatological data, State of (Arizona)

(8) ------
1979. National summary, climatic data.

(9) ------
1979. Hourly rainfall data, State of (Arizona).

Chapter 2. A MODEL TO ESTIMATE SEDIMENT YIELD FROM FIELD-SIZED AREAS: SELECTION OF PARAMETER VALUES

G. R. Foster, L. J. Lane, and J. D. Nowlin[1/]

INTRODUCTION

The erosion/sediment yield component of CREAMS discussed in this chapter is for use by planners and managers who select practices to control nonpoint pollution due to sediment coming from field-sized agricultural areas. This model combines new modeling concepts with such commonly accepted relationships as the Universal Soil Loss Equation (USLE) to provide a flexible, powerful model requiring a reasonable number of inputs. The model computes erosion, sediment yield, and particle composition of the sediment on a storm-by-storm basis. Long-term effects are evaluated by simulating over a long record. Main inputs are rainfall erosivity and runoff for each storm and erosion-sediment transport characteristics of the area. Effects of spatial variability in a downslope direction can be analyzed.

The model is based on the fundamental concept that if sediment available from detachment is less than transport capacity, detachment controls sediment yield. Conversely, if sediment load exceeds transport capacity, transport capacity controls sediment yield.

MODEL STRUCTURE

The model is structured around three basic elements: overland flow; concentrated (channel) flow; and an impoundment (pond). The study area is represented by a sequence of these elements. The overland flow element is called first, followed by a channel or pond element, or both, if these additional elements are required.

IMPLEMENTATION OF THE MODEL

The model is programed in standard FORTRAN. The main program, which is a control program, calls subprograms that read data, calculate erosion and sediment yield, and display the output.

1/The authors are, respectively, hydraulic engineer, USDA-SEA, Agricultural Engineering Department, Purdue University, West Lafayette, Ind.; hydrologist, USDA-SCA, Southwest Rangeland Research Watershed Center, Tucson, Ariz.; and computer programer, Agricultural Engineering Department, Purdue University, West Lafayette, Ind.

PROGRAM FLOW

The program operates over a series of storms but takes each storm individually. The program uses two data input files. One file contains the hydrologic input data, that is, storm erosivity (EI), volume of runoff, and peak runoff rate. The second input file contains inputs that characterize the erosion and sediment transport characteristics of the area (for example, soil erodibility, hydraulic roughness, slope shape). Sediment yield is computed by the program calling the elements in the sequence defined by the user. Output is sediment load and concentration of each particle type.

Factors that change with time are updated periodically. The date for each storm is compared with the date entered for the erosion parameter values. If the date for the storm exceeds the last date that current parameter values apply, new values are read for parameters that change. If updating is unnecessary, the program proceeds with computations for the next storm by reusing values from the previous storm.

SUBPROGRAMS

Main

The main subprogram is actually the control section of the overall program. It calls subroutines for input, output, and erosion and transport computations for the elements.

Input

Subroutines read data from the input files and convert all input data to units of feet, seconds, and pounds. This reduces confusion in using common variables among different subroutines. The unit for length variables is feet anywhere in the program, for example, except for input or output where the variables are in customary units for the user's convenience.

Output

Subprograms print out detailed results, as requested by the user.

Overland Flow

The overland flow subprogram computes interrill-rill (sheet-rill) erosion and sediment transport by overland flow. A modified version of the USLE has separate terms for detachment caused by flow and detachment caused by the impact of rain drops. The relationship uses input values for storm EI, volume of runoff, and peak discharge rate for the storm, and it uses USLE factors for soil erodibility, cover-management, and contouring.

The Yalin sediment transport equation is used to compute transport capacity. Rate of deposition is assumed to be directly proportional to the difference between transport capacity and sediment load. The model uses size and density of particles to estimate selective deposition. Hydraulic roughness for the overland flow surface characterizes the effect of roughness and vegetation on transport capacity.

The subprogram calls the subroutine PROFILE, which constructs a concave, convex, or a complex slope from slopes at the beginning, midsection, and end of the hillslope profile supplied as input. The program defines three slope segments for any convex slope shape, 10 segments for any concave slope shape, and a single segment for any uniform section.

The subprogram merges coordinates for these segments, along with coordinates where soil erodibility, cover-management, contouring, and hydraulic roughness change, into a single array of coordinates. Even when a uniform slope is specified, the user can consider changes in soil erodibility and the other factors along a slope. Computations proceed downslope segment by segment. The amount of sediment produced by detachment from interrill erosion is calculated and added to that arriving from upslope segments. The sum (potential sediment load) is compared with transport capacity. If transport capacity exceeds the potential sediment load, no deposition occurs and detachment by flow occurs at a capacity rate or a rate that will just fill transport capacity. If transport capacity is less than the potential sediment load, however, deposition occurs.

Sediment composed of up to five particle types can be considered. For most soils, particles are eroded as both aggregates and primary particles. Primary particles are sand, silt, and clay, and aggregates are conglomerates of primary particles and organic matter. A high percentage of sediment for silt loam soils is aggregated. The user supplies information on particles (density and diameter), or the model will estimate a distribution from the distribution of primary particles of the soil mass.

Principal output from the overland flow component is sediment load and concentration for each particle type for the storm. These values are final if overland flow is the only element called in the sequence. Otherwise, overland flow output is input for downstream elements.

Concentrated (Channel) Flow

The channel subprogram represents detachment and sediment transport in terrace channels; waterways; and small intermittent streams. Flow concentrations also include areas through the middle of a field where overland flow concentrates due to natural topography. Flow also may concentrate along field boundaries where a ridge on the outside of the field causes overland flow to collect along the edge of the field. Grass or a ridge at the field outlet also may slow the flow, causing deposition in the backwater.

The initial section of this subprogram sets up increments along the channel equal to 0.1 of the channel's effective length, the length of the channel if it were long enough to have zero flow at its upper end with the assumed lateral inflow. Some channels begin with an initial flow rate where overland flow area is above the entrance to the channel. Additional increments are defined if changes, such as in cover, occur along the channel.

The program selects from a variety of dimensionless curves to compute friction slope. This selection is based on channel slope and outlet control. If outlet control causes backwater, one curve from a group approximates the decrease in slope of the energy gradeline. If critical flow controls at the outlet, the program selects one curve from three applicable curves. Three curves pertain to a channel having zero slope, but friction slope may be assumed to equal channel slope.

The channel is assumed to be triangular (user supplies sideslope) or rectangular (user supplies channel width). A third option is for the program to compute eroded widths for a rectangular channel.

Computations proceed downslope as in the overland flow subprogram. Exactly the same concept of detachment or transport limiting is used to route the sediment downstream.

Pond

The pond subprogram estimates deposition of sediment in impoundment terraces having controlled pipe outlets. Deposition in shallow natural impoundments caused by a ridge around a field, heavy vegetation at the outlet to the field, or a pipe culvert is analyzed with the channel element and a backwater curve. The deposition relationship for the pond element is basically an exponential decay function with parameter values related to volume of runoff, geometrical characteristics of the impounded area, discharge rate from the impounded area, infiltration rate over the pond area, and size and density density of sediment particles.

MAJOR ASSUMPTIONS

This model, like any other model, is based on many assumptions. The user should be aware of the most significant assumptions because in some applications the model is invalid.

Profile

The curved portions of a land profile are assumed to be described by a quadratic equation where the end slopes are the same as the adjoining uniform slopes. The actual field slope may not be duplicated, but the essential effects of concavity, convexity, and complexity are included.

Discharge

Discharge at any point in the watershed is assumed to be directly proportional to the drainage area above that point. Overland flow discharge at a location is computed, therefore, as a product of length of slope to that point and maximum excess rainfall rate which is attenuated for nonuniform rainfall rates and travel time. This attenuated peak discharge is used as a characteristic discharge for the runoff event.

Erosion and Transport on Overland Flow Areas

The relationship to estimate detachment is on a storm basis, while transport is estimated on an instantaneous discharge basis. Sediment concentration in the flow is assumed to be the average concentration for the storm. Concentration for detachment is determined by dividing the amount of sediment detached for the storm for a segment by total amount of runoff per unit area. The characteristic discharge multiplied by the concentration gives rate of soil loss (per unit time) at the characteristic discharge. Transport also is computed with the characteristic discharge rate so that detachment and transport will be on the same basis.

An assumption is necessary to deal with simultaneous deposition and detachment. Whether the flow is detaching or depositing, the model always assumes that interrill erosion adds sediment to the flow. On a given segment, the potential sediment load is computed by adding detachment from interrill erosion to the incoming sediment load from the next upslope segment. If this potential sediment load exceeds transport capacity on the segment, deposition occurs on the segment. If deposition occurs, no rill erosion is allowed. If transport capacity exceeds this sediment load, two other possibilities exist. The first is that rill erosion can occur at its capacity rate and still give a total sediment load less than the transport capacity. The second possibility is that if rill erosion were to occur at its capacity rate, total sediment load at the end of the segment would exceed the transport capacity. In this situation, rill erosion is limited to that which would just fill transport capacity. This concept is more realistic than assuming that rill erosion occurs at its capacity rate even when deposition occurs. To allow simultaneous rill erosion and deposition at the same time is a conceptual inconsistency for erosion and transport over cohesive agricultural soils.

The capacity of overland flow to transport sediment is estimated using the Yalin bedload sediment transport equation. If desired, the user can increase a constant in the equation to account for both bedload and suspended load. Of several sediment transport equations considered, the Yalin equation appeared to be as good or better than most for transport by overland flow, especially when small particles and particles having specific gravities less than that of sand are considered.

The Yalin equation is modified to consider particle mixtures. If the sediment load of each particle type exceeds the transport capacity of the respective particle type, sediment transport capacity is distributed among the particle types based on transportability of the particles. If sediment load of a particular type is less than the transport capacity for that type, its excess transport capacity is shifted to particles having a deficit transport capacity. This modification prevents a small load of a particular particle type from having more than its share of the total transport capacity. When deposition occurs, rate of deposition is assumed to be directly proportional to the difference between transport capacity and sediment load. The proportionality constant is assumed to be directly proportional to fall velocity of the particle divided by the product of flow velocity and flow depth. This gives an exponential decay for rate of deposition as a function of distance.

Any deposited particles are assumed to become reattached immediately to the soil mass, that is, deposited particles are unavailable as detached particles for subsequent transport without being redetached. Likewise, tillage is not assumed to produce a supply of detached particles that is depleted over time by transport. Increased erosion from tillage is analyzed by adjusting the USLE soil loss ratio.

Erosion and Transport in Channel Flow

Input to the channel is a uniform lateral inflow of runoff and sediment from an overland flow or another channel element. The characteristic discharge is used to compute detachment, sediment transport, deposition, and sediment concentration in the channel elements.

Outlet conditions for the channel are assumed to be controlled by a downstream uniform flow, critical depth, or a structure having a known rating curve (for example, a flume restriction in an experimental watershed or a field boundary). Subcritical flow is assumed unless the option is specified that slope of the energy gradeline (friction slope) equals the channel slope.

Since many channels in farm fields may be approximated as triangular channels, a triangular channel with 5:1 sideslopes was used to develop the friction slope curves. Therefore, the actual channel must be approximated by a triangular channel to compute the friction slope. Remaining channel computations are made assuming a triangular, rectangular, or eroding channel section. The triangular and rectangular channel sections may have cover, but the eroding channel section is assumed to be bare with no cover. Enlargement occurs in the eroded channel and section properties remain fixed in the triangular channel. Width of the rectangular channel increases once the computed eroded width exceeds the initial width read as input.

Concepts for detachment and transport in the channel are exactly the same as those for overland flow. Lateral inflow of sediment in the channels is equivalent to interrill erosion, and channel erosion is equivalent to rill erosion. Relationships for the detachment capacity of channel erosion are computed using expressions developed from an experimental and analytical rill erosion study by Lane and Foster (vol. III, ch. 11). The algorithm considers the influence of a nonerodible boundary at some depth below the bottom of the channel. When a channel erodes to the nonerodible boundary, the channel widens and erosion rate decreases with time. This frequently occurs in many midwestern fields at planting time in areas where grass waterways should be installed.

The effect of tillage on erosion by channel flow is modeled by assuming that tillage greatly decreases the critical shear stress for detachment to begin. Tillage is assumed to loosen, (that is, decrease the critical shear stress) down to a given depth. No erosion is allowed below this depth. Critical shear stress increases after tillage as the soil consolidates from traffic, wetting and drying, and other processes. Values for critical shear stress and soil erodibility by channel flow have not been validated substantially.

Channel erosion does not occur throughout the duration of a storm. Detachment occurs only when shear stress exceeds critical shear stress. This time is estimated by assuming that the shear stress is linearly distributed in time. Detachment is assumed to occur at a rate based on the characteristic discharge for the period that shear stress exceeds the critical shear stress.

Transport Through Impoundments

These relationships for the impoundment component were derived from a detailed model (3) based on settling theory in still water. Output from the detailed model fit observed experimental data well. Regression analyses were used to fit the relationships used in this model to output from the detailed model.

This component applies to impoundment terraces that drain completely after a runoff event and to other small impoundments where discharge is controlled by an orifice restriction in an outlet pipe. Although several other impoundments occur behind ridges around fields, pipe culverts, and farm ponds, the impoundment component generally should not be applied to these situations. The channel element with backwater is recommended for impoundments by ridges and culverts. Relationships for standard reservoir deposition may be applied to farm ponds, using the model output to estimate runoff reaching the pond.

MODEL INPUTS AND PARAMETERS

The model inputs are the hydrologic variables rainfall storm erosivity (EI), volume of runoff, and characteristic peak excess rainfall rate (peak runoff rate at the outlet divided by area). These generally are obtained from the hydrology component of CREAMS or from observed data. Table II-11 shows the hydrology pass file variables, format, and sample data for the input to the erosion and sediment yield program. Figure II-14 a represents card image format for the pass file from the hydrology model. The model parameters characterize the erosion-sediment transport-deposition features of the area. These values come from a variety of sources. Tables II-12 and II-13 identify inputs and parameters, possible sources, and indications of the quality of a parameter estimate.

PREPARATION OF INPUT DATA FILES

This section shows how to assemble input data files by briefly describing the parameters and their location in the data set. For a more comprehensive definition of the parameters and a method of selecting values, refer to the following section of this publication.

The model reads input from two separate files: a hydrology file and a parameter file. Unless specifically stated otherwise in the card description, all numeric input formats consist of 8-column fields. Integers are read with I8 formats, and real numbers are read with F8.0 formats. Integers must be right justified in columns 1-8, 9-16, 17-24, ..., 73-80. Real numbers must be contained within these same columns, and the decimal point must be entered in the number. A sample value is given after each parameter is defined. If the sample has a decimal, the parameter is real; otherwise, it is an integer. The alphanumeric input is read with A4 formats. Specific instructions are given whenever alphanumeric input is required. A schematic representation of the parameter data deck is shown in figure II-15. Figure II-16 is a schematic data deck with sample data for 3 years on watershed P-2 at Watkinsville, Ga.

Blank cards or blank entries are used on some data cards to indicate a zero entry. If your computer does not read blanks as zeroes, enter zeroes instead of leaving those parameters blank. Some computers can be set using control statements to read blanks as zeroes.

If some updateable parameter does not change when others change, the last value read in for the parameter will be used by the program, if desired. When this option is used, data cards for these parameters are omitted. This is discussed in greater detail in the sections on updateable parameters.

Table II-11.—Hydrology pass file description and data for input to the erosion/sediment yield model

A. Storm/Hydrology Data File

Card 1. SDATE, RNFALL, RUNOFF, EXRAIN, EI, DP, PERCOL, AVGTMP, AVGSWC, ACCPEV, POTPEV, ACCSEV, POTSEV

SDATE	Date of storm (Julian date), e.g. 73146
RNFALL	Volume rainfall (in), e.g. 4.27
RUNOFF	Volume of runoff (in), e.g. 1.58
EXRAIN	Characteristic excess rainfall rate (in/hr), e.g. 4.13
EI	Wischmeier English EI for the given storm, e.g. 67.41
DP	Number of days since the last storm when percolation occured, e.g. 1
PERCOL	Percolation below the root zone (in), e.g. 1.015
AVGTMP	Average temperature between storms (Degrees F.), e.g. 72.8
AVGSWC	Average soil water between storms (in/in), e.g. 0.3239
ACCPEV	Actual EP (evaporation from plants) for the period between storms (in), e.g. 0.022|0.056
POTPEV	Potential EP for the period between storms (in), e.g. 0.022|0.056
ACCSEV	Actual ES (evaporation from soil) for the period between storms (in), e.g. 0.000|0.000
POTSEV	Potential ES for the period between storms (in), e.g. 0.000|0.000

Card 1 is repeated for each rainfall event. The last card in the file should be blank to indicate the end of data. The Hydrology program creates a file called "HYDPAS" specifically for use as this file. The values in the Storm/Hydrology file from DP to POTSEV are only read into the Erosion program so they can be passed through to the Chemicals program. If a file for the Chemicals program isn't going to be created then the only values required for the Storm/Hydrology file are SDATE, RNFALL, RUNOFF, EXRAIN, and EI.

Table II-11.—Hydrology pass file description and data for input to the erosion/sediment yield model--continued

A small sample of a typical Storm/Hydrology Data file follows to illustrate the file structure.

Format(I6,F6.2,F6.2,F6.2,F6.2,I2,F6.2,F6.2,F6.4,F6.3,F6.3,F6.3,F6.3)

73148	4.2700	1.8686	3.7387	51.9727	1	.770	76.1221	.3378	1.3266	2.9891
73156	.2800	0	0	.7860	1	.002	77.1801	.3577	1.9393	5.2168
73157	1.2200	.1583	.4580	7.5645	1	.585	77.8851	.3672	2.0836	5.4975
73159	.6000	.0128	.0540	2.5388	2	.302	78.0943	.3653	2.3687	6.0597

Table II-12. Model input

Variable	Source
Runoff Volume (V)	Estimated by a model.
Characteristic runoff rate (σ_p)	Estimated by a model.
Storm erosivity (EI)	Estimated from volume of rainfall and maximum 30-min intensity or volume of rainfall alone.

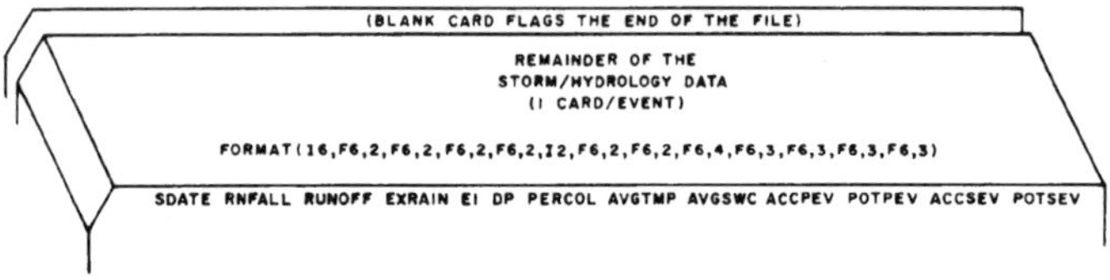

Figure II-14.—Sample format and card image arrangement for hydrology pass file, from either hydrology model option.

Some cards, such as card 13, contain unnecessary information. If a rating curve is specified, for example, the first four parameters on card 13 are unnecessary. These may be left blank, assigned zero, or assigned an obviously incorrect value, such as 999. This latter entry could help trace input errors.

Since all data files must be input in English units, values on the sample card decks are shown with English units. The model is written for variables with English units.

SELECTION OF INPUT VALUES

Input values generally can be selected from readily available information from such sources as a soil survey, topographic maps, aerial photographs, soil description, cropping history, and a site visit. Note the following input requirements and assemble the required source materials.

Obviously, the model is more sensitive to some parameters, as discussed in volume I, chapter 6. The sensitive variables require more careful selection. If sediment yield is primarilly controlled by detachment, overall the detachment parameters are more important whereas transport parameters are more important when deposition primarily controls sediment yeild. However, for specific locations, detachment may limit sediment yield for some storms, while transport will limit for other storms. Detachment may limit on one part of the watershed while transport will limit on another part. Detachment may control sediment yield for the fines while at the same time transport will control the yield of coarse particles. The result is mixed control between detachment and transport parameters, preventing general statements on the sensitivity of particular par-

Table II-13.—Erosion model parameters, definitions, and sources and quality of estimates

Parameter	Definition	Source of estimate	Quality of Estimate
ν	Kinematic viscosity	Handbook	Excellent. However, only parameter expressing temperature effect. Quality for expressing that effect unknown.
n_{bov}	Manning's n for overland flow over bare smooth soil (fine seedbed).	Model manual	Good but subjective.
n_{bch}	Manning's n for channel flow over bare, smooth soil (fine seedbed).	Model manual	Good but subjective.
ρ_{soil}	Weight density of soil mass.	Soil survey and experience.	Good.
K_{rch}	Soil erodibility factor for channel erosion.	Model manual	Poor. May require calibration.
C_{yal}	Constant in Yalin sediment transport equation.	Model manual	Good. Supposedly fixed, but may require calibration.
Sand, silt, clay.	Primary particle distribution of original soil mass.	Soil survey, soil tests experience.	Very good.
Particle characteristics.	Particle size class and density of particle.	Model manual and soil survey information or calculated from model equations using primary clay, silt and sand.	Good for most midwestern silt loam soils; unknown for most other soils.
λ_{ov}	Overland flow slope length.	Maps, soil survey, field observation.	Good, but problem of choosing representative length.
$\bar{S}$	Average overland flow slope steepness.	Maps, soil survey, field observation.	Good, but problem of choosing representative length.
S_b	Slope at beginning of overland flow profile.	Maps, soil survey field observation.	Good, but problem of choosing representative steepness.
S_m	Slope at middle of overland flow profile.	Maps, soil survey, field observation.	Good, but problem of choosing representative steepness.
S_e	Slope at end of overland flow profile.	Maps, soil survey, field observation.	Good, but problem of choosing representative steepness.
x_3, y_3 x_4, y_4	Coordinates of mid-uniform slope section.	Maps, soil survey, field observation.	Good, but problem of choosing representative section.
A_{ov}	Overland flow area	Map	Very good.
K	Soil erodibility factor (rill-interrill erosion).	Model manual; also USLE Handbook.	Good, based on extensive plot data.
C	Cover-management factor (rill-interrill erosion).	Model manual; also USLE Handbook.	Good, based on extensive plot data.
P	Contouring factor (rill-interrill erosion).	Model manual; also USLE Handbook.	Poor; value poorly defined for individual storms.
n_{cov}	Manning's n for overland flow over a covered soil surface.	Model manual	Good, but subjective.

Table II-13.—Erosion model parameters, definitions, and sources and quality of estimates--continued

Parameter	Definition	Source of estimate	Quality of Estimate
Shape	Channel shape	Experience and field observation.	Good, but subjective.
λ_{ch}	Channel length	Map, field observation.	Good, but can be quite subjective.
A_{chup}	Drainage area draining into upper end of channel.	Map	Very good.
A_{chlo}	Area drained by channel	Map	Very good.
Outlet control.	Outlet control parameters including channel width, sideslope, longitudinal slope, Manning's n, rating curve.	Experience, field observation, model manual.	Poor and highly subjective.
Slope	Slope along channel	Map, field observation.	Very good.
n_{ch}	Manning's n for channel with cover.	Model manual, handbooks provided; n_{bch} selected from same handbook.	Good, but subjective.
τ_{cr}	Critical shear stress which erosion begins in channel.	Model manual, experience.	Poor, values not known for many agricultural soils and management effects not known.
τ_{cov}	Critical shear stress for cover breakup.	Model manual, experience.	Fair for nonincorporated residue, poor for incorporated.
d_{ne}	Depth to nonerodible layer in channel.	Model manual, experience, field observation.	Fair, but subjective.
d_{side}	Depth to nonerodible layer at side of channel.	Model manual, experience, field observation.	Poor and highly subjective.
w_{ch}	Channel width	Model manual, field observation photo.	Fair.
Z	Channel sideslope	Model manual, field observation map.	Fair to good.
F_s, B	Coefficients for pond surface area vs depth intake rate.	Field survey model manual, map	Excellent with field survey, good with other means of estimating.
i	Intake rate	Soil survey, experience.	Good.
d_{or}	Diameter of orifice in outlet pipe.	Design notes, field observation experience.	Excellent or good if based on experience.
A_{pd}	Drainage area above pond.	Map	Excellent.

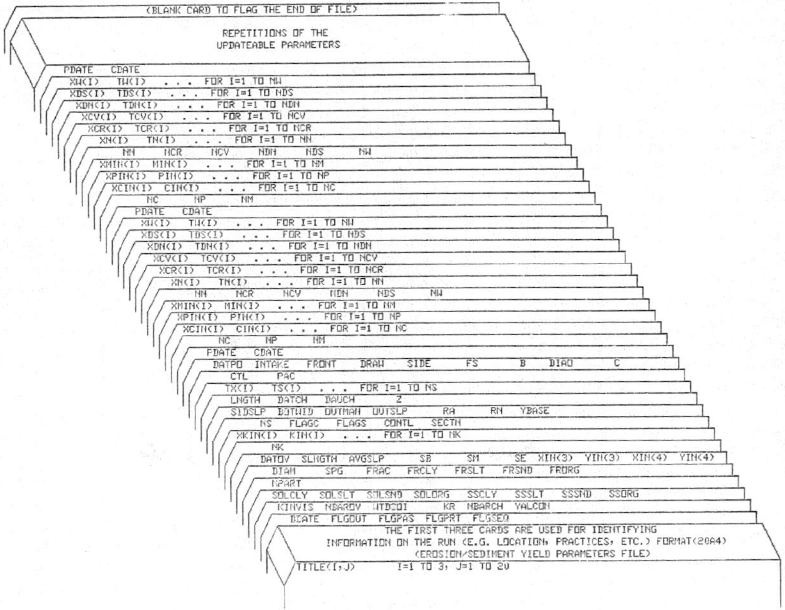

Figure II-15.--Sample format and input parameter card deck arrangement for the erosion model.

NOTE: Shaded card image represents a required blank

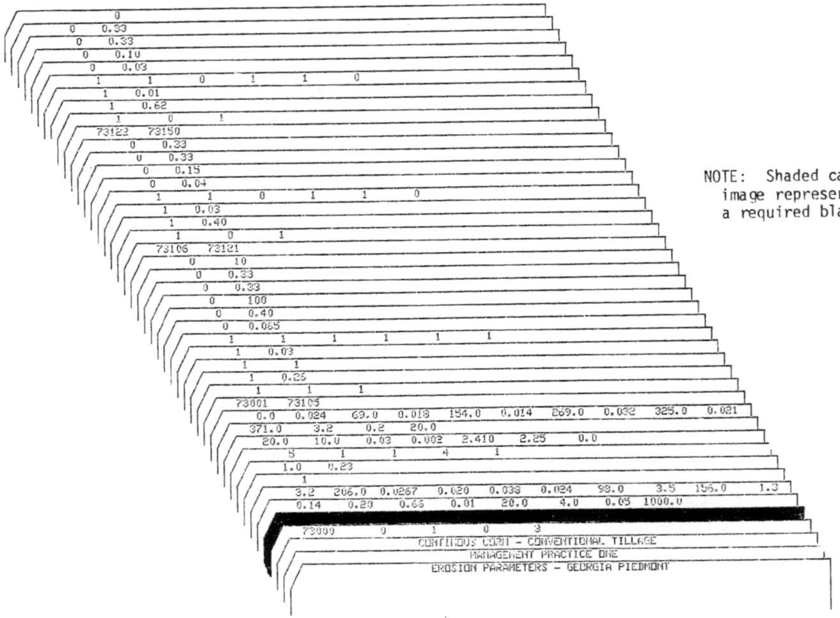

Figure II-6.--Sample deck arrangement of input data for the erosion model on watershed P2, Watkinsville, Ga.

ameters. Generally, a sensitivity analysis is advisable for each specific problem.

Discussion of selection of input values generally parallels the layout of the input data cards. Some input variables are not discussed because selection of a value is obvious.

Storm Hydrology Input

Hydrology Pass File

SDATE--The Julian calendar in table II-14 may be used to convert an ordinary calendar date. This date is given by specifying the last two digits of the year (for example, 78) followed by the Julian date (for example, 094 for April 4, or 78094).

RNFALL--Rainfall volumes for a series of storms are available from rainfall records of the National Weather Service and other agencies that collect weather data in your locale. Breakpoint data are most desirable, although hourly or daily data may be used.

RUNOFF--This is runoff volume per unit watershed area for the storm. Runoff is assumed to be uniform over the drainage area.

EXRAIN--The characteristic peak excess rainfall rate for the storm is obtained by dividing peak discharge at the watershed outlet by watershed area. If it is computed by subtracting infiltration rate from rainfall rate, it must be attenuated to account for nonuniform rainfall rates and time of travel. In the erosion/sediment yield component of CREAMS, characteristic peak runoff rate at any point in the watershed is taken as directly proportional to the drainage area above that point.

EI--The EI variable, as defined by Wischmeier and Smith (12), is a measure of rainfall erosivity. If a rainfall hyetograph of the given storm is available, EI for the storm may be estimated from the following procedure. Divide the rainfall hyetograph into periods so that rainfall intensity may be assumed to be constant for a period. For each period, calculate the unit rainfall energy per unit of rainfall from

$$e = 916 + 331 \log_{10} i \qquad [II-5]$$

where e = unit rainfall energy (ft-tons/acre-in of rain) and i = rainfall intensity (in/hr). Multiply the unit energy by the rainfall amount in the period to obtain the energy for that period. Add these incremental energies for all periods to obtain the total rainfall energy for the storm. The storm energy multiplied by the storm's maximum 30-min intensity divided by 100 gives EI in Wischmeier's English EI units.

Where the rainfall hyetograph is unavailable, total storm energy, E, may be estimated by computing e (unit rainfall energy) using the maximum 30-min storm intensity and multiplying by volume of rainfall. This value multiplied by maximum 30-min intensity is an estimate of EI. If the maximum 30-min intensity is not known but the maximum 60-min intensity is available, multiply the maximum 60-min intensity by 1.6 to estimate the maximum 30-min intensity.

These detailed data may be unavailable for a given storm. The best available information may be hourly or daily rainfall amounts, which are, by themselves, poor estimates of rainfall erosivity (6). A good estimate of intensity is required for good erosion estimates. Given only rainfall amount, however, storm EI may be estimated from:

$$EI = 8.00\ V_r^{1.51} \qquad [II\text{-}6]$$

where V_r = volume of rainfall (in). Since the coefficient of determination (R^2) for this equation is 0.54, EI values from the equation are subject to considerable error for any given specific storm.

DP, PERCOL, AVGTMP, AVGSWC, ACCPEV, POTPEV, ACCSEV, and POTSEV--These fields are not used by the erosion/sediment yield component. They may be left blank unless the erosion program is used to construct an input data file for the chemical program.

Table II-14.—Julian Calendar[1/]

			* * * * * January * * * * *			
1/ 1	2/ 2	3/ 3	4/ 4	5/ 5	6/ 6	7/ 7
8/ 8	9/ 9	10/10	11/11	12/12	13/13	14/14
15/15	16/16	17/17	18/18	19/19	20/20	21/21
22/22	23/23	24/24	25/25	26/26	27/27	28/28
29/29	30/30	31/31				
			* * * * * February * * * * *			
32/ 1	33/ 2	34/ 3	35/ 4	36/ 5	37/ 6	38/ 7
39/ 8	40/ 9	41/10	42/11	43/12	44/13	45/14
46/15	47/16	48/17	49/18	50/19	51/20	52/21
53/22	54/23	55/24	56/25	57/26	58/27	59/28
For leap years, add 1 day to Julian date following February 28.						
			* * * * * March * * * * *			
60/ 1	61/ 2	62/ 3	63/ 4	64/ 5	65/ 6	66/ 7
67/ 8	68/ 9	69/10	70/11	71/12	72/13	73/14
74/15	75/16	76/17	77/18	78/19	79/20	80/21
81/22	82/23	83/24	84/25	85/26	86/27	87/28
88/29	89/30	90/31				
			* * * * * April * * * * *			
91/ 1	92/ 2	93/ 3	94/ 4	95/ 5	96/ 6	97/ 7
98/ 8	99/ 9	100/10	101/11	102/12	103/13	104/14
105/15	106/16	107/17	108/18	109/19	110/20	111/21
112/22	113/23	114/24	115/25	116/26	117/27	118/28
119/29	120/30					
			* * * * * May * * * * *			
121/ 1	122/ 2	123/ 3	124/ 4	125/ 5	126/ 6	127/ 7
128/ 8	129/ 9	130/10	131/11	132/12	133/13	134/14
135/15	136/16	137/17	138/18	139/19	140/20	141/21
142/22	143/23	144/24	145/25	146/26	147/27	148/28
149/29	150/30	151/31				

Table II-14.--Julian calendar--continued

		* * * * * June * * * * *				
152/ 1	153/ 2	154/ 3	155/ 4	156/ 5	157/ 6	158/ 7
159/ 8	160/ 9	161/10	162/11	163/12	164/13	165/14
166/15	167/16	168/17	169/18	170/19	171/20	172/21
173/22	174/23	175/24	176/25	177/26	178/27	179/28
180/29	181/30					
		* * * * * July * * * * *				
182/ 1	183/ 2	184/ 3	185/ 4	186/ 5	187/ 6	188/ 7
189/ 8	190/ 9	191/10	192/11	193/12	194/13	195/14
196/15	197/16	198/17	199/18	200/19	201/20	202/21
203/22	204/23	205/24	206/25	207/26	208/27	209/28
210/29	211/30	212/31				
		* * * * * August * * * * *				
213/ 1	214/ 2	215/ 3	216/ 4	217/ 5	218/ 6	219/ 7
220/ 8	221/ 9	222/10	223/11	224/12	225/13	226/14
227/15	228/16	229/17	230/18	231/19	232/20	233/21
234/22	235/23	236/24	237/25	238/26	239/27	240/28
241/29	242/30	243/31				
		* * * * * September * * * * *				
244/ 1	245/ 2	246/ 3	247/ 4	248/ 5	249/ 6	250/ 7
251/ 8	252/ 9	253/10	254/11	255/12	256/13	257/14
258/15	259/16	260/17	261/18	262/19	263/20	264/21
265/22	266/23	267/24	268/25	269/26	270/27	271/28
272/29	273/30					
		* * * * * October * * * * *				
274/ 1	275/ 2	276/ 3	277/ 4	278/ 5	279/ 6	280/ 7
281/ 8	282/ 9	283/10	284/11	285/12	286/13	287/14
288/15	289/16	290/17	291/18	292/19	293/20	294/21
295/22	296/23	297/24	298/25	299/26	300/27	301/28
302/29	303/30	304/31				
		* * * * * November * * * * *				
305/ 1	306/ 2	307/ 3	308/ 4	309/ 5	310/ 6	311/ 7
312/ 8	313/ 9	314/10	315/11	316/12	317/13	318/14
319/15	320/16	321/17	322/18	323/19	324/20	325/21
326/22	327/23	328/24	329/25	330/26	331/27	332/28
333/29	334/30					
		* * * * * December * * * * *				
335/ 1	336/ 2	337/ 3	338/ 4	339/ 5	340/ 6	341/ 7
342/ 8	343/ 9	344/10	345/11	346/12	347/13	348/14
349/15	350/16	351/17	352/18	353/19	354/20	355/21
356/22	357/23	358/24	359/25	360/26	361/27	362/28
363/29	364/30	365/31				

[1/] Date to right of slash (/) is day of month.

Initial Inputs

The following description of parameter inputs is given in the same order as that of the input cards shown in table II-15

Table II-15.—Parameter file for erosion/sediment yield component

-----	Initial General Parameter Inputs	
Card 1-3.	TITLE()	
	TITLE	Three lines of 80 Characters each for alphanumeric information to be printed at the beginning of the output. format (20A4)
Card 4.	BDATE, FLGOUT, FLGPAS, FLGPRT, FLGSEQ	
	BDATE	The beginning date for simulation. It must be less than the first storm date (SDATE). (Julian date), e.g. 73000
	FLGOUT	0 for annual summary output 1 for monthly and annual summary output 2 for storm by storm and both types of summary output 3 for a single storm and detailed output by segments
	FLGPAS	0 if no file should be created for the Chemicals program 1 if the program should create a file for use by the Chemicals program
	FLGPRT	0 for the particle specifications to be computed with default values 1 for the particle specifications to be read in
	FLGSEQ	Execution sequence of erosion submodels: 1 - overland 2 - overland-pond 3 - overland-channel 4 - overland-channel-channel 5 - overland-channel-pond 6 - overland-channel-channel-pond

FLGSEQ is used to decide whether certain groups of cards should be read in. Cards 9-11 are always read, and only once. Cards 12-15 are only read when FLGSEQ is greater than or equal to 3, and they are repeated for a second channel if FLGSEQ is 4 or 6. Cards 16 and 17 are read if FLGSEQ is 2,5, or 6, and they are never read more than once.

Card 5 KINVIS, NBAROV, WTDSOI, KR, NBARCH, YALCON

If a default value is to be used, leave that position on the card blank. Otherwise enter the desired value. If all defaults are assumed, insert a blank card.

Table II-15.—Parameter file for erosion/sediment yield component--continued

	KINVIS	Kinematic viscosity (ft^2/sec), e.g. default 1.21E-05
	NBAROV	Manning's n for overland flow over bare soil, e.g. default 0.01
	WTDSOI	Weight density of soil (lbs/ft^3), e.g. default 96.0
	KR	Soil erodibility for erosion by concentrated flow ((lbs/ft^2 sec)(1/lbs/ft^2)$^{1.05}$) e.g. default 0.135
	NBARCH	Manning's n for channel flow over bare soil e.g. default 0.03
	YALCON	Yalin constant for sediment transport, e.g. default 0.635
Card 6.	SOLCLY, SOLSLT, SOLSND, SOLORG, SSCLY, SSSLT, SSSND, SSORG	
	SOLCLY	Fraction of clay in the original surface soil layer exposed to erosion, e.g. 0.14
	SOLSLT	Fraction of silt in the original surface soil layer exposed to erosion, e.g. 0.20
	SOLSND	Fraction of sand in the original surface soil layer exposed to erosion, e.g. 0.66
	SOLORG	Fraction of organic matter in the original surface soil layer exposed to erosion, e.g. 0.01
	SSCLY	Specific surface area of clay particles ($meters^2$/gram of soil), e.g. 20.0
	SSSLT	Specific surface area of silt particles ($meters^2$/gram of soil), e.g. 4.0
	SSSND	Specific surface area of sand particles ($meters^2$/gram of soil), e.g. 0.05
	SSORG	Specific surface area of organic matter particles ($meters^2$/gram of organic carbon), e.g. 1000.0 (organic carbon = organic matter/1.73)

The fractions of clay, silt, and sand should total 1.0, with the organic matter being a fraction, of the total of organic matter and soil particles.

If the specific surface area values are left blank the model defaults to 20.0, 4.0, 0.05, and 1000.0 for clay, silt, sand, and organic matter respectively.

If the particle specifications flag (FLGPRT, card 4) is 0 then no card 7 or card 8's will be read, and the number of particle types (NPART, card 7) will be calculated.

Card 7. NPART

NPART	The number of particle types, e.g. 5

Card 8. DIAM, SPG, FRAC, FRCLY, FRSLT, FRSND, FRORG

[Repeat card 8 for each particle (NPART, card 7)]

DIAM	Particle diameter (mm), e.g. 0.030
SPG	Specific gravity of particle (g/cm^3) , e.g. 1.8
FRAC	Fraction of sediment detached that is made up of this particular particle type, e.g. 0.50
FRCLY	Fraction of particle made up of clay, e.g. 0.3
FRSLT	Fraction of particle made up of silt, e.g. 0.5
FRSND	Fraction of particle made up of sand, e.g. 0.2
FRORG	Fraction of particle made up of organic matter, e.g. 0.02

The sum of the fractions for clay, silt, and sand should equal 1.0, with the organic matter being a fraction of the total organic matter and soil particles.

----- Initial Overland Flow Inputs

Card 9. DATOV, SLNGTH, AVGSLP, SB, SM, SE, XIN(3), YIN(3), XIN(4), YIN(4)

DATOV	Area represented by overland flow profile (acres), e.g. 3.2
SLNGTH	Slope length of representative overland flow profile (ft), e.g. 206.0
AVGSLP	Average slope of representative overland flow profile (ft/ft), e.g. 0.027
SB	Slope at the upper end of profile, e.g. 0.020
SM	Slope of mid-section, e.g. 0.0380
SE	Slope at the lower end of profile, e.g. 0.024

	XIN(3)	Distance from top of slope where mid-uniform section begins (ft), e.g. 98.0
	YIN(3)	Elevation above lowest point where mid-uniform section begins (ft), e.g. 3.5
	XIN(4)	Distance from top of slope where mid-uniform section ends (ft), 156.0
	YIN(4)	Elevation above lowest point where mid-uniform section ends (ft), e.g. 0.0

When simulating a uniform slope SB = SM = SE = AVGSLP; XIN(3) = XIN(4) = SLNGTH; YIN(3) = YIN(4) = 0.0

Card 10.	NK	
	NK	Number of slope segments differentiated by changes in soil erodibility factor, e.g. 1
Card 11.	XKIN(I), KIN(I), . . . for I=1 to NK (card 10)	
	XKIN(I)	Relative horizontal distance from the top of the slope to the bottom of segment I, e.g. 1.0
	KIN(I)	Soil erodibility factor for slope segment just above XKIN(I) (tons/acre/English EI) e.g. 0.23

The order of the following cards depends on the execution sequence (FLGSEQ, card 4). In some cases the following cards (12-17) won't be used, e.g. FLGSEQ = 1, or there may be two sets of channel inputs (12-15) and a pond (16,17), e.g. FLGSEQ = 6.

----- Initial Channel Inputs

Card 12.	NS, FLAGC, FLAGS, CONTL, SECTN	
	NS	Number of channel segments differentiated by changes in slope, e.g. 5
	FLAGC	Flag that indicates channel shape: 1 - Triangular channel 2 - Rectangular channel 3 - Naturally eroded channel
	FLAGS	1 for program to use curves for slopes of energy grade-line. (friction slope) 2 for program to assume friction slope equals channel slope.

	CONTL	1 if critical depth controls depth in outlet channel 2 if uniform flow controls in the outlet channel 3 if the program should use the maximum of 1 and 2 4 if the program should use a rating curve for control depth at outlet. $Q = RA\ (Y - YBASE)^{RN}$ $Q(ft^3/sec)$, Y and YBASE (ft),
	SECTN	1 if the shape of the outlet channel is triangular 2 if the shape is rectangular
Card 13	SIDSLP, BOTWID, OUTMAN, OUTSLP, RA, RN, YBASE	
	SIDSLP	Side slope of a cross-section of the outlet control channel, expressed as horizontal to vertical, e.g. 20.0
	BOTWID	Bottom width of the outlet control channel (ft), e.g. 10.0
	OUTMAN	Manning's N for the outlet control channel, e.g. 0.030
	OUTSLP	Slope of the outlet control channel, e.g. 0.002
	RA	Coefficient in the rating curve equation e.g. 2.41
	RN	Exponent in the rating curve equation, e.g. 2.25
	YBASE	Minimum depth for flow to begin (ft), e.g. 0.0
Card 14.	LNGTH, DATCH, DAUCH, Z	
	LNGTH	Channel length (ft), e.g. 371.0
	DATCH	Total drainage area of channel at lower end of channel (acres), e.g. 3.2
	DAUCH	Drainage area above upper end of channel (acres), e.g. 0.2
	Z	Sideslope of channel cross-section, expressed as horizontal to vertical, e.g. 20.0
	If the channel shape flag (FLAGC, card 12) is a 2 or 3, enter the value for Z that most closely approximates the channel shape.	
Card 15	TX(I), TS(I), . . . for I=1 to NS (card 12)	
	TX(I)	Distance from lower end of the channel to the bottom of segment I (ft), e.g. 0.0
	TS(I)	Slope of segment directly above TX(I), e.g. 0.024

----- Initial Pond Inputs

Card 16. CTL, PAC

CTL 1 for pipe outlet control as typical of impoundment type terraces
3 when the orifice coefficient (C, card 17) is read in

PAC 1 for program to calculate coefficients for pond surface area-depth relationship from user supplied parameters for impoundment basin slopes.
2 for user supplied coefficients SA = FC(Y^B), Where SA = Surface area (FT^2), Y = depth (ft)

Card 17 DATPO, INTAKE, FRONT, DRAW, SIDE, FS, B, DIAO, C

DATPO Total drainage area above the pond (acres), e.g. 3.2

INTAKE Soil water intake rate within the pond, in/hr, e.g. 0.2

FRONT Embankment front slope, e.g. 0.2

DRAW Slope along channel draining into pond, e.g. 0.024

SIDE Slope of land at pond toward draw, e.g. 0.01

FS depth relationship, e.g. 9500.0

B depth relationship, e.g. 1.73

DIAO diameter of pipe orfice (in), e.g. 3.0

C Orifice coeficient, e.g. 3000.0

----- Updateable General Parameter Inputs

The remaining inputs to the Erosion program are updateable. The program checks the dates (SDATE, card 1) from the hydrology file against the parameters control date (CDATE, card 18). If the control date is less than the date of the storm, the program reads in a new set of the updateable parameters. If the program reads a blank in place of the control date (CDATE, card 18) the program stops executing. The execution sequence flag (FLGSEQ, card 4) is used to determine whether or not cards in this section are read as in the Initial Inputs section. There are no updateable Pond parameters. The Overland flow parameters are on cards 19-22, and the Channel parameters are on cards 23-29.

Card 18. PDATE, CDATE

PDATE First date that the following erosion parameters are valid (Julian), e.g. 73138

The program doesn't read in the value for PDATE. PDATE is only used as an aid in putting together the data file.

CDATE — Last date that the following erosion parameters are valid (Julian), e.g. 73105

NOTE: A card 18. should always be the first card in a set of updateable parameters.

----- Updateable Overland Flow Inputs

Card 19. NC, NP, NM

NC — Number of slope segments differentiated by changes in cropping management factor, e.g. 1

NP — Number of slope segments differentiated by changes in contouring factor, e.g. 1

NM — Number of slope segments differentiated by changes in Manning's N, e.g. 1

On the initial pass through the program, each of the "N"'s should be at least 1 in order to read initial values for the parameters. In subsequent passes, a blank "N" indicates no change in the corresponding parameter from the previous update. To skip reading a parameter, for example Manning's n, read in a blank NM. If no new overland flow parameters are to be read, card 19 should be left blank. Input cards for a parameter should not be included in the data file when it's "N" is left blank.

Card 20 XCIN(I), CIN(I), . . . for I=1 to NC (card 19)

XCIN(I) — Relative horizontal distance from top of slope to the bottom of segment I , e.g. 1.0

CIN(I) — Cropping management factor for slope segment just above XCIN(I), e.g. 0.26

Card 21 XPIN(I), PIN(I), . . . for I=1 to NP (card 19)

XPIN(I) — Relative horizontal distance from top of slope to the bottom of segment I , e.g. 1.0

PIN(I) — Contouring factor for slope segment just above XPIN(I), e.g. 1.0

Card 22 XMIN(I), MIN(I), . . . for I=1 to NM (card 19)

XMIN(I) — Relative horizontal distance from top of slope to the bottom of segment I , e.g. 1.0

MIN(I) Manning's N value for slope segment just above XMIN(I), e.g. 0.03

----- Updateable Channel Inputs

Card 23. NN, NCR, NCV, NDN, NDS, NW

NN Number of channel segments differentiated by changes in Manning's N, e.g. 1

NCR Number of channel segments differentiated by changes in critical shear stress, e.g. 1

NCV Number of channel segments differentiated by changes in shear stress for cover, e.g. 1

NDN Number of channel segments differentiated by changes in depth from channel middle to the nonerodible layer, e.g. 1

NDS Number of channel segments differentiated by changes in depth from the channel side to the nonerodible layer, e.g. 1

NW Number of channel segments differentiated by changes in width, e.g. 1

On the initial pass through the program, each of the "N"'s should be at least 1 in order to read initial values for the parameters. In subsequent passes, a blank "N" indicates no change in the corresponding parameter from the previous update. To skip reading a parameter, for example channel width, read in a blank NW. If no new channel parameters are to be read, card 23 should be left blank. Input cards for a parameter should not be included in the data file when it's "N" is left blank.

Card 24. XN(I), TN(I), . . . for I=1 to NN (card 23)

XN(I) Distance from the lower end of the channel to the bottom of segment I (ft). e.g. 0.0

TN(I) Manning's n of channel directly above XN(I), e.g. 0.065

Card 25. XCR(I), TCR(I), . . . for I=1 to NCR (card 23)

XCR(I) Distance from the lower end of the channel to the bottom of segment I (ft). e.g. 0.0

TCR(I) Critical shear stress of channel directly above XCR(I), (lbs/ft^2), e.g. 0.40

Table II-15.—Parameter file for erosion/sediment yield component--continued

Card 26.	XCV(I), TCV(I), . . . for I=1 to NCV (card 23)	
	XCV(I)	Distance from the lower end of the channel to the bottom of segment I (ft). e.g. 0.0
	TCV(I)	Shear stress for cover stability for channel directly above XCV(I), (lbs/ft^2), e.g. 100.0
Card 27.	XDN(I), TDN(I), . . . for I=1 to NDN (card 23)	
	XDN(I)	Distance from the lower end of the channel to the bottom of segment I (ft), e.g. 0.0
	TDN(I)	Depth to the nonerodible layer in the middle of channel directly above XDN(I) (ft), e.g. 0.33
Card 28.	XDS(I), TDS(I), . . . for I=1 to NDS (card 23)	
	XDS(I)	Distance from the lower end of the channel to the bottom of segment I (ft), e.g. 0.0
	TDS(I)	Depth to the nonerodible layer along the side of channel directly above XDS(I), e.g. 0.33
Card 29.	XW(I), TW(I), . . . for I=1 to NW (card 23)	
	XW(I)	Distance from the lower end of the channel to the bottom of segment I (ft), e.g. 0.0
	TW(I)	Width of channel directly above XW(I) (ft), e.g. 10.0
	If the channel shape flag (FLAGC, card 12) is a 1 or 3, enter the value for TW that most closely approximates the channel shape.	

Cards 19 and 23 must be included, depending on the execution sequence (FLGSEQ, card 4), every time the updateable parameters are repeated. Cards 20-22 and 24-29 are included only if indicated on cards 19 and 23.

Table II-15.—Parameter file for erosion/sediment yield component--continued

A sample partial data file for the Control Parameters follows. It wil' help demonstrate the file structure.

CARD NO	EROSION PARAMETER DATA									
1	EROSION PARAMETERS - GEORGIA PIEDMONT									
2	MANAGEMENT PRACTICE ONE									
3	CONTINOUS CORN - CONVENTIONAL TILLAGE									
4	73000	0	1	0	3					
5	0.000	0.000	0.000	0.000	0.000	0.000				
6	0.140	0.200	0.660	0.010	20.000	4.000	0.0501000.000			
9	3.200	206.000	0.027	0.020	0.038	0.024	98.000	3.500	156.000	0.000
10	1									
11	1.000	0.230								
12	5	1	1	4	1					
13	20.000	10.000	0.030	0.002	2.410	2.250	0.000			
14	371.000	3.200	0.200	20.000						
15	0.000	0.024	69.000	0.018	154.000	0.014	269.000	0.032	325.000	0.021
18		73105								
19	1	1	1							
20	1.000	0.260								
21	1.000	1.000								
22	1.000	0.030								
23	1	1	1	1	1	1				
24	0.000	0.065								
25	0.000	0.400								
26	0.000	100.000								
27	0.000	0.330								
28	0.000	0.330								
29	0.000	10.000								
18		73121								
19	1	0	1							
20	1.000	0.400								
22	1.000	0.030								
23	1	1	0	1	1	0				
24	0.000	0.040								
25	0.000	0.150								
27	0.000	0.330								
28	0.000	0.330								

General Parameter Values

Starting Date--Set this value to zero if the model is used for a single storm. For multiple storms, the date should be less than that of the first storm (1 day less is sufficient). The date is given by first specifying the last two digits of the year (for example, 78) followed by the Julian date (for example, 094 for April 4, or 78094).

FLGOUT--This input determines whether the model runs for a single storm or for a series of storms. A 0 selects multiple storms, but the output is limited to annual summaries. A 1 selects multiple storms and gives output as monthly and annual summaries. A 2 gives output for each storm as well as the summaries. A 3 is used when the model is run for a single storm. It gives additional output of soil loss for each segment on the overland flow and channel elements. This indicates areas in the watershed where intense erosion or deposition occurs.

FLGPRT--If the particle distribution is computed, it is computed from the primary particle size distribution of the original soil mass. Management and other factors affecting aggregate sizes are not considered.

FLGPAS--Set to 0 if erosion/sediment yield estimates are not needed in other computations outside of the erosion/sediment yield component. Set to 1, the model writes date (Julian), volume of rainfall (in), volume of runoff (in), enrichment ratio (specific surface area of sediment and organic matter to that of original soil mass), sediment yield per unit area (tons/acre), and values input into the program for DP, PERCOL, AVGTMP, AVGPEV, POTPEV, ACCSEV, and POTSEV. These data are written into file 7, named PASS.

Sequence--The watershed is represented by a combination of such elements as overland flow, channel, and pond, and the calling sequence of these elements. Table II-16 gives the permissible sequences.

Table II-16.—Elements and their sequence numbers to represent main watershed features

Sequence number	Sequence of elements
1	Overland.
2	Overland-pond.
3	Overland-channel.
4	Overland-channel-channel.
5	Overland-channel-pond.
6	Overland-channel-channel-pond.

Before selecting the element sequence number, identify major features in the watershed that affect erosion and sediment yield. An aerial photograph and a site visit are especially useful. USGS topographic maps are generally too coarse for this application. A representative overland flow profile, channel(s), and impoundment are used to characterize the watershed elements. This characterization is discussed in later sections.

All watersheds are assumed to be composed of an overland flow element. Natural topography causes overland flow to converge into major flow concentrations on many farm fields. These few concentrations are readily distinguishable from the many rills that may exist on a field. The definition of a rill becoming a gully when a rill can no longer be obliterated by tillage is not workable, nor is the definition workable that a rill becomes a gully when it exceeds a certain size. The critical factor is how rills behave hydraulically. Removing a single rill has a negligible effect on the hydrologic-hydraulic response of the watershed, whereas removal of a single flow concentration has a major effect. Flow concentrations are easily identifiable with a site visit to a field tilled immediately before a major storm. In fact, site visits to typical fields before using the model are very helpful.

Other flow concentrations may exist besides these natural ones. Examples include terrace channels, grass waterways, and diversion ditches. A ridge develops around many fields, which often collects overland flow and causes a flow concentration along the edge of the field.

Impoundment terraces obviously pond water and are represented by a pond element. Other types of ponds are formed by natural depressions, roadways with pipe culverts, and other structures. A ridge and dense grass around the edge of many fields may pond runoff, causing considerable deposition near the edge of the field.

Obtain a map of the area to be modeled and identify the watershed boundary. Next, identify the channel and pond elements within the watershed. Only a single overland flow element may be called, which always is called first, and only a single pond element may be called, which always is called last. Flow through a series of ponds cannot be modeled. Typical examples are:

1. If an estimate of erosion on the overland flow areas alone is needed or if the field area is a simple overland flow area adjacent to a stream, only the overland flow element is called (CSEQ = 1) (Fig. II-17a).

2. The study area may be a simple watershed with a single concentration of flow down the middle, an overland flow section draining down a ridge formed by a field boundary that directs the flow along the field edge as concentrated flow, or an overland flow section cut off by a diversion ditch (CSEQ = 2) (fig. II-17b).

3. The study area may be a watershed with a major main flow concentration with several lateral flow concentrations feeding it, or it may be a series of terrace channels feeding an outlet channel (CSEQ = 4) (fig. II-17c).

4. If backwater is at the outlet for any of these situations, a second channel with backwater outlet control is added to the sequence (example 1 would become CSEQ = 3, or example 2 would be CSEQ = 4 (fig. II-17e)).

5. For impoundment terraces, two options are available for delivering flow to the impoundment. Overland flow goes directly to the impoundment (CSEQ = 2) (fig. II-17d), or overland flow first goes to a channel and then goes to the pond. Two channels may be involved, one in the draw, and one along the terrace. Select parameters based on the

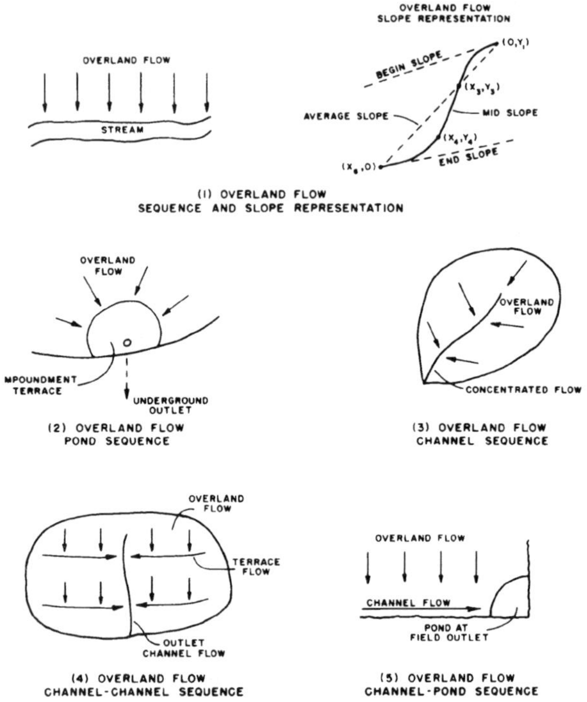

Figure II-17.—Schematic representation of typical field systems in the field-scale erosion/sediment yield model.

one delivering the most flow. The model does not permit a combination of both overland flow and channel delivery to the pond. The model can be run assuming different delivery systems and averaging results.

6. The pond outlet is assumed to be outside the study area, which prohibits analyzing ponds in a series.

7. If the sequence changes during the simulation period, the simulation must be run in parts, stopping when the sequence changes and then

restarting.

Kinematic Viscosity--The model defaults to a kinematic viscosity of 1.21 x 10^{-5} ft^2/s, the value for a temperature of 60° F. The value is assumed to be constant for the duration of the simulation period. The value was chosen assuming that most highly erosive storms occur in April and May. The value should be selected according to the temperature when most erosive storms occur if the model is being run for multiple storms. If it is being run for a single storm, a value appropriate for the temperature at the time of the storm should be selected. Table II-17 gives kinematic viscosity values for a range of temperatures.

Table II-17.—Kinematic viscosity for water over a typical range of temperatures

Temperature (°F)	Kinematic viscosity ($ft^2/s \times 10^5$)
40	1.67
50	1.41
60	1.21
70	1.05
80	0.90
90	0.82
100	0.74

Manning's n for overland flow over bare soil--The default value is 0.01, which represents smooth areas where broad overland flow has deposited sediment (5). Although the overland flow surface may be much rougher, a value for a smooth surface must be input because it is used to compute the portion of the flow's total shear stress that acts on the soil to cause detachment and transport.

Weight density of soil mass--This input is for the weight density of the soil mass in areas of flow concentrations. Although tillage, soil, management, time of year, and other factors significantly affect the density of the soil mass, a constant density is assumed over the simulation period. The default is 96 lb/ft^3 (1.54 g/cm^3 bulk density), which is larger than typical for surface soils. Flow concentrations typically occur in low areas and if erosion has occurred, much of the original surface soil is gone, leaving a more dense subsurface soil exposed. Compaction from farm equipment also is assumed to be greater in these areas. Therefore, bulk densities typical of tilled surface soils, especially after tillage, may be too small. The bulk density of the B horizon is probably a good value. On 36 of the Indiana soils used in the soil erodibility study by Wischmeier and others (13), bulk density of the B horizon ranged from 0.97 to 1.70 g/cm^3 with a mean of 1.37 g/cm^3 and a standard deviation of 0.15 g/cm^3. The mean weight density was 85 lb/ft^3. Table II-18 is recommended for selecting a value.

Soil Erodibility Factor for Erosion by Concentrated Flow--Some soils are much less susceptible to erosion by flow. Little information is available in the literature that may be used to estimate soil erodibility due to flow. A value of 0.135 (lb/ft^2/s)(ft^2/lb)$^{1.05}$ was obtained experimentally in a rill erosion study on a tilled silt loam soil (vol. III, ch. 11). For most applications, the default value is recommended. If the factor is varied, estimate the first approximation of K from the soil erodibility nomograph of Wischmeier and others (13) and multiply by 0.39. This assumes K_{rch} = 0.135 (lb/ft^2/s) for a first approximation of K = 0.35 (ton/ac/EI). Use the default value for sandy soils. Soil structure and permeability in the nomograph of Wischmeier and others (13) are considered nonapplicable to erosion by concentrated flow.

Table II-18.—Bulk and weight densities in areas of concentrated flow

Condition	Bulk density	Weight density
	(g/cm^3)	(lb/ft^3)
Loose.	1.20	75
Not subject to compaction and tilled regularly with primary tillage equipment.	1.37	85
Subject to compaction and tilled regularly with primary tillage equipment.	1.54	96
Not subject to compaction and not tilled regularly with primary tillage equipment.	1.54	98
Subject to compaction and not tilled regularly.	1.65	103

Manning's n for Channel Flow over Bare Soil--The default value is 0.03, which seems typical for nonvegetated earth channels, such as those channels where flow concentrates in farm fields (1, 8). This value is also consistent with Manning's n estimated from rill erosion experiments (vol. III, ch. 11). This n represents the roughness for flow over a seedbed or a relatively smooth soil that has eroded down to a nonerodible layer. Although the channel being analyzed is rough, covered with crop residue, or vegetated, a value for bare soil must be input because it is used to compute the portion of the flow's total shear stress that acts on the soil to cause detachment and transport.

Yalin's Constant--Yalin's sediment transport equation contains a constant equal to 0.635, which Yalin (14) obtained by fitting his equation to approximate sediment transport data from natural stream channels. When the equation was tested against overland flow data, the constant had to be increased to 0.88 to give good results for sand and it had to be decreased to 0.47 for coal particles having a 1.6 specific gravity (2). This equation is for bedload transport and may underpredict when a significant quantity of sediment is transported as suspended load. The constant can be increased to account for the suspended load transport capacity. No values are suggested, however. Use 0.635 unless other validated values are available (vol. III, ch. 10).

Particle Description

Particle Distribution of Residual Soil

The action of clay, silt, sand, and organic matter are for the original soil mass in the upper layer exposed to erosion and are based on the standard USDA classification. Use soil survey information or soil tests to estimate these values. The fractions are expressed so that the soil mineral fractions total 1.00. The fraction of organic matter is expressed as part of the total. These data are used to compute the distribution of particles of detached sediment if that option (FLGPRT = 0) is selected. With FLGPRT = 0 or 1, these data are used to compute enrichment ratios for the eroded sediment. (See vol. I, ch. 3 for procedures.)

Description of Sediment Particles

Two options are available in describing detached sediment particles and organic matter. The first option is to use the assumed relationships described in volume I, chapter 3. The second option is to input detailed information on the particles. Required information includes particle distribution (diameter, specific gravity, and fraction) of the sediment as it is detached, composition of each particle type, the specific surface area of clay, silt, sand, and organic matter; and relation of organic matter to clay in the eroded sediment.

Only limited information is given for choosing input values for the second option. A user that chooses the second option must research the required information on his own.

Sediment mixtures composed of up to twenty particle types are allowed with the second option. A type is a specific combination of size and density. Although clay-sized particles can be specified, flocculation or dispersion is not considered. If water or sediment chemistry causes flocculation or dispersion, identify this potential, enter appropriate particle size and density, and interpret the model results accordingly.

Many soils erode as aggregates, which are conglomerates composed of primary sand, silt, and clay particles having specific gravities less than the density of primary particles. Particle sizes are functions of soil properties, management, cover, and detachment by raindrop impact vs. detachment by runoff. Based on Young's analysis[2/] of available data, the following particle types are suggested in table II-19.

Although the classes in table II-19 are rather broad, do not deviate greatly from them except where the soil is poorly aggregated and aggregate stability is low. Since most agricultural soils erode as aggregates, especially in the midwest, size distribution of primary particles should be used directly as the eroded particle distribution only if the soil is totally nonaggregated when it erodes.

[2/] Personal communication with R. A. Young, USDA-SEA-AR Morris, Minn.

Table II-19.—Soil particle types

Condition	Size	Density	Fraction in detached sediment	Particle
	(mm)	(g/cm^3)		
Soils with ratio of silt to sand and clay > 0.5 (sa 15%, si 60%, cl 5%).	a. 0.002	2.60	0.05	Primary clay.
	b. .010	2.65	.08	Primary silt.
	c. .020	1.80	.50	Small aggregate.
	d. .500	1.60	.31	Large aggregate.
	e. .200	2.65	.06	Primary sand.
High clay soils (sa 10%, si 40%, cl 50%).	a. .002	2.60	.10	Primary clay.
	b. .010	2.65	.06	Primary silt.
	c. .075	1.80	.57	Small aggregate.
	d. 1.000	1.60	.25	Large aggregate.
	e. .200	2.65	.02	Primary sand.
High sand soils (sa 75%, si 15%, cl 10%).	a. .002	2.60	.02	Primary clay.
	b. .010	2.65	.02	Primary silt.
	c. .030	1.80	.16	Small aggregate.
	d. .200	1.60	.20	Large aggregate.
	e. .200	2.65	.60	Primary sand.

Factors that reduce rill erosion in relation to interrill erosion seem to increase the amount of primary particles and small aggregates in the fine size range. Particles from rough surfaces, vegetated surfaces, and flatter slopes tend to be smaller.

Data for these effects are limited. The figures and the discussion on their use are given only to indicate the effects, which will help interpret results. To adjust values in table II-19 for slope, read off a factor value from figure II-18. Multiply the fractions for the large particles, (that is, large aggregate and primary sand) by the factor and increase the other fraction in equal proportions to account for the reduction so that the total of the fractions is one.

Use figure II-19 to adjust for cover and the effect of consolidation. When soil lays exposed to consolidating traffic, it becomes more resistant to rill erosion and particle size is believed to decrease. The consolidation curve represents the change over the growing season or the effect of notillage, which is assumed not to change over time. Primary tillage lies in between. These curves do not account for roughness from primary tillage and its effect on transport capacity, which must be considered in the hydraulic roughness in put. Figures II-18 and II-19 are extrapolations from Young's review[2/] of avail-available data on particle size, and should be used carefully. In most problems, the effects described in these figures can be neglected.

2/ Op Cit

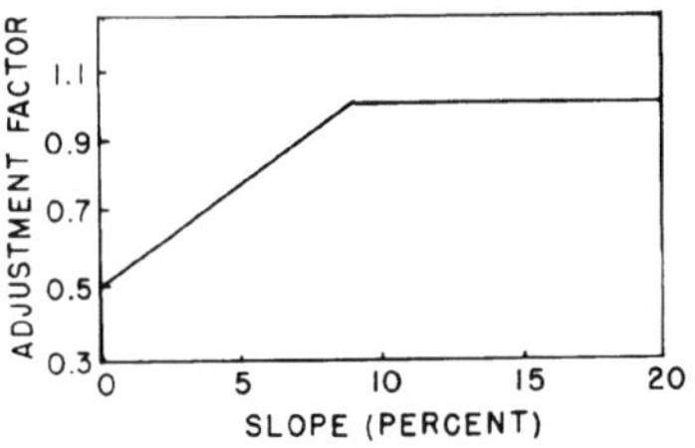

Figure II-18.—Adjustment factor for multiplying fractions of large particles (large aggregates and sand) to account for effect of slope on particle size. (Note: Curves are speculative.)

Only limited information is given on selection of information for particle characteristics to help the user obtain detailed information. Information on specific surface area is available in texts on soil physics. Note that specific surface area is used for organic carbon, rather than organic matter. If values are not input for specific surface areas, the model devaults to 20, 4, 0.05, and 1,000 m^2/g, respectively, for clay, silt, sand, and organic carbon. The specific surface area of 20 m^2/g is typical of kaolionitic clay. Montmorillonite clay may be as high as 800 m^2/g.

Particle composition is used to compute specific area and enrichment ratio of specific surface area. Aggregates are made up of organic matter, clay, silt, and sand. The specific surface area of the aggregate depends on the composition of these basic components. Little information on aggregate composition is available. The equations given in volume I, chapter 3 may be used as initial approximations.

Particle composition is not used to compute transport capacity and deposition of the particle types. Consequently, sediment yield estimates are accurate regardless of the accuracy of the input of particle composition.

Overland Flow

The overland flow element represents typical overland flow conditions on the watershed. After a representative land profile is selected, values must be identified for the erosion variables and their relative locations along the representative profile. The model uses averages in the direction perpendicular to the downslope direction, that is, along the contour but not along the slope. Thus, spatial effects along the profile can be considered with the model. Variations in cropping practices over the field cannot be analyzed with the model except for changes in cropping practice along the profile, such as strip cropping and grass buffer strips. The model assumes uniformity along the contour.

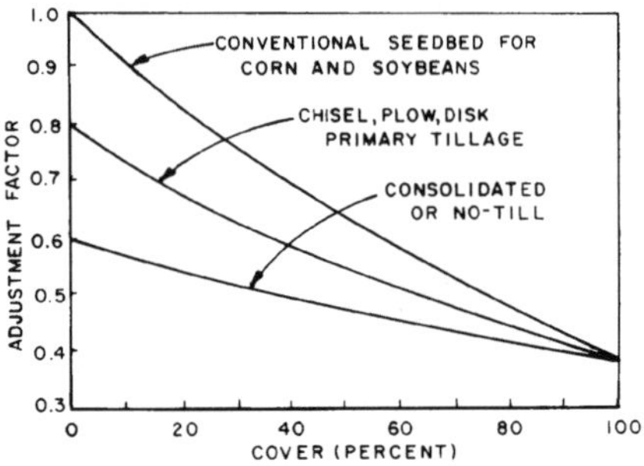

Figure II-19.—Adjustment factor for multiplying fraction of large particles (large aggregates and sand) to account for effect of cover and "consolidation" (soil disturbance) on particle size. (Note· Curves are speculative.)

Location on the slope of such features as highly erodible soils dramatically can effect sediment yield. A sediment delivery ratio concept or a P factor concept from the USLE, except for contouring, is not used. Both sediment delivery ratio and P factors are highly variable from storm to storm.

Nonupdateable Parameters

Overland Flow Area--This area of the watershed is represented by the overland flow element. It usually equals the total watershed area.

A modification of the USLE is used to compute detachment on the overland flow element. The modified equation uses soil erodibility, crop stage soil loss ratio, and contouring factors from the USLE without change. The next step is to assign values for these parameters and other detachment-transport parameters along the land profile.

Slope length--On simple rectangular areas, slope length is the distance from the point that overland flow originates to where it reaches concentrated flow. On typical midwestern fields, slope lengths seldom exceed 300 ft unless flow is constrained by tillage marks, crop row ridges, graded furrows, or formed channels. Do not use USGS contour maps to estimate slope length; they usually give excessively long slope lengths.

On simple areas, such as between terraces, the slope length is the typical distance between terraces. On more complex areas, the method of Williams and Berndt (11) may be used. This contour-extreme point method requires a contour map with the closest contour intervals available.

When a flow concentration crosses a contour, the contour comes to a point generally in the direction of the watershed divide (fig. II-20). These are called extreme points because they are local maxima in an uphill direction. Three contours, LC_{25}, LC_{50}, and LC_{75} are located at 25, 50, and 75% of the total watershed relief, and their lengths are determined. Next, the length is measured around the base of the LC contour (LB in fig. II-20). Slope length is given by:

$$\lambda = (LC_{50} \cdot LB)/(2EP \cdot (LC_{50}^2 - LB_{50}^2)^{1/2}) \quad [II-7]$$

where EP = number of extreme points.

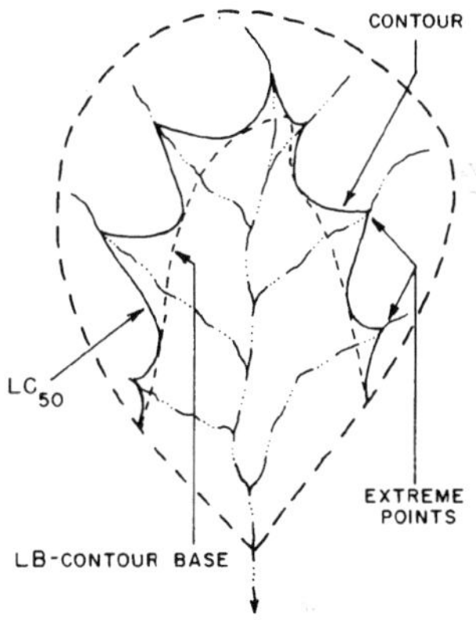

Figure II-20.—Sample watershed showing contour extreme points and base contour.

The resulting slope length should be inspected to determine if it is reasonable. This method breaks down as the watershed approaches a simple plane such as the area between terraces.

A subjective approach also may be used to determine such characteristics of slope as length and steepness. The watershed is divided into 10 to 15 areas approximately equal in area. Flow (stream) lines are drawn perpendicular to the contour lines. Slope length for each stream line is the distance from where overland flow originates to where it reaches concentrated flow, such as that in waterways or drainageways in farm fields. The slope length used in the model is the average of these lengths. Slope parameters to be discussed also can be evaluated for each flow line and averaged.

Average slope steepness of representative overland flow profile--Determining steepness is simple for simple land forms. The methods of Williams and Berndt's (11) are recommended for complex watersheds:

$$S = 0.25\ Z\ (LC_{25} + LC_{50} + LC_{75})/DATWS \qquad [II\text{-}8]$$

where Z = difference between an elevation of the highest point in the watershed and the elevation of the outlet and DATWS = total area of the watershed. The result should be inspected for consistency and reasonableness. The preceding subjective approach may be used as an alternate.

Profile midsection--A typical overland flow profile is identified for the watershed. The length and steepness of its slope have been determined. The following steps fill in its shape, soil, and cover characteristics.

Field profiles occur in a variety of shapes, such as those shown in figure II-21. The elements of slope profile used by the model to represent the actual field profile are identified in the complex, convex-concave slope in figure II-21. Read from left to right to identify slope elements and to name a slope. A slope is complex if it has convex and concave elements. It is a complex, convex-concave slope if the first curved element is convex. Values are assigned one at a time to the variables used to describe slope shape until all variables required to describe the slope have assigned values. Any remaining variables are assigned the last value appropriate for that type of variable.

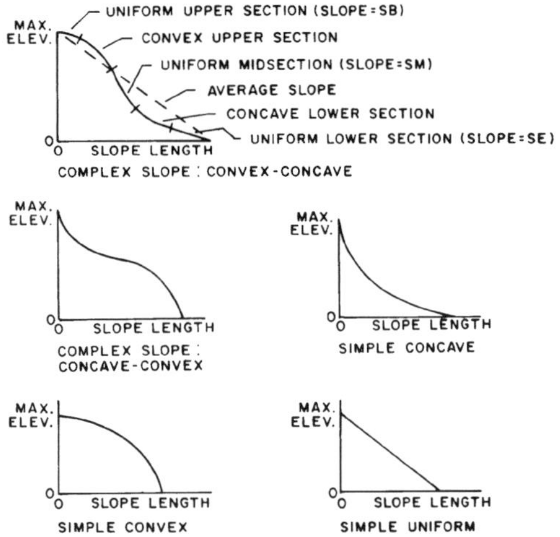

Figure II-21.—Slope shapes that can be analyzed with the erosion/sediment yield model.

In the coordinate system, x = 0 at the origin of overland flow where y = maximum elevation and y = 0 at x = slope length.

If a midsection exists, it must be located and its coordinates must be determined. If an upper convex section immediately changes to a concave section, no miduniform section exists. In this situation, coordinates of the lower end of the convex portion and the upper end of the concave portion are set equal to each other and equal to the coordinates where the two curves meet. The slope at this point must be specified later as SM. For a simple slope, coordinates of the midsection will be those at the upper and lower ends of the miduniform segment, if it exists. If the miduniform section does not exist, the coordinates of both ends of the miduniform section are set equal to the coordinates of the end of the slope, that is, x = slope length; y = 0.

Slope at upper end--The slope at the upper end, SB, is the slope of the upper end of the uniform segment, if it exists. If the uniform segment does not exist, SB is the slope at x = 0.

Slope of midsection--The slope of the midsection, SM, is the slope of the miduniform segment, if it exists. If not, it equals slope of the land profile where the upper and lower sections meet. On simple convex or concave slope, it is slope of the land profile at x = slope length.

Slope at lower end--The slope at lower end, SE, is the slope of the lower end of the uniform segment, if it exists. If not, SE is the slope of the land profile at the end of the profile where x = slope length. For uniform slopes, SB = SM = SE. For simple concave and convex slopes, SM = SE. On convex (simple or complex) slopes, SM is greater than the average slope and less than the average slope for concave (simple or complex) slopes. The upper slope, SB, is less than the average slopes for convex slopes and greater than the average slope for concave slopes. On complex, convex-concave slopes, SE is less than the average slope, while on complex, concave-convex slopes, SE is greater than the average slope. Failure to satisfy these conditions may cause fatal errors during the program execution (for example, dividing by zero or raising negative numbers to a power).

Soil erodibility--Some soils are more erodible than others. The erodibility of a soil is expressed by the soil erodibility factor. Values for this factor may be estimated from a soil erodibility nomograph (fig. II-22) (12). Using this nomograph requires a mechanical analysis of the soil to determine sand (0.1-2.0 mm), very fine sand, silt, clay (USDA classification), and organic matter fractions. Soil survey classification for soil structure and permeability also is required. Local offices of the USDA-Soil Conservation Service usually can provide soil erodibility values for local soils.

Identify the relative position (distance from top of slope/slope length) along the slope where the factor changes. The factor is assumed to be constant over the slope segment just above the point of change. If the entire slope has a single factor value, 1.0 is entered for relative distance.

To illustrate input, assume a slope length of 200 ft and K = 0.4 tons/acre/EI for the first 150 ft, and 0.2 ton/acre/EI for the last 50 ft. The

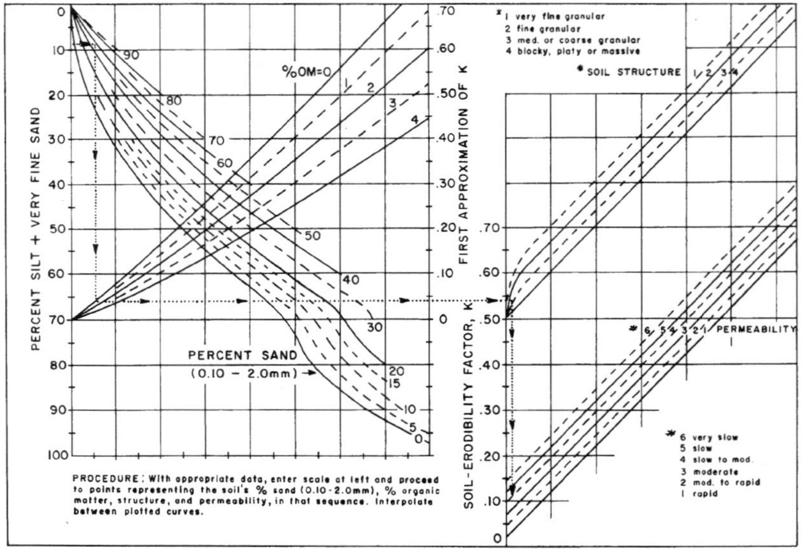

NOTE: K has units of (tons/acre) per EI unit where an EI unit is 100(ft-tons/acre)(in/hr).

Figure II-22.—The soil-erodibility nomograph. Where the silt fraction does not exceed 70 percent, t equation is $100\ K = 2.1\ M^{1.14}\ (10^{-4})\ (12 - a) + 3.25\ (b - 2) + 2.5\ (c - 3)$ where M = (percent si vfs) (100 percent c), a = percent organic matter, b = structure code, and c = profile permeabili class. From Wischmeier and Smith (12).

Table II-20.—Soil loss ratios (SLR) to describe the effects of cropping management (12)

Line No.	Cover, crop sequence, and management[1]	Spring residue[2]	Cover after plant[3]	Soil loss ratio[4] for cropstage period and canopy cover[5]: F	SB	1	2	3:80	90	96	4L[6]
		Lb	Pct	Pct	Pct	Pct	Pct	Pct	Pct	Pct	Pct
	CORN AFTER C, GS, G OR COT IN MEADOWLESS SYSTEMS										
	Moldboard plow, conv till:										
1	RdL, sprg TP	4,500		31	55	48	38			20	23
2		3,400	—	36	60	52	41	—	24	20	30
3		2,600		43	64	56	43	32	25	21	37
4		2,000		51	68	60	45	33	26	22	47
5	RdL, fall TP	HP[7]		44	65	53	38	—		20	—
6		GP		49	70	57	41	—	24	20	
7		FP		57	74	61	43	32	25	21	—
8		LP		65	78	65	45	32	26	22	
9	RdR, sprg TP	HP		66	74	65	47			22	56
10		GP	—	67	75	66	47	—	27	23	62
11		FP	—	68	76	67	48	35	27		69
12		LP	—	69	77	68	49	35		—	74
13	RdR, fall TP	HP		76	82	70	49			22	—
14		GP	—	77	83	71	50		27	23	
15		FP		78	85	72	51	35	27	—	—
16		LP		79	86	73	52	35	—		—
17	*Wheeltrack pl, RdL, TP*[8]	4,500	—	—	31	27	25		—	18	23
18		3,400		—	36	32	30		22	18	30
19		2,600	—		43	36	32	29	23	19	37
20		2,000	—		51	43	36	31	24	20	47
21	*Deep off-set disk or*	4,500	10	—	45	38	34	—		20	23
22	*disk plow*	3,400	10	—	52	43	37	—	24	20	30
23		2,600	5		57	48	40	32	25	21	37
24		2,000	—	—	61	51	42	33	26	22	47
25	*No-till plant in crop residue*[9]	6,000	95	—	2	2	2	—		2	14
26		6,000	90	—	3	3	3			3	14
27		4,500	80		5	5	5			5	15
28		3,400	70		8	8	8		8	6	19
29		3,400	60		12	12	12	12	9	8	23
30		3,400	50	—	15	15	14	14	11	9	27
31		2,600	40	—	21	20	18	17	13	11	30
32		2,600	30		26	24	22	21	17	14	36
	Chisel, shallow disk, or fld cult, as only tillage:										
33	On moderate slopes	6,000	70		8	8	7			7	17
34			60		10	9	8			8	17
35			50		13	11	10			9	18
36			40		15	13	11			10	19
37			30		18	15	13			12	20
38			20		23	20	18			16	21
39	D	1,500	70		9	8	7			7	18
40			60		12	10	9			8	18
41			50		14	13	11			9	19
42			40		17	15	13			10	20
43			30		21	18	15			13	21
44			20		25	22	19			16	22

Line No.	Cover, crop sequence, and management[1]	Spring residue[2]	Cover after plant[3]	Soil loss ratio[4] for cropstage period and canopy cover[5]: F	SB	1	2	3:80	90	96	4L[6]
		Lb	Pct	Pct	Pct	Pct	Pct	Pct	Pct	Pct	Pct
	CORN AFTER WC OF RYEGRASS OR WHEAT SEEDED IN C STUBBLE										
	WC reaches stemming stage:										
79	No-till pl in killed WC	4,000			7	7	7	—	7	6	(13)
80		3,000	—		11	11	11	11	9	7	
81		2,000	—	—	15	15	14	14	11	9	
82		1,500	—		20	19	18	18	14	11	
	Strip till one-fourth row space										
83	Rows U/D slope	4,000		—	13	12	11	—	11	9	(13)
84		3,000	—	—	18	17	16	16	13	10	
85		2,000	—		23	22	20	19	15	12	
86		1,500	—	—	28	26	24	22	17	14	
87	Rows on contour[11]	4,000	—	—	10	10	10	—	10	8	(13)
88		3,000	—	—	15	15	15	15	12	9	
89		2,000	—		20	20	19	19	15	12	
90		1,500	—	—	25	24	23	22	17	14	
91	TP, conv seedbed	4,000	—	36	60	52	41	—	24	20	(13)
92		3,000	—	43	64	56	43	31	25	21	
93		2,000	—	51	68	60	45	33	26	22	
94		1,500	—	61	73	64	47	35	27	23	
	WC succulent blades only:										
95	No-till pl in killed WC	3,000	—		11	11	17	23	18	16	(13)
96		2,000	—		15	15	20	25	20	17	
97		1,500	—	—	20	20	23	26	21	18	
98		1,000	—	—	26	26	27	27	22	19	
99	Strip till one-fourth row space	3,000	—		18	18	21	25	20	17	(13)
100		2,000	—		23	23	25	27	21	18	
101		1,500	—	—	28	28	28	28	22	19	
102		1,000	—		33	33	31	29	23	20	
	CORN IN SOD-BASED SYSTEMS										
	No-till pl in killed sod:										
103	3 to 5 tons hay yld				1	1	1	—	1	1	1
104	1 to 2 tons hay yld		—		2	2	2	2	2	2	2
	Strip till, 3-5 ton M:										
105	50 percent cover, tilled strips				2	2	2	—	2	2	4
106	20 percent cover, tilled strips	—		—	3	3	3	—	3	3	5
	Strip till, 1-2 ton M:										
107	40 percent cover, tilled strips	—		—	4	4	4	4	4	4	6
108	20 percent cover, tilled strips	—	—	—	5	5	5	5	5	5	7
	Other tillage after sod:			(14)	(14)	(14)	(14)	(14)	(14)	(14)	(14)
	CORN AFTER SOYBEANS										
109	*Sprg TP, conv till*	HP	—	40	72	60	48	—		25	29
110		GP	—	47	78	65	51	—	30	25	37
111		FP	—	56	83	70	54	40	31	26	44
112	*Fall TP, conv till*	HP	—	47	75	60	48	—	—	25	—
113		GP	—	53	81	65	51	—	30	25	—
114		FP	—	62	86	70	54	40	31	26	—

Table II-20.—Soil loss ratios (SLR) to describe the effects of cropping management--continued (12)

45	Do.	3,400	60		13	11	10		10	8	20
46			50	-	16	13	12	-	12	9	24
47			40	- -	19	17	16	.	14	11	25
48			30	-	23	21	19	.	17	14	26
49			20		29	25	23		21	16	27
50			10		36	32	29		24	20	30
51	Do.	2,600	50		17	16	15	15	13	10	29
52			40	- -	21	20	19	19	15	12	30
53			30	- -	25	23	22	22	18	14	32
54			20	-	32	29	28	27	22	17	34
55			10	- -	41	36	34	32	25	21	37
56	Do.	2,000	40	-	23	21	20	20	15	12	37
57			30		27	25	24	23	19	15	39
58			20	-	35	32	30	28	22	18	42
59			10		46	42	38	33	26	22	47
	On slopes > 12 percent.										
60	Lines 33-59 times factor of:	–	- -	-	1.3	1.3	1.1	1.0	1.0	1.0	1.0
	Disk or harrow after spring chisel or fld cult:										
	Lines 33-59 times factor of:										
61	On moderate slopes	--	--	—	1.1	1.1	1.1	1.0	1.0	1.0	1.0
62	On slopes > 12 percent	—	—	--	1.4	1.4	1.2	1.0	1.0	1.0	1.0
	Ridge plant:[10]										
	Lines 33-59 times factor of:										
63	Rows on contour[11]	—		-	.7	.7	.7	.7	.7	.7	.7
64	Rows U/D slope < 12 percent	--		—	.7	.7	1.0	1.0	1.0	1.0	1.0
65	Rows U/D slope > 12 percent	—	—	--	.9	.9	1.0	1.0	1.0	1.0	1.0
	Till plant:										
	Lines 33-59 times factor of:										
66	Rows on contour[11]	—	--	--	.7	.85	1.0	1.0	1.0	1.0	1.0
67	Rows U/D slope < 7 percent	--	--	—	1.0	1.0	1.0	1.0	1.0	1.0	1.0
	Strip till one-fourth of row spacing:										
68	Rows on contour[11]	4,500	[12]60	---	12	10	9		—	8	23
69		3,400	50	--	16	14	12	--	11	10	27
70		2,600	40	--	22	19	17	17	14	12	30
71		2,000	30	- -	27	23	21	20	16	13	36
72	Rows U/D slope	4,500	[12]60	---	16	13	11		-	9	23
73		3,400	50	-	20	17	14	-	12	11	27
74		2,600	40	- -	26	22	19	17	14	12	30
75		2,000	30	—	31	26	23	20	16	13	36
	Vari-till:										
76	Rows on contour[11]	3,400	40	---	13	12	11		--	11	22
77		3,400	30	--	16	15	14	14	13	12	26
78		2,600	20	—	21	19	19	19	16	14	34
115	Fall & sprg chisel or cult	HP	[12]30		40	35	29			23	29
116		GP	25	- -	45	39	33		27	23	37
117		GP	20	-	51	44	39	34	27	23	37
118		FP	15	- -	58	51	44	36	28	23	44
119		LP	10		67	59	48	36	28	23	54
120	No-till pl in crop res'd	HP	[12]40	—	25	20	19	--	14	11	26
121		GP	30	--	33	29	25	22	18	14	33
122		FP	20	-	44	38	32	27	23	18	40
	BEANS AFTER CORN										
123	Sprg TP, RdL, conv till	HP	-	33	60	52	38	--	20	17	(16)
124		GP	- -	39	64	56	41	- -	21	18	
125		FP	—	45	68	60	43	29	22	---	
126	Fall TP, RdL, conv till	HP		45	69	57	38	--	20	17	(16)
127		GP	--	52	73	61	41	--	21	18	
128		FP	- -	59	77	65	43	29	22	—	
	Chisel or fld cult:			(17)	(17)	(17)	(17)	(17)	(17)	(17)	(18)
	BEANS AFTER BEANS			(15)	(15)	(15)	(15)	(18)	(15)	(15)	(16)
	GRAIN AFTER C, G, GS, COT[19]										
129	In disked residues:	4,500	70	—	12	12	11	7	4	2	(20)
130		3,400	60	--	16	14	12	7	4	2	
131			50	--	22	18	14	8	5	3	
132			40	—	27	21	16	9	5	3	
133			30	—	32	25	18	9	6	3	
134			20	- -	38	30	21	10	6	3	
135	Do.	2,600	40	- -	29	24	19	9	6	3	(20)
136			20	—	43	34	24	11	7	4	
137			10	—	52	39	27	12	7	4	
138	Do.	2,000	30	--	38	30	23	11	7	4	(20)
139			20	- -	46	36	26	12	7	4	
140			10	- -	56	43	30	13	8	5	
141	In disked stubble, RdR	- -	—	- -	79	62	42	17	11	6	(20)
142	Winter G after fall TP, RdL	HP	--	31	55	48	31	12	7	5	(20)
143		GP	-	36	60	52	33	13	8	5	
144		FP	--	43	64	56	36	14	9	5	
145		LP	—	53	68	60	38	15	10	6	
	GRAIN AFTER SUMMER FALLOW										
146	With grain residues	200	10		70	55	43	18	13	11	(21)
147		500	30	--	43	34	23	13	10	8	
148		750	40	- -	34	27	18	10	7	7	
149		1,000	50	---	26	21	15	8	7	6	
150		1,500	60	--	20	16	12	7	5	5	
151		2,000	70	—	14	11	9	7	5	5	
152	With row crop residues	300	5	- -	82	65	44	19	14	12	(21)
153		500	15	—	62	49	35	17	13	11	
154		750	23	--	50	40	29	14	11	9	
155		1,000	30	—	40	31	24	13	10	8	
156		1,500	45	--	31	24	18	10	8	7	
157		2,000	55	—	23	19	14	8	7	5	
158		2,500	65	- -	17	14	12	7	5	4	
	POTATOES										
159	Rows with slope	---	-	43	64	56	36	26	19	16	
160	Contoured rows, ridged when canopy cover is about 50 percent[11]			43	64	56	18	13	10	8	

Table II-20.—Soil loss ratios (SLR) to describe the effects of cropping management--continued (12)

[1] Symbols: B, soybeans; C, corn; conv till, plow, disk and harrow for seedbed; cot, cotton; F, rough fallow; fld cult, field cultivator; G, small grain; GS, grain sorghum; M, grass and legume meadow, at least 1 full year; pl, plant; RdL, crop residues left on field; RdR, crop residues removed; SB, seedbed period; sprg, spring; TP, plowed with moldboard; WC, winter cover crop; —, insignificant or an unlikely combination of variables.

[2] Dry weight per acre after winter loss and reductions by grazing or partial removal: 4,500 lbs represents 100 to 125 bu corn; 3,400 lbs, 75 to 99 bu; 2,600 lbs, 60 to 74 bu; and 2,000 lbs, 40 to 59 bu; with normal 30-percent winter loss. For RdR or fall-plow practices, these four productivity levels are indicated by HP, GP, FP and LP, respectively (high, good, fair, and low productivity). In lines 79 to 102, this column indicates dry weight of the winter-cover crop.

[3] Percentage of soil surface covered by plant residue mulch *after* crop seeding. The difference between spring residue and that on the surface after crop seeding is reflected in the soil loss ratios as residues mixed with the topsoil.

[4] The soil loss ratios, given as percentages, assume that the indicated crop sequence and practices are followed consistently. One-year deviations from normal practices do not have the effect of a permanent change. Linear interpolation between lines is recommended when justified by field conditions.

[5] Cropstage periods are as defined on p. 18. The three columns for cropstage 3 are for 80, 90, and 96 to 100 percent canopy cover at maturity.

[6] Column 4L is for all residues left on field. Corn stalks partially standing as left by some mechanical pickers. If stalks are shredded and spread by picker, select ratio from table 5-C. When residues are reduced by grazing, take ratio from lower spring-residue line.

[7] Period 4 values in lines 9 to 12 are for corn stubble (stover removed).

[8] Inversion plowed, no secondary tillage. For this practice, residues must be left and incorporated.

[9] Soil surface and chopped residues of *matured* preceding crop undisturbed except in narrow slots in which seeds are planted.

[10] Top of old row ridge sliced off, throwing residues and some soil into furrow areas. Reridging assumed to occur near end of cropstage 1.

[11] Where lower soil loss ratios are listed for rows on the contour, this reduction is in addition to the standard field contouring credit. The *P* value for contouring is used with these reduced loss ratios.

[12] Field-average percent cover; probably about three-fourths of percent cover on undisturbed strips.

[13] If again seeded to WC crop in corn stubble, evaluate winter period as a winter grain seeding (lines 132 to 148). Otherwise, see table 5-C.

[14] Select the appropriate line for the crop, tillage, and productivity level and multiply the listed soil loss ratios by sod residual factors from table 5-D.

[15] Spring residue may include carryover from prior corn crop.

[16] See table 5-C.

[17] Use values from lines 33 to 62 with appropriate dates and lengths of cropstage periods for beans in the locality.

[18] Values in lines 109 to 122 are best available estimates, but planting dates and lengths of cropstages may differ.

[19] When meadow is seeded with the grain, its effect will be reflected through higher percentages of cover in cropstages 3 and 4.

[20] Ratio depends on percent cover. See table 5-C.

[21] See item 12, table 5-B.

Table II-21.—Approximate soil loss ratios for cotton (12)

Expected final canopy percent cover:		65	80	95
Estimated initial percent cover from defoliation + stalks down:		30	45	60
Practice Number	Tillage operation(s)	Soil loss ratio[1]		
COTTON ANNUALLY:		Percent		
1	*None:*			
	Defoliation to Dec. 31	36	24	15
	Jan. 1 to Feb. or Mar. tillage:			
	Cot Rd only	52	41	32
	Rd & 20 percent cover vol veg[2]	32	26	20
	Rd & 30 percent cover vol veg	26	20	14
2	*Chisel plow soon after cot harvest:*			
	Chiseling to Dec. 31	40	31	24
	Jan. 1 to sprg tillage	56	47	40
3	*Fall disk after chisel:*			
	Disking to Dec. 31	53	45	37
	Jan. 1 to sprg tillage	62	54	47
4	*Chisel plow Feb-Mar, no prior tillage:*			
	Cot Rd only	50	42	35
	Rd & 20 percent vol veg	39	33	28
	Rd & 30 percent vol veg	34	29	25
5	*Bed ("hip") Feb-Mar, no prior tillage:*			
	Cot Rd only	100	84	70
	Rd & 20 percent vol veg	78	66	56
	Rd & 30 percent vol veg	68	58	50
	Split ridges & plant after hip, or Disk & plant after chisel (SB):			
	Cot Rd only	61	54	47
	Rd & 20 percent vol veg	53	47	41
	Rd & 30 percent vol veg	50	44	38
	Cropstage 1:			
	Cot Rd only	57	50	43
	Rd & 20 percent vol veg	49	43	38
	Rd & 30 percent vol veg	46	41	36
	Cropstage 2	45	39	34
	Cropstage 3	40	27	17
6	*Bed (hip) after 1 prior tillage:*			
	Cot Rd only	110	96	84
	Rd & 20 percent veg	94	82	72
	Rd & 30 percent veg	90	78	68
	Split ridges after hip (SB):			
	Cot Rd only	66	61	52
	Rd & 20 to 30 percent veg	61	55	49
	Cropstage 1:			
	Cot Rd only	60	56	49
	Rd & 20 to 30 percent veg	56	51	46
	Cropstage 2	47	44	38
	Cropstage 3	42	30	19
7	*Hip after 2 prior tillages:*			
	Cot Rd only	116	108	98
	Rd & 20-30 percent veg	108	98	88
	Split ridges after hip (SB)	67	62	57
8	*Hip after 3 or more tillages:*	120	110	102
	Split ridges after hip (SB)	68	64	59
9	*Conventional moldboard plow and disk:*			
	Fallow period	42	39	36
	Seedbed period	68	64	59
	Cropstage 1	63	59	55
	Cropstage 2	49	46	43
	Cropstage 3	44	32	22
	Cropstage 4 (See practtices 1, 2, and 3)			

COTTON AFTER SOD CROP:

For the first or second crop after a grass or grass-and-legume meadow has been turnplowed, multiply values given in the last five lines above by sod residual factors from table 5-D.

COTTON AFTER SOYBEANS:

Select values from above and multiply by 1.25.

See footnotes at right.

input card would be:

0.75 0.4 1.0 0.2.

Updateable Parameters

Selection of inputs thus far has been discussed in the same order of the inputs. Inputs to this point are fixed for the simulation period. If they must be updated, the run is stopped. The input file is changed to the new values, and a new run is started.

Initial constants for the channel and pond elements are read before the updateable inputs for the overland area are read. The discussion continues with the updateable overland flow variables.

PDATE, CDATE (Card 19)--This date, expressed as a Julian date, is the first and last date that a set of parameters is valid, including those of both overland flow and channel elements. Once the storm date exceeds CDATE, the program reads the next set of parameter values. If zero is entered for the number of data points for a parameter, the program uses the most recent value for that parameter. That is, only new values are required for the parameters that change.

[1] *Alternate procedure for estimating the soil loss ratios:*

The ratios given above for cotton are based on estimates for reductions in percent cover through normal winter loss and by the successive tillage operations. Research is underway in Mississippi to obtain more accurate residue data in relation to tillage practices. This research should provide more accurate soil loss ratios for cotton within a few years.

Where the reductions in percent cover by winter loss and tillage operations are small, the following procedure may be used to compute soil loss ratios for the preplant and seedbed periods: Enter figure 6 with the percentage of the field surface covered by residue mulch, move vertically to the upper curve, and read the mulch factor on the scale at the left. Multiply this factor by a factor selected from the following tabulation to credit for effects of land-use residual, surface roughness and porosity.

Productivitty level	No tillage	Rough surface	Smoothed surface
High	0.66	0.50	0.56
Medium	.71	.54	.61
Poor	.75	.58	.65

Values for the bedded period on slopes of less than 1 percent should be estimated at twice the value computed above for rough surfaces.

[2] Rd, crop residue; vol veg, volunteer vegetation.

Table II-22.—Soil loss ratios for conditions not evaluated in table II-20 (12)

COTTON:

See table 5-A.

CROPSTAGE 4 FOR ROWCROPS:

Stalks broken and partially standing: Use col. 4L.

Stalks standing after hand picking: Col. 4L times 1.15.

Stalks shredded without soil tillage: See table 5-C.

Fall chisel: Select values from lines 33-62, seedbed column.

CROPSTAGE 4 FOR SMALL GRAIN:

See table 5-C.

DOUBLE CROPPING:

Derive annual C value by selecting from table 5 the soil loss percentages for the successive cropstage periods of each crop.

ESTABLISHED MEADOW, FULL-YEAR PERCENTAGES:

Grass and legume mix, 3 to 5 t hay		0.4
Do.	2 to 3 t hay	.6
Do.	1 t hay	1.0
Sericea, after second year		1.0
Red clover		1.5
Alfalfa, lespedeza, and second-year sericea		2.0
Sweetclover		2.5

MEADOW SEEDING WITHOUT NURSE CROP:

Determine appropriate lengths of cropstage periods SB, 1, and 2 and apply values given for small grain seeding.

PEANUTS:

Comparison with soybeans is suggested.

PINEAPPLES:

Direct data not available. Tentative values derived analytically are available from the SCS in Hawaii or the Western Technical Service Center at Portland, Oreg. (Reference 5).

SORGHUM:

Select values given for corn, on the basis of expected crop residues and canopy cover.

SUGARBEETS:

Direct data not available. Probably most nearly comparable to potatoes, without the ridging credit.

SUGARCANE:

Tentative values available from sources given for pineapples.

SUMMER FALLOW IN LOW-RAINFALL AREAS, USE GRAIN OR ROW CROP RESIDUES:

The approximate soil loss percentage after each successive tillage operation may be obtained from the following tabulation by estimating the percent surface cover after that tillage and selecting the column for the appropriate amount of initial residue. The given values credit benefits of the residue mulch, residues mixed with soil by tillage, and the crop system residual.

Percent cover by mulch	Initial residue (lbs/A)			
	>4,000	3,000	2,000	1,500
90	4	—	—	—
80	8	[1]8	—	—
70	12	13	[1]14	—
60	16	17	[1]18	[1]19
50	20	22	24	[1]25
40	25	27	30	32
30	29	33	37	39
20	35	39	44	48
10	47	55	63	68

[1] For grain residue only.

WINTER COVER SEEDING IN ROW CROP STUBBLE OR RESIDUES:

Define cropstage periods based on the cover seeding date and apply values from lines 129 to 145.

Table II-23.—Soil loss ratios (pct) for cropstage 4 when stalks are chopped and distributed without soil tillage (12)

Mulch cover[1]	Corn or Sorghum		Soybeans		Grain Stubble[4]
	Tilled seedbed[2]	No-till	Tilled seedbed[2]	No-till in corn rd[3]	
20	48	34	60	42	48
30	37	26	46	32	37
40	30	21	38	26	30
50	22	15	28	19	22
60	17	12	21	16	17
70	12	8	15	10	12
80	7	5	9	6	7
90	4	3	—	—	4
95	3	2	—	—	3

[1] Part of a field surface directly covered by pieces of residue mulch.

[2] This column applies for all systems other than no-till.

[3] Cover after bean harvest may include an appreciable number of stalks carried over from the prior corn crop.

[4] For grain with meadow seeding, include meadow growth in percent cover and limit grain period 4 to 2 mo. Thereafter, classify as established meadow.

Table II-24.— Factors to credit residual effects of turned sod[1] (12)

Crop	Hay yield	Factor for cropstage period:				
		F	SB and 1	2	3	4
	Tons					
First year after mead:						
Row crop or grain	3-5	0.25	0.40	0.45	0.50	0.60
	2-3	.30	.45	.50	.55	.65
	1-2	.35	.50	.55	.60	.70
Second year after mead:						
Row crop	3.5	.70	.80	.85	.90	.95
	2-3	.75	.85	.90	.95	1.0
	1-2	.80	.90	.95	1.0	1.0
Spring grain	3-5	—	.75	.80	.85	.95
	2-3	—	.80	.85	.90	1.0
	1-2	—	.85	.90	.95	1.0
Winter grain	3-5	—	.60	.70	.85	.95
	2-3	—	.65	.75	.90	1.0
	1-2	—	.70	.85	.95	1.0

[1] These factors are to be multiplied by the appropriate soil loss percentages selected from table 5. They are directly applicable for sod-forming meadows of at least 1 full year duration, plowed not more than 1 month before final seedbed preparation.

When sod is fall plowed for spring planting, the listed values for all cropstage periods are increased by adding 0.02 for each additional month by which the plowing precedes spring seedbed preparation. For example, September plowing would precede May disking by 8 months and 0.02(8—1), or 0.14, would be added to each value in the table. For nonsod-forming meadows, like sweetclover or lespedeza, multiply the factors by 1.2. When the computed value is greater than 1.0, use as 1.0.

Cover-management--Cover, tillage, stage of crop growth, and previous management history greatly affect erosion. A grass-covered slope hardly erodes, while erosion on a bare slope may be excessive. Similarly, a freshly prepared, finely tilled seedbed for corn is much more susceptible to erosion than it is immediately following harvest.

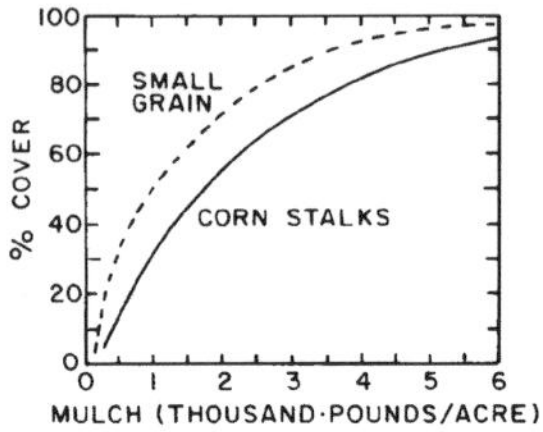

Figure II-23.—Relation of percentage of cover to dry weight of uniformly distributed residue mulch. [From Wischmeier and Smith (12).]

Values of the USLE crop-stage-soil-loss ratio (SLR) appropriate for the history and present conditions on the field are used to describe this factor. Tables II-20 through II-24 and Figure II-23 for soil loss ratios (SLR) were taken from Wischmeier and Smith (12), and should be used to select cover-management factor values. The C factor in the USLE is not the same as the soil loss ratio. The C factor is an average annual value that integrates the variable soil loss ratio and variable rainfall erosivity over the year. This model does not use the average annual C factor.

The soil loss ratio describes erosion characteristics at specific times during the cropping cycles. The soil loss ratios in the tables of Wischmeier and Smith are step functions. In many situations (but not all cases, such as at harvest), they are continuous. A continuous function in the model may be approximated by reading new soil loss ratios at frequent intervals.

Enter SLR values as a fraction rather than as a percentage, as shown in the tables. Values are entered as (1) relative location where the soil loss ratio changes and (2) the soil loss ratio just upslope of the point of change. The following examples illustrate.

Example 1: Entire field is in continuous conventional corn at seedbed:

X_*	SLR
1.0	0.8

Example 2: A 20-ft grass buffer strip is at the toe of a 200 ft slope with corn at seedbed time. Total slope length is 220 ft. The relative location for the first change is 200/220, or 0.91. The entered values might be:

x_*	SLR	x_*	SLR
0.91	0.80	1.00	0.03 (soil loss ratio for the grass).

Example 3: The field is strip cropped with alternating strips of corn and oats. Assume corn is 0 to 75 ft, oats is 75 to 125 ft, corn is 125 to 200 ft,

and oats is 200 to 250 ft. The entries could be:

$x_{\star}$	SLR	$x_{\star}$	SLR	$x_{\star}$	SLR	$x_{\star}$	SLR
0.3	0.80	0.5	0.10	0.8	0.80	1.0	0.10

where $x_{\star}$ = relative distance and SLR = soil loss ratio.

Contouring--Direction of tillage may significantly influence erosion and sediment yield. Contouring may store most runoff from small storms, greatly reducing sediment yield. For large storms that cause breakovers, however, sediment yield can be greater than from uphill and downhill tillage. Following breakovers, sediment yield from small storms increases. Unfortunately, contouring factor values are defined poorly on a storm-by-storm basis. The values in Table II-25 are taken from the USLE and represent long-term averages. Future refinements are needed greatly for this factor. Contouring loses its effectiveness for long slopes. For slope lengths beyond those shown in table II-25, assign a contouring factor of 1.0.

Table II-25.—Contour factor values and slope-length limits for contouring
[From Wischmeier and Smith (12)]

Land slope	Contour factor value	Maximum length[1/]
(%)		(ft)
1 to 2	0.60	400
3 to 5	.50	300
6 to 8	.50	200
9 to 12	.60	120
13 to 16	.70	80
17 to 20	.80	60
21 to 25	.90	50

[1/] Limit may be increased by 25 % if residue cover after crop seedling regularly will exceed 50 %.

Values in table II-25 are for contouring typical of conventional farming practices where row ridges are formed during cultivation after crops emerge. Ridges are typically 4 to 6 in above the row middles.

Rows in many fields follow field boundaries rather than on the contour. In one part of the watershed, rows may be directly uphill and downhill while in another part they may be on the contour and somewhere in between for the rest of the watershed. Effectiveness of contouring is partially due to deposition in the middle, where the gradient of flow around the slope is low and transport capacity is small. Transport capacity increases rapidly, however, when slope steepens. Deposition in row middles consequently decreases rapidly as the row deviates slightly from the contour. Use figure II-24 to adjust nonlinearly the contouring factor where rows are neither on the contour nor directly uphill and downhill. This adjusted factor also should be weighted for the watershed area, as well.

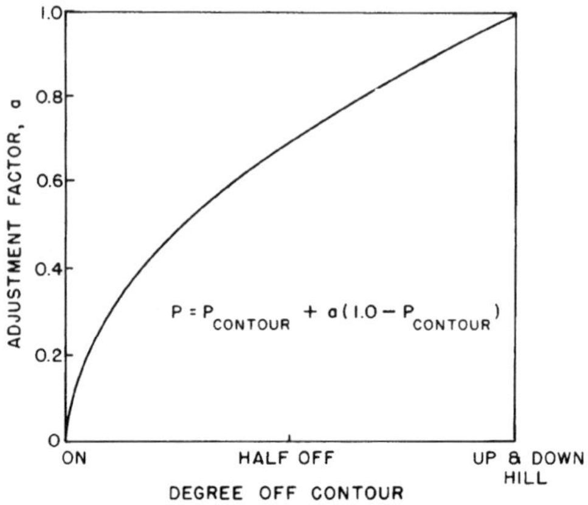

Figure II-24.—Adjustment factor for being off contour with tillage.

Furrows of some contouring systems lead runoff to one of several flow concentrations in small draws. If excess rainfall exceeds storage capacity, the ridges overtop in the draws. Erosion in these overtopped areas may be analyzed as concentrated flow erosion. Select parameter values according to instructions for a concentrated flow element that requires estimating the depth of soil from the bottom of the channel to the nonerodible layer. Consider the nonerodible layer to be at the depth of secondary tillage. Add to this depth one-third of the difference in height from the top of the row ridge to the bottom of the row middles to obtain a total depth. Use this value as the depth of the soil beside the channel. Assume a naturally eroded channel shape. If this approach is used, the overland flow land profile is taken along the rows leading to the flow concentration.

Graded row middles may act as individual channel systems. These may be analyzed by assuming that the row ridges are the overland flow area. The average steepness for the row sideslope is used for slope steepness, and slope length is the distance from the row ridge to the water's edge in the row middle. The row middle is described with the model's channel element. Be aware that this technique and the technique for row breakover have not been validated for the model.

Hydraulic roughness--Cover and roughness on the soil surface slow overland flow and reduce its transport capacity. The reduction in velocity depends on the cover material and its density and the degree of surface roughness. Table II-26 may be used to select Manning's n values, which the model uses to estimate the transport capacity of overland flow. The ratio of n from table II-26 to that assumed for overland flow over bare soil is a key value. The values in table II-26 are based on n = 0.01 for overland flow over bare soil. If that value is increased, the values in table II-26 should be changed to maintain the same ratio of n for cover to n for bare soil. Conversely, if the values in table II-26 are adjusted as a whole, similarly adjust the n for bare soil.

Table II-26. Estimates of Manning's n for overland flow and soil covers[1/]

Treatment	Manning's n	Treatment	Manning's n
Cornstalk residue applied to fallow surface:		Crushed stone mulch	
1 ton/acre	0.020	15 tons/acre	0.012
2 tons/acre	.040	60 tons/acre	.023
4 tons/acre	.070	135 tons/acre	.046
		240 tons/acre	.074
Cornstalk residue disk-harrow incorporated:		375 tons/acre	.074
		Grass	
1 ton/acre	0.012	Sparse	0.015
2 tons/acre	.020	Poor	.023
4 tons/acre	.023	Fair	.032
		Good	.046
Wheat straw mulch:		Excellent	.074
		Dense	.150
0.25 ton/acre	0.015	Very dense	.400
0.5 ton/acre	.018		
1 ton/acre	.032	Rough surface depressions	
2 tons/acre	.070		
4 tons/acre	.074	4 to 5 in deep	0.046
		2 to 4 in deep	.023
		1 to 2 in deep	.014
		No surface depressions	.010

Small grain (20% to full maturity)	Across slope	Upslope & Downslope
Poor stand	0.018	0.012
Moderate stand	.023	.015
Good stand	.032	.023
Dense	.046	.032

1/Based on data form Lane and others (5) and Neibling and Foster (7).

Overland flow transport is related to Manning's n for bare soil, the ratio of n with bare soil to that with cover or roughness, Yalin's constant, particle characteristics, and the deposition reaction coefficent. With the exception of the reaction coefficient, which can be changed only by an internal program modification, all these variables must be considered during any optimization of transport capacity.

Strip cropping and grass buffer strips reduce yield of sediment because close growing vegetation slows the runoff, greatly reducing its transport capacity to where deposition occurs. If these practices are not on the contour, causing flow to move along the strips, or if flow concentrations submerge the grass, their effectiveness is reduced greatly. Flow along the upper edge of a strip of grass or other dense vegetation may be treated as a naturally eroded channel with a slope equal to that along the upper edge of the strip. A triangular channel may be assumed to pass concentrated flow through a strip.

If the problem does not lend itself to treating concentrated flow through the grass as a channel element, decrease Manning's n to account for a lesser reduction in flow velocity with the deeper flow. The relation of Manning's n for flow through grass to the product of velocity and hydraulic radius is not built into the model. Choose a value that best represents the grass and the given runoff event.

Channel Element

The channel element describes erosion and sediment transport in flow concentrations within farm fields. These are not necessarily defined channels unless they happen to be grassed waterways, terrace channels, or diversions. On many farm fields, flow concentrates in natural, small draws. A heavy rain on freshly prepared seedbed can cause major erosion in these concentrations. The soil often will erode down to the depth of secondary or primary tillage. Ridges and field borders can cause flow concentrations at the edge of a field. The channel element that describes these situations is unsuitable for hydraulic design of terraces or diversions. Channel sides are assumed to be sufficiently high to contain the flow.

Non-updateable Parameters

Like the overland flow parameters, there are two sets of nonupdateable parameters: (1) those that remain constant for the simulation period and (2) those that are updateable on a storm-by-storm basis. If the nonupdateable parameters change, the simulation is stopped and restarted. The order in which the following variables are discussed is changed slightly from the input order.

<u>Channel shape</u>--The user may specify a triangular, rectangular, or naturally eroded channel. The triangular channel with equal side slopes is used for most terrace and grass waterway channels although they are usually parabolic. If the channel is too parabolic for a triangular section, assume a rectangular channel. Assume a naturally eroded channel when channel dimensions depend strongly on previous erosion. Hydraulic calculations are completely independent of erosion for the triangular channel. However, erosion rate is limited by a nonerodible layer and extent of previous erosion. For the rectangular channel, erosion rate is also limited by a nonerodible layer and extent of previous erosion. If the calculated eroded width exceeds the specified channel width, channel width is reset to the eroded width. For the naturally eroded channel, erosion rate and channel dimensions depend on extent of previous erosion and the existence of a nonerodible layer.

The channel element might be applied to small stream channels (less than 5 to 10 ft wide). Refer to volume I, chapter 3, for a description of the method used to estimate erosion to consider whether this method is satisfactory in relation to other known methods.

The model should not be applied to gully erosion. Many features of gully erosion, such as headcutting and sloughing of the sidewalls, are not included in the channel component.

A single typical flow concentration is chosen to represent flow concentrations of a given stream order. The flow concentration analyzed by the model may not exist, or it may be one of a number in the watershed, for example, one out of five in a system of terraces. The output sediment concentration from the typical flow concentration is assumed to equal the average concentration for all flow concentrations represented by the chosen channel.

Slope of the energy gradeline (friction slope)--Flow in most channels is dynamic and spatially varied. For many field-sized watersheds, dynamic terms in the flow momentum equation may be dropped (that is, kinematic assumption). At your option, normal flow may be assumed to set the friction slope equal to the slope of the channel. The relatively flat 0.1 to 1/2% slope of many terrace channels may invalidate this kinematic assumption. Roughness from vegetation or a ridge at the field's edge may cause backwater not considered by the kinematic assumption. In terrace channels where the outlet is unrestricted, flow accelerates near the outlet, producing higher shear stress than the normal flow assumption would calculate. At the upper end, shear stresses are lower than those from the normal flow assumptions.

The normal flow assumption will not work for a zero grade terrace channel unless slope of the energy gradeline is used as slope input. The normal slope assumption gives no backwater effect for a restricted outlet unless the slopes are input as slope of the energy gradeline.

The other option is to use the built-in nonuniform flow curves. These were developed by normalizing the spatially varied flow equation and solving it for a range of typical parameter values. Regression analyses were used to fit polynominal curves to the solutions. Although flow is unsteady, the analysis assumes steady flow and uses the peak runoff rate as a characteristic discharge for hydraulic computations. Refer to table II-27 as a guide to when to use the two options.

Outlet Control--Four types of control that may be specified are: (1) critical depth, (2) uniform flow in a downstream control channel, (3) the greater of (1) or (2), and (4) a structure or control having a known rating curve. Use (1) for terrace, diversion, or other channels when the depth of flow in the outlet channel has no restricting effect. Use (2) when a reach at the lower end of the channel sets the depth (for example, a heavy vegetation at the channel outlet). Use (3) when the model is to choose the greater of (1) or (2). Use (4) when a control structure (for example, weir, ridge that acts as a weir, or flume) controls flow depth according to a known rating curve.

The outlet control is used only when the friction slope curves are used. This control determines depth at the channel outlet, which is a parameter in

Table II-27.—Guidelines on using the built-in friction slope curves

Condition	Friction slope assumption
Supercritical flow all along the channel and at the outlet.	Kinematic (normal) flow.
Small discharge; very flat channel gradient (0.001) to 0.005); critical depth at outlet (for example, channel flow in a row middle).	Kinematic (normal) flow.
Restricted outlet giving backwater.	Friction slope curves.
Critical depth at the outlet of diversions and conventional terrace channels.	Friction slope curves.
Zero grade channel (for example, level terraces).	Friction slope curves.

the friction slope curves. If (1), (2), or (3) is specified, a triangular channel section and its sideslope must be selected (ignore any reference to a rectangular control section). The preferred sideslope is that of the channel element. If another sideslope is specified, do not use a flatter slope than that for the channel. When uniform flow control is selected, slope of the outlet reach and its Manning's n is specified. Refer to a later section for values of Manning's n. The form of the rating curve is:

$$Q = RA(Y - YBASE)^{RB}. \qquad [II\text{-}9]$$

Specify values for the coefficient RA and exponent RB. The term Y is flow depth, and YBASE is the minimum depth for flow to begin. Refer to weir and flume rating tables to determine the parameter values.

Channel length--This is the length of channel from its outlet to its origin. Channel origin is the point where overland flow has converged to the point that it can be considered concentrated flow.

Drainage area at the outlet--If the flow concentration being analyzed is the main stem in the watershed, the drainage area at the outlet is the total area in the watershed. If the channel is one of several terrace channels, the area is that drained by the representative channel chosen for analysis. This area will be smaller than the overland flow area since the channel only drains a part of the total overland flow area. Figure II-25 illustrates this area and the area at the upper end of the channel.

Drainage area at the upper end--If overland flow converges to form the origin of the channel, drainage area at the upper end equals the overland flow area draining into the upper end of the channel. This variable is used to compute discharge at the upper end of the channel. Channels for terrace outlet usually originate in the middle of the field with a drainage area at the upper end. Usually, no drainage area is at the upper end of a terrace channel because of zero discharge at the upper end (fig. II-25).

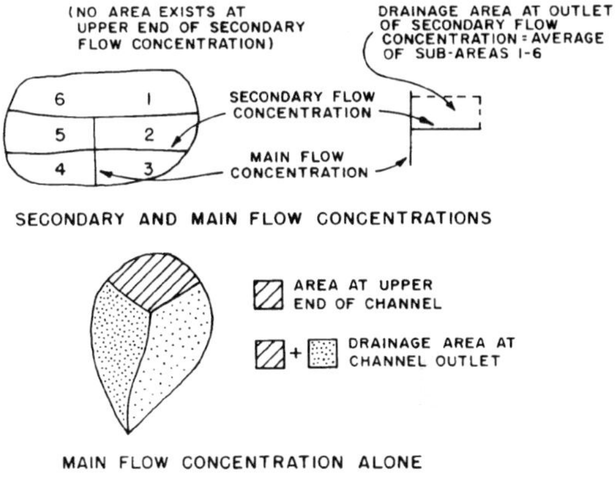

Figure II-25.—Channel areas.

Channel section sideslope--Use a sideslope of 5 for terrace channels and grass waterways unless more specific information is available. Use 10 or a value indicated by more specific information for concentrated flow in an area regularly tilled but susceptible to major erosion. Use 20 for flow concentrations caused by ridges along field boundaries. Even if a rectangular or naturally eroded channel is specified, approximate and enter the sideslope for the channel. Sideslope is used by the model to compute the friction slope. An approximate value for sideslope for a rectangular channel is:

$$Z = 3\ B^{5/3}/Q^{2/3} \qquad [II-10]$$

where B = bottom width of the rectangular channel and Q = discharge. Use a weighted discharge based on the square of the discharges for all storms or equalize the flow area for a depth weighted toward the larger events.

Distance along the channel--For input, x = 0 at the channel outlet and x increases going upstream in the channel. Internally, the model inverts this order and reassigns x = 0 to the upper end of the effective channel length. Effective channel length is defined from the following relationships.

Discharge rate Q_{low} at the lower end of the channel is:

$$Q_{low} = \sigma_p A_{low} \qquad [II-11]$$

where σ_p = characteristic peak excess rainfall rate and A_{low} = drainage area above the lower end of the channel. Discharge rate q_{up} of the upper end of the channel is:

$$Q_{up} = \sigma_p A_{up} \qquad [II-12]$$

where A_{up} = drainage area above the upper end of the channel. Lateral inflow rate q_i is:

$$q_i = (Q_{low} - Q_{up})/\lambda_{ch} \qquad [II-13]$$

where λ_{ch} = channel length. Discharge at any point along the channel is:

$$Q = Q_{low} - x_{ch}\, q_i \qquad [II-14]$$

where x_{ch} is the distance upstream from the lower end of the channel. Effective channel length λ_{cheff} is defined as the x_{ch} where Q = 0, or from the preceding equation:

$$\lambda_{cheff} = Q_{low}/q_i\ , \qquad [II-15]$$

which is the same as:

$$\lambda_{cheff} = \lambda_{ch}/(1.0 - A_{up}/A_{low}). \qquad [II-16]$$

Channel slope--Channel slope at a point (that is, at the specified x's rather than for a segment) is input. These slopes may be estimated from a plot of the profile of the channel. A minimum of five points is needed to represent a curved channel profile. Since the model assumes that the channel profile is curved, abrupt changes in slope can only be approximated. Enter slope values and locations on either side of the break for best approximations. The locations can be a minimum of 1 ft apart.

The model defines channel segments equal to 0.1 of the effective length. If the actual length is short in relation to the effective length, less than three channel segments may exist. If this occurs, "fool" the model to have at least three segments (five or more preferred) by specifying a slightly different Manning's n (for example, 0.03001 and 0.03000) at x's when segment ends are desired. This technique also may be used to "fool" the model to obtain a finer definition of uniform channel segments.

Updateable Channel Parameters

Channel parameters are read with an x where the parameter changes and the value of the parameter is just upslope of the change. Input x's are specified as distance upstream from the channel outlet. When the parameters are updated, the number of changes is read for each parameter along the channel. If a parameter does not change from the previous storm, updating of that parameter is not required. A zero is read for the number of changes along the channel, and the model uses the parameter value from the last storm.

Changes along a channel--The model can describe changes in all parameters along the channel, but it primarily represents end or beginning of grass within the length of the channel. Figure II-26 shows a typical grassed waterway ending within a field where the channel flattens to where erosion in the waterway probably would not occur. The channel from 0 to 100 ft would be tilled. At seedbed time, a sample parameter data set would be:

	Tilled portion		Grassed waterway	
	Distance from END	Parameter value	Distance from END	Parameter value
Manning's n - - - - - - - - - - -	0.0	0.03	100	0.13
Critical soil shear - - - - - -	.0	.10	100	.60
Critical cover shear- - - - - -	.0	100	100	100
Depth to nonerodible- - - - - -	.0	.33	Same for entire channel length	
Depth at side of channel- - - -	.0	.33		
Channel width - - - - - - - - -	.0	20		

Refer to following sections for discussion of selection of these parameter values.

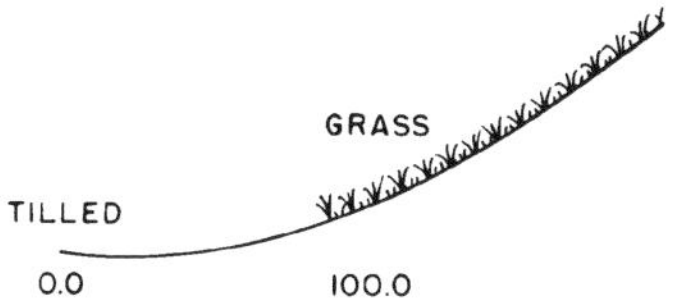

Figure II-26.—Representation of a channel having a change in cover along channel.

Manning's n--Different covers have different values for hydraulic roughness, depending on their density, height, and type of vegetation. Their hydraulic roughness also depends on the rigidity of the vegetation and degree of submergence by the flow. Although Manning's n may vary over a considerable range of the product of velocity and hydraulic radius, the model assumes that n does not change with discharge. Table II-28 gives estimated n's for several covers. Chow (1), Ree and Crow (8), and hydraulics handbooks of USDA's Soil Conservation Service provide additional information. Do not enter a Manning's n less than the value entered for the Manning's n for bare soil.

The Manning's n's in table II-28 are moderate values for a range of the product, velocity times hydraulic radius. These values generally are for V · R of 1.0 to 1.5. For high flows that definitely submerge the cover, the n's are too high.

Refer to the overland flow, Manning's n section, for discussion on adjustment of Manning's n from those in table II-28.

Critical shear stress of the soil--Some soils and soil conditions are more susceptible to detachment by flow than are others. Although considerable information exists on critical shear stress, it is contradictory and generally does not apply to agriculture. For this model, estimate a base critical shear stress (lb/ft^2) modified from Smerdon and Beasley's equation (10):

$$\tau_{cr} = 0.213/d_r^{0.63} \qquad [II-17]$$

Table II-28.—Manning's n for typical soil covers[1/]

Cover	Cover density	Manning's n
Smooth, bare soil; roughness elements.	Less than 1 in deep	0.030
	1-2 in deep	.033
	2-4 in deep	.038
	4-6 in deep	.045
Corn stalks (assumes residue stays in place and is not washed away).	1 ton/acre	.050
	2 tons/acre	.075
	3 tons/acre	.100
	4 tons/acre	.13
Wheat straw (assumes residue stays in place and is not washed away).	1 ton/acre	.060
	1.5 tons/acre	.100
	2 tons/acre	.15
	4 tons/acre	.25
Grass (assumes grass is erect and as deep as the flow).	Sparse	.04
	Poor	.05
	Fair	.06
	Good	.08
	Excellent	.13
	Dense	.20
	Very dense	.30
Small grain (20% to full maturity-rows with flow).	Poor, 7 in rows	.13
	Poor, 14 in rows	.13
	Good, 7 in rows	.30
	Good, 14 in rows	.20
(Rows across flows)	Good	.30
Sorghum and cotton	Poor	.07
	Good	.09
Sudangrass	Good	.20
Lespedeza	Good	.10
Lovegrass	Good	.15

[1/]Does not include effects of submergence or product of velocity-hydraulic radius.

where d_r = dispersion ratio. Dispersion ratio is the ratio expressed as a percentage of suspension percentage divided by percentage of silt plus percentage of clay. Typical values for dispersion ratio range from 5 to 25 with most about 10 to 12. The base critical shear stress applies to a finely pulverized seedbed, which usually occurs when the soil is most susceptible to erosion.

A value of 0.05 lb/ft^2 may be used for this base value if no better information is available. The soil gradually consolidates and becomes less erodible. Critical shear stress seems to decrease as tillage more finely pulverizes the soil. Use figure II-27 to estimate a factor to multiply the base critical shear stress for an estimate of apparent critical shear stress. Table II-29 may be used for critical shear stress if differences in types of soil are not considered.

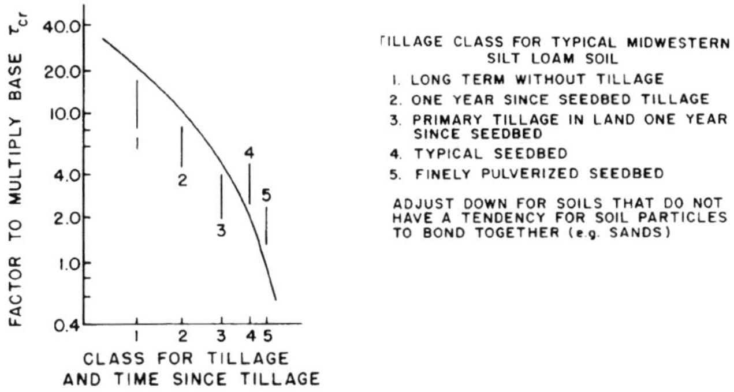

Figure II-27.—Effect of tillage on critical shear stress.

Table II-29.—Critical shear stress values as a function of tillage and consolidation for moderately erodible soils

Tillage-consolidation condition	Critical shear stress
	(lb/ft^2)
Moldboard plowed	0.20
Chisel or disk for primary tillage	.15
Disking for common seedbed for corn or cultivation of crop	.10
Finely pulverized seedbed	.05
1 month after last tillage of common seedbed	.20
2 months after last tillage of common seedbed	.30
3 months after last tillage of common seedbed	.40
Long term, undisturbed	.60

When flow bends vegetation over so that it lies flat on the channel, the vegetation effectively armors the soil and prevents erosion. The model cannot directly handle this problem. If this effect is suspected, increase the critical shear stress for the soil to prevent erosion, and decrease n to prevent deposition, which is unlikely with the flattened vegetation.

Shear stress at failure of crop residue--As in conservation tillage, shear stress for concentrated flow through and over crop residue may exceed a critical shear stress at which the cover may begin to move. The model assumes that if shear stress on the cover exceeds a critical shear stress for the particular type and rate of a mulch, the cover fails and shear stress is computed as if no cover exists. The failed cover assumption remains in effect until a new set of Manning's n is read. Estimates for this critical shear stress are in table II-30. Assign a value of 100.0 to the variable if cover failure is not allowed.

Table II-30.—Critical shear stress value for corn stalk and wheat stalk en-masse movement[1/]

Type of mulch	Rate	Critical shear stress
	(tons/acre)	(lb/ft^2)
Corn stalks	1	0.051
(not incorporated).	2	.105
	3	.156
	4	.210
Wheat straw	1	.064
(not anchored)	1.5	.140
	2	.232
	4	.841

[1/]Stress acting on mulch; does not include stress acting on soil.

Depth of soil from channel bottom to nonerodible layer--When concentrated flow in a farm field erodes, it often erodes through the tilled layer until it strikes a nontilled layer and then it rapidly widens. The nonerodible layer is frequently at the bottom of the surface layer of secondary tillage, which typically is 0.3 to 0.4 ft deep. Although primary tillage disturbs the soil to a greater depth than secondary tillage, the large soil chunks turned over by primary tillage are much less erodible than the surface soils that have been exposed to secondary tillage. The large soil chunks may act like grade control structures.

In a natural channel, a rock layer or an armor layer acts as a nonerodible layer. Large flows can destroy the armor layer, however, and the channel will deepen again until a new armor layer develops. A pond or a larger stream also may control like a nonerodible layer. The model cannot describe, however, the development of a concave channel profile upslope from a control like a pond.

Depths to the nonerodible layer are shown in figure II-28. Whenever tillage occurs, this value should be reset. If it is not reset, the model uses the depth that the channel has eroded to during the previous storm. If the effect of the nonerodible layer is to be neglected, assign a large value, for example, 1000.0, to this parameter.

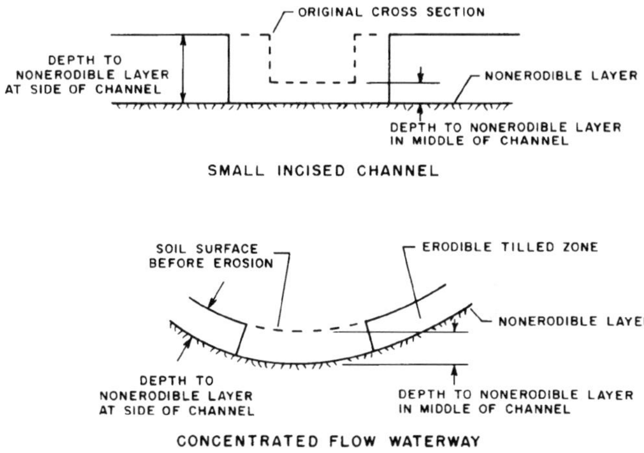

Figure II-28.—Defining sketch for depths to nonerodible layer for a small, encised channel and for a concentrated-flow waterway.

Depth to nonerodible layer at the side of the channel--If the channel is a flow concentration through the field in a regularly tilled area, use the same value as the depth to nonerodible layer in the middle of the channel. For more defined, incised channels, use the height of the effective channel wall that moves horizontally as the channel widens.

Channel width--Specify channel width when a rectangular channel section is assumed. When a triangular or naturally eroded section is analyzed, specify the width of the rectangular channel that most closely approximates the channel. Do not leave this parameter blank. In some situations of no erosion, the model defaults to a rectangular section.

Pond (Impoundment) Element

Relationships for the pond element were derived from analysis of output from a simulation model (3) supported with field observations from impoundment terraces (4).

The pond element is primarily meant to describe deposition in impoundment terraces with pipe outlets, which drain between storms. The pond element can describe deposition in ponds and impounded water behind ridges and culverts, but it should not be applied to those situations unless discharge is controlled by a pipe outlet or the equivalent orifice coefficient is known and the impoundment drains between storms.

The pond element makes now allowance for short circuiting where the sediment load enters near the outlet or the entrance is connected directly with a flow path to the outlet. The pond is assumed to drain completely after the runoff event and to pass all runoff except for infiltrated water. If some storm water is retained, the effect on sediment yield must be accounted for outside the model.

Initial Parameters

If any pond parameters change with time, the simulation must be stopped when the parameters change and restarted with new parameter values.

Control--The user may specify one of two possible controls: pipe outlet-riser as typical in parallel tile outlet terraces, or control where the equivalent orifice coefficient is available. Refer to volume I, chapter 3 for a definition of equivalent orifice. This option may be unreliable unless it is for the impoundment terrace type of drain.

Surface area-depth--Values are required for the coefficient and exponent for the surface area-depth relationship given by:

$$SA = F_s\, y^B \qquad [II-18]$$

where SA = surface area (ft^2), F_s = coefficient, B = exponent, and y = water depth in the pond (ft). The coefficint and exponent depend on topography within the ponded area. These values sometimes can be determined from design-construction surveys. Table II-31 shows typical values for some impoundment terraces.

Table II-31.—Coefficient and exponent for surface area-depth relationships observed for typical impoundment terraces

Terrace location	Coefficient F_s	Exponent B
Eldora, Iowa[1/] - - - - - - - - - - - - -	8,247	1.10
Charles City, Iowa[1/] - - - - - - - - - -	9,465	1.73
Guthrie Center, Iowa[1/] - - - - - - - - -	4,485	1.28
Marvyn, Ala.[2/] - - - - - - - - - - - - -	7,950	1.77

[1/] From Laflen (4).

[2/] From Rochester and Busch (10).

Values for front, draw, and side slopes may be used by the model to estimate F_s and B if values for them are unavailable. These slopes are front (embankment front slope), draw (slope at the pond along draw draining into pond), and side (slope of land at the pond toward the draw). The exponent B is assumed to be 2 and coefficent F_s is calculated from (9):

$$F_s = [(f + d)/f]^2/(d \cdot s) \qquad [II-19]$$

where f = front slope, s = side slope, d = draw slope.

Drainage area--This land area drains into the pond. It is generally the watershed area since the pond is assumed to be the last element.

Intake rate--Intake rate is the infiltration rate within the pond and not on the watershed. It depends on type of soil, sealing, and tillage through the impoundment area. Refer to the soil survey of USDA-SCS for an indication of permeability with adjustments for sealing and tillage. A typical value for a silt loam soil with good intake would be 0.4 in/hr.

Diameter of orifice in outlet pipe--An orifice of a small diameter delays passage of the runoff through the impoundment and increases depositon. Diameter of the orifice usually is selected based on volume of impoundment, runoff rates and volumes, and time to drain. Consult designers of these terraces (usually USDA-SCS) in your local area to determine actual sizes for the given terraces or an estimate of typical sizes. If no value can be found, use 3.0 in.

Orifice coefficient--The model actually requires an estimate of C in the equation:

$$C = 3600\ Q/Y^{1/2} \qquad [II-19]$$

where Q = peak discharge (ft^3/s) out of the pond and Y = depth (ft) of water above control. If a value for this coefficient is known, it may be input directly into the model. Otherwise, the model estimates it from the diameter of the pipe orifice.

OUTPUT

The user gets basic output from the model describing basic parameter values for the watershed (fig. II-29). The user can select additional output in various levels of detail. The first option is an annual summary for each year in the simulation period, giving sediment yield from the most downstream element in the sequence. Sediment yield is for all the types of particle and for each type individually. Totals for the entire simulation period, also are given. Figure II-30 illustrates a summary output.

The second option provides monthly and annual summaries, as shown in figure II-31.

The third option summarizes information for each storm and for each element in addition to sediment yield. Figure II-32 illustrates this output.

The fourth option is output from a single storm where loss or deposition of soil is given for each segment in each element. Figure II-33 illustrates

NONPOINT SOURCE POLLUTION MODEL (EROSION/SEDIMENT YIELD)

EROSION PARAMETERS - GEORGIA PIEDMONT
MANAGEMENT PRACTICE ONE
CONTINOUS CORN - CONVENTIONAL TILLAGE

INITIAL CONSTANTS

STARTING DATE FOR THIS RUN	74000	JULIAN DATE
WT. DENSITY SOIL (IN PLACE)	96.0	LBSF/FT**3
WT. DENSITY WATER	62.4	LBSF/FT**3
MASS DENSITY WATER	1.94	SLUGS/FT**3
ACC. DUE TO GRAVITY	32.2	FT/SEC**2
KINEMATIC VISCOSITY	0.121E-04	FT**2/SEC
MANNING N BARE SOIL (OVER)	0.010	
MANNING N BARE SOIL (CHAN)	0.030	
CHANNEL ERODIBILITY FACTOR (LBS/FT**2 SEC)/(LBS/FT**2)**1.05	0.135	
YALIN CONSTANT (ALL PART.)	0.635	
MOMENTUM COEFF. FOR NONUNIFORM VELOCITY IN CROSS SECTION	1.56	(NO UNITS)

DISTRIBUTION OF PRIMARY PARTICLES
AND ORGANIC MATTER IN THE ORIGINAL SOIL MASS

TYPE	FRACTION	SPECIFIC SURFACE
		(M**2/G OF SOIL)
CLAY	0.140	20.000
SILT	0.200	4.000
SAND	0.660	0.050
		(M**2/G OF ORGANIC CARBON)
ORGANIC MATTER	0.010	1000.000

(ORGANIC CARBON = ORGANIC MATTER/1.73)

INDEX OF SPECIFIC SURFACE 9.38 M**2/G OF TOTAL SOIL

Figure II-29.—Basic input values for the erosion model.

PARTICLE SPECIFICATIONS

TYPE NO.	DIA. MM	EQSAND DIA. MM	FALL VEL. FT/SEC	SPGRAV. GM/CM**3	FRAC. IN DETACH. SED.
1	0.002	0.002	0.102E-04	2.60	0.03
2	0.010	0.010	0.263E-03	2.65	0.03
3	0.030	0.020	0.125E-02	1.80	0.23
4	0.280	0.158	0.542E-01	1.60	0.27
5	0.200	0.201	0.759E-01	2.65	0.45

PARTICLE COMPOSITION

TYPE NO.	PRIMARY PARTICLE FRACTIONS CLAY	SILT	SAND	ORGANIC MATTER
1	1.000	0.000	0.000	0.071
2	0.000	1.000	0.000	0.000
3	0.412	0.588	0.000	0.029
4	0.070	0.153	0.777	0.005
5	0.000	0.000	1.000	0.000

OVERLAND INPUTS

OVERLAND AREA	3.2000	ACRES
SLOPE LENGTH	205.00	FT
MAXIMUM ELEVATION	5.50	FT
AVERAGE SLOPE	0.0267	
SLOPE OF UPPER END	0.0200	
SLOPE OF MID SECTION	0.0380	
SLOPE OF LOWER END	0.0240	

THE SLOPE IS A CONVEX CONCAVE

LOCATION OF UNIFORM SECTION
DISTANCE,ELEVATION 98.0, 3.5
DISTANCE,ELEVATION 156.0, 1.3
DISTANCE MEASURED FROM THE UPPER END
ELEVATION MEASURED ABOVE LOWEST POINT

Figure II-29.—Basic input values for the erosion model--continued.

```
                CHANNEL INPUTS
                --------------

CHANNEL LENGTH              371.00    FT
DRAINAGE AREA UPPER END       0.2000  ACRES
EFFCT. LENGTH UPPER END      24.73    FT
DRAINAGE AREA LOWER END       3.2000  ACRES
EFFCT. LENGTH LOWER END     395.73    FT
MANNING N FOR BARE SOIL       0.030
SOIL ERODIBILITY FACTOR       0.135

    A TRIANGULAR SHAPED CHANNEL

          ENERGY GRADELINE
  JSES THE ENERGY GRADELINE CURVES

       RATING CURVE CONTROL

        Q = RA*(Y-YBASE)**RN
          RA    =     2.410
          RN    =     2.250
          YBASE =     0.00
```

Figure II-29.—Basic input values for the erosion model--continued.

ANNUAL SUMMARY FOR 1974

67 STORMS PRODUCED 40.26 IN. OF RAINFALL
13 STORMS PRODUCED 3.49 IN. OF RUNOFF

QUANTITY OF ERODED SEDIMENT

PART. TYPE	SOIL LOSS LBS.	CONCENTRATIONS (SOIL/WATER) LBSF/FT**3	LBSF/LBSF	PPM (WT)
1	1354.	0.0334	0.0005	535.
2	1213.	0.0299	0.0005	480.
3	10337.	0.2550	0.0041	4087.
4	8016.	0.1978	0.0032	3170.
5	11046.	0.2726	0.0044	4368.
TOTAL	31966.	0.7887	0.0126	12640.

AVERAGE SOIL LOSS FOR AREA 5.00 TONS/ACRE
(AREA = 3.2000 ACRES)

DISTRIBUTION OF PRIMARY PARTICLES
AND ORGANIC MATTER IN THE ERODED SEDIMENT

TYPE	FRACTION
CLAY	0.193
SILT	0.265
SAND	0.540
ORGANIC MATTER	0.014

INDEX OF SPECIFIC SURFACE 12.86 M**2/G OF TOTAL SEDIMENT

ENRICHMENT RATIO OF SPECIFIC SURFACE 1.371

Figure II-30.—Annual and total summaries from the erosion/sediment yield model.

ANNUAL SUMMARY FOR 1975

71 STORMS PRODUCED 48.25 IN. OF RAINFALL
26 STORMS PRODUCED 7.49 IN. OF RUNOFF

THE QUANTITY OF ERODED SEDIMENT

PART.	SOIL LOSS	CONCENTRATIONS (SOIL/WATER)		
TYPE	LBS.	LBSF/FT**3	LBSF/LBSF	PPM (WT)
1	1621.	0.0186	0.0003	298.
2	1454.	0.0167	0.0003	268.
3	12215.	0.1404	0.0022	2249.
4	7771.	0.0893	0.0014	1431.
5	8254.	0.0948	0.0015	1520.
TOTAL	31315.	0.3598	0.0058	5766.

AVERAGE SOIL LOSS FOR AREA 4.90 TONS/ACRE
(AREA = 3.2000 ACRES)

DISTRIBUTION OF PRIMARY PARTICLES
AND ORGANIC MATTER IN THE ERODED SEDIMENT

TYPE	FRACTION
CLAY	0.230
SILT	0.314
SAND	0.456
ORGANIC MATTER	0.016

INDEX OF SPECIFIC SURFACE 15.26 M**2/G OF TOTAL SEDIMENT

ENRICHMENT RATIO OF SPECIFIC SURFACE 1.628

Figure II-30.—Annual and total summaries from the erosion/sediment yield model--continued.

NONPOINT SOURCE POLLUTION MODEL (EROSION/SEDIMENT YIELD)

EROSION PARAMETERS - GEORGIA PIEDMONT
MANAGEMENT PRACTICE ONE
CONTINOUS CORN - CONVENTIONAL TILLAGE

STORM SUMMARY

138 STORMS PRODUCED 88.51 IN. OF RAINFALL
39 STORMS PRODUCED 10.98 IN. OF RUNOFF

THE QUANTITY OF ERODED SEDIMENT IN RUNOFF
VALUES FOR ALL STORMS

PART. TYPE	SOIL LOSS LBS.	CONCENTRATIONS (SOIL/WATER) LBSF/FT**3	LBSF/LBSF	PPM (WT)
1	2975.	0.0233	0.0004	374.
2	2668.	0.0209	0.0003	335.
3	22552.	0.1768	0.0028	2833.
4	15787.	0.1238	0.0020	1983.
5	19300.	0.1513	0.0024	2425.
TOTAL	63281.	0.4961	0.0080	7950.

TOTAL SOIL LOSS FOR AREA 9.90 TONS/ACRE
(AREA = 3.2000 ACRES)

DISTRIBUTION OF PRIMARY PARTICLES
AND ORGANIC MATTER IN THE ERODED SEDIMENT

TYPE	FRACTION
CLAY	0.211
SILT	0.290
SAND	0.495
ORGANIC MATTER	0.015

INDEX OF SPECIFIC SURFACE 14.05 M**2/G OF TOTAL SEDIMENT

ENRICHMENT RATIO OF SPECIFIC SURFACE 1.498

Figure II-30.—Annual and total summaries from the erosion/sediment yield model--continued.

MONTHLY SUMMARY FOR MAY, 1974

10 STORMS PRODUCED 5.42 IN. OF RAINFALL
1 STORMS PRODUCED 0.64 IN. OF RUNOFF

THE QUANTITY OF ERODED SEDIMENT

PART. TYPE	SOIL LOSS LBS.	CONCENTRATIONS (SOIL/WATER) LBSF/FT**3	LBSF/LBSF	PPM (WT)
1	386.	0.0520	0.0008	834.
2	357.	0.0482	0.0008	773.
3	3088.	0.4169	0.0067	6680.
4	2315.	0.3125	0.0050	5008.
5	3289.	0.4439	0.0071	7114.
TOTAL	9435.	1.2736	0.0204	20410.

AVERAGE SOIL LOSS FOR AREA 1.48 TONS/ACRE
(AREA = 3.2000 ACRES)

DISTRIBUTION OF PRIMARY PARTICLES
AND ORGANIC MATTER IN THE ERODED SEDIMENT

TYPE	FRACTION
CLAY	0.193
SILT	0.268
SAND	0.539
ORGANIC MATTER	0.014

INDEX OF SPECIFIC SURFACE 12.85 M**2/G OF TOTAL SEDIMENT

ENRICHMENT RATIO OF SPECIFIC SURFACE 1.370

Figure II-31.—Sample of monthly summaries from the erosion/sediment yield model.

MONTHLY SUMMARY FOR JUN, 1974

4 STORMS PRODUCED 5.29 IN. OF RAINFALL
1 STORMS PRODUCED 1.25 IN. OF RUNOFF

THE QUANTITY OF ERODED SEDIMENT

PART. TYPE	SOIL LOSS LBS.	CONCENTRATIONS (SOIL/WATER) LBSF/FT**3	LBSF/LBSF	PPM (WT)
1	725.	0.0498	0.0008	798.
2	672.	0.0461	0.0007	739.
3	5827.	0.4000	0.0064	6411.
4	4979.	0.3418	0.0055	5478.
5	7536.	0.5173	0.0083	8291.
TOTAL	19739.	1.3551	0.0217	21716.

AVERAGE SOIL LOSS FOR AREA 3.09 TONS/ACRE
(AREA = 3.2000 ACRES)

DISTRIBUTION OF PRIMARY PARTICLES
AND ORGANIC MATTER IN THE ERODED SEDIMENT

TYPE	FRACTION
CLAY	0.176
SILT	0.246
SAND	0.578
ORGANIC MATTER	0.013

INDEX OF SPECIFIC SURFACE 11.74 M**2/G OF TOTAL SEDIMENT

ENRICHMENT RATIO OF SPECIFIC SURFACE 1.252

Figure II-31.—Sample of monthly summaries from the erosion/sediment yield model--continued.

THE FOLLOWING PARAMETERS ARE VALID BETWEEN THE DATES (JULIAN)
74000 - 74105

POINTS OF CHANGE ALONG THE OVERLAND PROFILE

DISTANCE FEET	DISTANCE NONDIM.	SLOPE	SOIL EROD. K FACTOR	CROPPING C FACTOR	CONTOUR P FACTOR	MANNINGS N
0.0	0.000	0.020	0.230	0.260	1.000	0.030
93.5	0.454	0.020	0.230	0.260	1.000	0.030
95.0	0.461	0.023	0.230	0.260	1.000	0.030
96.5	0.469	0.029	0.230	0.260	1.000	0.030
98.0	0.476	0.035	0.230	0.260	1.000	0.030
156.0	0.757	0.038	0.230	0.260	1.000	0.030
157.4	0.764	0.037	0.230	0.260	1.000	0.030
158.9	0.771	0.036	0.230	0.260	1.000	0.030
160.3	0.778	0.034	0.230	0.260	1.000	0.030
161.7	0.785	0.033	0.230	0.260	1.000	0.030
163.1	0.792	0.032	0.230	0.260	1.000	0.030
164.6	0.799	0.030	0.230	0.260	1.000	0.030
166.0	0.806	0.029	0.230	0.260	1.000	0.030
167.4	0.813	0.027	0.230	0.260	1.000	0.030
168.9	0.820	0.026	0.230	0.260	1.000	0.030
170.3	0.827	0.025	0.230	0.260	1.000	0.030
206.0	1.000	0.024	0.230	0.260	1.000	0.030

POINTS OF CHANGE ALONG THE CHANNEL

DISTANCE FEET	DISTANCE NONDIM.	SLOPE	MANN. N	WIDTH FEET	DEPTH MIDDLE FEET	DEPTH SIDE FEET	SHEAR CRIT. LB/FT**2	SHEAR COVER LB/FT**2
24.7	0.063	0.021	0.065	10.000	0.330	0.330	0.400	100.000
39.6	0.100	0.021	0.065	10.000	0.330	0.330	0.400	100.000
79.1	0.200	0.023	0.065	10.000	0.330	0.330	0.400	100.000
118.7	0.300	0.030	0.065	10.000	0.330	0.330	0.400	100.000
158.3	0.400	0.027	0.065	10.000	0.330	0.330	0.400	100.000
197.9	0.500	0.021	0.065	10.000	0.330	0.330	0.400	100.000
237.4	0.600	0.015	0.065	10.000	0.330	0.330	0.400	100.000
277.0	0.700	0.016	0.065	10.000	0.330	0.330	0.400	100.000
316.6	0.800	0.018	0.065	10.000	0.330	0.330	0.400	100.000
356.2	0.900	0.021	0.065	10.000	0.330	0.330	0.400	100.000
395.7	1.000	0.024	0.065	10.000	0.330	0.330	0.400	100.000

Figure II-32.—Sample output from the erosion/sediment yield model showing storm-by-storm and element-by-element data.

STORM INPUTS

DATE	74037	JULIAN DATE
RAINFALL	1.70	INCHES
RUNOFF VOLUME	0.26	INCHES
EXCESS RAINFALL	0.90	INCHES/HR
EI	16.78	WISCHMEIER ENGL. UNITS

VALUES FOR STORM 74037 FROM OVERLAND FLOW

THE QUANTITY OF ERODED SEDIMENT IN RUNOFF

PART. TYPE	SOIL LOSS LBS.	CONCENTRATIONS (SOIL/WATER) LBSF/FT**3	LBSF/LBSF	PPM (WT)
1	34.	0.0112	0.0002	180.
2	31.	0.0104	0.0002	166.
3	204.	0.0679	0.0011	1088.
4	74.	0.0246	0.0004	394.
5	0.	0.0001	0.0000	1.
TOTAL	343.	0.1142	0.0018	1830.

AVERAGE SOIL LOSS FOR AREA 0.05 TONS/ACRE

Figure II-32.—Sample output from the erosion/sediment yield model showing storm-by-storm and element-by-element data--continued.

VALUES FOR STORM 74037 FROM CHANNEL ONE

PEAK DISCHARGE UPPER END	0.182	FT**3/SEC
PEAK DISCHARGE LOWER END	2.914	FT**3/SEC
CONTROL DEPTH	1.088	FT

THE QUANTITY OF ERODED SEDIMENT IN RUNOFF

PART.	SOIL LOSS	CONCENTRATIONS (SOIL/WATER)		
TYPE	LBS.	LBSF/FT**3	LBSF/LBSF	PPM (WT)
1	34.	0.0112	0.0002	180.
2	31.	0.0103	0.0002	166.
3	201.	0.0670	0.0011	1073.
4	49.	0.0164	0.0003	263.
5	0.	0.0001	0.0000	1.
TOTAL	315.	0.1051	0.0017	1684.

AVERAGE SOIL LOSS FOR AREA 0.05 TONS/ACRE

DISTRIBUTION OF PRIMARY PARTICLES
AND ORGANIC MATTER IN THE ERODED SEDIMENT

TYPE	FRACTION
CLAY	0.380
SILT	0.497
SAND	0.122
ORGANIC MATTER	0.027

INDEX OF SPECIFIC SURFACE 25.05 M**2/G OF TOTAL SEDIMENT

ENRICHMENT RATIO OF SPECIFIC SURFACE 2.672

STORM INPUTS

DATE	74038	JULIAN DATE
RAINFALL	0.20	INCHES
RUNOFF VOLUME	0.00	INCHES
EXCESS RAINFALL	0.00	INCHES/HR
EI	0.66	WISCHMEIER ENGL. UNITS

*** NO RUNOFF - NO LOSSES ***

Figure II-32.—Sample output from the erosion/sediment yield model showing storm-by-storm and element-by-element data--continued.

```
                    STORM INPUTS
                    ------------

         DATE               74178    JULIAN DATE
         RAINFALL            4.26    INCHES
         RUNOFF VOLUME       1.25    INCHES
         EXCESS RAINFALL     3.40    INCHES/HR
         EI                 67.17    WISCHMEIER ENGL. UNITS


VALUES FOR THE SEGMENT   93.5 FT. FROM THE PROFILE TOP
     PARTICLE       NET SOIL LOSS     RILL SOIL LOSS
       TYPE             (TONS/ACRE OF SEGMENT)
       ----             -------           -------
         1               0.03              0.01
         2               0.03              0.00
         3               0.27              0.04
         4               0.32              0.05
         5               0.55              0.08

       TOTAL             1.20              0.18

VALUES FOR THE SEGMENT   95.0 FT. FROM THE PROFILE TOP
     PARTICLE       NET SOIL LOSS     RILL SOIL LOSS
       TYPE             (TONS/ACRE OF SEGMENT)
       ----             -------           -------
         1               0.04              0.01
         2               0.04              0.01
         3               0.35              0.10
         4               0.41              0.11
         5               0.70              0.20

       TOTAL             1.54              0.43

VALUES FOR THE SEGMENT   96.5 FT. FROM THE PROFILE TOP
     PARTICLE       NET SOIL LOSS     RILL SOIL LOSS
       TYPE             (TONS/ACRE OF SEGMENT)
       ----             -------           -------
         1               0.05              0.02
         2               0.05              0.02
         3               0.44              0.15
         4               0.51              0.17
         5               0.88              0.29

       TOTAL             1.93              0.65

VALUES FOR THE SEGMENT   98.0 FT. FROM THE PROFILE TOP
     PARTICLE       NET SOIL LOSS     RILL SOIL LOSS
       TYPE             (TONS/ACRE OF SEGMENT)
       ----             -------           -------
         1               0.07              0.03
         2               0.06              0.03
         3               0.56              0.22
         4               0.65              0.26
         5               1.11              0.45

       TOTAL             2.45              0.99
```

Figure II-33.—Sample output from the erosion/sediment yield model for successive segments

VALUES FOR THE SEGMENT 156.0 FT. FROM THE PROFILE TOP

PARTICLE TYPE	NET SOIL LOSS (TONS/ACRE OF SEGMENT)	RILL SOIL LOSS
1	0.09	0.05
2	0.08	0.04
3	0.73	0.38
4	0.86	0.45
5	1.46	0.76
TOTAL	3.23	1.68

VALUES FOR THE SEGMENT 157.4 FT. FROM THE PROFILE TOP

PARTICLE TYPE	NET SOIL LOSS (TONS/ACRE OF SEGMENT)	RILL SOIL LOSS
1	0.10	0.06
2	0.10	0.06
3	0.83	0.49
4	0.98	0.57
5	1.67	0.97
TOTAL	3.68	2.15

VALUES FOR THE SEGMENT 158.9 FT. FROM THE PROFILE TOP

PARTICLE TYPE	NET SOIL LOSS (TONS/ACRE OF SEGMENT)	RILL SOIL LOSS
1	0.10	0.06
2	0.09	0.05
3	0.80	0.46
4	0.94	0.54
5	1.60	0.93
TOTAL	3.54	2.05

VALUES FOR THE SEGMENT 160.3 FT. FROM THE PROFILE TOP

PARTICLE TYPE	NET SOIL LOSS (TONS/ACRE OF SEGMENT)	RILL SOIL LOSS
1	0.09	0.05
2	0.09	0.05
3	0.76	0.43
4	0.89	0.51
5	1.52	0.87
TOTAL	3.36	1.91

VALUES FOR THE SEGMENT 161.7 FT. FROM THE PROFILE TOP

PARTICLE TYPE	NET SOIL LOSS (TONS/ACRE OF SEGMENT)	RILL SOIL LOSS
1	0.09	0.05
2	0.08	0.05
3	0.72	0.40
4	0.84	0.47
5	1.44	0.80
TOTAL	3.18	1.77

Figure II-33.—Sample output from the erosion/sediment yield model for successive segments--continued.

VALUES FOR THE SEGMENT 163.1 FT. FROM THE PROFILE TOP

PARTICLE TYPE	NET SOIL LOSS	RILL SOIL LOSS
	(TONS/ACRE OF SEGMENT)	
1	0.08	0.05
2	0.08	0.04
3	0.68	0.37
4	0.80	0.44
5	1.36	0.74
TOTAL	3.00	1.64

VALUES FOR THE SEGMENT 164.6 FT. FROM THE PROFILE TOP

PARTICLE TYPE	NET SOIL LOSS	RILL SOIL LOSS
	(TONS/ACRE OF SEGMENT)	
1	0.08	0.04
2	0.07	0.04
3	0.64	0.34
4	0.75	0.40
5	1.28	0.68
TOTAL	2.83	1.51

VALUES FOR THE SEGMENT 166.0 FT. FROM THE PROFILE TOP

PARTICLE TYPE	NET SOIL LOSS	RILL SOIL LOSS
	(TONS/ACRE OF SEGMENT)	
1	0.07	0.04
2	0.07	0.04
3	0.60	0.31
4	0.71	0.37
5	1.21	0.63
TOTAL	2.67	1.39

VALUES FOR THE SEGMENT 167.4 FT. FROM THE PROFILE TOP

PARTICLE TYPE	NET SOIL LOSS	RILL SOIL LOSS
	(TONS/ACRE OF SEGMENT)	
1	0.07	0.04
2	0.07	0.03
3	0.57	0.29
4	0.67	0.34
5	1.14	0.57
TOTAL	2.51	1.27

Figure II-33.__Sample output from the erosion/sediment yield model for successive segments--continued.

VALUES FOR THE SEGMENT 168.9 FT. FROM THE PROFILE TOP

PARTICLE TYPE	NET SOIL LOSS (TONS/ACRE	RILL SOIL LOSS OF SEGMENT)
1	0.07	0.03
2	0.06	0.03
3	0.53	0.26
4	0.62	0.31
5	1.06	0.52
TOTAL	2.35	1.15

VALUES FOR THE SEGMENT 170.3 FT. FROM THE PROFILE TOP

PARTICLE TYPE	NET SOIL LOSS (TONS/ACRE	RILL SOIL LOSS OF SEGMENT)
1	0.06	0.03
2	0.06	0.03
3	0.50	0.24
4	0.58	0.28
5	1.00	0.47
TOTAL	2.20	1.04

VALUES FOR THE SEGMENT 206.0 FT. FROM THE PROFILE TOP

PARTICLE TYPE	NET SOIL LOSS (TONS/ACRE	RILL SOIL LOSS OF SEGMENT)
1	0.06	0.03
2	0.06	0.03
3	0.49	0.23
4	0.57	0.27
5	0.98	0.46
TOTAL	2.16	1.02

VALUES FOR STORM 74178 FROM OVERLAND FLOW

THE QUANTITY OF ERODED SEDIMENT IN RUNOFF

PART. TYPE	SOIL LOSS LBS.	CONCENTRATIONS (SOIL/WATER) LBSF/FT**3	LBSF/LBSF	PPM (WT)
1	372.	0.0255	0.0004	409.
2	345.	0.0237	0.0004	380.
3	3013.	0.2068	0.0033	3315.
4	3531.	0.2424	0.0039	3885.
5	6022.	0.4134	0.0066	6626.
TOTAL	13284.	0.9120	0.0146	14615.

AVERAGE SOIL LOSS FOR AREA 2.08 TONS/ACRE

Figure II-33.—Sample output from the erosion/sediment yield model for successive segments--continued.

VALUES FOR THE SEGMENT 39.6 FT. FROM THE CHANNEL TOP

PARTICLE TYPE	NET SOIL LOSS	CHAN SOIL LOSS
	(LBS/FT OF CHANNEL SEGMENT)	
1	1.02	0.08
2	0.95	0.08
3	8.28	0.67
4	9.71	0.79
5	16.56	1.34
TOTAL	36.52	2.95

VALUES FOR THE SEGMENT 79.1 FT. FROM THE CHANNEL TOP

PARTICLE TYPE	NET SOIL LOSS	CHAN SOIL LOSS
	(LBS/FT OF CHANNEL SEGMENT)	
1	1.24	0.30
2	1.15	0.28
3	10.03	2.41
4	11.75	2.83
5	20.04	4.83
TOTAL	44.21	10.64

VALUES FOR THE SEGMENT 118.7 FT. FROM THE CHANNEL TOP

PARTICLE TYPE	NET SOIL LOSS	CHAN SOIL LOSS
	(LBS/FT OF CHANNEL SEGMENT)	
1	1.80	0.86
2	1.67	0.80
3	14.57	6.95
4	17.07	8.15
5	29.12	13.90
TOTAL	64.22	30.65

VALUES FOR THE SEGMENT 158.3 FT. FROM THE CHANNEL TOP

PARTICLE TYPE	NET SOIL LOSS	CHAN SOIL LOSS
	(LBS/FT OF CHANNEL SEGMENT)	
1	2.29	1.35
2	2.13	1.26
3	18.57	10.96
4	21.77	12.84
5	37.13	21.91
TOTAL	81.89	48.32

Figure II-33. Sample output from the erosion/sediment yield model for successive segments--continued.

VALUES FOR THE SEGMENT 197.9 FT. FROM THE CHANNEL TOP

PARTICLE TYPE	NET SOIL LOSS (LBS/FT OF CHANNEL SEGMENT)	CHAN SOIL LOSS
1	2.31	1.37
2	2.15	1.27
3	18.72	11.11
4	21.94	13.02
5	37.42	22.20
TOTAL	82.54	48.97

VALUES FOR THE SEGMENT 237.4 FT. FROM THE CHANNEL TOP

PARTICLE TYPE	NET SOIL LOSS (LBS/FT OF CHANNEL SEGMENT)	CHAN SOIL LOSS
1	1.93	0.99
2	1.79	0.92
3	15.64	8.03
4	18.33	9.41
5	31.27	16.05
TOTAL	68.97	35.40

VALUES FOR THE SEGMENT 277.0 FT. FROM THE CHANNEL TOP

PARTICLE TYPE	NET SOIL LOSS (LBS/FT OF CHANNEL SEGMENT)	CHAN SOIL LOSS
1	1.80	0.86
2	1.67	0.80
3	14.57	6.96
4	17.08	8.15
5	29.13	13.91
TOTAL	64.24	30.67

Figure II-33. Sample output from the erosion/sediment yield model for successive segments--continued.

VALUES FOR THE SEGMENT 316.6 FT. FROM THE CHANNEL TOP

PARTICLE TYPE	NET SOIL LOSS	CHAN SOIL LOSS
	(LBS/FT OF CHANNEL SEGMENT)	
1	1.65	0.71
2	1.53	0.66
3	13.35	5.74
4	15.65	6.73
5	26.69	11.48
TOTAL	58.88	25.31

VALUES FOR THE SEGMENT 356.2 FT. FROM THE CHANNEL TOP

PARTICLE TYPE	NET SOIL LOSS	CHAN SOIL LOSS
	(LBS/FT OF CHANNEL SEGMENT)	
1	2.35	1.41
2	2.18	1.31
3	19.06	11.44
4	22.34	13.41
5	38.09	22.88
TOTAL	84.02	50.46

VALUES FOR THE SEGMENT 395.7 FT. FROM THE CHANNEL TOP

PARTICLE TYPE	NET SOIL LOSS	CHAN SOIL LOSS
	(LBS/FT OF CHANNEL SEGMENT)	
1	1.99	1.05
2	1.81	0.94
3	14.86	7.25
4	-29.33	-38.26
5	-74.18	-89.40
TOTAL	-84.85	-118.42

igure II-33. Sample output from the erosion/sediment yield model for successive segments--continued.

```
VALUES FOR STORM 74178 FROM CHANNEL ONE
---------------------------------------

PEAK DISCHARGE UPPER END    0.686 FT**3/SEC
PEAK DISCHARGE LOWER END   10.981 FT**3/SEC
CONTROL DEPTH               1.962 FT

THE QUANTITY OF ERODED SEDIMENT IN RUNOFF

PART.    SOIL LOSS        CONCENTRATIONS (SOIL/WATER)
TYPE       LBS.      LBSF/FT**3    LBSF/LBSF     PPM (WT)
----     -------       ------        -----       --------
  1         725.       0.0498       0.0008          798.
  2         672.       0.0461       0.0007          739.
  3        5827.       0.4000       0.0064         6411.
  4        4979.       0.3418       0.0055         5478.
  5        7536.       0.5173       0.0083         8291.

TOTAL     19739.       1.3551       0.0217        21716

       AVERAGE SOIL LOSS FOR AREA   3.09 TONS/ACRE


          DISTRIBUTION OF PRIMARY PARTICLES
        AND ORGANIC MATTER IN THE ERODED SEDIMENT
        -----------------------------------------
                TYPE            FRACTION
            ------------        --------
                CLAY             0.176
                SILT             0.246
                SAND             0.578
            ORGANIC MATTER       0.013

INDEX OF SPECIFIC SURFACE    11.74 M**2/G OF TOTAL SEDIMENT

       ENRICHMENT RATIO OF SPECIFIC SURFACE   1.252
```

Figure II-33. Sample output from the erosion/sediment yield model for successive segments--continued.

this output. The soil loss values for the overland flow or channel segment are the net loss or gain of sediment from the segment, that is, net loss (or deposition) = [sediment out - sediment in + lateral contribution + flow detachment (or deposition)]/[area (or length)]. Negative values indicate deposition, but positive values do not necessarily indicate flow detachment. A positive value indicates a net loss value, which simply means that more sediment left the lower end of the segment than entered the upper end. This increase could be from lateral inflow or lateral inflow plus flow detachment. An increase in the soil loss per unit watershed area from the overland flow element to the channel element indicates net channel erosion, while a decrease indicates net channel deposition.

The option chosen depends on the type of information needed to reach a management decision. If long-term averages are important, annual summaries are adequate. If storm-to-storm variability is needed, the third option is chosen. The fourth option is selected to identify critical areas in the watershed where rates of erosion or deposition are large and when sediment yield for a design storm is needed.

MODEL APPLICATION

The intended application of the model is to evaluate sediment yield and the particle composition of the sediment as influenced by rainfall and runoff, soil, topography, and management practices. For a given site, management practices would be studied to identify management schemes that limit total sediment yield and yield of clay to some tolerable level. Table II-32 summarizes values of sediment yield abstracted from simulation runs for 14 storms on 17 different management practices.

In some situations, makeup of the sediment is as important as the total amount. Figure II-30 shows how some situations affect the particle fractions in the sediment yield.

Interpretation of the results indicates several important considerations. If the tolerable sediment yield for the simulation time period is 3 tons/acre, nine practices would be acceptable. Concave slopes, especially those less than 0.5% over an extended distance at the toe, significantly reduce sediment yield by inducing deposition. Sediment yield from terraces depends on their grade. Erosion was calculated in the 1% terrace grade, while deposition was calculated in the 0.25% grade. All terraces are not equally effective in controlling sediment yield.

Table II-32 shows that delivery ratio is not constant for all storms and a single value, such as terraces, cannot be used for a management system. The delivery ratios in table II-32 are from model output. The model does not use delivery ratio to compute sediment yield.

While deposition reduces sediment yield, it segregates the sediment enriching the fines (fig. II-30). Since the composition depends on rainfall and runoff characteristics, a single design storm is inadequate to evaluate the effectiveness of best management practices to control pollution.

The breadth of the conditions in table II-32 indicates the ability of the model to consider such watershed conditions as slope shape, restricted outlets, eroded drainageways, and a broad range of management practices. Model parameter values are readily available without calibration. Accuracy of the results is believed to equal or exceed that of most available models.

Table II-32.—Typical best management practices that can be analyzed with the model and typical estimates for sediment yield

Practice	All 14 storms		Small storm EI = 3.6, Runoff = 0.11 in		Large storm EI = 45.4, Runoff = 1.74 in	
	Sediment yield	Computed delivery ratio	Sediment yield	Computed delivery ratio	Sediment yield	Computed delivery ratio
	(tons/acre)		(tons/acre)		(tons/acre)	
1. Conventional	13.18	1/ 1.00	0.21	1/ 1.00	10.63	1/ 1.00
2. Conventional, complex slope with concave at toe.	2.29	1/ .17	.01	1/ .05	2.12	1/ .20
3. Strip cropping, grass buffer strip.	.78	1/ .06	.00	1/ .00	.72	1/ .07
4. Conventional, concentrated flow.	16.33	1/ 1.24	.21	1/ 1.00	12.68	1/ 1.19
5. Conventional, concentrated flow, restricted outlet.	12.42	1/ .94	.14	1/ .67	10.04	1/ .94
6. Conventional, grass waterway.	5.40	1/ .41	.05	1/ .24	4.70	1/ .44
7. Conventional, 40 ft terrace interval, 1% grade.	9.86	1/ .75	.10	1/ .48	8.88	1/ .84
8. Conventional, 40 ft terrace interval 0.8% grade.	7.00	1/ .53	.10	1/ .48	6.11	1/ .57
9. Conventional, 40 ft terrace interval, 0.5% grade.	4.62	1/ .35	.10	1/ .48	3.78	1/ .36

Table II-32.—Typical best management practices that can be analyzed with the model and typical estimates for sediment yield--continued.

Practice	All 14 storms		Small storm EI = 3.6, Runoff = 0.11 in		Large storm EI = 45.4, Runoff = 1.74 in	
	Sediment yield	Computed delivery ratio	Sediment yield	Computed delivery ratio	Sediment yield	Computed delivery ratio
	(tons/acre)		(tons/acre)		(tons/acre)	
10. Conventional, 40 ft terrace interval, 0.25% grade.	2.86	1/0.22	0.05	1/0.24	2.36	1/0.22
11. Conventional, impoundment.	.20	1/.02	.01	1/.03	.13	1/.01
12. Chisel, 4500 lb/acre, 50% cover.	2.33	2/.00	.02	2/1.00	2.16	2/1.00
13. Chisel, 2000 lb/acre, 20% cover.	5.87	2/1.00	.07	2/1.00	5.31	2/1.00
14. No-till, 4500 lb/acre, 80% cover.	.92	2/1.00	.01	2/1.00	.83	2/1.00
15. No-till in killed sod.	.15	2/1.00	.00	2/1.00	.14	2/1.00
16. Chisel, 2000 lb/acre, 20% cover, 40 ft terrace, 0.5% grade.	2.41	3/.41	.01	3/.09	2.17	3/.41
17. No-till, 4500 lb/acre, 80% cover, 40 ft terrace, 0.5% grade.	1.30	3/1.41	.00	3/.08	1.22	3/1.47

1/Ratio of sediment yield at outlet to sediment yield from uniform slope, conventional management.

2/Ratio of sediment yield with practice to same practice on uniform slope.

3/Ratio of sediment yield at terrace outlet to sediment yield from uniform slope with no terraces. Slope length and steepness = 160 ft and 6 pct, respectively. Corn at seedbed time.

Table II-33.—Analysis of several best management practices for the sample Piedmont watershed

Practice	Sediment yield (SY)[1/]	Enrichment ratio (ER) for specific surface area	Product SY*ER
	(tons/acre)		
1. Continous corn, moldboard plow, disk, cultivate, unprotected waterway.	6.9	1.8	12.4
2. Same as (1) except grassed waterway.	2.4	2.7	6.5
3. Same as (1) except chisel plow, no cultivation, and a grassed waterway.	1.2	2.3	2.8
4. Same as (1) except conventional terraces on a 0.2% grade and a grass outlet channel.	1.7	2.8	4.8
5. Same as (1) except impoundment at lower end of waterway.	.7	4.2	2.9

[1/]Total for approximately 1-2/3 yr of record.

Practice (1) for the Piedmont watershed reflects sediment yield from the field where the waterway outlet is restricted, causing ponding and deposition. Sediment yield from overland flow was estimated at 8.1 tons/acre. This erosion is calculated but is not printed out when channel elements are used. It was obtained by rerunning the model and deleting the channel component.

Installing a grass waterway reduces sediment yield by 65%. Although some of this reduction is due to elimination of erosion in the waterway, much of the reduction is due to deposition in the waterway as shown by the increased enrichment ratio. Deposition in the waterway, however, may cause difficult maintenance problems.

Chisel plowing limited sediment yield by reducing erosion on the field surface. Terraces and the impoundment control sediment yield by inducing deposition. Practices that reduce sediment by deposition increase enrichment due to an increase in the fractions for fines and organic matter.

Table II-34.—Analysis of 3 best management practices for the sample Delta watershed

Practice	Sediment yield (SY)[1/]	Enrichment ratio (ER) for specific surface area	Product SY*ER
	(tons/acre)		
1. Continuous cotton, fall tillage, multiple spring tillage, grassed field ditch.	15.8	2.5	39.5
2. Same as (1) except no fall tillage, winter cover, and a 20-ft grassed buffer strip along edge of field.	9.8	2.7	26.5
3. Same as (2) except limited spring tillage.	8.7	2.8	24.4

[1/]Total for 3 yr of record.

The grass buffer strip cut sediment yield by about one-third for the Delta watershed. Winter cover and reduced tillage somewhat reduced sediment yield. Even with these practices, the soil is relatively bare for a significant portion of the year. The enrichment ratios are high because much deposition occurs in the row middles which significantly enriches the clays. Practically no sand leaves the field even for the poorest protection.

Grassed waterways significantly reduced sediment yield for the West Tennessee watershed as well as the Piedmont watershed.

To establish the contribution of overland flow, runs were made with an overland element alone for both the complex slope shape and a uniform shape having the same slope as the average slope of the complex slope. Sediment yield was 17.8 tons/acre from the complex overland flow profile and 76.7 tons/acre from the uniform slope. The difference between 76.7 and 17.8 is due to slope shape. Much of the difference was due to deposition on the concave portion on the lower part of the complex slope. The difference between 17.8 and 34.3 tons/acre in table II-35 for practice 1 is due to erosion by concentrated flow. Even though grassed waterways controlled erosion by concentrated flow, erosion on the steeper portions of the overland flow slope was excessive. A practice such as practice 4 in table II-35 which controls both sheet and rill erosion, erosion by concentrated flow, and sediment yield is desireable.

The effect of most practices on the west Tennessee watershed was similar to that for the Piedmont watershed. As expected, the enrichment ratio generally increased as sediment yield decreased. Scatter is great in the relation-

Table II-35.—Analysis of several best management practices for the sample West Tennessee watershed

Practice	Sediment yield (SY)[1/]	Enrichment ratio (ER) for specific surface area	Product SY*ER
	(tons/acre)		
1. Continuous corn, moldboard plow, disk, cultivate, unprotected waterways.	34.3	1.9	65.2
2. Same as (1) except grassed waterways.	12.9	2.8	36.1
3. Permanent pasture.	.12	2.2	.2
4. Same as (1) except no-till (3400 lb/acre 60% cover), grassed waterway.	4.9	2.7	13.2
5. Same as (1) except impoundments at end of tributary waterways.	6.5	4.2	27.3

[1/] Total for 3 yr of record.

ship, however. Note the low enrichment ratio for the pasture practice on the west Tennessee watershed. For this situation, sediment yield was controlled by detachment, which was limited by surface cover. Conversely, note the large enrichment ratios for large sediment yield rates for the Delta watershed. These enrichment ratios are large because deposition controlled sediment yield. Using this type of model is advantageous in that the model can represent complex interactions.

The product of sediment yield and enrichment ratio is a pollution index for the sediment in that it measures the amount and fineness of sediment. Viewed in that perspective of the cropping system analyzed, the best are the chisel system on the Piedmont watershed, the limited tillage and grass buffer strip system on the Delta watershed, and the no-till system on the West Tennessee watershed. Depending on the selected tolerance level, one or more practices might be acceptable. Of practices that give sediment loads meeting the tolerance level, the farmer selects the one that best fits his total farming operation

CALIBRATION

Obviously, model results can be improved by calibration. However, if calibration is used, the following cautions should be observed.

Carefully inspect the quality of the observed data. Especially make sure that deposition at the flume did not reduce the data to a measure of transport capacity through the flume.

Keep the calibrated parameters minimal. On areas of overland flow, the parameters most likely to be in error are soil erodibility factor (up to a factor of 2 to 3) and Manning's n (a factor of 2 to 3). The soil loss ratios represent well the influence of management practices, although in a given situation they might be off by a factor of 2. For overland flow, therefore, the calibrated variables should be limited to soil erodibility and Manning's n.

The main calibration factors in the channels, are soil erodibility (off by a factor of 2 to perhaps 5), critical shear stress (off by a factor of 2 to perhaps 5), outlet control characteristics, and Manning's n. Manning's n is reasonably well defined for channels, but the model does not allow it to vary with discharge or flow depth. If the outlet control is calibrated, use a rating-curve control.

Although a representative particle size could be selected by calibration, input the primary particle size and use the default distribution. Distribution of primary particles of the soil does not represent the distribution of sediment particles for most agricultural soils.

Intake rate, shape parameters, and orifice coefficients are calibratable parameters for the pond.

Overall, peak runoff rate should be considered to be a calibratable parameter. The user should recognize the magnitude of errors likely in estimating peak runoff rate. Also, variation of Manning's n values, discussed above, could affect peak runoff rates.

Use calibration sparingly. Calibration for one management practice will not insure an adequate evaluation of an alternate management practice on the same watershed. The model parameter values have been given to minimize the need for calibration.

REFERENCES

(1) Chow, V. T.
1959. Open-channel hydraulics. McGraw-Hill Book Company, Inc., New York, N.Y., 680 pp.

(2) Davis, S. S.
1978. Deposition of nonuniform sediment by overland flow on concave slopes. MS Thesis, Purdue University, West Lafayette, Ind., 137 pp.

(3) Laflen, J. M., H. P. Johnson, and R. O. Hartwig.
1978. Sedimentation modeling of impoundment terraces. Transactions of the American Society of Agricultural Engineers 21(6):1131-1135.

(4) ------, H. P. Johnson, and R. C. Reeve.
1972. Soil loss from tile outlet terraces. Journal Soil and Water Conservation 27(2):74-77.

(5) Lane, L. J., D. A. Woolhiser, and V. Yevjevich.
1975. Influence of simplification in watershed geometry in simulation of surface runoff. Hydrology Paper No. 81, Colorado State University, Fort Collins, Colo., 50 pp.

(6) Lombardi, F.
1979. Universal Soil Loss Equation (USLE), runoff erosivity factor, slope length exponent, and slope steepness exponent for individual storms. PhD Thesis, Purdue University, West Lafayette, Ind.

(7) Neibling, W. H., and G. R. Foster.
1977. Estimating deposition and sediment yield from overland flow processes. In: D. T. Kao (ed.), Proceedings of the International Symposium on Urban Hydrology, Hydraulics and Sediment Control. UKY-BU114. University of Kentucky, Lexington, Ken., pp 75-86.

(8) Ree, W. O., and F. R. Crow.
1977. Friction factors for vegetated waterways of small slope. U. S. Department of Agriculture, Agricultural Research Service, Southern Region, ARS-S-151, 56 pp. (Series discontinued; Agricultural Research Service is now Science and Education Adminstration-Agricultural Research.)

(9) Rochester, E. W., and C. D. Busch.
1974. Hydraulic design for impoundment terraces. Transactions of the American Society of Agricultural Engineers 17(4):694-696, 700.

(10) Smerdon, E. T., and R. P. Beasley.
1959. Tractive force theory applied to stability of open channels in cohesive soils. Agricultural Experiment Station, University of Missouri, Research Bulletin No. 715. Columbia, Mo., 36 pp.

(11) Williams, J. R., and H. D. Berndt.
1977. Determining the Universal Soil Loss Equation's length-slope factor for watersheds. In: Soil Erosion: Prediction and Control. Soil Conservation Society of America, Ankeny, Iowa. Special Publication No. 21, pp. 217-225.

(12) Wischmeier, W. H., and D. D. Smith.
1978. Predicting rainfall erosion losses. U. S. Deparment of Agriculture, Agriculture Handbook No. 537, 58 pp.

(13) ------, C. B. Johnson, and B. V. Cross.
1971. A soil erodibility nomograph for farmland and construction sites. Journal of Soil and Water Conservation 26(5):189-193.

(14) Yalin, Y. S.
1963. An expression for bed-load transportation. Journal of the Hydraulics Division, Proceedings of the American Society of Civil Engineers 89(HY3):221-250.

Chapter 3. NUTRIENT SUBMODEL

M. H. Frere and J. D. Nowlin[1/]

INTRODUCTION

This nutrient model was developed to provide the user with estimates of nitrogen and phosphorus losses from fields. With the model, the user can simulate the effects of such best management practices as erosion control practices or timing and method of nutrient applications. The results of these simulations can be analyzed to determine if any proposed practice increases losses or which practices most effectively control nutrient losses.

The model was developed with a minimum amount of information needed for a reasonable or acceptable prediction. Most of the relations used, therefore, are simple and do not require many parameters which are frequently unavailable. The simple relations are solved sequentially rather than simultaneously, which reduces computer time.

Since the variability of physical and chemical parameters across the field is often ± 20%, our goal for overall accuracy was ± 40% over several years when average measured parameter values are used. Since the error in predicting individual storm events can be considerably greater, a wide variety of climatic conditions (10-30 yr), should be used to generate information for probability-type analysis.

Submodel Structure

A main program calls both pesticide and nutrient subprograms. The six nutrient subprograms are NUTRIN, NUT208, NUTEND, NUTRES, NUTANN, and NUTMON. Two other subroutines, ANNPCP and THEEND, are called to print annual and end-of-record headings with rainfall and runoff summaries.

Subroutine NUTRIN is called to read in values of the parameters and initial conditions for operating the model. This subroutine also prints these values at the beginning of the simulation results to document the values used.

Subroutine NUT208 is the main subprogram that calculates the movement of nutrients between compartments and subsequent losses of nutrients. The first section under "initial conditions" establishes the initial conditions for several variables and calculates the value of some parameters. These calculations

---

1/ Soil scientist, USDA-SEA-AR, Southern Region Office, New Orleans, La., and computer programmer, Agricultural Engineering Department, Purdue University, West Lafayette, Ind., respectively.

are bypassed, except when new conditions are introduced, NEWNT > 0. The rest of the section calculates the transpiration ratio for the period since the last storm, TR, the amount of nitrogen added in the current rainfall, RN, and updates or initializes some variables.

When percolation occurred during a period, denitrification (DNI) and nitrate leaching (NL) are assumed to have occurred. These values cannot exceed the current value of nitrate in the root zone, NO3. Burns' estimation of leaching (BNL) also is calculated although it only has significance at the end of the year.

Following leaching, nutrients from fertilizer, wastes, and residues are added to the soluble N and P compartments, SOLN and SOLP, and the soil nitrate, NO3, on the day of fertilization, DATE F or DF (ID). If loss of sediment occurred, SOLOSS > 0, N and P losses with the sediment, SEDN and SEDP, are calculated.

The fraction of rainfall that does not appear as runoff infiltrates the soil and leaches some N and P out of the surface layer. The leached nitrate is added to the nitrate in the root zone, which is subject to further changes. Because of the buffering capacity of the soil, the phosphate level is not permitted to be reduced below the initial soil level.

When there is runoff, RUNOFF > 0, some N and P are lost in the runoff, RON and ROP, in proportion to the concentration in the surface layer, CN and CP. These runoff losses cannot exceed the amount available in the surface layer.

Under "mineralization", the average temperature, ATP, is used to modify the rate constant, TK. The amount mineralized, MN, is calculated from the amount of potentially mineralizable nitrogen, POTM, and the soil water correction, WK, equal to the ratio of average water content to field capacity.

Under "uptake", the plant only takes up nitrogen from the nitrate in the soil between the date of emergence, DEMERG, and the harvest date, DHRVST . The value of OPT determines which option will be used for calculating nitrogen uptake.

In option 1, the amount of dry matter, DM, is calculated from the yield potential, YP, and the ratio of actual transpiration, ACTSP, to potential water use, PWU. The fraction of the total growth expected is calculated using the fraction of remaining potential transpiration, 1-SWU/PWU. The concentration of nitrogen in the plant material changes with growth and is the minimum value calculated from two power equations. The amount of nitrogen currently in the plant is the product of the dry matter and the concentration in the dry matter. Uptake, UP, is the difference between the current and the previous value.

In option 2, the time since emergence, T, is used to compute normalized probability variate, X, from the mean, DOM, and standard deviation, SD, both in days. The fraction of potential uptake is calculated using a fourth order polynominal representation of the probability curve. The actual nitrogen in the plant material is calculated using the amount of potential uptake, PU, and the transpiration ratio, TR, to account for water stress.

The amount of uptake from either option cannot be higher than the amount of nitrate in the root zone, NO3. Finally, the total runoff and sediment losses of N and P are accumulated.

The subroutine NUTRES prints out the losses of N and P after each storm. If no runoff occurs, only the uptake, mineralization, and drainage losses are printed. The average concentration of N and P in the runoff waters is PPMN and PPMP.

This information is useful in identifying when and, hence, under what conditions, the highest losses occur. It can be used to select management practices that might be effective. Since some storms are more frequent than others, the information printed out by the subroutine can be used to develop probability of occurrence graphs. These graphs are most useful in comparing the effects of management practices.

Subroutine NUTMON is called at the end of each month to print out monthly summaries of nutrient losses. The monthly summaries provide a convenient means of reviewing losses on a seasonal basis relative to rainfall and management regimes.

The subroutine ANNNUT is called at the end of the year to print out the accumulated nutrient losses during that year. This annual summary is useful because nutrient problems tend to be chronic rather than acute. Therefore, annual loads leaving a field probably are a better reflection of impact than any single storm event.

The subroutine NUTEND is called at the end of the simulation period to print out the total accumulated nutrient losses. This final summary is the best single value for evaluating best management practices for controlling nutrient pollution. Since nutrient problems are long term, the losses accumulated over many storms and years better reflect the average effects.

SELECTION OF VALUES FOR INPUT DATA

Storm/Hydrology/Erosion Data File

This file is created in the hydrology and erosion components of the model passed from the erosion component. Table II-36 shows the card format, variable name, and variable definition for the data from the erosion pass file. A sample card image arrangement for the pass file is shown in figure II-34. This file would not need to be recreated if various fertilization practices were evaluated or if the nutrient model itself was evaluated.

SDATE is the Julian date of the storm, including both the last two digits of the year and the day number, for example, 74123.

RNFALL is the inches of rainfall occurring in a storm on that date. It is converted to millimeters for use in the nutrient model.

RUNOFF is the inches of runoff from the storm and is converted to millimeters for the nutrient model.

Table II-36.—Chemistry model input

Storm/Hydrology/Erosion Data File

Card 1. SDATE, RNFALL, RUNOFF, SOLOSS, ENRICH, DP, PERCOL, AVGTMP, AVGSWC, ACCPEV, POTPEV, ACCSEV, POTSEV

SDATE	Date of storm (Julian date), e.g. 73148
RNFALL	Volume rainfall (in|cm), e.g. 4.27|10.85
RUNOFF	Volume of runoff (in|cm), e.g. 1.58|4.01
SOLOSS	Amount of eroded sediment (tons/acre|kg/ha), e.g. 4.32|9674.0
ENRICH	The sediment enrichment ratio computed with particle size distribution information, e.g. 1.30
DP	Number of days since the last storm when percolation occured, e.g. 1
PERCOL	Percolation below the root zone (in|cm), e.g. 1.015|2.58
AVGTMP	Average temperature between storms (Degrees F.|C.), e.g. 72.8|22.7
AVGSWC	Average soil water between storms (in/in), e.g. 0.3239
ACCPEV	Actual EP (evaporation from plants) for the period between storms (in|cm), e.g. 0.022|0.056
POTPEV	Potential EP for the period between storms (in|cm), e.g. 0.022|0.056
ACCSEV	Actual ES (evaporation from soil) for the period between storms (in|cm), e.g. 0.000|0.000
POTSEV	Potential ES for the period between storms (in|cm), e.g. 0.000|0.000

Card 1 is repeated for each rainfall event. The last card on the file should be blank to indicate the end of data. The Erosion program creates a file called "SEDPAS" specifically for use as this file. The values in the Storm/Hydrology/Erosion file are in English units when it is created with the Erosion program. If the file isn't created with the Erosion program then either English or Metric units may be used.

A small sample of a typical Storm/Hydrology/Erosion Data file follows. It will illustrate the file structure.

Format(I6,F6.2,F6.2,F6.2,F6.2,I2,F6.2,F6.2,F6.4,F6.3,F6.3,F6.3,F6.3)

EROSION PASS FILE EXAMPLE

```
73139  0.48  0.00  0.00  0.00 0  0.00 70.730.3171 0.002 0.002 0.000 0.000
73143  0.52  0.00  0.00  0.00 0  0.00 71.530.3138 0.010 0.010 0.000 0.000
73144  0.23  0.00  0.00  0.00 0  0.00 72.180.3191 0.004 0.004 0.000 0.000
73148  4.27  1.58  4.34  1.31 1  1.01 72.810.3239 0.022 0.022 0.000 0.000
73156  0.28  0.00  0.00  0.00 0  0.00 74.230.3509 0.065 0.065 0.000 0.000
73157  1.22  0.12  0.11  2.63 1  0.38 75.220.3664 0.008 0.008 0.000 0.000
73159  0.60  0.01  0.01  4.15 1  0.13 75.530.3625 0.017 0.017 0.000 0.000
73160  0.50  0.03  0.02  2.77 1  0.20 75.830.3664 0.009 0.009 0.000 0.000
73161  0.25  0.00  0.00  0.00 0  0.00 76.030.3654 0.009 0.009 0.000 0.000
73164  0.78  0.03  0.03  2.91 1  0.19 76.400.3597 0.026 0.026 0.000 0.000
```

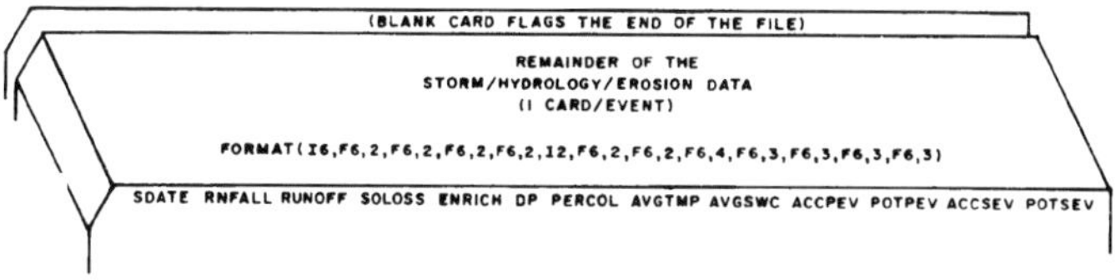

Figure II-34.—Schematic representation of a sample card deck arrangement and format for the erosion/sediment yield pass file.

SOLOSS is the tons/acre of soil loss in that storm. It is converted to kilograms/hectare and called SED in the nutrient model.

ENRICH is an enrichment factor computed in the erosion model but not used currently in the nutrient model.

DP is the number of days since the last storm when percolation occurred.

PERCOL is the inches of percolation below the root zone since the last storm. It is converted to millimeters and called PERC in the nutrient model.

AVGTMP is the average Farenheit temperature between storms. It is converted to Celsius and called ATP in the nutrient model. The temperature of the soil in the top 2 ft is preferred, but air temperature is used as an approximation.

AVGSWC is the average volumetric soil water content of the root zone between storms and is called AWC in the nutrient model.

ACCPEV is the inches of actual transpiration between storms. It is converted to millimeters and called ACTSP in the nutrient model.

POTPEV is the inches of potential transpiration between storms. It is converted to millimeters and called ACPTSP in the nutrient model.

ACCSEV is the inches of actual soil evaporation between storms. It is converted to millimeters in the nutrient model.

POTSEV is the inches of potential soil evaporation between the storms and is converted to millimeters in the nutrient model.

Nutrient Parameter File

The chemistry component of CREAMS contains the plant nutrient submodel as well as the pesticide submodel. The complete chemistry parameter listing, given in table II-37, is used here and again in chapter 4 to prevent confusion

Table II-37.—Chemistry model input parameters file

----- Initial General Parameter Inputs

Card 1-3.	TITLE()	
	TITLE	Three lines of 80 Characters each for alphanumeric information to be printed at the beginning of the output. format (20A4)
Card 4.	BDATE, FLGOUT, FLGIN, FLGPST, FLGNUT	
	BDATE	The beginning date for simulation. It <u>must</u> be less than the first storm date (SDATE). (Julian date), e.g. 73138
	FLGOUT	0 for annual summary output 1 for annual and monthly summary output 2 for individual storm and all summary output
	FLGIN	0 if the Storm Hydrology input is in English units and will need to be converted to metric 1 if the values are already in metric units
	FLGPST	0 if there will be no Pesticide inputs 1 if there will be Pesticide simulation
	FLGNUT	0 if there will be no Plant Nutrient inputs 1 if there will be Plant Nutrient simulation
Card 5.	SOLPOR, FC, OM	
	SOLPOR	Soil porosity (cc/cc), e.g. 0.41
	FC	Field capacity (cc/cc), e.g. 0.32
	OM	Organic matter available for denitrification (% of soil mass), e.g. 0.65

----- Initial Pesticide Inputs

Card 6.	NPEST, PBDATE, PEDATE	
	NPEST	Number of pesticides, e.g. 2, MAX of 10 If blank the pesticides portion of the model is bypassed.
	PBDATE	Date the model begins to consider pesticides (Julian date), e.g. 74120

Table II-37.—Chemistry model input parameters file--continued

	PEDATE	Date the model stops considering pesticides (Julian date), e.g. 75365
-----	Initial Plant Nutrient Inputs	
Card 7.	OPT	
	OPT	1 for option one nitrogen uptake 2 for option two nitrogen uptake
Card 8.	SOLN, SOLP, NO3, SOILN, SOILP, EXKN, EXKP, AN, BN, AP	
	SOLN	Soluble nitrogen (kg/ha), e.g. 0.2
	SOLP	Soluble phosphorous (kg/ha), e.g. 0.2
	NO3	Nitrate (kg/ha), e.g. 20.0
	SOILN	Soil nitrogen (kg/kg), e.g. 0.00035
	SOILP	Soil phosphorous (kg/kg), e.g. 0.00018
	EXKN	Extraction coefficient for nitrogen, e.g. 0.0576
	EXKP	Extraction coefficient for phosphorous, e.g. 0.07
	AN	Enrichment coefficient for nitrogen, e.g. 16.8
	BN	Enrichment exponent for nitrogen, e.g. -0.16
	AP	Enrichment coefficient for phosphorous, e.g. 11.2
Card 9.	BP, RCN	
	BP	Enrichment exponent for phosphorous, e.g. -0.146
	RCN	Concentration of nitrogen in rainfall (mg/l), e.g. 0.8
-----	Updateable General Parameter Inputs	

The rest of the input to the Chemicals program is updateable. The program checks the dates (SDATE, card 1) from the Storm/Hydrology/Erosion file against the parameters control date (CDATE, card 10). If the control date is less than the date of the storm, the program reads in a new set of the updateable parameters. If the program reads a blank in place of the control date (CDATE, card 10) the program stops executing.

Card 10.	PDATE, CDATE	

Table II-37.—Chemistry model input parameters file--continued

	PDATE	First date that the following chemical parameters are valid (Julian), e.g. 73138
		The program doesn't read in a value for PDATE. PDATE is only used as an aid in putting together the data file.
	CDATE	Last date that the following chemical parameters are valid, for example one day before the next pesticide application or one day before a change in the plant nutrients parameters (Julian), e.g. 73131
		NOTE: A card 10. should always be the first card in a new set of updateable parameters.

----- Updateable Pesticide Inputs

Card 11.	APDATE	
	APDATE	Date the pesticide is applied (Julian date), e.g. 73121
		if blank, cards 12-14 are not read
Card 12.	PSTNAM()	
	PSTNAM	The pesticide name, up to 24 characters, format (6A4), e.g. ATRAZINE
Card 13.	APRATE, DEPINC, EFFINC, FOLFRC, SOLFRC, FOLRES, SOLRES, WSHFRC, WSHTHR	
	APRATE	Rate of application (kg/ha), e.g. 3.36
	DEPINC	Depth of incorporation (cm), e.g. 1.0
	EFFINC	Efficiency of incorporation, e.g. 1.0
	FOLFRC	Fraction of pesticide applied to the foliage, e.g. 0.0
	SOLFRC	Fraction applied to the soil, e.g. 1.0
	FOLRES	Amount of pesticide residue on the foliage prior to this application (ug/g), e.g. 0.0
	SOLRES	Amount on the soil prior to this application (ug/g), e.g. 0.0
	WSHFRC	Fraction on the foliage available for rainfall washoff, e.g. 0.0
	WSHTHR	Rainfall threshold for foliage washoff (cm), e.g. 0.0

Card 14. SOLH2O, HAFLIF, EXTRCT, DECAY, KD

SOLH2O	Water solubility (PPM), e.g. 33.0
HAFLIF	Foliar residue half life (days), e.g. 0.0
EXTRCT	Extraction ratio, e.g. 0.1
DECAY	Decay constant, e.g. 0.10
KD	KD, e.g. 2.0

Cards 11-14 are repeated for each pesticide (NPEST, card 6). If the application date (APDATE, card 11) is blank then cards 12-14 are omitted for that pesticide and the old values, including APDATE, are retained. This is usefull when one of the pesticides is to be reapplied but others are not. If more than one pesticide is applied and the pesticides are applied on different dates, blank cards must be inserted at the appropriate places in the file for each pesticide not being applied with this update. The following example is given for clarification.

Assume 3 pesticides (NPEST, card 6) are applied with the following application dates:

Atrazine - 3/20/74 (74079), 3/27/75 (75086)
2,4-D - 4/15/74 (74105), 4/12/75 (75102)
Parathion - 6/13/74 (74164), 7/05/74 (74186),
6/20/75 (75171), 7/21/75 (75202)

The following cards 10-14 would be used:

Card 10: 74000 74104
Card 11: 74079
Card 12: Atrazine
Card 13 and 14: appropriate data
2 blank card 11's for 2,4-D and Parathion
Card 10: 74105 74163
1 blank card 11 for Atrazine
Card 11: 74105
Card 12: 2,4-D
Card 13 and 14: appropriate data
1 blank card 11 for Parathion
Card 10: 74164 74185
2 blank card 11's for Atrazine and 2,4-D
Card 11: 74164
Card 12: Parathion
Card 13 and 14: appropriate data
Card 10: 74186 75085
2 blank card 11's for Atrazine and 2,4-D
Card 11: 74186
Card 12: Parathion

Card 13 and 14: appropriate data
Card 10: 75086 75101
Card 11: 75086
Card 12: Atrazine
Card 13 and 14: appropriate data
2 blank card 11's for 2,4-D and Parathion
etc...

NOTE: Some computers read a blank card as undefined or some other type of illegal data that will result in an execution error. A zero punched in the data fields on blank cards will prevent this from occuring

----- Updateable Plant Nutrient Inputs

Card 15. NF, DEMERG, DHRVST

NF	number of fertilizer applications, e.g. 2
DEMERG	Date of plant emergence (Julian date, no year), e.g. 141
DHRVST	Date of plant harvesting (Julian date, no year), e.g. 305

When no new Plant Nutrient values are to be read card 15 should be left blank. The program will then skip reading the remaining Plant Nutrient parameters.

----- For Option One Nitrogen Uptake

Card 16. RZMAX, YP, DMY, POTM, AWU, PWU

RZMAX	Maximum depth of the root zone (mm), e.g. 450.0
YP	Potential yield (kg/ha), e.g. 5700.0
DMY	Dry matter yield ratio, e.g. 2.5
POTM	Potential mineralizable nitrogen (kg/ha), e.g. 47.0
AWU	Actual water use (mm), e.g. 570.0
PWU	Potential water use (mm), e.g. 780.0

Card 17 C1, C2, C3, C4

C1,C3	Cubic coefficients, e.g. 0.0209, 0.0128
C2,C4	Cubic exponents, e.g. -0.157, -0.415

Table II-37.—Chemistry model input parameters file--continued

----- For Option Two Nitrogen Uptake

Card 16. RZMAX, YP, DMY, POTM, DOM, SD, PU

RZMAX	Maximum depth of the root zone (mm), e.g. 450.0
YP	Potential yield (kg/ha), e.g. 5700.0
DMY	Dry matter yield ratio, e.g. 2.5
POTM	Potential mineralizable nitrogen (kg/ha), e.g. 47.0
DOM	Date of mid point in nitrogen uptake cycle (days), e.g. 73.0
SD	Standard deviation of DOM (days), e.g. 30.0
PU	Potential nitrogen uptake (kg/ha), e.g. 250.0

----- Both Options Continue

Card 18. DF(1)

DF	Date of fertilizer application (Julian date), e.g. 73131

Card 19. FN(1), FP(1), FA(1)

FN	Nitrogen applied (kg/ha), e.g. 28.0
FP	Phosphorous applied (kg/ha), e.g. 28.0
FA	Surface fraction of application, e.g. 0.1

Cards 17 and 18 are repeated for each application of fertilizer (NF, card 15). A maximum of 20 applications can be read in one update.

Table II-37.—Chemistry model input parameters file--continued

---

A sample data file for the Control Parameters for the plant nutrients model follows. It will help demonstrate the file structure

```
CARD
 NO                                   CHEMISTRY PARAMETER DATA
----   -----------------------------------------------------------------------------------------
  1                         NUTRIENTS PARAMETERS - GEORGIA PIEDMONT
  2                                 MANAGEMENT PRACTICE ONE
  3                          CONTINOUS CORN - CONVENTIONAL TILLAGE
  4       73138         1         0         0         1
  5       0.410     0.320     0.650
  7           2
  8       0.200     0.200    20.000 0.00035 0.00018 0.05760 0.07000 16.8000 -0.1600 11.2000
  9      -0.146     0.800
 10                 73305
 15           2       141       305
 16     450.0005700.000     2.500    47.000    73.000    30.000 250.000
 18       73131
 19      28.000    28.000     0.100
 18       73174
 19     112.000     0.000     1.000
 10                 74305
 15           2       129       305
 16     450.0005700.000     2.500    47.000    73.000    30.000 250.000
 18       74119
 19      28.000    28.000     0.100
 18       74162
 19     112.000     0.000     1.000
 10                 75305
 15           2       151       305
 16     450.0005700.000     2.500    47.000    73.000    30.000 250.000
 18       75141
 19      28.000    28.000     0.100
 18       75176
 19     112.000     0.000     1.000
 10                 75365
 15           0       151       305
 16     450.0005700.000     2.500    47.000    73.000    30.000 250.000
 10                     0
```

---

of overall program input. Only the nutrient parameter file will be discussed in this chapter. The chemistry component can be run on the computer with only nutrient data if the user desires.

General Parameters

Some parameters do not change significantly during a simulation period, although it is recognized that such changes may be gradual over time. If an intensive management system significantly changes the organic matter content, for example, simulation should be stopped and started again with better values for those parameters.

SOLPOR, the soil porosity, is the fraction of the soil that can be filled with water or air. The value for a soil can be calculated from the bulk density, BD, the oven dry weight of a known volume of soil. Assuming the solid density is 2.65 g/cm^3:

$$SOLPOR = 1-(BD/2.65) \qquad [II-21]$$

Values of porosity in the range of 0.3 to 0.5 are often available in reports by the SCS. This value is used as POR in the NUT208 program and POROS in the hydrology models.

FC, field capacity, is the fraction of the soil volume filled with water after a day's drainage or in equilibrium with tensions of 0.1 to 0.3 bar. If measurements are impossible, some values in the range of 0.2 to 0.4 may be found in SCS reports. The value used here must be compatible with the variable FUL used in the hydrology models.

OM, organic matter, is the percentage of the soil that is composed of biological residues. OM is 1.724 times the percent total organic carbon in the soil. Values in the range of 0.1 to 2 for OM or values for total organic carbon are often given in reports from SCS. The value for OM <u>must not</u> be the same as that used for PEROG in the erosion model, since OM is the average in the root zone. If good information is unavailable, OM can be set, realistically, as one half of PEROG.

Initial Parameters

The user can select the method of nitrogen uptake calculations as given in volume I, chapter 4. Plant nutrients in the surface soil layer and root zone at the beginning of simulation can be measured, or estimated if measurements are unavailable. These nutrient contents change with fertilizer, waste, and residue applications as well as from plant uptake, leaching, denitrification, and washoff. The model provides an accounting during the simulation, and only initial values are needed.

OPT is 1 for nitrogen uptake to be simulated by plant growth and nitrogen content. OPT is 2 when the normal probability curve is used to describe the nitrogen uptake.

SOLN and SOLP are the kilograms/hectare of soluble N and P in the surface centimeter of soil. The initial value for these parameters is best estimated by determining the equilibrium nitrate and phosphate concentrations in samples of the soil (CREAMS, vol. III, ch. 15). The next best estimate is obtained using measured data for several storms by fitting the relation PPMN=EXKN * CN where PPMN is the parts per million concentration in the runoff, EXKN is the unitless extraction coefficient, and CN is the parts per million concentration in the pore water of the surface centimeter of soil. CN is related to SOLN by

$$SOLN = 1/10\ CN * POR \qquad [II\text{-}22]$$

where POR is the porosity.

Similar relations exist for phosphate. A default value in the range of 0.01 to 0.4 for SOLN, SOLP, EXKN, and EXKP can be obtained from information in CREAMS, volume III, chapter 14 and chapter 15. Accuracy of the value for SOLP is most important, because the model assumes that SOLP never drops below this value. The presence of residues on the soil surface at the beginning of the simulation is accounted for by having a nutrient addition on day 0.

NO3 is the kilograms of Nitrate/hectare in the root zone. The initial value should come from laboratory analysis of soil samples taken from the root zone. A default value of 20 kg/ha can be used with only a small effect for a long simulation period because this variable is dynamic.

SOILN and SOILP are the contents of total nitrogen and total phosphorus in the surface soil, kilograms of nutrient per kilogram of soil. These values are available or can be estimated from SCS reports and soil test results at State experiment stations. They range from 0.0005 to 0.003 for N and 0.0001 to 0.0013 for phosphorus.

EXKN and EXKP are unitless extraction coefficients whose estimation is discussed in the preceding paragraph in connection with estimating values for SOLN and SOLP.

AN and AP are enrichment coefficients, and

BN and BP are enrichment exponents for calculating the degree of N and P enrichment in the sediment. These must be calculated from measured values of N and P in sediments. Default values are 7.4 for the coefficients and -0.2 for the exponents.

RCN is the nitrogen concentration in rainfall in parts per million. The concentration varies from slightly less to slightly more than 1 ppm. A map in the description of the nutrient model shows how the nitrogen input in rainfall varies across the country (CREAMS, vol. I, ch 4, fig. I-16).

General Updateable Parameters

The nutrient model is structured such that dates of applicability are specified by the user. Such date specification results in the program reading at the appropriate time updateable information such as fertilizer additions.

PDATE is the first date (year and Julian day) on which the updateable parameters are valid.

CDATE is the last date (year and Julian day) on which the updateable parameters are valid. CDATE would be on a day prior to fertilizer application as an example.

Updateable Parameters

Updateable parameters permit specification of the information that changes with crop or for year-to-year changes for the same crop. Some parameters are applicable to both options for nitrogen uptake. Parameters RZMAX, YP, DMY, and POTM are required by both options.

NF is the number of nutrient additions (fertilizer, wastes, residues, and so forth) that are made during the year.

DEMERG is the Julian date of plant emergence when nitrogen uptake starts.

DHRVST is the Julian date of harvest when nitrogen uptake stops.

Nitrogen uptake option 1--Nitrogen uptake by plants is calculated in this option by using the ratio of actual plant evaporation to potential plant evaporation, AWU/PWU, and cubic coefficients to estimate the nitrogen content in the crop dry matter.

RZMAX is the maximum depth of the potential root zone in millimeters. This value is best obtained from field observations because many fields have layers or conditions that limit root growth below normal values for crops given in table I-13 (CREAMS, vol. I, ch. 4) of the model documentation. The value used here must be compatible with depths used in the hydrology models.

YP is the kilograms/hectare potential yield of grain (seed cotton in the case of cotton) for the crop grown under ideal conditions. Values can be obtained from table I-11.

DMY is the ratio of total dry matter yield (grain + stover + roots) to the dry matter yield of grain.

POTM is the kilograms/hectare of potentially mineralizable nitrogen in the root zone, which should be measured with laboratory tests. Default values can be estimated from carbon or organic matter contents, using tables in CREAMS, volume III, chapter 13. Care must be taken because values of carbon or organic matter in SCS reports are for well managed soils and may be considerably higher than those for poorly managed soils. Over estimation of this parameter can cause over estimation of nitrate leaching. A low value is 50 kg/ha. POTM is included in the updateable parameters to allow resetting to account for residue added after harvest.

AWU is the millimeters of actual water used by the crop and is the actual transpiration accumulated for the year. Values of this parameter are obtained from the output of the hydrology model.

PWU is the millimeters of potential water use by the crop and is the total potential transpiration for the year. Preliminary runs of the hydrology model provides estimates of this parameter.

C1, C2, C3, and C4 are coefficients relating the nitrogen content of the crop to its stage of growth as reflected in its amount of dry matter. These coefficients for corn, sorghum, wheat, cotton, and soybeans are given in table 3 of Smith and others (vol. III, ch. 13).

<u>Nitrogen uptake option 2</u>--The previously described updateable parameters RZMAX, YP, DMY, and POTM are used with the option 2 method of estimating nitrogen uptake by the crop. Nitrogen uptake calculations in this option are based upon the number of days to reach 50% uptake, DOM, and the number of days between 50% and 84% uptake, SD, determined from the normal distribution curve.

DOM is the number of days after emergence that half the nitrogen is taken up and is equivalent to the mean of the probability distribution.

SD is the number of days required after 50% uptake to reach 84% uptake and is equivalent to the standard deviation of the probability distribution. Estimates of DOM and SD for four crops are given in CREAMS, volume III, chapter 13, table 5.

PU is the potential uptake of nitrogen, in kilograms/hectare, by the crop under ideal conditions. Values are determined best from field studies, but estimates can be made as they are for YP.

The user can specify dates and rates of fertilization and depths of incorporation. The previously described parameter, NF, number of fertilizer applications, permit the user to make split applications. Fertilizer may be incorporated at planting time and a top-dress application of nitrogen may be added later, for example.

DF is the Julian date that nutrients are applied to the field. If residues are on the field at the start of the simulation, a date of 0 can be used.

FN and FP are the kilograms/hectare of nutrients applied to the field on each of the dates, DF. The content of nutrients in residues and manures is given in tables I-11 and I-12 of the nutrient model.

FA is the application factor that is the reciprocal of the depth of application. Surface application is given a value of 1, while an application that is mixed into the top 10 cm is given a value of 1/10.

Figure II-35 schematically represents a data deck arrangement. The plant nutrient and pesticide models are both included in the same computer program.

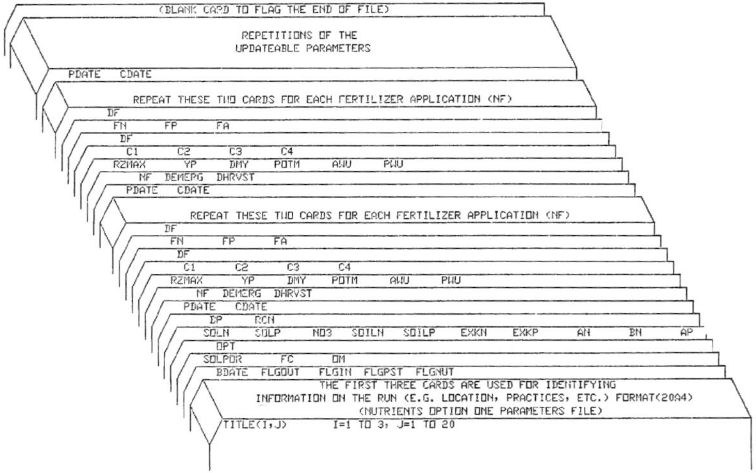

Figure II-35.—Sample format and data deck arrangement for the nutrient model.

Table II-38 is a list of parameters and definitions used in the nutrient model, and it also gives the source and relative quality of estimates.

OUTPUT

Optional output is available to the user and is specified as FLGOUT, input card 4. If only annual summaries are desired, FLGOUT = 0. A sample of an annual summary for plant nutrients is shown in figure II-36. This summary shows the total number of storms and total rainfall for the year, as well as the number of runoff-producing storms and total runoff. Unit nutrient losses are shown for the elements and the values are accumulated for the year. Total nutrient losses are not added. For example, total phosphorus would be the sum of phosphorus in runoff and phosphorus with sediment. Total nitrogen loss would include nitrogen in runoff and nitrogen with sediment. Other elements in figure II-36 include nitrogen uptake and mineralization, nitrate remaining in the soil, rainfall nitrogen, nitrate leached, and denitrification. Maximum and minimum values are given for nitrate leaching. Since leaching is difficult to estimate, the extremes are given and actual leaching would be somewhere between these values.

For storm values of nutrient losses, FLGOUT is set to 1. Figure II-37 shows a sample of nutrient losses for a storm. Summarized input data from the erosion pass file are shown at the top of this figure. The output data include the type of data used for the annual summary, as well as soluble N and P available in the surface layer. Figure II-38 shows output for a storm that did not cause runoff. The data are abbreviated since runoff and erosion did not occur. It is possible to have percolation from a storm that did not produce runoff, and therefore, nitrate leaching is included. Storm output will help the user to consider nutrient losses that might occur from storms shortly after applying fertilizer. Concentrations of nitrogen and phosphorus in runoff (fig. II-37) are averages for the storm. Storm losses also are useful in considering seasonal losses. They would be helpful in analyzing fertilizer-use efficiency as well as nonpoint source pollution.

Table II-39.—Inputs and parameters for pesticide submodel

Parameter	Definition	Source of estimate	Quality of estimate[1/]
R,[2/] ARATE.[3/] - -	Pesticide application rate.	Recommendations on label, farm records, table II-40.	Good, but may vary depending on application equipment and operator care.
ID, DEPINC. - -	Depth of pesticide incorporation.	Application recommendation, experience.	Good, but may vary depending on soil conditions.
EF, EFFINC. - -	Efficiency factor for incorporation.	Measurement, experience.	Fair to good, depending on soil conditions.
FF, FOLFRC. - -	Fraction on foliage	Model manual, experience, observations.	Fair to good, depending on source of estimate.
SF, SOLFRC. - -	Fraction on soil	Model manual, experience, observations.	Fair to good, depending on source of estimate.
FOLRES - - -	Initial foliar residue	Experience, measurement.	Unknown, depends on source of estimate.
SOLRES - - -	Initial soil residue	Measurement, inferred from past management and pesticide persistence.	Good if measured; poor if inferred.
WSHFRC - - -	Fraction of foliar pesticide washed off.	Model manual, literature.	Good for limited number of pesticides; fair to unknown for others.
THRWSH - - -	Rainfall threshold for washoff.	Judgment based on canopy.	Probably fair, subjective.
H2OSOL - - -	Pesticide solubility in water.	Handbooks, table II-40, and table II-41.	Good to excellent for most pesticides.
$C_{1/2}$, HAFLIF. - -	Foliar pesticide half-life.	Model manual, literature, measurement.	Fair to Good for limited pesticides, but is site- and condition-specific.
k_s, DECAY. - -	Dissipation rate from soil surface.	Model manual, literature, measurement.	Fair to good, but site- and condition-specific, estimates from bulk soil. Measurements often underestimate.
B, EXTRCT. - -	Extraction ratio, ratio of soil:water in mixing zone.	Model manual	Fair based on model performance, but subjective.
K_d, KD. - - - -	Distribution coefficient.	Model manual, literature, measurement.	Fair to good, but laboratory value may poorly describe field behavior.

1/ Excellent - known to be within few percent; Good - errors of 50% possible; Fair - error by factor of 2 possible; Poor - error by factor in excess of 2 possible.
2/ Notation used in documentation.
3/ Notation used in computer program.

```
            ANNUAL SUMMARY FOR 1974
            -----------------------

67 STORMS PRODUCED    102.26 CM. OF RAINFALL
14 STORMS PRODUCED      8.91 CM. OF RUNOFF


          THE PLANT NUTRIENT LOSSES
          -------------------------

NITROGEN IN RUNOFF               0.3214 KG/HA
PHOSPHORUS IN RUNOFF             0.3117 KG/HA
NITROGEN WITH SEDIMENT          17.6607 KG/HA
PHOSPHORUS WITH SEDIMENT         6.7851 KG/HA
ACCUMULATED DRAINAGE           151.16   MM
MINERALIZED N                   14.2020 KG/HA
N UPTAKE                       144.3028 KG/HA
SOIL NITRATE                     2.4268 KG/HA
RAINFALL NITRATE                 8.1808 KG/HA
ESTIMATE 1 NITRATE LEACHED      12.3446 KG/HA
BURNS ESTIMATE                  19.2652 KG/HA
ACCUMULATED DENITRIFICATION     23.8776 KG/HA
```

Figure II-36.—Sample output of annual summary from the plant nutrient component.

```
            STORM INPUTS
            ------------

DATE               74178    JULIAN DATE
RAINFALL           10.82    CM.
RUNOFF VOLUME       3.07    CM.
SOIL LOSS        4970.30    KG/HA
ENRICH. RATIO       1.55
PERCOLATION         3.55    CM.
AVG. TEMP.         25.71    DEGREES C.
AVG. SOIL WATER      .35    VOL/VOl
ACCUMULATED ET     33.93    CM.
POTENTIAL ET       77.12    CM.



THE QUANTITY OF PLANT NUTRIENTS IN RUNOFF AND LEACHED
               VALUES FOR STORM 74178

NITROGEN IN RUNOFF                       0 KG/HA
NITROGEN IN RUNOFF                       0 PPM
PHOSPHORUS IN RUNOFF                 .1074 KG/HA
PHOSPHORUS IN RUNOFF                  3500 PPM
NITROGEN WITH SEDIMENT              7.4875 KG/HA
PHOSPHORUS WITH SEDIMENT            2.8920 KG/HA

DRAINAGE THIS STORM                35.46    MM
ACCUMULATED DRAINAGE              156.69    MM
MINERALIZED N                        .9600 KG/HA
N UPTAKE                                 0 KG/HA
NITRATE LEACHED THIS STORM          9.2111 KG/HA
SOIL NITRATE                      124.3780 KG/HA
SOLUBLE N                                0 KG/HA
SOLUBLE P                            .0926 KG/HA
DENITRIFICATION                    12.4201 KG/HA
```

Figure II-37.—Sample output of nutrient data for a runoff-producing storm.

```
                STORM INPUTS
                ------------

  DATE                74171    JULIAN DATE
  RAINFALL             1.22    CM.
  RUNOFF VOLUME           0    CM.
  SOIL LOSS               0    KG/HA
  ENRICH. RATIO        2.00
  PERCOLATION             0    CM.
  AVG. TEMP.          25.00    DEGREES C
  AVG. SOIL WATER       .36    VOL/VOL
  ACCUMULATED ET      31.87    CM.
  POTENTIAL ET        72.12    CM.


     ***   NO RUNOFF - NO LOSSES   ***

DRAINAGE THIS STORM                    0     MM
ACCUMULATED DRAINAGE              121.23     MM
MINERALIZED N                     1.3656 KG/HA
N UPTAKE                               0 KG/HA
NITRATE LEACHED THIS STORM             0 KG/HA
SOIL NITRATE                     51.2943 KG/HA
SOLUBLE N                        92.6894 KG/HA
SOLUBLE P                          .2000 KG/HA
DENITRIFICATION                        0 KG/HA
```

Figure II-38. Sample output for plant nutrient data when runoff and nutrient losses did not occur.

Chapter 4. THE PESTICIDE SUBMODEL

R. A. Leonard and J. D. Nowlin[1/]

This submodel provides procedures to assess the effects of management options on potential pesticide losses in runoff. Its applicability is in making relative comparisons among options. It is not designed to provide predictions of pesticide concentrations in runoff to be used as an absolute value in making water quality assessments. The model is for field-scale application and will provide estimates of pesticide mass and storm-mean concentrations at the edge of the field. The percentage of this quantity actually reaching and impacting a body of water or stream is not addressed, and will depend on such factors as location of the field with respect to receiving waters and properties of the particular pesticide.

The impact of pesticides is caused largely by their concentrations in water rather than their total mass. Pesticide concentrations are determined by their rate of loss with respect to rate of runoff water and sediment and volume of downstream receiving water. The submodel developed here does not describe rate of pesticide transport within a single storm event. Experiments on small plots have shown that pesticide concentration in runoff may decrease several fold from the beginning of runoff to the end, depending on storm duration and mode of pesticide transport. Other experiments on field-sized or small, complex-slope watersheds have shown that distinct within-storm concentration patterns are unusual except for pesticides that are transported by sediment. Concentrations within storms usually range between certain limits in an apparently random way by factors of 2 or more, even up to 10. An explanation for this behavior is that the runoff material reaching the field edge originates at different locations within the field and has different times of travel to the measuring point. In these situations, even accurate measurement of total storm losses is difficult and representation of within-storm concentrations by models is impossible without tremendous detail. In this submodel, use of daily or storm totals generated by the hydrology and erosion submodels precludes any within-storm description.

Details of model development are in CREAMS, volume I, chapter 5. Other supporting documentation is provided in volume III, chapters 17, 18, and 19. A potential user should consult this material for general familiarization with the model.

1/ Soil scientist, Southeast Watershed Research Program, Athens, Ga., and computer programer, Agricultural Engineering Department, Purdue University, West Lafayette, Ind., respectively.

MODEL STRUCTURE

The model is structured to account for multiple applications of the same pesticide applied to soil or foliage. Different rates of dissipation or decay can be used, if necessary, for that part of the chemical residing on foliage as compared to that in soil. Movement of pesticide below the soil surface and out of the runoff-active zone as a result of infiltrating water also is estimated for potentially mobile compounds. Concentrations of pesticides in solution and in sediment are computed, as well as total mass transported by each vehicle. The initial pesticide residue concentration in the soil or on foliage, if any, at the beginning of the modeling period is initially specified. Pesticide residue remaining at the soil surface after each storm is computed, and the residue and storm pesticide runoff are printed in the output.

MAJOR ASSUMPTIONS AND SIMPLIFICATIONS

In developing the model, many simplifying assumptions were required to reduce the description of complex systems and processes to a concept that could be represented by simple mathematical expressions. The model user must be aware of these assumptions and inherent limitations to avoid misapplication or overinterpretation of the significance of the model outputs. Many assumptions and limitations imposed and summarized here are discussed at length elsewhere (CREAMS, vols. I and III).

Source of Pesticide in Runoff

The immediate soil surface is most active in supplying pesticide to runoff. In interrill areas, extraction of pesticide may occur from a soil zone only a few millimeters deep. Pesticide may be extracted, however, by active rill erosion from a zone several centimeters deep. Extraction also may occur as runoff water seeps through surface irregularities and furrows.

This model assumes an effective pesticide source zone of 1 cm deep at the soil surface. Runoff concentrations are assumed to be proportional to pesticide concentrations in this soil layer, expressed in units of micrograms per gram (ppm). Initial surface concentrations after application are computed from the rate of application and depth of incorporation. To specify a surface concentration, pesticides applied as a surface spray are assumed to be mixed uniformly with the 0- to 1-cm soil depth increment. Incorporated pesticide is assumed to be uniformly or nonuniformly mixed throughout the depth of incorporation.

Rate of Pesticide Dissipation from Soil Surface

Pesticide is assumed to dissipate from, or decay in, the surface 0- to 1-cm zone at a rate proportional to the amount present, as described by a simple exponential function commonly known as a first-order rate expression. A single parameter, referred to as the "decay constant," k_s, is used in the function to compute surface concentration as a function of time. This is a "lumped" parameter for degradation, volatilization, and other processes contributing to pesticide dissipation from the soil surface. Assumptions and limitations involved are discussed in detail in volume III. This simplification tends to underestimate dissipation rates immediately after application and overestimate dissipation rates after several weeks of pesticide contact with the soil. The model provides no direct or prescribed way of incorporating the time variable decay rate. However, the decay constant may be entered several times throughout a model application period as an updateable parameter.

Rate of Pesticide Dissipation from Foliage

The dissipation of pesticide from foliage also is assumed describable by a simple exponential decay function. In the model, foliar dissipation is described by the parameter "half-life in days" or more correctly called half-concentration time, which is related to a decay constant, k_f by: Half-life = $= 0.693/k_f$.

Mechanism of Foliar Washoff

How washoff of pesticide from foliage contributes to observed runoff is not well understood. For some pesticides, that part residing on foliage has been described experimentally in terms of a fraction that can be dislodged and a fraction that cannot be dislodged. Rainfall can remove part of the pesticide described as "dislodgeable," depending on the pesticide and, probably, leaf characteristics and time after application. In the model, a fraction of that remaining as computed from the decay function is specified as "washoff fraction." This part of the remaining pesticide is assumed to be moved from the foliage to the soil surface when rainfall exceeds a "washoff threshold," which is approximated by the amount of rainfall in centimeters that the plant canopy can intercept and store as droplets on the surface of the leaf.

Vertical Transport of Pesticide from the Soil Surface

Pesticides that are mobile in soil, that is, soluble and not strongly adsorbed, can be leached from the soil surface by infiltrating rainfall. Before runoff concentrations are estimated for a storm, surface concentrations of pesticide are reduced, depending on the amount of rainfall in excess of runoff and initial wetting as a measure of flux through the surface layer of the soil. For pesticides with solubilities greater than 1 ppm, a pesticide distribution coefficient, K_d, is assumed to describe the availability of the pesticide for transport by the infiltrating water. This procedure is approximate compared with other more exact procedures that require detailed within-storm information. This simple method, therefore, may either overestimate or underestimate vertical transport within a storm, depending on rainfall intensity and beginning of runoff relative to rainfall.

Distribution of Pesticide Between Solution and Soil Phases

As indicated, a coefficient, K_d, is assumed to describe the distribution of pesticide between the water or solution phase and the adsorbed phase. This coefficient is defined as the ratio of the concentration in soil (μg/g) to the concentration in solution at equilibrium (μg/ml). Values of K_d normally are assigned from equilibrium experiments in the laboratory using soil suspensions containing added pesticide. In the model, the most serious assumptions regarding the use of K_d are:

(1) K_d is independent of pesticide concentration. This assumption is discussed in detail in volume III, chapter 19. Where this assumption is violated, the affinity of the soil or sediment for pesticide generally would be underestimated at very low concentrations.

(2) Adsorption-desorption processes in soil are reversible, and equilibrium is achieved rapidly. Many runoff experiments have shown that, with time of contact in soil, pesticide becomes more difficult to displace in water; that is, the apparent K_d increases. This observation may be related to both irreversible adsorption and to the dependence of K_d concentration. Equilibration in the dynamic runoff stream probably is never achieved. If the desorption rate is slow in relation to changes in the ratio of water to soil in the runoff-active zone, solution extraction and transport will be overestimated.

Serious errors in applying and interpreting the model can be avoided if K_d values are used to distinguish behavioral differences between major pesticide classes (weakly adsorbed, moderately adsorbed, and so forth) as reflected by K_d differences of an order of magnitude. When the model is used for relative comparisons, and when used in this manner, smaller differences in K_d may be significant.

Pesticide Extraction Into Runoff

In developing the function relating concentrations of pesticide in runoff water to concentration in the soil, assuming a value was necessary for the ratio of soil:water in the mixing zone at the surface. Otherwise, it must be assumed that the runoff water equilibrates with the pore water or extracts pesticide from a mass of soil represented by the sediment yield. The total mass of pesticide at the soil surface can be computed as a potential runoff source from concentration and assumed depth. During a runoff event, however, all of this pesticide does not react, or is not mixed, with runoff water. The extraction ratio parameter, B, as defined, represents the effective soil:water ratio in the mixed zone during a runoff event. The value of this parameter cannot be measured directly and should be related to storm intensity, slope, and other factors. A limited range of values for B is required, however, for satisfactory predictions. Insufficient data are available to relate B or another representation of the mixing zone to site and storm characteristics.

The model assumes that as the soil is mixed with runoff water at the soil: water ratio specified, a distribution of pesticide between the solution and soil phase is approached as approximated by K_d. Approximate equilibrium conditions must be assumed. The solution concentration predicted at the field edge is assumed to be the same as determined above. In the mixing zone, however, the absorbed phase concentration computed is for the soil, not sediment. The concentration in the sediment delivered at the field edge is assumed to be increased by an enrichment factor or ratio reflecting preferential removal and transport of clay and organic matter. An enrichment ratio is computed in the erosion model based on particle characteristics of the sediment compared to characteristics of the original soil.

The model has no mechanism for limiting the maximum mass of pesticide in runoff during a single event. The surface concentration is reduced by vertical transport before runoff. Since the surface concentration is reduced by the amount in runoff only at the end of a runoff event, however, total runoff mass may be overestimated in unusually large storms, that is, > 5-8 cm of runoff.

MODEL INPUTS AND PARAMETERS

Hydrologic inputs required are rainfall and runoff volume. These are obtained from the hydrology model or input as observed data. Sediment yield is obtained from the erosion model, experimental observations, or other estimates. A hydrology pass file is used to generate an erosion pass file, which also contains the hydrology data needed in the chemistry model. Figure II-39 schematically represents the card deck from the erosion pass file. The figure was given in the previous chapter for plant nutrients, but is repeated here for user reference. The erosion model estimates enrichment factor for pesticide transported by sediment. Table II-39 identifies additional pesticide model parameters and inputs required with suggested sources of estimate and expected quality of the estimate.

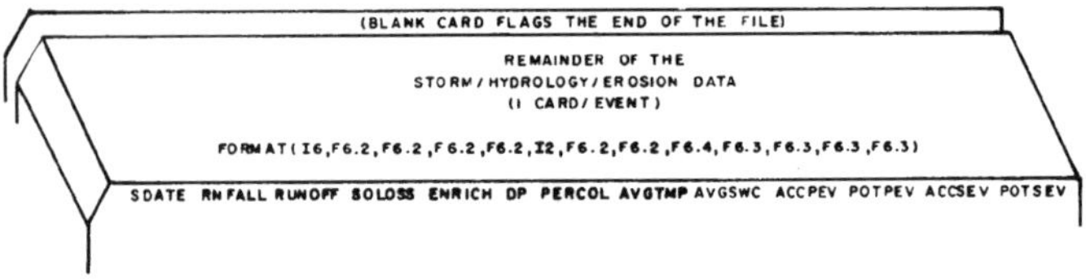

Figure II-39.—Schematic representation of a sample card deck arrangement and format for the erosion/sediment yield pass file.

Table II-38.—Nutrient model parameters

Parameter	Definition	Source of estimate	Quality of estimate	
SOILN	Soil nitrogen	Soil survey data; lab analysis; literature.	± 40% ± 20% ± 100%	Good for sampled soil series.
SOILP	Soil phosphorus	Soil survey data; lab analysis; literature.	± 40% ± 20% ± 100%	Dependent upon sampling scheme for unsurveyed soils.
EXKN	Extraction coefficient for nitrogen.	Analysis of runoff data; literature.	± 100% ± 300%	Do.
EXKP	Extraction coefficient for phosphorus.	Analysis of runoff data; literature.	± 100% ± 300%	Do.
AN	Enrichment coefficient for nitrogen.	Analysis of erosion data; literature.	± 30% ± 300%	Do.
BN	Enrichment exponent for nitrogen.	Analysis of erosion data; literature.	± 30% ± 300%	Do.
AP	Enrichment coefficient for phosphorus.	Analysis of erosion data; literature.	± 30% ± 300%	Do.
BP	Enrichment exponent for phosphorus.	Analysis of erosion data; literature.	± 30% ± 300%	Do.
FC	Field capacity	Soil survey data; lab studies.	± 30% ± 15%	Excellent for point samples; fair to poor for variability in space.
POR	Porosity	Soil survey data; lab studies.	± 30% ± 15%	Do.
POTM	Potential mineralization for nitrogen.	Lab analysis; literature.	± 20% ± 100%	Do.
RCN	Concentration of nitrogen in rainfall.	Lab analysis; literature.	± 10% ± 100%	Do.
RZMAX	Maximum depth of root zone.	Field study; soil survey.	± 20% ± 100%	Good for cultivated crops; poor for weeds, rangelands.
DOM	Date of miduptake	Local information; general information.	± 15% ± 30%	Generally not available on a local basis.
SD	Standard deviation of uptake.	Local information; general information.	± 15% ± 30%	Do.
PU	Potential nitrogen uptake.	Local information; general information.	± 15% ± 30%	Do.
YP	Yield potential	Local information; general information.	± 15% ± 30%	Occasionally available locally.
C_1; C_2; C_3; C_4	Plant nitrogen uptake coefficients.	Manual		Good for crops measured.

Selection of Input Values

The following discussion provides a guide to estimating input values and parameters. Much of this information has been extracted from comprehensive reviews and analyses in volume III of CREAMS. Where extensive data tables are required, see volume III with suggestions, if appropriate, on how to use these data. The discussions in volume III, help show how values were derived, possible errors, and how values may vary depending on site and condition. If available, use additional site-specific information from other sources rather than average or generalized information in this publication.

Table II-40 summarizes solubilities and application rates for some commonly used herbicides. Table II-41 gives solubility for some common insecticides. More complete tabulations as can be found in the handbooks referenced in these tables.

Table II-40.—Water solubility (SOLH2O) and application rate (ARATE) of commonly used herbicides[1/]

Pesticide trade name	Pesticide common name	Water solubility	Application rates[2/]
		(ppm)	(lb/acre)
AMEX 820	A-820	1.0	1 - 5
Lasso	ALACHLOR	242	1 - 4
EVIK	AMETRYNE	185	2 - 8
Amitrol-T	AMITROLE	280,000	2 - 10
Dessicant	ARSENIC ACID	Freely	1.5
AA trex	ATRAZINE	33	2 - 4
Balan	BENEFIN	<1	1.12 - 1.5
Basagran	BENTAZON	5%	0.5 - 1.5
Hyvar-X	BROMACIL	815	1.5 - 24
Machete	BUTACHLOR	23	1.5 - 4
Sutan	BUTYLATE	45	3 - 4
Bromex	CHLORBROMURON	50	0.75 - 4
Norex	CHLOROXURON	2.7	2 - 8
	2,4-D	900	0.25 - 4
DOWPON	DALAPON	Very soluble	0.75 - 20
Banvel	DICAMBA	4,500	0.06 - 10
COBEX	DINITRAMINE	1	1/3 - 2/3
DYMID, ENIDE	DIPHENAMID	260	2 - 6
Karmex	DIURON	42	0.6 - 48
Urab	FENURON	3,850	18 - 27
Cotoran	FLUOMETURON	90	0.5 - 4
Roundup	GLYPHOSATE	1.2%	1 - 4
PAARLAN	ISOPROPALIN	0.11	1 - 2
Sencor	METRIBUZIN	1,220	0.25 - 1.0
Daconate, Weed-Hoe	MSMA	Very soluble	2 - 3.8
Telvar	MONURON	230	4 - 48
Planavin	NITRALIN	0.6	0.5 - 1.5
Ryzelan	ORYZALIN	2.4	0.75 - 1.75
Ortho Paraquat	PARAQUAT	Completely	0.25 - 1
Tordan	PICLORAM	430	1 - 8
Tolban	PROFLURALIN	0.1	0.5 - 1.5
Pramitol	PROMETONE	750	10 - 60
Caparol	PROMETRYNE	48	0.48 - 2.75
Ramrod	PROPACHLOR	580	3 - 6
Milogard	PROPAZINE	8.6	1 - 4
Pyramin	PYRAZON	400	2 - 4
2,4,5-TP	SILVEX	140	0.75 - 16
Princep	SIMAZINE	5	2 - 4
2,4,5-T	2,4,5-T	238	0.5 - 16
Randox	TCBE	2	2.6
Treflan	TRIFLURALIN	1	0.5 - 2

1/ Hilton, H. L., R. W. Bovey, H. M. Hull, W. R. Mullison, and R. E. Talbert. 1974. Herbicide Handbook of the Weed Science Society of America. Third edition. Champaign, Ill. 430 pp.

2/ Range for active ingredient.

Table II-41.—Water solubility (SOLH2O) of commonly used insecticides[1/] [2/]

Insecticide trade name	Insecticide common name	Water solubility
Orthene	ACEPHATE	65%
Guthion	AZINPHOSMETHYL	29
Bux	BUFENCARB	low
Sevin	CARBARYL	40
Furadan	CARBOFURAN	700
Lorsban	CHLORPYPIFOS	2
Spectracide, Diazinon	DIAZINON	0.004%
Di-Syston	DISULFOTON	25
Dasanit	FENSULFOTHION	1,600
Cythion	MALATHION	145
Supracide	METHIDATHION	240
Lannate, Nudrin	METHOMYL	58,000
Metacide	METHYL PARATHION	50 - 60
Methyl Parathion		
Niran, Bladan	PARATHION	24
Thimet, Phorate-10G	PHORATE	50
Toxaphene	TOXAPHENE	3

1/ Berg, G. L., (ed). 1979. Farm Chemicals Handbook, Section D - Pesticide Dictionary, Merster Pub. Co., Willoughby, Ohio. 316 pp.
2/ Lawless, E. W., T. L. Ferguson, and A. F. Meiners. 1975. Guidelines for the disposal of small quantities of unused pesticides. U.S. Environmental Protection Agency Technology Series. EPA-670/2-75-057. 331 pp.

Table II-42 describes the pesticide input parameters and format, and a schematic representation of an input data deck is shown in Figure II-40.

Table II-42.—Chemistry model input parameter file

----- Initial General Parameter Inputs

Card 1-3. TITLE()

TITLE — Three lines of 80 Characters each for alphanumeric information to be printed at the beginning of the output. format (20A4)

Card 4. BDATE, FLGOUT, FLGIN, FLGPST, FLGNUT

BDATE — The beginning date for simulation. It must be less than the first storm date (SDATE). (Julian date), e.g. 73138

FLGOUT — 0 for annual summary output
1 for annual and monthly summary output
2 for individual storm and all summary output

FLGIN — 0 if the Storm Hydrology input is in English units and will need to be converted to metric
1 if the values are already in metric units

FLGPST — 0 if there will be no Pesticide inputs
1 if there will be Pesticide simulation

FLGNUT — 0 if there will be no Plant Nutrient inputs
1 if there will be Plant Nutrient simulation

Card 5 SOLPOR, FC, OM

SOLPOR — Soil porosity (cc/cc), e.g. 0.41

FC — Field capacity (cc/cc), e.g. 0.32

OM — Organic matter available for denitrification (% of soil mass), e.g. 0.65

----- Initial Pesticide Inputs

Card 6. NPEST, PBDATE, PEDATE

NPEST — Number of pesticides, e.g. 2, MAX of 10
If blank the pesticides portion of the model is bypassed.

PBDATE — Date the model begins to consider pesticides (Julian date), e.g. 74120

Table II-42.—Chemistry model input parameter file--continued

	PEDATE	Date the model stops considering pesticides (Julian date), e.g. 75365

----- Initial Plant Nutrient Inputs

Card 7.	OPT	
	OPT	1 for option one nitrogen uptake 2 for option two nitrogen uptake
Card 8.	SOLN, SOLP, NO3, SOILN, SOILP, EXKN, EXKP, AN, BN, AP	
	SOLN	Soluble nitrogen (kg/ha), e.g. 0.2
	SOLP	Soluble phosphorous (kg/ha), e.g. 0.2
	NO3	Nitrate (kg/ha), e.g. 20.0
	SOILN	Soil nitrogen (kg/kg), e.g. 0.00035
	SOILP	Soil phosphorous (kg/kg), e.g. 0.00018
	EXKN	Extraction coefficient for nitrogen, e.g. 0.0576
	EXKP	Extraction coefficient for phosphorous, e.g. 0.07
	AN	Enrichment coefficient for nitrogen, e.g. 16.8
	BN	Enrichment exponent for nitrogen, e.g. -0.16
	AP	Enrichment coefficient for phosphorous, e.g. 11.2
Card 9.	BP, RCN	
	BP	Enrichment exponent for phosphorous, e.g. -0.146
	RCN	Concentration of nitrogen in rainfall (mg/l), e.g. 0.8

----- Updateable General Parameter Inputs

The rest of the input to the Chemicals program is updateable. The program checks the dates (SDATE, card 1) from the Storm/Hydrology/Erosion file against the parameters control date (CDATE, card 10). If the control date is less than the date of the storm, the program reads in a new set of the updateable parameters. If the program reads a blank in place of the control date (CDATE, card 10) the program stops executing.

Card 10. PDATE, CDATE

	PDATE	First date that the following chemical parameters are valid (Julian), e.g. 73138 The program doesn't read in a value for PDATE. PDATE is only used as an aid in putting together the data file.
	CDATE	Last date that the following chemical parameters are valid, for example one day before the next pesticide application or one day before a change in the plant nutrients parameters (Julian), e.g. 73131 NOTE: A card 10. should always be the first card in a new set of updateable parameters.
-----	Updateable Pesticide Inputs	
Card 11.	APDATE	
	APDATE	Date the pesticide is applied (Julian date), e.g. 73121 if blank, cards 12-14 are not read
Card 12.	PSTNAM()	
	PSTNAM	The pesticide name, up to 24 characters, format (6A4), e.g. ATRAZINE
Card 13.	APRATE, DEPINC, EFFINC, FOLFRC, SOLFRC, FOLRES, SOLRES, WSHFRC, WSHTHR	
	APRATE	Rate of application (kg/ha), e.g. 3.36
	DEPINC	Depth of incorporation (cm), e.g. 1.0
	EFFINC	Efficiency of incorporation, e.g. 1.0
	FOLFRC	Fraction of pesticide applied to the foliage, e.g. 0.0
	SOLFRC	Fraction applied to the soil, e.g. 1.0
	FOLRES	Amount of pesticide residue on the foliage prior to this application (ug/g), e.g. 0.0
	SOLRES	Amount on the soil prior to this application (ug/g), e.g. 0.0
	WSHFRC	Fraction on the foliage available for rainfall washoff, e.g. 0.0
	WSHTHR	Rainfall threshold for foliage washoff (cm), e.g. 0.0

Card 14. SOLH2O, HAFLIF, EXTRCT, DECAY, KD

SOLH2O	Water solubility (PPM), e.g. 33.0
HAFLIF	Foliar residue half life (days), e.g. 0.0
EXTRCT	Extraction ratio, e.g. 0.1
DECAY	Decay constant, e.g. 0.10
KD	KD, e.g. 2.0

Cards 11-14 are repeated for each pesticide (NPEST, card 6). If the application date (APDATE, card 11) is blank then cards 12-14 are omitted for that pesticide and the old values, including APDATE, are retained. This is usefull when one of the pesticides is to be reapplied but others are not. If more than one pesticide is applied and the pesticides are applied on different dates, blank cards must be inserted at the appropriate places in the file for each pesticide not being applied with this update. The following example is given for clarification.

Assume 3 pesticides (NPEST, card 6) are applied with the following application dates:

Atrazine - 3/20/74 (74079), 3/27/75 (75086)
2,4-D - 4/15/74 (74105), 4/12/75 (75102)
Parathion - 6/13/74 (74164), 7/05/74 (74186),
6/20/75 (75171), 7/21/75 (75202)

The following cards 10-14 would be used:

Card 10: 74000 74104
Card 11: 74079
Card 12: Atrazine
Card 13 and 14: appropriate data
2 blank card 11's for 2,4-D and Parathion
Card 10: 74105 74163
1 blank card 11 for Atrazine
Card 11: 74105
Card 12: 2,4-D
Card 13 and 14: appropriate data
1 blank card 11 for Parathion
Card 10: 74164 74185
2 blank card 11's for Atrazine and 2,4-D
Card 11: 74164
Card 12: Parathion
Card 13 and 14: appropriate data
Card 10: 74186 75085
2 blank card 11's for Atrazine and 2,4-D
Card 11: 74186
Card 12: Parathion

Card 13 and 14: appropriate data
Card 10: 75086 75101
Card 11: 75086
Card 12: Atrazine
Card 13 and 14: appropriate data
2 blank card 11's for 2,4-D and Parathion
etc...

NOTE: Some computers read a blank card as undefined or some other type of illegal data that will result in an execution error. A zero punched in the data fields on blank cards will prevent this from occuring.

----- Updateable Plant Nutrient Inputs

Card 15. NF, DEMERG, DHRVST

NF	number of fertilizer applications, e.g. 2
DEMERG	Date of plant emergence (Julian date, no year), e.g. 141
DHRVST	Date of plant harvesting (Julian date, no year), e.g. 305

When no new Plant Nutrient values are to be read card 15 should be left blank. The program will then skip reading the remaining Plant Nutrient parameters.

----- For Option One Nitrogen Uptake

Card 16. RZMAX, YP, DMY, POTM, AWU, PWU

RZMAX	Maximum depth of the root zone (mm), e.g. 450.0
YP	Potential yield (kg/ha), e.g. 5700.0
DMY	Dry matter yield ratio, e.g. 2.5
POTM	Potential mineralizable nitrogen (kg/ha), e.g. 47.0
AWU	Actual water use (mm), e.g. 570.0
PWU	Potential water use (mm), e.g. 780.0

Card 17. C1, C2, C3, C4

C1,C3	Cubic coefficients, e.g. 0.0209, 0.0128
C2,C4	Cubic exponents, e.g. -0.157, -0.415

Table II-42.—Chemistry model input parameter file--continued

----- For Option Two Nitrogen Uptake

Card 16. RZMAX, YP, DMY, POTM, DOM, SD, PU

RZMAX	Maximum depth of the root zone (mm), e.g. 450.0
YP	Potential yield (kg/ha), e.g. 5700.0
DMY	Dry matter yield ratio, e.g. 2.5
POTM	Potential mineralizable nitrogen (kg/ha), e.g. 47.0
DOM	Date of mid point in nitrogen uptake cycle (days), e.g. 73.0
SD	Standard deviation of DOM (days), e.g. 30.0
PU	Potential nitrogen uptake (kg/ha), e.g. 250.0

----- Both Options Continue

Card 18. DF(1)

DF	Date of fertilizer application (Julian date), e.g. 73131

Card 19. FN(1), FP(1), FA(1)

FN	Nitrogen applied (kg/ha), e.g. 28.0
FP	Phosphorous applied (kg/ha), e.g. 28.0
FA	Surface fraction of application, e.g. 0.1

Cards 17 and 18 are repeated for each application of fertilizer (NF, card 15). A maximum of 20 applications can be read in one update.

Table II-42.—Chemistry model input parameter file--continued

A sample data file for the Control Parameters for pesticides follows. It will help demonstrate the file structure.

```
CARD
 NO                          CHEMISTRY PARAMETER DATA
----    ---------------------------------------------------------------------------
 1                        PESTICIDES PARAMETERS - GEORGIA PIEDMONT
 2                                MANAGEMENT PRACTICE ONE
 3                         CONTINOUS CORN - CONVENTIONAL TILLAGE
 4          73138        0        0        1        0
 5          0.410    0.320    0.650
 6              2    74120    75365
10                   73131
11          73121
12       ATRAZINE
13          3.360    1.000    1.000    0.000    1.000    0.000    0.000    0.000    0.000
14           33.0      0.0   0.1000   0.1000      2.0
11              0
10                   74120
11              0
11          73132
12       PARAQUAT
13          2.049    1.000    1.000    0.000    1.000    0.000    0.000    0.000    0.000
14       500000.0      0.0   0.1000   0.0070100000.0
10                   74131
11          74121
12       ATRAZINE
13          3.360    1.000    1.000    0.000    1.000    0.000    0.000    0.000    0.000
14           33.0      0.0   0.1000   0.1000      2.0
11              0
10                   75120
11              0
11          74132
12       PARAQUAT
13          2.049    1.000    1.000    0.000    1.000    0.000    0.000    0.000    0.000
14       500000.0      0.0   0.1000   0.0070100000.0
10                   75131
11          75121
12       ATRAZINE
13          3.360    1.000    1.000    0.000    1.000    0.000    0.000    0.000    0.000
14           33.0      0.0   0.1000   0.1000      2.0
11              0
10                   75366
11              0
11          75132
12       PARAQUAT
13          2.049    1.000    1.000    0.000    1.000    0.000    0.000    0.000    0.000
14       500000.0      0.0   0.1000   0.0070100000.0
10                       0
```

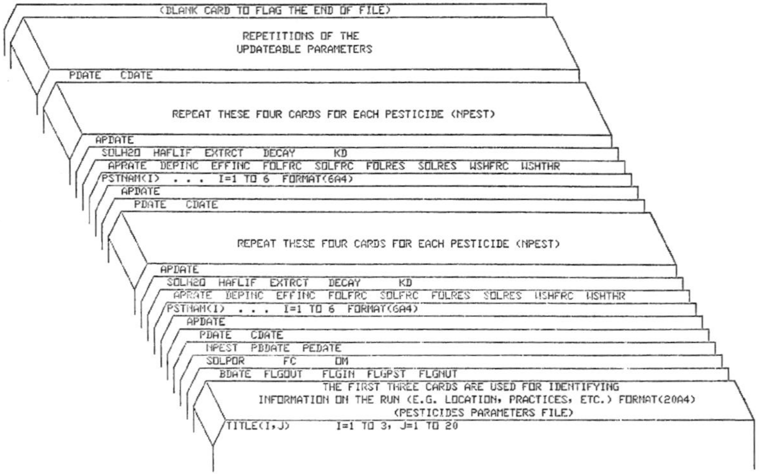

Figure II-40.—Schematic representation of a data deck for the pesticide model.

Application Rate

The desired pesticide rate for a given application usually is specified within certain limits by the registration data on the label or is obtained as recommendations from the supplier or extension specialists. The number of applications for some pesticides, particularly foliar-applied insecticides, will depend on extent of insect infestation or established spray schedules. Application rate is input in units of kilograms/hectare. See table II-40 for ranges of application of some common herbicides.

Depth of Pesticide Incorporation

Pesticides often are incorporated by double-disking, rotary tillers, and other equipment for harrowing or smoothing the soil surface. Depth of pesticide incorporation will depend on the type of tillage equipment used and soil conditions. Depth of incorporation normally ranges from about 8 to 15 cm (3 to 6 in). When the pesticide is incorporated, select the depth based on the tillage equipment used. For surface-applied chemicals, a value of 1 cm is input as the incorporation depth since the surface is defined arbitrarily as having a depth of 1 cm.

Efficiency Factor for Incorporation

Most incorporation devices do not mix the applied pesticide uniformly throughout the entire depth. The concentration remaining at the surface may be significantly higher than at lower depths. Injected pesticides may have a low surface concentration due to their placement below the surface. The efficiency factor can be used to adjust the surface concentration based on known patterns of incorporation. If an incorporation device leaves a concentration in the surface of twice that achieved by uniform mixing, for example, an efficiency factor equal to half the incorporation depth could be used. For injected pesticides, an efficiency factor of less than one will reduce the surface concentration in proportion. Since this type of information usually is unavailable, a value of 1 would be input with the assumption that uniform mixing was achieved.

Fraction on Soil and Foliage

When crops are treated with pesticides applied to the plant canopy, some of the application, depending on degree of canopy closure, will reach the surface of the soil directly, some will remain on the foliage, and the rest will be lost by drift and volatilization. At full canopy, about 75 ± 20% and 50 ± 20% of the ground and aerial applications, respectively, reach the canopy (CREAMS, vol. III, ch. 18). If the amounts reaching soil directly are assumed negligible at full canopy, about 25 to 50% can be lost by drift and volatilization during application. For incomplete canopy, the fraction reaching soil should be somewhat proportional to the extent of ground cover although insufficient information is available to provide any functional relationship. The actual distribution between soil, foliage, and off-target loss will be highly variable and dependent on atmospheric conditions, path of application, and canopy characteristics. If site-specific information is unavailable, at full canopy closure use 0.4 to 0.6 on foliage for aerial applications and 0.7 to 0.8 on foliage for ground applications. Assume an insignificant fraction reaching the soil. For less than full closure, use a fraction for soil interception in proportion to exposed ground surface. For example, suppose an aerial application is made to cotton that, on projection, covers 50% of the ground surface. The fraction on foliage would be 0.3 and the fraction on soil would be 0.3, with the rest, 0.4, assumed as off-target losses.

Initial Foliar Residues

Pesticides normally dissipate from foliage such that a residue will not be present at the beginning of a model application period. This option, is provided, however, so that the model can be applied on any date. To estimate an initial residue from a previous application, assume interception fraction, as was suggested, and use equations given in volume III, chapter 18, to estimate dissipation with time. Rates of foliar dissipation are discussed in a following section. The value input should be in units of milligrams of pesticide per square meter of ground surface. Initial residue can be determined best by direct measurement, but this procedure usually is not practical except for research.

Initial Soil Residue

As for foliar residue, the amount of pesticide present in soil at the beginning of a model application period is best determined by sampling and analysis. Little residue of nonpersistent pesticides would be expected at the beginning of a growing season. When persistent pesticides, such as organochlorines, have been used for several years on a site, however, a significant residue will be present. If sampling and analysis cannot be accomplished, published data should be sought on residues in the soils of the area. The input value should be in units of micrograms per gram (ppm). If the initial residue cannot be determined by measurement or cannot be estimated from published information such as that found in the Pesticide Monitoring Journal, levels of initial residue may be estimated by using the values in volume III, chapter 17, if past application history is known. First-order decay may be assumed, or an equation of best fit may be used, such as that in volume III, chapter 17 on observed pesticide persistence.

The initial residue parameter also provides a device for updating the concentration of pesticide in the surface of the soil as a result of redistribution caused by major tillage. Persistent pesticide may accumulate at the soil surface during an application season. This accumulated residue would be predicted as output from the model. At the time of tillage, a new value for the concentration at the surface of the soil can be computed, based on the accumulated residue and tillage depth, and can be entered as an initial soil residue for a new model application period.

Foliar Washoff Threshold

This parameter estimates the amount of rainfall required to exceed the capacity of the canopy to intercept and retain rainfall as droplets on the leaf surfaces. Once this amount of rainfall is exceeded, pesticide washoff is assumed. The value of this parameter probably ranges from about 0.1 cm to 0.3 cm for a dense crop canopy.

Washoff Fraction

Little information is available on extent and patterns of pesticide washoff from foliage. The efficiency of the washoff process may be related to several factors. Information in volume III, chapter 18, suggests that rainfall can remove about 60% of the dislodgeable residue of most pesticides. Organochlorines, and possibly other pesticides, however, are exceptions. Less than 10% of these compounds is removed by rainfall. Values of 0.6 to 0.7 are suggested, therefore, for all except the organochlorines, where values in the range of 0.05 to 0.1 should be used.

Water Solubility

Pesticide solubilities can be found in many handbooks on pesticide properties. In the model, solubility serves two functions. If solubility is $<$ 1 ppm, the vertical transport computation is bypassed. Secondly, the predicted runoff concentration in solution is compared to solubility. If solubility is exceeded, the solution concentration is limited to the water solubility. Solubility is, therefore, a critical input parameter only for the relatively insoluble pesticides. Solubilities of some common pesticides are given in tables II-40 and II-41.

Foliar Residue Half-Life

Consult volume III, chapter 18, for half-life values of pesticides on foliage. Pesticides generally are not as persistent on foliage as in soil.

Extraction Ratio

This parameter describes the efficiency of the runoff stream in removing or extracting pesticide. Conceptually, it is the ratio of soil:water in the mixing zone. Tests with the model indicate that values in the range of 0.05 to 0.2 are needed--the higher values for conditions of excessive runoff and erosion. Predicted runoff concentrations of those pesticides transported entirely in solution vary in direct proportion to the value of the extraction ratio. As sediment transport becomes more significant, sensitivity to this parameter decreases. A value of 0.1 gives adequate prediction in most situations.

Soil Decay Constant

Values of rate constants, k_s, are tabulated in volume III, chapter 17 for the assumed expoential decay function applied to several pesticides and conditions. Because dissipation rates are affected by climatic factors, the results of individual experiments also should be reviewed before making a final selection (vol. III, ch. 17). Note that decay constants for surface and subsurface (bulk soil) are given in volume III, chapter 17, if data were available. Many pesticides dissipate more rapidly at the surface of the soil than from the soil bulk. The k_s values for surface dissipation are more appropriate for runoff prediction, but more results have been reported on persistence in the soil bulk. Where k_s values are given for soil bulk but not for surface, differences reported for similar compounds may be used in making a subjective judgment on how the surface k_s might differ from the reported bulk soil k_s.

Additional information is provided in volume III, chapter 17, on how k_s values can be estimated based on properties of the pesticides and their environment. In addition to a better perspective of factors influencing dissipation rates, methods are provided by which k_s values can be estimated where little experimental data are available.

In some instances, the first-order decay equation poorly describes dissipation of a pesticide. Another alternative is suggested in chapter 17, volume III, whereby pesticide concentration as a function of time can be obtained from equations fitted to experimental data. No direct method is provided in the present model for substituting these equations for the first-order decay equation. The k_s values can be updated, however, using different values for different times after application. A best-fit equation could be used to compute k_s values for shorter time segments of the linear log c vs. t relationship assumed. Since all equations of best fit are not incorporated in this version of the model, a user should consult the author of chapter 17, volume III when neccessary.

Distribution Coefficient K_d

Chapter 19, volume III, discusses how K_d is determined, the factors affecting its value for different pesticides and soils, and how to estimate K_d for a specific situation. Tables 1 through 4 list mean K_d with standard deviations for several pesticides. These tables also provide for estimating K_d as a function of soil texture and organic matter content, thus tying k_d to both the pesticidal properties and controlling site-specific characteristics of the soils. Additional relationships for estimating K_d are based on observed soil thin-layer chromatography and pesticide solubility.

Some assumptions are discussed for using K_d to predict distribution of pesticide between solution and adsorbed phases. Figure 1 (volume III, chapter 19), shows how the apparent K_d can vary with pesticide concentration if the adsorption relationship or isotherm is nonlinear. Users should compare potential errors due to linearity and other assumptions in relation to the accuracy of required output to achieve the objectives of their simulation. Since the effect of these assumptions on the validity of model output is uncertain, K_d values for an order of magnitude might be warranted when distinguishing major behavioral differences. Expressing K_d values explicitly as per reference may be useful to analyze certain problems or situations, using model simulations to compare effects of different management alternatives on the same site.

OUTPUT

The user can specify the type of output from the pesticide model by the input on card 4: FLGOUT = 0 for annual summary only, FLGOUT = 1 for monthly and annual summary, or FLGOUT = 2 for storm output as well as monthly and annual summaries. This enables users to select the best output for their problems. If a potential toxicity problem exists, storm output would be needed, whereas an overall assessment of the pesticide losses could be determined from the annual summary.

Figure II-41 shows an annual summary output for a situation where six pesticides were applied and a seventh pesticide was applied in previous years. A storm summary of rainfall and runoff for the year is shown at the top of the figure which gives the total mass of pesticide in water and with sediment. Loss of pesticide as a percentage of that applied is shown also. Only the total mass is shown for toxaphene, which was not applied during the year and the percentage of application shows residue. Figure II-42 shows sample output of monthly summaries for the same seven pesticides.

Figure II-43 shows output for a single storm event when there was no runoff or pesticide loss. Figure II-44 shows the model output for a storm event that resulted in runoff, erosion, and pesticide loss. The pesticide numbers in figure II-44 correspond to order of input and the order for the annual summary (fig. II-41). Concentrations in water and sediment are averages for the storm.

```
                 ANNUAL SUMMARY FOR 1974
                 -----------------------

    107 STORMS PRODUCED     179.60 CM. OF RAINFALL
     48 STORMS PRODUCED      63.28 CM. OF RUNOFF


                   THE PESTICIDE LOSSES
                   --------------------

    PESTICIDE         TOTAL MASS        PERCENT OF
      NAME               G/HA           APPLICATION
    ---------         ----------        -----------
  FLUOMETURON             8.26                 .55
  TRIFLURALIN              .05                 .01
  MSMA                  268.14               17.88
  DIURON                  55                   28
  METHYL PARATHION        .66                  01
  EPN.                  199.71                3.99
  TOXAPHENE             296.31             RESIDUL
```

Figure II-41.—Sample output of annual summary of pesticide component where six pesticides were applied and the seventh pesticide carried over from previous years.

MONTHLY SUMMARY FOR JUL. 1974

7 STORMS PRODUCED 14.83 CM. OF RAINFALL
2 STORMS PRODUCED 1.85 CM. OF RUNOFF

THE PESTICIDE LOSSES

PESTICIDE NAME	TOTAL MASS G/HA	PERCENT OF APPLICATION
FLUOMETURON	.04	RESIDUE
TRIFLURALIN	.05	RESIDUE
MSMA	.10	RESIDUE
DIURON	.54	.27
METHYL PARATHION	.13	.01
EPN	27.90	1.66
TOXAPHENE	0	RESIDUE

MONTHLY SUMMARY FOR AUG. 1974

13 STORMS PRODUCED 16.56 CM. OF RAINFALL
3 STORMS PRODUCED 1.46 CM. OF RUNOFF

THE PESTICIDE LOSSES

PESTICIDE NAME	TOTAL MASS G/HA	PERCENT OF APPLICATION
FLUOMETURON	.00	RESIDUE
TRIFLURALIN	.00	RESIDUE
MSMA	.00	RESIDUE
DIURON	.01	RESIDUE
METHYL PARATHION	.25	.01
EPN	75.45	3.77
TOXAPHENE	0	RESIDUE

Figure II-42.— Sample output of monthly summaries for the pesticide component where six pesticides were applied and the seventh pesticide carried over from previous years.

STORM INPUTS

DATE	74192	JULIAN DATE
RAINFALL	53	CM.
RUNOFF VOLUME	0	CM.
SOIL LOSS	0	KG/HA
ENRICH. RATIO	2.00	
PERCOLATION	0	CM
AVG. TEMP.	27.37	DEGREES C
AVG. SOIL WATER	26	VOL/VOL
ACCUMULATED ET	55.30	CM.
POTENTIAL ET	90.77	CM.

*** NO RUNOFF - NO LOSSES ***

Figure II-43.—Sample output from the pesticide component for a single storm event that did not produce runoff.

STORM INPUTS

DATE	74190	JULIAN DATE
RAINFALL	5.97	CM.
RUNOFF VOLUME	1.31	CM.
SOIL LOSS	726.95	KG/HA
ENRICH. RATIO	2.49	
PERCOLATION	0	CM.
AVG. TEMP.	27.14	DEGREES C.
AVG. SOIL WATER	.28	VOL/VOL
ACCUMULATED ET	53.61	CM.
POTENTIAL ET	89.29	CM

THE QUANTITY OF PESTICIDE IN RUNOFF
VALUES FOR STORM 74190

PEST. NO	CONC. AVA. RESIDUE UG/G	CONC. IN WATER UG/ML	MASS IN WATER G/HA	CONC. IN SEDIMENT UG/G	MASS IN SEDIMENT G/HA	TOTAL MASS G/HA	REMAIN. RESIDUE UG/G
1	0	0	0	0	0	0	0
2	0	0	0	0	0	0	0
3	00	0004	.0478	0001	0000	.0479	0
4	.00	.0026	5575	.0012	.0009	.5582	0
7	0	0	0	0	0	0	0

Figure II-44.—Sample output from the pesticide component for a runoff-producing storm.

MODEL APPLICATION

Selection of best management practices rarely will hinge around solving only a single potential problem affecting the soil resource or water quality or both. A balance will be sought among total production, production efficiency, net profits, protection of the resource base, and need and potential for improving downstream water quality. For most constituents in water draining from agricultural fields, including pesticides, no absolute standards or criteria have been set as goals or requirements that must be met. Model output cannot be compared, therefore, for selecting management options that meet a fixed set of criteria. If this were possible, predictions and criteria also must deal with probabilities of occurrence and the permissible or reasonable level of environmental risks.

Whenever a toxin is used widely, some risk is incurred to at least part of the environment. Acute toxicity problems are identified more easily by their effects than long-term chronic exposure. Dangers of long-term exposure to very low levels of chemicals in the environment are not well understood, nor is there general agreement on the extent of danger. Therefore, the basic question of how much pesticide runoff consititutes a problem and how much it should be reduced cannot be answered at this time, and is beyond the scope of this discussion. The general philosophy in the environmental community is that reduction of all off-target losses of pesticides to some practical minimum is desirable. This is not to say, however, that some management option shows potential for reducing pesticide runoff should necessarily be selected over another option. All factors must be considered, including reduction of soil and plant nutrient losses, effects on production, costs, and net return; and potential problems caused by the pesticide. In using nonpoint source pollution models, therefore, the planner for land use and water quality must examine and rate management options with uncertainties of the issue as well as uncertainties in the model outputs.

An example of how the model could be used is to compare relative losses of pesticides under different management schemes designed to limit sediment yield. Table II-43 shows results of a simulation for a hypothetical situation where it was assumed that 3 cm rainfall occurred on days 5, 10, 15, 20, and 25 after pesticide was applied on the surface at the rate of 3 kg/ha. Each storm was assumed to produce 1 cm runoff and 500 kg/ha sediment yield. Pesticides with K_d's of 5 and 5,000 were considered. Both pesticides were assumed to have decay constants, k_s, of 0.10, and a sediment enrichment factor of 2.0 was assumed in all events. In terms of total mass, the pesticide with a K_d = 5 was transported almost totally in the water phase, whereas the pesticide with a K_d of 5,000 was transported by sediment. At the assumed level of sediment production, total losses of the sediment-transported pesticide were less than those predicted for the water-transported pesticide. Pesticide losses would be similar in each situation if sediment production was increased by a factor of about 4. Total loss of the pesticide with a K_d of 5 would not be changed significantly, however, by increased or decreased sediment yield unless the volume of runoff also was changed. Sample model runs on actual situations are given in chapter 5 of this volume and may be examined for additional illustrations of model use.

Table II-43.—Pesticide in runoff predicted for hypothetical situation of 3 cm rainfall, 1 cm runoff, and 500 kg/ha sediment yield on days indiated[1/]

Days after application	Concentration in water	Concentration in sediment	Total mass in runoff	Percent in water
	(ppb)	(ppm)	(grams)	(%)
K_d = 5				
5	670	6.70	70.3	95
10	335	3.35	35.3	95
15	170	1.70	17.8	95
20	85	0.85	8.9	95
25	43	0.43	4.5	95
K_d = 5,000				
5	2.4	24.2	12.4	2
10	1.5	14.7	7.4	2
15	0.9	9.0	4.6	2
20	0.5	5.4	2.8	2
25	0.3	3.3	1.7	2

1/ R = 3.0 kg/ha, k_s = 0.10, enrichment factor = 2, extraction coefficient = 0.1.

Chapter 5. EXAMPLE APPLICATIONS FOR TYPICAL FIELD SITUATIONS

G. R. Foster, M. H. Frere, W. G. Knisel, R. A. Leonard, A. D. Nicks, J. D. Nowlin, R. E. Smith, and J. R. Williams[1/]

INTRODUCTION

This chapter cites three typical field situations to show how parameter values are obtained for a real-world problem. Limited interpretive information will help the user understand the significance of specific aspects of the CREAMS model and its parameters. The three typical field situations represent different physiographic areas of Georgia Piedmont, Mississippi Delta, and western Tennessee. These examples typify gently rolling topography; flat, landformed topography; and steep slopes with long, slender fields.

The management practices for sample computer runs may not be recommended by the SCS or acceptable by farmers, but the procedures are valid and should help the user understand the model operations. Two management practices are considered for each location. Table II-44 shows the three locations, management practices (MP1 and MP2), and model components.

DESCRIPTION OF APPLICATION SITES

Georgia Piedmont

The topography of the Georgia Piedmont field (fig. II-45) is typical of Piedmont cropland. Drainage from the field is restricted at the fence line and causes in some temporary ponding of runoff. The soil is Cecil sandy loam with a depth of 24 in to the B2 horizon. Continuous corn is assumed for the crop.

Two management practices in table II-44 are (1) MP1, conventional tillage with rows running across the drainage, more or less on the contour in the upper end and (2) MP2, modified tillage with a grass waterway extending approximately two-thirds of the field length. Conventional tillage consists of spring moldboarding, disking twice, planting, and cultivating twice. Modified tillage consists of chiseling, disking, planting, and not cultivating. Plant nutrient application consists of the locally customary application of 140 kg/ha of

---

1/ Hydraulic engineer, USDA-SEA-AR, Lafayette, Ind.; soil scientist, USDA-SEA-AR, New Orleans, La.; hydraulic engineer, USDA-SEA-AR, Tucson, Ariz.; soil scientist, USDA-SEA-AR, Athens, Ga.; agricultural engineer, USDA-SEA-AR, Chickasha, Okla.; computer technician, Purdue University, West Lafayette, Ind.; hydraulic engineer, USDA-SEA-AR, Fort Collins, Colo.; and hydraulic engineer, USDA-SEA-AR, Temple, Tex.

Table II-44.—Typical field situations for sample model runs

Model Component	Component Option	Location		
		Georgia Piedmont	Mississippi Delta	Western Tennessee
Hydrology	Option 1			MP1[1/] MP2
	Option 2	MP1	MP1	
Erosion			MP1 MP2	MP1
Nutrients	Method 1			MP1
	Method 2	MP1		
Pesticides			MP1	

1/ MP1 is management practice 1; MP2 is management practice 2.

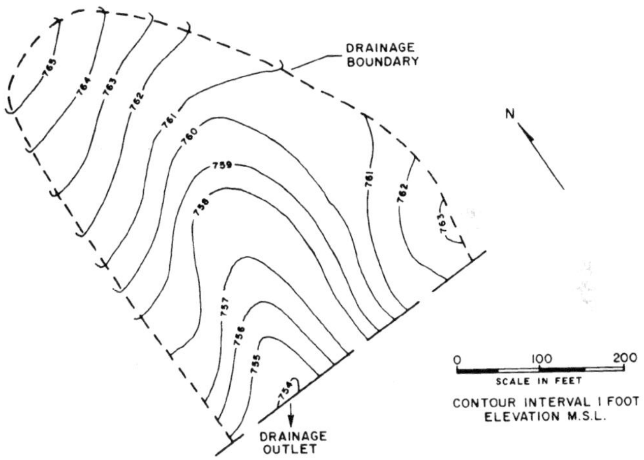

Figure II-45.—Topographic map for Georgia Piedmont field.

nitrogen and 28 kg/ha of phosphorus. At planting time, 28 kg/ha of both N and P are applied and incorporated by disking. The remaining nitrogen is surface applied in June. Atrazine is surface applied at planting time at the rate of 3.36 kg/ha. Planting is assumed to occur on May 1 each year. The second nitrogen application is assumed to occur on June 11. These same dates and rates for applying nutrients and pesticide are used in both management practices. Table II-45 gives dates of tillage operations for MP1 and dates and rates for applying fertilizer and pesticides.

Table II-45.—Tillage operations and applications of fertilizer and pesticide on the Georgia Piedmont field[1/]

Date	Field operation	Fertilizer		Pesticide	
		N	P	Name	Rate
		(kg/ha)	(kg/ha)		(kg/ha)
74105[2/]	Moldboard plow	--	--	--	--
74122	Disk/plant/fertilize	28	28	Atrazine	3.36
74150	Cultivate	--	--	--	--
74162	Fertilize	112	--	--	--
74165	Cultivate	--	--	--	--
74274	Harvest/shred stalks	--	--	--	--

1/ Management practice MP1 (continuous corn with conventional tillage). Operations are assumed to be the same each year of simulation.

2/ 74105 is Julian day 105 in calendar year 1974.

Mississippi Delta

The Mississippi Delta farmland has flat slopes on poorly drained soils with a relatively high water table that fluctuates considerably during the growing season. Farmers in the Delta is form land to obtain a uniform field slope. Rows are run in the direction of the slope to provide good drainage along the furrows. A field drain provides drainage from the ends of the rows. This drain is relatively broad and flat, normally is grassed, and is used as a turnrow for farm equipment. Figure II-46 shows a typical field in the Mississippi Delta. The field is roughly rectangular in shape, with a 32-acre drainage area. Row length is 1,300 ft on a 0.4% grade. The rows drain directly into a triangular-shaped channel that has a longitudinal slope of 0.1%, channel side slopes of 10:1, and a bermuda grass cover. The drainage channel flows through a culvert into a larger drainage ditch. Backwater occurs at the culvert entrance, but free outfall occurs at the culvert outlet. The soils in the field are Commerce silt loam. A phreatic water table in the Mississippi Delta fluctuates from year to year and within the year, depending upon rainfall amounts and time of occurrence. These fluctuations cause rooting depths to vary from year to year. A maximum rooting depth of 40 in was estimated represent normal or average conditions.[2/]

Figure II-46 also shows a typical row cross-section. In this figure overland flow is assumed to occur on the row ridges and concentrated flow in the furrow is assumed to be channel 1 for the erosion model. Channel 2 for the erosion model is the field drain.

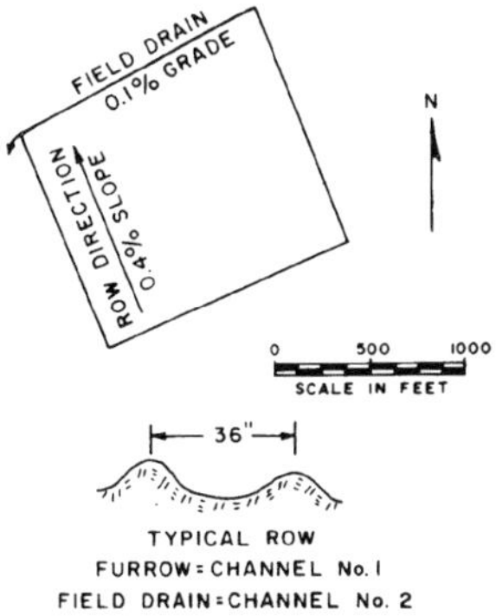

Figure II-46.—Representative field, Mississippe Delta.

Typical field operations consist of several diskings between cotton harvest in the fall and seedbed preparation the following spring. Several cultivations are made during the cotton-growing season. These field operations are shown in table II-46 for the conventional management practice (MP1) along with fertilization and pesticide applications. A spectrum of pesticides is used to control weeds and insects in the clean-till system.

Few optional management practices are practical, acceptable by the farmer, and effective for controlling erosion in the Delta. The second management system (MP2) considers maximum erosion control under the clean-till cotton production. This system includes ryegrass as a winter cover crop. Ryegrass would be seeded after cotton is harvested to provide protection from erosion during the winter. It would be disked before preparing the seedbed and applying herbicides and fertilizer. Field operations for management practice MP2 are shown in table II-47.

Western Tennessee

The western Tennessee area considered for application has by steep slopes with severely eroded soils. Erosion rates are high with continuous row cropping. Considerable interrill and rill erosion occurs on the steep slopes with deposition on the toe of the slope or in the concentrated flow where slopes are much flatter. Most of the land is class V because of slopes and erosion and normally would not be recommended for farming by the Soil Conservation Service. Terraces constructed on these slopes would be about 50 ft apart and would be objectionable to the farmer. This site represents a special problem and demonstrates what could be expected under two extreme management practices: (1) continuous corn with conventional tillage and (2) permanent fescue harvested annually for seed (a cash crop for possible replacement of corn). Complete

2/ Personal communication with G. H. Willis, soil scientist, USDA-SEA-AR, Baton Rough, La.

Table II-46.—Field operations for Mississippi Delta management practice 1 for 1974

Date	Field operation	Fertilizer N	Fertilizer P	Pesticide Name	Pesticide Rate
		(kg/ha)	(kg/ha)		(kg/ha)
74035[1/]	Disk	--	--	--	--
74042	Disk/herbicide	--	--	Fluometuron	1.5
74064	Disk/bed/fertilize	100.	--	--	--
74109	Rebed/knock down/plant/herbicide	--	--	Trifluralin	1.0
74127	Cultivate	--	--	--	--
74133	Cultivate	--	--	--	--
74143	Cultivate	--	--	--	--
74149	Herbicide	--	--	MSMA	.5
74154	Herbicide	--	--	MSMA	.5
74165	Cultivate	--	--	--	--
74172	Fertilize/herbicide	90.	--	MSMA	.5
74182	Cultivate/herbicide	--	--	Diuron	.2
74198	Insecticide	--	--	Methyl Para-thion/EPN	.5/.5
74203	Insecticide	--	--	Methyl Para-thion/EPN	.5/.5
74210	Insecticide	--	--	Methyl Para-thion/EPN	.5/.5
74218	Insecticide	--	--	Methyl Para-thion/EPN	.5/.5
74225	Insecticide	--	--	Methyl Para-thion/EPN	.5/.5
74232	Insecticide	--	--	Methyl Para-thion/EPN	.5/.5
74239	Insecticide			Methyl Para-thion/EPN	.5/.5
74247	Insecticide	--	--	Methyl Para-thion/EPN	.5/.5
74253	Insecticide	--	--	Methyl Para-thion/EPN	.5/.5
74260	Insecticide	--	--	Methyl Para-thion/EPN	.5/.5
74320	Harvest	--	--	--	--
74324	Cut and shred stalks	--	--	--	--
74325	Disk	--	--	--	--

1/ 74035 is Julian day 35 in calendar year 1974.

Table II-47.—Field operations for Mississippi Delta management practice 2 for 1974

Date	Field operation	Fertilizer N	Fertilizer P	Pesticide Name	Pesticide Rate
		(kg/ha)	(kg/ha)		(kg/ha)
74078[1/]	Disk	--	--	--	--
74107	Disk/bed/knock down/fertilize/ Plant cotton/herbicide	100	--	Trifluralin/ Fluometuron	1.0 1.5
74127	Cultivate	--	--	--	--
74143	Cultivate	--	--	--	--
74149	Herbicide	--	--	MSMA	.5
74154	Herbicide	--	--	MSMA	.5
74165	Cultivate	--	--	--	--
74172	Fertilize/herbicide	90	--	MSMA	.5
74182	Herbicide	--	--	Diuron	.5
74198	Insecticide	--	--	Methyl Para- thion/EPN	.5/ .5
74203	Insecticide	--	--	Methyl Para- thion/EPN	.5/ .5
74210	Insecticide	--	--	Methyl Para- thion/EPN	.5/ .5
74218	Insecticide	--	--	Methyl Para- thion/EPN	.5/ .5
74225	Insecticide	--	--	Methyl Para- thion/EPN	.5/ .5
74232	Insecticide	--	--	Methyl Para- thion/EPN	.5/ .5
74239	Insecticide	--	--	Methyl Para- thion/EPN	.5/ .5
74247	Insecticide	--	--	Methyl Para- thion/EPN	.5/ .5
74253	Insecticide	--	--	Methyl Para- thion/EPN	.5/ .5
74260	Insecticide	--	--	Methyl Para- thion/EPN	.5/ .5
74293	Harvest/shred stalks	--	--	--	--
74294	Disk/plant ryegrass	--	--	--	--

1/ 74078 is Julian day 78 in calendar year 1974.

information is unavailable for the western Tennessee site.[3/]

Figure II-47 is a soils and drainage map of the selected field. This field is long and slender, which significantly attenuates runoff peak rate at the field outlet. Soil on the hilltops and steep slopes is Loring silt loam with a hardpan approximately 26 in below the surface. Drainage from the hillsides concentrates in the alluvial Collins silt loam, which has a slope of 0 to 2%. Since a topographic map is unavailable, the soils map was used to generate the representative overland flow profile (fig. II-48). The average slope from the soils map represented each slope segment, that is, a "B" slope ranges from 2 to 5%. Thus, the average 3.5% was used. A "D" slope ranges from 8 to 12%, and the average 10% was used. The Collins silt loam in the alluvial valley has a 0 to 2% slope. The side slope from the toe of the Loring was assumed to have a 1% slope, whereas the slope in the direction of concentrated flow is assumed as 2%.

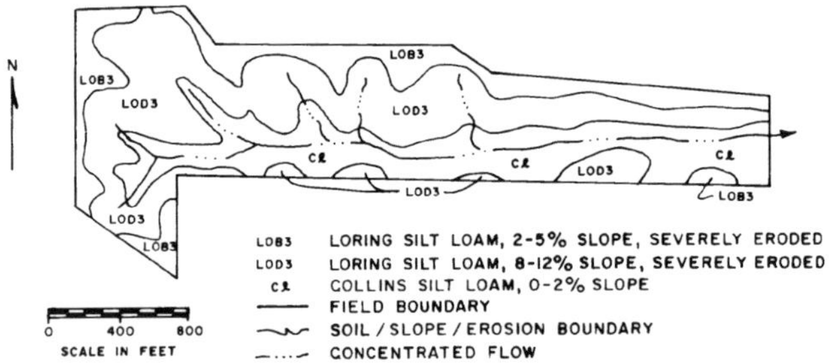

Figure II-47.—Soils-drainage map of western Tennessee field.

The lack of a topographic map will cause inaccurate estimates of runoff and erosion even if the model was exact. This is of little consequence since the model is for comparing management practices. Absolute accuracy is not as important as the relative magnitudes between practices.

Table II-48 shows field operations and dates of herbicide and fertilizer application for management practice 1. These operations are normal for the area with normal planting and harvest dates. For the second management practice (MP2) with permanent fescue grass, the fescue is assumed to be established at the beginning of simulation. This assumption is in keeping with the previous description of extreme conditions for hydrology and erosion.

3/ Data on soils and general information were abstracted by SCS personnel from a University of tennessee master's thesis.

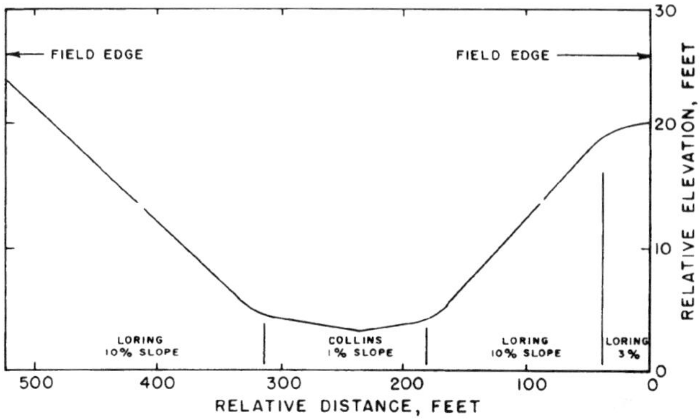

Figure II-48.—Representative overland flow profile estimated from soil map and average slopes.

Table II-48.—Field operations for western Tennessee, management practice 1, continuous corn with conventional tillage

Date	Field Operation	Fertilizer		Pesticide	
		N	P	Name	Rate
		(kg/ha)	(kg/ha)		(kg/ha)
74092[1/]	Moldboard	--	--	--	--
74119	Disk	--	--	--	--
74122	Disk/plant/fertilize/ apply herbicide.	140	20	Atrazine	3.36
74154	Cultivate	--	--	--	--
74177	Cultivate	--	--	--	--
74198	Cultivate	--	--	--	--
74309	Harvest/shred	--	--	--	--

1/ 74092 is Julian day 92 in calendar year 1974.

APPLICATIONS AND PARAMETER ESTIMATES

The rest of this chapter is presented by components. That is, hydrology is presented for the applications in one section so the user can see these applications without having to read through erosion, plant nutrients, and pesticides. All erosion applications are in a single section so the user can see the different conditions represented. Parameter values are discussed sufficiently to help the user understand the estimation process.

HYDROLOGY

Application of both hydrology options in this chapter will help the user select parameter values. Hydrology option 1 is given for both management alternates on the western Tennessee site. Hydrology option 2 is used for the Georgia Piedmont and Mississippi Delta.

Hydrology Option One

Western Tennessee

The process for obtaining parameter values is designed to be as objective as possible while maintaining general applicability. Judgment is required, however, in estimating missing data, crop production, soil characteristic variations, and so forth. To demonstrate this process, each parameter estimate is decribed in detail for management practice 1. Only parameter estimates that change values in converting from MP1 to MP2 are presented for MP2.

Management Practice 1 -- The source of information is given for each parameter value.

Card 1 to 3:

TITLE = DAILY HYDROLOGY PARAMETERS - WESTERN TENNESSEE
MANAGEMENT PRACTICE 1
CONTINUOUS CORN - CONVENTIONAL TILLAGE (HYDONE)

Alphanumeric information describing the problem.

Card 4:

BDATE = 74001 Beginning date of simulation (year and Julian day).

FLGOUT = 1 Specifies output for each storm and for annual summaries.

FLGPAS = 1 Creates a hydrology file for use by the erosion model.

FLGOPT = 1 Indicates that the hydrology option 1 model will be used.

Card 5:

DACRE = 69.2 Drainage area of the field measured from drainage map, acres.

RC = 0.10 Effective saturated conductivity of the predominant soil, Loring silt loam, in/hr, obtained from description of soils given by Holtan and others (1).

FUL = 0.80 Fraction of plant-available water storage filled at field capacity. FUL = (Field capacity - BR15)/(Porosity - BR15) where field capacity is estimated as the volumetric content at 0.1 bar tension and BR15 is the volumetric water content at 15 bars tension obtained from Holtan and others (1).

BST = 0.5 — Fraction of plant-available water storage filled when simulation begins. Since this simulation begins on January 1, the plant-available water storage is assumed to be half full. If the simulation began in the fall after harvest, BST would be estimated much lower (0.5 - 0.1) because the crop normally uses most water in the root zone.

CONA = 3.5 — Soil evaporation parameter. A value of 3.5 is satisfactory for most soils. The suggested range is from 3.3 for sands to 4.5 for loams.

POROS = 0.48 — Soil porosity in the root zone, cm^3/cm^3, obtained from Holtan and others (1).

Card 6:

SIA = 0.2 — Initial abstraction coefficient for SCS runoff equation. The 0.2 value generally is recommended.

CN2 = 86 — Two condition SCS curve number (5, tables 7.1 and 9.1).

CHS = 0.01 — Channel slope, ft/ft, determined by dividing the elevation difference by the distance along the main drainageway from the field outlet to the most distant point.

WLW = 5.4 — Field length-width ratio determined by dividing the square of the channel length as defined in CHS by the drainage area. For this calculation, channel length is expressed in miles and drainage area in square miles.

RD = 27 — Root depth, inches, estimated as the depth to the hardpan. If no hardpan exists, the root depth for most crops is about 36 in. This depends on the type of crop, soil, and production level.

Card 7:

UL(1-7) = 0.27, 1.15, 1.44, 1.28, 1.03, 0.93
Plant-available soil water storage, inches, for each of seven storages. UL = Porosity - (BR15)(RD)(D), D = 1/36 for top storage, 5/36 for second storage, and 1/6 for 5 lower storages, obtained from Holtan and others (1).

Cards 8, 9:

TEMP(1-12) = 41.2, 46.0, 53.5, 60.4, 68.5, 74.6, 79.4, 77.7, 68.6, 59.6, 48.2, 40.6
Average monthly temperature at Nashville, Tenn., obtained from the Climatic Atlas of the United States (6).

Cards 10, 11:

RAD(1-12) = 149, 228, 322, 432, 503, 551, 473, 403, 308, 208, 150. Average monthly solar radiation, langleys/day, at Nashville, Tenn., obtained from the Climatic Atlas of the United States (6) or table II-7.

Card 12:

GR = 1.0 Winter cover factor. Essentially no winter cover is assumed for continuous corn with conventional tillage.

Cards 13 to 25: Number of cards varies for different crops.

1	0.0
132	0.0
167	0.09
182	0.19
196	0.23
211	0.49
225	1.16
240	2.97
254	3.00
269	2.72
283	1.83
309	0.00
366	0.00

Julian day and leaf area index for the crop grown during the first year, or simulation (table II-8).

Rainfall data--Daily rainfall data were used from USDA-SEA-AR records near Clarksdale, Miss. One year's data are placed on 37 cards.

Since management practice 1 specifies continuous corn, the leaf area index and winter cover factor remain the same thoughout the simulation. New values of these variables must be input each year if the crop changes.

Management practice 2--Only three input changes are required in converting from management practice 1 to management practice 2. Since these parameters (CN2, GR, and LAI) are the primary indicators of management changes, adjusting other parameters usually is unnecessary when considering management practices.

The two condition SCS curve number is 79 (5, tables 7.1 and 9.1). The winter cover factor is GR = 0.5 because the fescue grass should provide an excellent winter cover.

Leaf area index data are unavailable for fescue grass. Information was obtained for growing period, approximate dates and rates of fertilization, dates of harvest, approximate yields, and recommended herbicide practices. Fescue grass is a cool-season grass that begins rapid growth about mid-February in western Tennessee. A balanced fertilizer is applied at that time at the rate of 60 lb/acre nitrogen, 20 lb/acre phosphorus, and 30 lb/acre potassium.

A light application of 2,4-D is made in late April or early May at the rate of 1.5 kg/ha. Growth rate begins to decrease the beginning of April and reaches a maximum by mid-May. Seed is ready for harvest by late June. Seed is harvested with a grain combine that leaves much plant material standing. A good yield is 1,000 lb/acre of seed and 5,000 lb/acre of dry matter. Fescue does not go completely dormant during the summer, but it grows little until temperatures drop in mid-September. The growth rate again declines by the first of November but, like winter small grain, a transpiring canopy exists throughout the winter. Table II-49 gives the leaf area index used in the hydrology model.

Table II-49.—Leaf area index for fescue grass, management practice 2 (MP2), western Tennessee[1/]

Date	Julian day	Leaf Area Index
1-1	001	0.35
2-15	046	.40
3-1	091	2.10
5-15	135	2.80
6-15	166	2.80
7-1	182	2.60
7-2	183	.20
9-15	258	.25
11-1	305	.35
12-31	366	.35

1/ Personal communication with S. R. Wilkinson, USDA-SEA-AR, Southern Piedmont Conservation Research Center, Watkinsville, Ga.

Interpretation of Results

Table II-50 compares results for the two simulations. Although the simulation period was only 3 yr, a wide variation in hydrologic conditions was observed. For example, rainfall ranged from 34.5 to 70.7 in. Average annual values in table II-50 shows that management practice 2 gives less surface runoff and more percolation than management practice 1. These results seem reasonable because the fescue grass increases the infiltration rate of soil considerably. Since evapotranspiration is essentially the same for the two management practices, percolation must be higher if infiltration is increased.

Hydrology Option 2

Many parameters necessary for hydrology option 2 are the same as for option 1. The two example applications show how parameters are selected where judgment is reflected. Parameters are discussed as listed in table II-5.

Georgia Piedmont Management Practice 1

The measured area (DACRE) of the field/watershed in figure II-44 is 3.2

Table II-50.—Results from hydrology option 1 for management practices 1 and 2 in western Tennessee

Year	Rainfall	Management Practice 1			Management Practice 2		
		Runoff	Evapotrans-piration	Perco-lation	Runoff	Evapotrans-piration	Perco-lation
	- - - - - - - -	- - - - - -	- - - - - -	-(in)-	- - - - - -	- - - - - -	- - - - - -
1974	70.71	24.17	35.72	8.61	16.04	35.45	16.98
1975	56.90	14.09	35.24	7.68	9.04	36.74	11.15
1976	34.53	6.77	27.07	4.94	4.41	26.99	5.66
Average Annual	54.05	15.01	32.68	7.08	9.83	33.06	11.26

acres. RC is set at 0.19 in/hr based on information on soils (1), SCS hydrologic soil group B (5), and guidelines in table II-9. A better procedure is to estimate RC and GA by best fit with infiltrometer measurements, but these are often unavailable. Management practice 2, with a grassed waterway, will have higher RC.

Volumetric saturation at field capacity, FUL, is estimated as 0.75. FUL will be higher for clay loams and lower for sandy soils. A corresponding value for available water content at the beginning of simulation, BST, is taken as 0.50, which is unknown, but simulation is not sensitive to this starting value. BST = 0.5 assumes that the soil contains half its capacity of stored water at the beginning of the year.

The soil evaporation parameter, CONA, is set at 3.75 according to guidelines in CREAMS, volume I, chapter 2 (3). Total porosity, POROS, is 0.41 as taken from data in Holtan and others (1). This value is a rough average of porosities reported for upper soil layer samples.

BR15, representing volume of immobile water, ranges from about 5 to 30%, depending on soil type and conditions. Measured values in Holtan and others (1) vary greatly from layer to layer but are generally low for sandy soils and high for clay soils. A mean value is 0.17 in/in for the upper layers of the Cecil soil used here.

Average monthly values of temperature and radiation are in the Climatic Atlas (6) (or CREAMS, vol. II, table II-7). Since this is a cropped watershed, winter cover is small and GR = 1.0. The leaf area index values for the corn crop are available from table II-8.

The depth of surface soil layer, DS, arbitrarily is taken as 2.0 in, and the root layer depth, DP, is assumed to be 22 in, making rooting depth, RD, equal to 24 in.

The infiltration parameter, GA, ordinarily is found for this soil (group B) from table II-9. The value used (13 in) is higher than the recommended range for a hydrologic group B soil but was chosen from comparisons to actual infiltrometer data. Tilled soils exhibit higher values of GA than untilled or undisturbed soils, which should be reflected when choosing a value for GA.

The remaining parameters reflect the hydraulic and topographic conditions governing overland flow and, in this model, the estimation of peak runoff rates. The Manning roughness value (RMN) of 0.03 is typical of flow along plowed furrows. Values for slope and distance of a "typical" overland flow path should reflect actual flow paths to make best estimates of runoff peaks. Flow often will follow furrows rather than the topographic "downhill" direction, which modifies slopes and lengths of flow.

The watershed in figure II-44 exhibits overland flow along furrows until the bottom of the swale is reached. Flow should move in a broad, rough channel to the outlet. Furrows in this example run east and west. The "channel" flow should be fairly rapid, and measured mean "furrow" distance is about 250 ft. A total estimated (weighted) flow length, XLP, is taken as 350 ft. The effective (weighted) field slope is estimated as 0.015 (1.5%). This value should put most weight on flow along furrows where the runoff water spends most of its time. The formal procedure for getting optimum mean slope (CREAMS, vol. I, ch. 2, eq. I-36) is not used since it applies to cascaded overland flow planes rather than a combination of planes and channels. In this example, furrow slope and swale slope are nearly equal.

Interpretation of Results

Results from hydrology option 2 consist of a summary of input parameters, a table of daily values of rainfall, runoff, and other water-balance information for all days on which rainfall occurs, plus the passfile created for subsequent model component simulations.

Simulation results for the Georgia Piedmont application are summarized in figure II-49. The fallow saturated-hydrologic conductivity in the program is 0.8 times the normal (cultivated) conductivity.

Results show that in each year total evapotranspiration is a large part of rainfall. For the lowest rainfall year, 1973, soil water never became large enough to cause deep percolation. Although it is unlikely that no soil water moved below the root zone, percolation was very small. The nature of the model's approximation to soil-water movement and simplification of the soil water system will insure that years with low percolation cannot be simulated accurately.

Mississippi Delta Management Practice 1

The Mississippi Delta is topographically much more straightforward to represent since it is rectangular and, therefore, has a uniform flow length for flow along the furrows. The soil parameters were obtained from Lund and Loftin (2). Table II-9 shows that the soil is in a low B or a high C group, with hydraulic conductivity (RC) of 0.16. A value of 11 for GA is consistent with this classification.

Parameters FUL, BST, POROS, and BR15 were assigned from values typical for

HYDROLOGY SUMMARY

DAILY HYDROLOGY PARAMETERS - GEORGIA PIEDMONT
MANAGEMENT PRACTICE ONE
CONTINOUS CORN - CONVENTIONAL TILLAGE

1974

MONTH	RAIN	RUNOFF	ET	PERC	AVG SW
JAN	2.700	0.008	2.157	0.000	2.514
FEB	4.110	0.436	2.270	1.380	2.830
MAR	1.920	0.008	1.618	0.106	2.544
APR	2.600	0.179	1.991	0.768	2.735
MAY	5.420	0.644	3.374	0.971	2.742
JUN	5.290	1.254	3.153	1.257	2.351
JUL	4.150	0.586	4.726	0.000	0.901
AUG	5.780	0.192	6.440	0.000	0.969
SEP	1.850	0.000	2.199	0.000	0.457
OCT	0.360	0.000	0.554	0.000	0.095
NOV	1.160	0.000	0.782	0.000	0.282
DEC	4.920	0.209	1.666	0.433	1.484
TOT	40.260	3.516	30.929	4.916	1.659

1975

MONTH	RAIN	RUNOFF	ET	PERC	AVG SW
JAN	5.020	0.490	2.321	2.522	2.977
FEB	7.170	1.035	2.460	3.611	2.920
MAR	9.780	2.961	2.930	3.682	2.864
APR	3.930	0.934	2.173	1.260	2.717
MAY	6.070	0.633	3.123	1.686	2.771
JUN	3.550	1.071	2.996	1.067	2.479
JUL	4.670	0.043	6.012	0.000	0.615
AUG	2.340	0.000	2.194	0.000	0.165
SEP	5.370	0.339	2.984	0.000	1.200
OCT	0.350	0.000	0.916	0.000	2.157
NOV	0.000	0.000	0.329	0.000	1.721
DEC	0.000	0.000	0.262	0.000	1.431
TOT	48.250	7.505	28.602	13.828	2.001

ANNUAL AVERAGES

MONTH	RAIN	RUNOFF	ET	PERC	AVG SW
JAN	3.860	0.249	2.239	1.261	2.745
FEB	5.640	0.735	2.365	2.496	2.875
MAR	5.850	1.485	2.274	1.894	2.704
APR	3.265	0.556	2.082	1.014	2.726
MAY	5.745	0.639	3.249	1.329	2.756
JUN	4.420	1.163	3.074	1.162	2.415
JUL	4.410	0.314	5.369	0.000	0.758
AUG	4.060	0.096	4.317	0.000	0.567
SEP	3.610	0.170	2.592	0.000	0.828
OCT	0.355	0.000	0.685	0.000	1.126
NOV	0.580	0.000	0.556	0.000	1.001
DEC	2.460	0.105	0.964	0.217	1.458
TOT	44.255	5.511	29.765	9.372	1.830

Figure II-49.—Annual summaries of simulation results using hydrology option 2, Ga. Piedmont application, MP1 soils of this type since measurements were unavailable. Although normal rooting depth for cotton can extend to 4 ft, 40 in reflect the expected effect of the fluctuating high water table in this area. The winter cover factor, GR, of 0.50 reflects the practice of leaving cotton plant residue on the soil over winter.

The relatively large row channels indicated a somewhat lower roughness value than ordinary furrows, and Manning's n (RMN) was chosen as 0.02. The effective slope of 0.003 is a weighted value combining the row slope of 0.004 and the channel slope of 0.001. Effect of the intercepting channel was included in the overall effective flow length of 2,000 ft, especially since it is a grass-lined channel. The value of XLP is, nevertheless, somewhat smaller than the total of the row and channel lengths in figure II-46.

Temperature and radiation data were from the nearest U.S. National Weather Service recording stations. In this situation, radiation data were from Shrevesport, La., and temperature data were from Jackson, Miss. Leaf area index data for cotton were taken from table II-8.

Interpretation of Results

Figure II-50 shows summary output information from the hydrology subroutines for HYDTWO for this example. Compared to the Georgia Piedmont example, total evapotranspiration is a lower percentage of the total rainfall but still is larger than the corresponding ET in Georgia. Rainfall here is much larger, although the 3 yr represent both a year with lower and higher than average rainfall. The lower RC intuitively causes a higher proportion of runoff. Since with the number of days of high soil-water content is large, percolation is also relatively large.

Not shown in either output HYDTWO data example is the predicted sequence of peak flows for the runoff events. These naturally will reflect the values chosen for surface response-related parameters, including RMN, SLOPE, and XLP. Management practices strongly are reflected in RMN and XLP values, and peak flows are affected considerably as a result.

EROSION

The western Tennessee and Mississippi Delta examples illustrate selection of parameter values and application of the erosion component of the model. The western Tennessee site is discussed first because it is typical of many cultivated fields. The Delta site shows a special application of the model.

Only the most significant cards and parameters are discussed. Refer to chapter 2 for card sequence and identification and definition of parameters. Several of the first cards of each input file are shown in accompanying figures.

Western Tennessee

Figure II-51 shows the initial part of the parameter file. Major entries

HYDROLOGY SUMMARY

BREAKPOINT HYDROLOGY PARAMETERS - MISSISSIPPI DELTA
MANAGEMENT PRACTICE ONE
CONTINUOUS COTTON - CONVENTIONAL TILLAGE

1974

MONTH	RAIN	RUNOFF	ET	PERC	AVG SW
JAN	8.880	0.912	1.256	4.672	3.904
FEB	3.470	0.400	1.152	2.173	4.158
MAR	1.920	0.039	1.082	0.681	4.098
APR	5.380	0.872	1.328	2.985	4.084
MAY	14.620	2.907	3.073	8.539	4.148
JUN	9.050	2.607	6.462	2.477	3.704
JUL	5.870	2.280	5.084	0.000	0.656
AUG	8.010	1.605	3.504	0.000	0.630
SEP	3.650	0.497	5.666	0.000	1.818
OCT	1.970	0.091	0.925	0.000	0.779
NOV	4.140	0.735	1.507	0.000	3.008
DEC	4.610	0.387	1.150	2.422	4.129
TOT	71.570	13.334	32.189	23.950	2.926

1975

MONTH	RAIN	RUNOFF	ET	PERC	AVG SW
JAN	4.160	0.223	1.128	2.720	4.226
FEB	4.870	0.652	1.470	3.008	4.274
MAR	7.170	0.808	1.842	4.493	4.197
APR	4.770	0.441	2.089	2.097	4.152
MAY	4.990	0.031	2.955	2.321	4.114
JUN	4.750	0.320	6.287	0.805	3.340
JUL	2.860	0.367	3.687	0.000	0.410
AUG	5.760	1.608	4.201	0.000	0.638
SEP	4.010	0.291	3.391	0.000	0.665
OCT	2.230	0.249	0.867	0.000	0.782
NOV	9.140	0.744	1.407	4.166	4.144
DEC	2.630	0.000	1.186	1.677	4.374
TOT	57.340	5.734	30.509	21.289	2.943

Figure II-50.—Summary results of simulation for the Miss. Delta MP1 using hydrology option 2.

1976

MONTH	RAIN	RUNOFF	ET	PERC	AVG SW
JAN	3.410	0.371	1.226	1.835	4.244
FEB	5.830	1.484	1.199	3.245	4.154
MAR	6.880	0.417	1.926	4.228	4.230
APR	1.220	0.022	1.277	0.233	4.082
MAY	1.490	0.000	1.458	0.261	3.973
JUN	4.800	0.149	5.661	0.000	3.280
JUL	1.220	0.062	3.666	0.000	0.664
AUG	0.140	0.000	0.285	0.000	0.056
SEP	3.400	0.352	2.885	0.000	0.470
OCT	3.140	0.000	1.614	0.000	0.551
NOV	1.900	0.000	1.086	0.000	1.919
DEC	1.380	-0.000	1.024	0.000	2.937
TOT	34.810	2.857	23.306	9.803	2.547

ANNUAL AVERAGES

MONTH	RAIN	RUNOFF	ET	PERC	AVG SW
JAN	5.483	0.502	1.203	3.076	4.125
FEB	4.723	0.846	1.273	2.809	4.195
MAR	5.323	0.421	1.617	3.134	4.175
APR	3.790	0.445	1.565	1.772	4.106
MAY	7.033	0.979	2.495	3.707	4.078
JUN	6.200	1.025	6.137	1.094	3.442
JUL	3.317	0.903	4.146	0.000	0.577
AUG	4.637	1.071	2.663	0.000	0.441
SEP	3.687	0.380	3.980	0.000	0.984
OCT	2.447	0.113	1.135	0.000	0.704
NOV	5.060	0.493	1.334	1.389	3.024
DEC	2.873	0.129	1.120	1.366	3.813
TOT	54.573	7.309	28.668	18.347	2.805

Figure II-50.—Summary results of simulation for the Miss. Delta MP1 using hydrology option 2--continued.

```
NO                              EROSION PARAMETER DATA
----      --------------------------------------------------------------------------------
 1                        EROSION PARAMETERS - WESTERN TENNESSEE
 2                                MANAGEMENT PRACTICE ONE
 3                        CONTINUOUS CORN - CONVENTIONAL TILLAGE
 4       74000         0         1         0         4
 5       0.000     0.000     0.000     0.000     0.000     0.000
 6       0.150     0.650     0.200     0.020     0.000     0.000     0.000     0.000
 9      69.210   276.000     0.074     0.030     0.120     0.010    61.000    18.400   174.000     4.840
10           1
11       1.000     0.400
12           3         1         1         2         1
13      20.000    10.000     0.035     0.010     2.000     2.000     0.000
14     568.000     6.500     0.800    20.000
15       0.000     0.010   284.000     0.020   568.000     0.030
12           5         1         1         4         1
13      20.000    20.000     0.035     0.005    31.000     2.000     0.000
14    4900.000    69.210     0.800    20.000
15       0.000     0.0051380.000     0.0061620.000     0.0102400.000     0.0153000.000     0.018
18       74001     74045
19           1         1         1
20       1.000     0.200
21       1.000     0.800
22       1.000     0.030
23           1         1         1
24       0.000     0.060
25       0.000     0.400
26       0.000   100.000
27       0.000     0.330
28       0.000     0.330
29       0.000    10.000
23           1         1
24       0.000     0.060
25       0.000     0.400
26       0.000   100.000
27       0.000     0.330
28       0.000     0.330
29       0.000    20.000
18       74046     74092
19           1         0         0
20       1.000     0.250
23           0         0         0         0         0         0
23           0         0         0         0         0         0
18       74093     74119
19           1         0         1
20       1.000     0.430
22       1.000     0.040
23           1         1         0         1         1         0
24       0.000     0.045
25       0.000     0.150
27       0.000     0.330
28       0.000     0.330
23           1         1         0                             0
24       0.000     0.045
25       0.000     0.100
27       0.000     0.330
28       0.000     0.330
```

Figure II-51.—Partial parameter file for the erosion component with application to western Tennessee, management practice 1.

by cards will be discussed.

Cards 1 to 3: The alphanumeric information identifies such items as site, management practice, and other important identification factors.

Card 4: BDATE = 74000. BDATE can be greater than the first PDATE in the parameter file, but it must be less than the first storm date, SDATE, in the hydrology file. FLGPRT = 0 so that the model will use the primary particle distribution of the soil to compute the sediment particle specifications from internal relationships in the model. FLGSEQ = 4 to accommodate a watershed with two stream orders (fig. II-52).

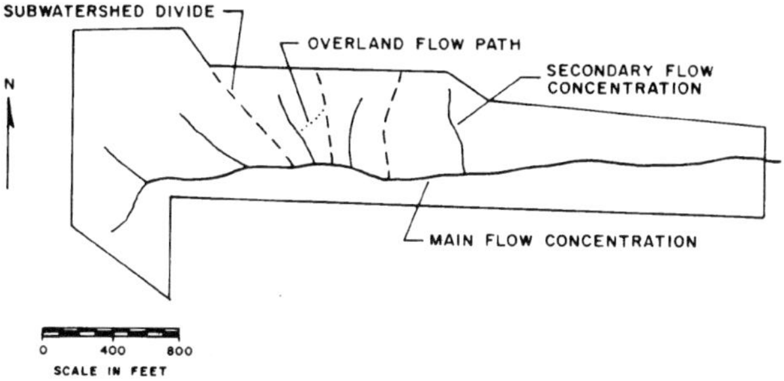

Figure II-52.—Field, subwatershed, channel, and overland flow definitions

Card 5: The defaults for variables on this card are used by leaving this card blank.

Card 6: The primary particle size and organic matter contents were estimated from SCS soil survey information. Specific surface area variables were left blank so that default values are used since better information was unavailable.

Cards 7 to 8: These cards are absent because default sediment particle specifications calculated within the model from primary particle size data are used. If FLGPRT = 3 on card 4, cards 7 and 8 must be present.

Card 9: This watershed is made up of several small subwatersheds. Average values for topographic factors must be estimated, or representative values must be chosen. Typical subwatersheds are shown in figure II-52. Since a contour map was unavailable, a typical overland flow profile was constructed from soil survey maps (fig. II-53). The parameter values are DATOV = 69.21 acres, which is the area of the total watershed. An average value for the subwatersheds could have been used. The only difference in the output would have been in total amount (lb) of sediment produced on the overland flow areas. DATOV does not affect the sediment yield per unit area and concentration of sediment in the runoff. SLNGTH, AVGSLP, SB, SM, SE, XIN(3), YIN(3), and XIN(4), YIN(4)

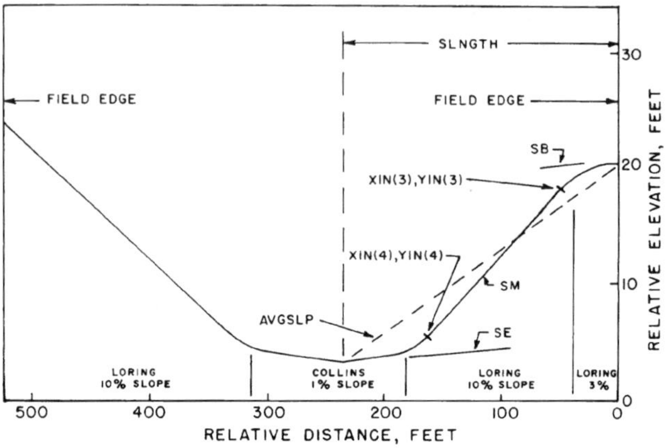

Figure II-53.—Representative overland flow profile constructed from soils slope data.

are, respectively, 276 ft, 0.074, 0.030, 0.120, 0.010, 61 ft, 18.4 ft, 174 ft, and 4.84 ft.

Cards 10 to 11: A soil erodibility of 0.4 ton/acre/EI was used for all locations along the representative overland flow profile. Values for soil erodibility are available from SCS.

Card 12: NS = 3: Three segments were used to describe the channel profile. FLAGC = 1 selected a triangular channel. FLAGS = 1 specified that the program use the equations for spatially varied flow. CONTL = 2 set uniform flow as control at the channel outlet. SECTN = 1 selected a triangular channel for the outlet control section.

Card 13: SIDSLP = 20 for the side slope, BOTWID = 10 ft, OUTMAN = 0.035 for Manning's n, and OUTSLP = 0.01 for slope of the outlet control channel. The rating parameters RA, RB, and YBASE are not used although values of 2, 2, and 0 are entered.

Card 14: LNGTH = 568 ft was the average length for the channel in each subwatershed. DATCH = 6.5 acres for the average drainage area in the subwatersheds above the channel outlet, and DAUCH = 0.8 acre for the average drainage area above the entrance to the channel in the subwatersheds. A side slope, Z = 20, for the secondary flow concentrations was used because the concentrations are farmed over.

Card 15: TX and TS are locations and slopes along the channel profile. Distances are referenced to the channel outlet. Therefore, TX(1) = 0 at the channel outlet. Slope along the channel was estimated from the soil survey map.

Cards 12 to 15: These cards repeat for the main channel of the watershed. FLAGS was set to 1 to use the energy gradeline curves to consider backwater for an assumed restricted outlet. CONTL = 4 was used to express control by a rating curve at the outlet. The rating coefficients RA = 31, RB = 2, and YBASE = 0 were selected to give estimated depths for assumed discharges. DATCH = 69.21 acres is the total watershed area. Channel slopes were estimated from the soil survey maps.

Cards 16 to 17: These cards are absent because no pond element is used.

Card 18: PDATE = 74000 and BDATE = 74045. These are the dates between which the following parameter values are valid. This period is one of winter stalk cover.

Card 19: NC, NP, and NM = 1 because of uniformity along the slope. In the first set of updateable parameter values, a nonzero value must be assigned to NC, NP, and NM to initialize the overland flow parameters.

Card 20: Since a uniform soil loss ratio is assumed for the slope, XCIN (1) = 1.0. At this crop stage, the soil loss ratio for about 1-1/2 tons/acre of surface residue is 0.20 from tables II-20 and II-23, and figure II-23. This represents an average C between standing stalks left by a cornpicker and stalks uniformly shredded with a shredder.

Card 21: PIN(1) = 0.80 for partial contouring, which results from the assumed "parallel to fence farming."

Card 22: Manning's n is a function of cover and roughness. A relative smooth surface and 1-1/2 ton/acre of cornstalks give MIN of 0.03 (table II-26).

Card 23: All variables on the card are set to 1 because of uniformity along the channels and to initialize the parameter values.

Card 24: The first TX, that is, TX(1), is always 0 because the reference is at the outlet end of the channel. TN = 0.06 (table II-28) is assigned to Manning's n. This value applies to the entire channel length.

Card 25: Critical shear stress, TCR, is set fairly high at 0.4 lb/ft^2 (table II-29) because the soil is assumed to have consolidated since the last cultivation in the summer.

Card 26: The effect of cover breakdown is ignored by setting TCV to the large number (100 lb/ft^2), which greatly exceeds values for the flow's shear stress.

Card 27: TDN, the depth to the nonerodible layer, is an initial value the first time it is read. A definite value for TDN is unknown except following tillage. Therefore, simulations are best started at the time of tillage, but this was inconvenient for the problem. A value of 0.33 ft approximates the depth of secondary tillage, the limiting depth for many fields.

Card 28: The nonerodible layer is assumed to follow the curvature of the surface soil, that is, it is parallel to the soil surface. Therefore, TDS is set equal to TDN.

Card 29: Although the channel is triangular, a width TW = 10 ft is specified because the model sometimes defaults to a rectangular channel.

Cards 23 to 29: These cards are repeated to describe the second channel.

Once the storm date exceeds CDATE, the model reads a new set of updateable parameters for the next crop stage. New values are required only for those parameters that change. The cards begin repeating at card 18.

Card 18: The new dates are 74046 and 74092. This is the end of the winter crop stage period. It is included to account for further decay of residue over the winter. The field is moldboard plowed on 74093.

Card 19: Contouring and Manning's n do not change from previous values. New input values are not read by setting NP and NM to 0. NC = 1 indicates a new soil loss ratio that is uniform along the slope.

Card 20: The new soil loss ratio is 0.25.

Cards 21 to 22: These are absent because NP and NM = 0.

Card 23: All values are set to zero to use previous values.

Cards 23 to 29: These do not appear because previous values are used for all parameters on the cards.

The next crop stage follows moldboard plowing on 74093. Plowing changes the soil loss ratio, and Manning's n for overland flow and channel flow reduces the critical shear stress and resets the depth to nonerodible layer. The next group of cards for 74093 and 74119 are inputs for these changes.

Interpretation of Results

The results of several management options for the West Tennessee site are discussed in a section late in chapter 2.

Mississippi Delta

Management practice 1--The Mississippi Delta site illustrates a special application of the model. This example is quite different because the watershed is flat and an unusual watershed representation is used.

The field is disked and bedded several times during the year. Well-defined row ridges and middles that form the flow patterns exist most of the year. The representation assumed is an overland area of row side slopes, channel 1 for a representative row middle, and channel 2 for the field ditch.

Figure II-54 partially lists the input parameters. The parameter values discussed differ significantly from the first example.

CARD NO	EROSION PARAMETER DATA									
1	EROSION PARAMETERS - MISSISSIPPI DELTA									
2	MANAGEMENT PRACTICE ONE									
3	CONTINUOUS COTTON - CONVENTIONAL TILLAGE									
4	74000	0	1	0	4					
5	0.000	0.000	0.000	0.000	0.050	0.000				
6	0.200	0.500	0.300	0.012	0.000	0.000	0.000	0.000		
9	0.090	1.500	0.200	0.200	0.200	0.200	1.500	0.000	1.500	0.000
10	1									
11	1.000	0.370								
12	1	2	2	2	1					
13	20.000	3.000	0.030	0.004	0.000	0.000	0.000			
14	1300.000	0.090	0.000	5.000						
15	0.000	0.004								
12	1	1	1	4	1					
13	5.000	3.000	0.020	0.010	15.000	0.500	0.000			
14	1270.000	32.000	0.000	5.000						
15	0.000	0.001								
18	74001	74035								
19	1	1	1							
20	1.000	0.540								
21	1.000	1.000								
22	1.000	0.050								
23	1	1	1	1	1	1				
24	0.000	0.060								
25	0.000	0.200								
26	0.000	100.000								
27	0.000	0.330								
28	0.000	0.330								
29	0.000	3.000								
23	1	1	1	1	1	1				
24	0.000	0.150								
25	0.000	0.700								
26	0.000	100.000								
27	0.000	100.000								
28	0.000	100.000								
29	0.000	5.000								
18	74035	74064								

Figure II-54.—Partial listing of parameter values for the Mississippi Delta, management practice 1.

Card 4: CSEQ = 4 designates overland flow (row side slopes) - channel (row middle) - channel (field ditch) for the watershed representation.

Card 5: The Manning's n for overland flow over bare soil is increased to 0.05 to ensure that no deposition is calculated on the row side slopes. The transport equations are intended for longer slopes. Increasing n for bare conditions increases computed transport capacity.

Card 9: A typical row side slope is the overland flow area. The rows are assumed to be 3.0 ft apart, and 1.5 ft is assumed for the width of a single row side slope, which is the length of overland flow. The overland flow area for a single row side slope for the 1,300 ft row·length is 0.0448 acre. Since two side slopes occur per row middle, a value of 0.090 acre is used for DATOV to obtain the total amount of sediment draining into the row middle. Slope length, SLNGTH, for the overland flow area is 1.5 ft. A uniform steepness of 0.2 is assumed. Therefore, AVGSLP, SB, SM, and SE = 0.2. Since the slope is uniform, XIN(3), YIN(3), and XIN(4), YIN(4) = 1.5, 0.0.

Card 12: Since the slope along the row middle is assumed to be uniform, NS = 1, that is, only one slope value is required. FLAGC = 2 for a rectangular channel in the row middle. Since FLAGS = 2 (table II-27), the slope of the energy gradeline (friction slope) equals the channel slope. Flow rates are small in row middles, and backwater effects do not extend beyond a few feet up the middles.

The parameters CONTL and SECTN are not used since FLAGS = 2, but dummy values are entered.

Card 13: Although values on this card are not used, the card must be present with nonzero dummy values, except for YBASE.

Card 14: LNGTH = 1300 ft for the length of the row middles. The drainage area DATCH for a single row is 0.090 acre. The drainage area DAUCH at the upper end of each middle is zero. The channel side slope Z is approximated at 5:1 (that is, 5.0).

Card 15: The slope along the row middle is 0.004. The entry on this card is 0.0, 0.004.

Cards 12 to 15: These cards repeat for the field ditch and are similar to other channel cards, except that backwater and a rating curve are assumed for outlet control.

Card 25: TCR for the second channel was set to 0.70 lb/ft^2, a relatively large critical shear stress representing the long period of consolidation since tillage.

Cards 27 to 28: TDN and TDS were set to large value, 100 ft, to ignore the effect of the nonerodible boundary, which was assumed not to exist.

Management Practice 2 - This management practice included a reduced number of tillage operations, winter cover, and a 20-ft grass strip at the end of the rows. Figure II-55 shows the first part of the input file for the problem.

Skip to card 23.

Card 23: The 20-ft grass strip at the end of the rows requires a different Manning's n, critical shear stress, and channel width from that used for the cultivated portion of the rows. Therefore, NN = 2, NCR = 2, NCV = 1, NDN = 1, NDS = 1, and NW = 2.

Card 24: A Manning's n of 0.10 is assumed for the grass and 0.06 for the cover crop, which begins 20 ft up the row. Entries on the card are 0.0, 0.15 (grass strip), 20.0, and 0.06 (tilled part of row).

CARD NO	EROSION PARAMETER DATA									
1	EROSION PARAMETERS - MISSISSIPPI DELTA									
2	MANAGEMENT PRACTICE TWO									
3	CONTINUOUS COTTON - MODIFIED TILLAGE - WINTER COVER									
4	74000	0	1	0	4					
5	0.000	0.000	0.000	0.000	0.050	0.000				
6	0.200	0.500	0.300	0.012	0.000	0.000	0.000	0.000		
9	0.090	1.500	0.200	0.200	0.200	0.200	1.500	0.000	1.500	0.000
10	1									
11	1.000	0.370								
12	1	2	2	2	1					
13	20.000	3.000	0.030	0.004	0.000	0.000	0.000			
14	1300.000	0.090	0.000	5.000						
15	0.000	0.004								
12	1	1	1	4	1					
13	5.000	3.000	0.020	0.010	15.000	0.500	0.000			
14	1270.000	32.000	0.000	5.000						
15	0.000	0.001								
18	74001	74046								
19	1	1	1							
20	1.000	0.200								
21	1.000	1.000								
22	1.000	0.050								
23	2	2	1	1	1	2				
24	0.000	0.150	20.000	0.060						
25	0.000	0.700	20.000	0.300						
26	0.000	100.000								
27	0.000	0.330								
28	0.000	0.330								
29	0.000	3.000	20.000	0.500						
23	1	1	1	1	1	1				
24	0.000	0.150								
25	0.000	0.700								
26	0.000	100.000								
27	0.000	100.000								
28	0.000	100.000								
29	0.000	5.000								
18	74047	74078								
19	1	0	0							
20	1.000	0.150								
23	2	2	0	0	0	0				
24	0.000	0.150	20.000	0.100						
25	0.000	0.700	20.000	0.400						
23	0	0	0	0	0	0				
18	74079	74107								
19	1	0	0							
20	1.000	0.580								
23	2	2	0	1	1	2				
24	0.000	0.150	20.000	0.040						
25	0.000	0.700	20.000	0.200						
27	0.000	0.330								
28	0.000	0.330								
29	0.000	3.000	20.000	0.500						
23	0	0	0	0	0	0				

Figure II-55.—Partial list of parameter values for the Mississippi Delta, management practice 2.

Card 25: The grass strip is not tilled, but the field is tilled above 20 ft from the end of the row. At this crop stage, the critical shear stress of the tilled soil is estimated to be 0.3 lb/ft^2 after subsoiling. The critical shear stress in the untilled grass strip is set to 0.7 lb/ft^2. The card is 0.0, 0.7, 20.0, and 0.3.

Card 26: Cover stability is assumed; therefore, TCV = 100 lb/ft^2.

Cards 27 to 28: The depth to the nonerodible layer is assigned 0.33 ft in the grass and tilled areas.

Card 29: The flow width in the grass is the entire row width, 3 ft, but it is 0.5 ft (20 ft up the row) in the row middles that remain after harvest.

Cards 23 to 29: Cards for the field ditch are the same as for example MP1.

Card 18: 74047, 74078. The cover is developed further in this crop stage. The field is disked on 078.

Card 19: NC = 1, NP and NM = 0. The soil loss ratio has decreased, but the contouring and Manning's n factors did not change.

Card 20: SLR = 0.15 for CIN.

Cards 21-22: These cards are absent since the parameters on them did not change.

Card 23: The only change is in the cover and consolidation of the soil. Therefore, only Manning's n and critical shear stress change. NN and NCR = 2. All other N's = 0.

Card 24: 0.0, 0.15, 20.0, 0.10. The Manning's n for the grass is not changed, but both n's must be reread since Manning's n changes for the tilled part of the channel. The n on the tilled part increases because resistance to the flow is assumed to increase from cover in the row middle.

Card 25: 0.0, 0.70, 20.0, 0.4. The critical shear stress for the tilled portion increases because of consolidation.

Cards 26 to 29: These cards do not exist because their parameters did not change. Since tillage did not occur, TDN and TDS are not reset. They are not reset until the next tillage.

On 74079, the field is tilled. This changes the soil loss ratio, Manning's n, critical shear stress, and depth to the nonerodible layer. The cards for 74079 to 74107 are for these changes.

Interpretation of Results

The results of three management practices for the Delta site are discussed in chapter 2.

PLANT NUTRIENTS

Two methods of nitrogen uptake by plants were given in chapter 3 and in volume I, chapter 4. Method 1 is applied on the western Tennessee management practice 1, and method 2 is applied on the Georgia Piedmont management practice 1.

Nitrogen Uptake Method 1

Western Tennessee Management Practice 1

It was stated in the "Description of Application Sites" section for western Tennessee that little information is available for the location. It was stated in chapter 3 that where soil-test data are not available, general information from soil surveys may be used in estimating parameter values. Research data are sometimes available on similar soils and sites. A combination of sources provided information for estimating parameter values for this application.

The following parameter values apply for the nutrient component with nitrogen uptake method 1.

Cards 1 to 3: WESTERN TENNESSEE MANAGEMENT PRACTICE 1
CONTINUOUS CORN, CONVENTIONAL TILLAGE
NITROGEN UPTAKE METHOD 1 (HYDONE PASSFILE)

Card 4: BDATE = 74001 The beginning date for simulation is January 1, 1974.

FLGOUT = 0 Code for type of output is user specified (coded for annual summary only).

FLGIN = 0 Code indicates input from hydrology pass file is in English units.

FLGPST = 0 Pesticides not included in this application.

FLGNUT = 1 Plant nutrients will be simulated.

Card 5: SOLPOR = 0.47 Porosity data for the specific site are unavailable. The soil survey data sheets give 1.4 for the bulk density for Loring silt loam. Porosity = 1 - (BD/2.65) = 0.47 cm^3/cm^3.

FC = 0.38 Field capacity, volumetric soil water content in cm^3/cm^3, was estimated from soil survey data sheets and personal communication with M. J. M. Romkens, USDA-SEA-AR, Oxford, Miss., who has conducted research on Loring soils in western Tennessee.

OM = 1.0 Percentage organic matter in the soil survey data sheets is for surface soil or plow layer. The organic matter content as used in the nutrient model is for calculations of denitrification in the root zone. Since data for the profile are unavailable, a reasonable value can be estimated as half the content in the surface soil. Since the surface value of 2.0% was obtained, half is 1.0%.

Card 7: OPT = 1 Nitrogen-uptake method 1 is used for this application.

Card 8: SOLN = 0.24 Initial soluble nitrogen in the top 1 cm of soil is generally unavailable unless soil tests have been made. For lack of data, an estimated 5 ppm of nitrogen in the soil water at saturation is reasonable. A porosity of 0.47 results in 0.235 kg/ha for soluble nitrogen.

SOLP = 0.09 Initial soluble phosphorus in the top centimeter of soil at saturation would not exceed 2 ppm. With a porosity of 0.47, 2 ppm would be 2×10^{-6} kg phosphorus/kg water, and 4.7×10^{4} kg/ha water at saturation, or 9.4×10^{-2} kg/ha soluble phosphorus.

NO3 = 20.0 Data are unavailable for nitrate in the root zone. Nitrate in soils at several locations is approximately 20 kg/ha. As indicated in chapter 4, the default value for NO3 is 20, which is used here.

SOILN = 0.0007 Soil survey data sheets for western Tennessee give approximately 0.07% nitrogen content in soil, which is in the range of 0.05% to 0.3% for nitrogen in the soil as given in volume I, chapter 4. This gives 0.0007 kg of nitrogen/kg of soil.

SOILP = 0.00035 The Loring soils in western Tennessee contain about 0.03% phosphorus. Conventionally, soil phosphorus is estimated as half of the soil nitrogen, resulting in a value of 0.00035 kg/kg for SOILP.

EXKN = 0.07 Extraction coefficient for nitrogen is an empirical coefficient relating nitrogen in the runoff to soluble nitrogen (SOLN) in the top centimeter of soil. Observations at several research locations have shown that the value of the coefficient should be in the range of 0.05 to 0.10.

EXKP = 0.07 Extraction coefficient for phosphorus is an empirical coefficient relating phosphorus in the runoff to soluble phosphorus (SOLP) in the top centimeter of soil. The coefficient for phosphorus should be about the same as for nitrogen.

AN = 7.4 The nitrogen enrichment coefficients for sediment were related to sediment transport in CREAMS, volume III, chapter 12. Since data are unavailable for Loring soils, the default value of 7.4 is used.

BN = -0.2 The nitrogen enrichment exponents for sediment were estimated in Vol. III, Chap. 12, also. Since data are not available for Loring soils, the default value, -0.2, is used here.

AP = 7.4 As for nitrogen, the phosphorus enrichment coefficient default value, 7.4, is used.

Card 9: BP = -0.2 The default value, -0.2, is used for phosphorus enrichment exponent.

RCN = 0.8 Nitrogen concentration in rainfall is read from the map in figure I-18 (CREAMS, vol. I, ch. 4).

Card 10: PDATE = 74001 PDATE is the Julian date on which the following parameters are valid, in this case, the beginning date of simulation.

CDATE = 74309 CDATE is the Julian date on which the model will stop using the following parameters. This is the day of harvest since potential mineralization (POTM) is reset due to increase of organic matter from decaying roots and stover.

Card 15: NF = 1 The number of fertilizer applications made until CDATE is reached.

DEMERG = 132 The Julian date of plant emergence is assumed to be 10 days after planting.

DHRVST = 309 Harvest is assumed at the end of October and is equivalent to the date when LAI goes to zero in the hydrology model.

Card 16: RZMAX = 660.0 Maximum rooting depth is estimated as 660 mm based on the depth of claypan (personal communication with M. J. M. Romkens, USDASEA-AR, Oxford, Miss.).

YP = 5000.0 The potential yield of corn grain under "ideal" conditions is about 9,400 kg/ha (CREAMS, vol. I, ch. 4, table I-11). Information from SCS indicates that 80 bu/acre (5,000 kg/ha) is a reasonable potential yield for the soils and slopes in this field.

DMY = 2.5 Dry matter ratio is the ratio of total dry matter yield to grain yield. The potential yield of grain and stover is 19480 kg/ha (CREAMS, vol. I, table I-11). Roots are considered 20% of the total dry matter production, giving a total potential of 24,350 kg/ha. DMY then is 24350/9400, or approximately 2.5.

POTM = 70.0 Potentially mineralizable nitrogen is calculated from the organic matter content in the root zone (CREAMS, vol. III, Ch. 13).

AWU = 299.0 Actual water use is the accumulated plant evaporation, in millimeters, for the growing season calculated in the hydrology component.

PWU = 299.0 Growing season potential plant evaporation, in millimeters, is calculated in the hydrology model. The value was calculated by HYDONE.

Card 17: C1 = 0.0209
C2 = -0.157
C3 = 0.0128
C4 = -0.415

The cubic coefficients and exponents for nitrogen uptake are given by crop (CREAMS, vol. III, Ch. 13, table 3).

Card 18: DF = 74122 The Julian date of fertilizer application is given in table II-48 (the same date is used each year).

Card 19: FN = 140.0 The amount of nitrogen fertilizer applied is 140 kg/ha (table II-48).

FP = 20.0 The amount of phosphorus fertilizer applied is 20 kg/ha (table II-48)

FA = 0.1 Fertilizer was incorporated by disking to a depth of 10 cm; therefore, the fraction of application in the surface centimeter is 1/10 or 0.1.

When day 74309 is reached in the simulation, card 10 is read for the new dates of applicability.

Card 10: PDATE = 74310 The Julian date when the new parameters are valid.

CDATE = 75309 The Julian date when the simulation ends with the following parameters; set at the harvest date in 1975.

Parameters on cards 15 through 19 are updateable to enable the user to specify actual and potential water use for different crops and different times, rates, and methods of fertilizer application. The nonupdateable parameters, such as SOLN, SOILN, NO3, and so forth, are updated automatically by accounting procedures in the computer program. Updating by the user is unnecessary. Multiple application of fertilizer during a year can be specified by repeating cards 18 and 19, as will be shown for the Georgia Piedmont location. Since updateable parameters in the western Tennessee application are repetitive for successive years, further discussion is unnecessary. The parameter values indicate a complete input file for the 3-yr application. Cards 15 through 19 for 1975 follow.

Card 15: NF = 1
DEMERG = 132
DHRVST = 309

Card 16: RZMAX = 660.0
YP = 5000.0
DMY = 2.5
POTM = 70.0
AWU = 283.0
PWU = 299.0

Card 17: C1 = 0.0209
C2 = -0.157
C3 = 0.0128
C4 = -0.415

Card 18: DF = 75122

Card 19: FN = 140.0
FP = 20.0
FA = 0.1

The following list shows cards 10 and 15 through 19 for calendar year 1976.

Card 10: PDATE = 75310
CDATE = 76366

Card 15: NF = 1

DEMERG = 132
DHRVST = 309

Card 16: RZMAX = 660.0
YP = 5000.0
DMY = 2.5
POTM = 70.0
AWU = 215.0
PWU = 299.0

Card 17: C1 = 0.0209
C2 = -0.157
C3 = 0.0128
C4 = -0.415

Card 18: DF = 76122

Card 19: FN = 140.0
FP = 20.0
FA = 0.1

Simulation is to terminate on the last day of 1976. A blank card is inserted following card 19. The blank card is read as a zero for CDATE, and simulation ceases.

Interpretation of Results

Results are summarized in table II-51 for the 3-yr simulation of plant nutrients for western Tennessee. This summary includes the water budget, sediment yield, and plant nutrient budget for the 28-ha area.

Nitrogen additions through fertilization, rainfall nitrogen, and nitrogen mineralization average approximately 185 kg/ha. Nitrogen uptake was relatively low for the simulation period due to the low estimated yield of 5,000 kg/ha. This yield may be low for the climate during the 3-yr period, but it is a realistic estimate for the conventional system on steep, eroded soil.

Denitrification is relatively high and probably is greater than expected. Since immobilization is not considered in the model process, mineralization and denitrification are higher than expected. The ratio of actual potential plant evaporation indicates a high soil-water content whereby denitrification is expected to be high. Nitrate leached, however, is not as high as expected for the large amounts of annual percolation and denitrification.

Annual summaries may be misleading for mass of pollutants. The sediment yield for 1974 (table II-51) was concentrated in two major storm events. Approximately 70% of the total annual soil loss resulted from these events. The first storm was the largest and occurred on January 10, which was well before application of fertilizer. Associated nutrient losses in runoff and sediment were low. The second largest storm occurred 13 days after fertilization, but runoff and soil loss were only half that of the January 10 storm. If the magnitudes of the two storms had been reversed, nutrient losses for the year would

Table II-51.—Annual summaries of erosion and water nutrient budgets for western Tennessee management practice 1

	1974	1975	1976	Annual Average
Rainfall (mm)	1,796.0	1,445.3	877.1	1,372.8
Runoff (mm)	613.9	357.8	172.0	381.2
Percolation (mm)	218.8	195.1	125.4	195.8
Actual evapotranspiration (mm)	907.3	895.1	687.5	830.0
Potential plant evaporation (mm)	299.8	299.8	299.8	299.8
Actual plant evaporation (mm)	299.8	283.2	214.6	265.9
Sediment yield (kg/ha)	51,446.	17,709.	6,501.	25,210.
Nitrogen fertilizer (kg/ha)	140.0	140.0	140.0	140.0
Nitrogen in rainfall (kg/ha)	14.35	11.56	7.02	10.98
Nitrogen mineralization (kg/ha)	42.85	39.85	34.22	38.97
Nitrogen in runoff (kg/ha)	8.89	4.14	2.45	5.16
Nitrogen in sediment (kg/ha)	.75	.23	.12	.37
Nitrogen uptake (kg/ha)	160.26	151.51	137.68	149.82
Nitrate leached (kg/ha)	4.92	4.16	5.53	4.87
Denitrification (kg/ha)	64.48	49.96	32.85	49.10
Phosphorus fertilizer (kg/ha)	20.	20.	20.	20.
Phosphorus in runoff (kg/ha)	.54	.31	.15	.33
Phosphorus in sediment (kg/ha)	.37	.11	.06	.18

have been significantly higher. Long-term simulation and analysis of the frequency of occurrence for selected periods of the year therefore are needed.

Nitrogen Uptake Method 2

Georgia Piedmont Management Practice 1

For the Georgia Piedmont example, much data by Smith and others (4) were used to select parameter values. Indications are given as to how values would have been assigned without specific published data. The following parameters are required for the nutrient model when nitrogen uptake is estimated by using method 2.

Cards 1 to 3: GEORGIA PIEDMONT MANAGEMENT PRACTICE 1
CONTINUOUS CORN, CONVENTIONAL TILLAGE
NITROGEN UPTAKE METHOD 2 (HYDTWO PASSFILE)

Card 4: BDATE = 74001 The beginning date for simulation is January 1, 1974.

FLGOUT = 0 Code for type of output desired (coded for annual summary).

FLGIN = 0 Code indicates input from hydrology pass file is

in English units.

FLGPST = 0 Pesticides not included in this application.

FLGNUT = 1 Code to indicate plant nutrient will be simulated.

Card 5: SOLPOR = 0.41 Soil porosity calculated from a bulk density of 1.56 (4).

FC = 0.32 Field capacity (4).

OM = 0.65 Percentage of soil organic matter content in root zone. Half of surface value is used as an estimate.

Card 7: OPT = 2 Nitrogen uptake method 2 is used for this application.

Card 8: SOLN = 0.20 Initial soluble N in top 1 cm of soil. Generally unknown unless soil is tested at beginning of simulation. A default value of 0.2 can be calculated assuming about 5 ppm in the soil solution at saturation.

NO3 = 20.0 Initial nitrate in root zone. Default value used (Ch. 4).

SOILN = 0.00035 Total nitrogen in root zone. Estimated from data (4). Units are kilograms of nitrogen per kilogram of soil.

SOILP = 0.00018 Total phosphorus in kilograms of phosphorus per kilogram of soil, often nearly half total nitrogen. Assigned value near those reported by Smith and others (4).

EXKN = 0.10 Extraction coefficient for soluble N in runoff. An empirical coefficient based on observations using data of Smith and others (4).

EXKP = 0.10 Extraction coefficient for soluble P in runoff. Value assigned by same rationale as for soluble N.

AN = 16.8 Coefficient for computing enrichment of N in sediment by methods in CREAMS, volume III, chapter 12. This value is computed using data of Smith and others (4).

BN = -0.16 Exponent for computing enrichment of N in sediment by methods in CREAMS, volume III, chapter 12. This value is computed using data of Smith and others (4). Without specific data, a default value of -0.2 would have been assigned.

AP = 11.2 Coefficient for computing enrichment of P in sediment by method in CREAMS, volume III, chapter 12. This value is computed using data of Smith and others (4). A default value of 7.4 would have been used without specific data.

Card 9: BP = -0.146 Exponent for computing phosphorus enrichment in sediment by method in CREAMS, volume III, chapter 12. Value computed using data of Smith and others (4).

RCN = 0.8 Nitrogen concentration in rainfall reported by Smith and others (4). Would have been estimated as 1.3 ppm from information in CREAMS, volume I, chapter 4, figure I-18.

Card 10: PDATE = 74001 The Julian date for the beginning of simulation when the following parameters are valid.

CDATE = 74255 The Julian date denoting when the following parameters are no longer valid (date of harvest to update POTM).

Card 15: NF = 2 The number of fertilizations during the year.

DEMERG = 132 Julian day of plant emergence. Assumed to be 10 days after planting (table II-45).

DHRVST = 255 Julian day of plant harvest, assumed to be on day leaf area index = 0.

Card 16: RZMAX = 610.0 Depth of potential root zone; taken here as depth to B2 horizon, in millimeters.

YP = 5700.0 Potential corn yield in kilograms per hectare under ideal conditions is 9,400 kilograms per hectare (CREAMS, vol. I, ch. 4, table I-11). Since conditions in the Georgia Piedmont are not ideal, potential is estimated as 5,700 kilograms per hectare.

DMY = 2.5 Dry matter yield ratio, the ratio of total dry matter production to grain production. DMY is about 2.5 for corn (CREAMS, vol. I, ch. 4, table II-11).

POTM = 47.0 Potentially mineralizable nitrogen. Estimated by method in CREAMS, volume III, chapter 13.

DOM = 60.0 Number of days after emergence until half of nitrogen is taken up (CREAMS, vol. III, ch. 13).

SD = 27.0 Standard deviation of DOM. The number of days between 50% and 84% uptake (CREAMS, vol. III, ch. 13).

PU = 250.0 Potential N uptake by the entire plant, kilograms per hectare. Based on actual uptake computed from data (4). Range recommended is 150 to 300 kilograms per hectare (CREAMS, vol. III, ch. 13).

Card 18: DF = 74122 Date of fertilization. The first fertilization was Julian day 122 (table II-45).

Card 19: FN = 28.0 Amount of nitrogen fertilizer applied, kilograms per hectare (table II-45).

FP = 28.0 Amount of phosphorus fertilizer applied, kilograms per hectare (table II-45).

FA = 0.1 Application factor for fertilizer. First application incorporated to 10 cm; therefore, FA = 0.1 for first application.

Table II-45 shows that fertilizer was applied twice in 1974; accordingly, NF = 2 on Card 15. Cards 18 and 19 must be repeated for each fertilization.

Card 18: DF = 74162 The second fertilizer application in 1974 was Julian day 162 (table II-45).

Card 19: FN = 112.0 The fertilizer rate was 112 kg/ha (table II-45).

FP = 0.0 Since the second fertilizer application consisted of ammonium nitrate, phosphorus was not applied.

FA = 1.0 The ammonium nitrate was surface applied, and all fertilizer is in the top centimeter of soil; thus, application factor is 1.0.

The parameter file is not shown for 1975. Only the updateable parameters are reset for following years.

Interpretation of Results

Georgia Piedmont is included in the application of nutrients to help the user select parameter values for nitrogen uptake method 2.

PESTICIDES

Mississippi Delta Management Practice 1

A pesticide application scheme was set up using the tillage/planting program (table II-46) for a pesticide program commonly used in the Mississippi Delta. Application dates were chosen not to coincide with rainfall, that is, no application during rainfall. Table II-52 shows pesticides, application dates, and assignment of parameter value for management practice 1 in 1974. For 1975 and 1976, only the application dates were changed as required to match changes in tillage/planting operation and to avoid pesticide application on rainy days. Hydrologic information was provided in pass files from the option 2 hydrology model.

Assignment of Parameter Values

Fluometuron--Applied at a rate of 1.5 kg/ha as a preplant incorporated herbicide. Incorporation was assumed uniform to a depth of 10 cm (4 in), that is, DEPINC = 10, EFFINC = 1. Since the herbicide is applied to soil, SOLFRC = 1, FOLFRC = 0. Since no initial residues were assumed, FOLRES and SOLRES = 0. Parameters for foliar washoff are not applicable or required. The program was written, however, so that zeros can be entered for WSHFRC, WSHTHR, and HAFLIF. Water solubility of fluometuron is 90 ppm (table II-40). A value of 0.1 was assumed for EXTRCT for fluometuron and all other pesticides (CREAMS, vol. I, Ch. 5). Persistence of fluometuron at the soil surface is described by DECAY =

Table II-52.—Pesticide application dates and parameters for Mississippi Delta application, management practice 1, 1974

PSTNAM	APDATE	APRATE	DEPINC	EFFINC	FOLFRC	SOLFRC	FOLRES	SOLRES	WSHTHR	WSHFRC	SOLH2O	HAFLIF	EXTRCT	DECAY	KD
		(kg/ha)	(cm)				(mg/m^2)	(ppm)	(cm)		(ppm)	(days)			
Fluometuron	74042	1.50	10	1.0	0	1.0	0	0	0	0	90	0	0.1	0.06	2
Trifluralin	74109	1.0	10	1.0	0	1.0	0	0	0	0	1.0	0	.1	.07	200
MSMA	74149	.50	1.0	1.0	0	1.0	0	0	0	0	100,000	0	.1	.07	4,000
	74154	.50	1.0	1.0	0	1.0	0	0	0	0	100,000	0	.1	.07	4,000
	74172	.50	1.0	1.0	0	1.0	0	0	0	0	100,000	0	.1	.07	4,000
Diuron	74182	.2	1.0	1.0	0	1.0	0	0	0	0	42	0	.1	.185	15
Methyl parathion	74198	.5	1.0	1.0	.4	.10	0	0	.2	.65	60	3	.1	.14	10
	74203	.5	1.0	1.0	.4	.10	0	0	.2	.65	60	3	.1	.14	10
	74210	.5	1.0	1.0	.4	.10	0	0	.2	.65	60	3	.1	.14	10
	74218	.5	1.0	1.0	.4	.10	0	0	.2	.65	60	3	.1	.14	10
	74225	.5	1.0	1.0	.5	0	0	0	.3	.65	60	3	.1	.14	10
	74232	.5	1.0	1.0	.5	0	0	0	.3	.65	60	3	.1	.14	10
	74239	.5	1.0	1.0	.5	0	0	0	.3	.65	60	3	.1	.14	10
	74247	.5	1.0	1.0	.5	0	0	0	.3	.65	60	3	.1	.14	10
	74253	.5	1.0	1.0	.5	0	0	0	.3	.65	60	3	.1	.14	10
	74260	.5	1.0	1.0	.5	0	0	0	.3	.65	60	3	.1	.14	10
EPN	74198	.5	1.0	1.0	.4	.10	0	0	.2	.65	.5	5	.1	.14	200
	74203	.5	1.0	1.0	.4	.10	0	0	.2	.65	.5	5	.1	.14	200
	74210	.5	1.0	1.0	.4	.10	0	0	.2	.65	.5	5	.1	.14	200
	74218	.5	1.0	1.0	.4	.10	0	0	.2	.65	.5	5	.1	.14	200
	74225	.5	1.0	1.0	.5	0	0	0	.3	.65	.5	5	.1	.14	200
	74232	.5	1.0	1.0	.5	0	0	0	.3	.65	.5	5	.1	.14	200
	74239	.5	1.0	1.0	.5	0	0	0	.3	.65	.5	5	.1	.14	200
	74247	.5	1.0	1.0	.5	0	0	0	.3	.65	.5	5	.1	.14	200
	74253	.5	1.0	1.0	.5	0	0	0	.3	.65	.5	5	.1	.14	200
	74260	.5	1.0	1.0	.5	0	0	0	.3	.65	.5	5	.1	.14	200
Toxaphene	74001	0	1.0	1.0	0	1.0	0	3.0	0	0	.4	0	.1	.0014	4,000

0.06, a midrange value (CREAMS, vol. III, ch. 17, table 1). The value of 2 for KD is the mean value for this component (CREAMS, vol. III, ch. 19, tables 2 or 4).

Trifluralin--This preplant-incorporated herbicide is applied at 1.0 kg/ha by procedures similar to those used for fluometuron; therefore, DEPINC = 10, EFFINC = 1, SOLFRC = 1, and FOLFRC = 0. Solubility is 1.0 ppm (table II-40), DECAY = 0.07, a midrange value (CREAMS, vol. III, ch. 17, table 1). A value of 200 was assumed for KD since some sediment transport of trifluralin is thought to occur (CREAMS, vol. III, ch. 16). Information in chapter 19 and other published information indicate, however, that KD for trifluralin would be about 20, making the sediment transport insignificant. Therefore, a value of 200 for KD is a compromise between somewhat conflicting observations.

MSMA--Applied three times as a directed postemergence spray at a rate of 0.5 kg/ha each. Since this herbicide is incorporated only by subsequent shallow cultivation, DEPINC = 1 and EFFINC = 1. Recall that 1 cm is the reference surface depth in the model for calculations of concentrations of surface-applied pesticides. No significant interception by foliage was assumed; therefore, SOLFRC = 1, and FOLFRC = 0, with remaining parameters pertaining to foliage set at 0. Since MSMA is very soluble, a large value (100,000 ppm) was assigned arbitrarily. DECAY = 0.07 and KD = 4000 were based on observations by R. D. Wauchope, U.S. Delta States Agricultural Research Center, Stoneville, Miss. (personal communication).

Diuron--One application applied as a direct postemergence spray. Parameter values are similar to those of MSMA, except SOLH2O = 42, DECAY = 0.185 (midrange value from CREAMS, vol. III, ch. 17, table 1), KD = 15 (mean value from CREAMS, vol. III, ch. 19, table 3).

Methyl Parathon/EPN--Applied together as aerial application to cotton foliage at rates of 0.5 kg/ha each per application for a total of 10 applications. DEPINC = 1 and EFFINC = 1 as 1 cm always is used as the reference depth for computing concentrations at the soil surface for pesticides that reach the soil surface and are not incorporated physically. Fifty percent of the intended application is assumed lost off-target by drift and volatilization (CREAMS, vol. III, ch. 18). The 50% intercepted by the target is distributed between the soil and foliage by FOLFRC = 0.4 and SOLFRC = 0.1 during the first five applications. Complete canopy closure is assumed at this stage so that FOLFRC = 0.5 and SOLFRC = 0. The washoff threshold, WSHTHR, was set at 0.2 and 0.3 cm rainfall, respectively, for these two periods. Organophosphates are removed readily by rainfall, so that WSHFRC for both methyl parathion and EPN was set at 0.65 (CREAMS, vol. III, ch.18). Foliar half-life values, HAFLIF, of 5 days for EPN and 3 days for methyl parathion were selected from CREAMS, volume III, chapter 18, table 2. Rates for soil decay, DECAY, 0.14 for both compounds were estimated from values in chapter 17 and from personal communications with L. L. McDowell, USDA Sedimentation Laboratory, Oxford, Miss. KD values for both compounds were estimated using solubility-KD relationships (CREAMS, vol. III, ch. 19, figure 6).

Toxaphene--Although toxaphene was not applied during the period 1974-76, it was assumed to be present as a residue in the soil at 3 ppm as a result of past use. Therefore, APRATE = 0 and SOLRES = 3. DECAY = 0.0014 and KD = 4000 were estimated by L. L. McDowell based on research studies on toxaphene in runoff.

A partial listing of the input parameter file is shown in figure II-56. The listing was continued to a point where all initial parameters and the first set of updateable parameters are given.

```
CARD
 NO                                  CHEMISTRY PARAMETER DATA
----     ---------------------------------------------------------------------------------
  1                          PESTICIDES PARAMETERS - MISSISSIPPI DELTA
  2                                  MANAGEMENT PRACTICE ONE
  3                          CONTINUOUS COTTON - CONVENTIONAL TILLAGE
  4         74000        0        0        1        0
  5         0.440    0.360    0.650
  6             7    74001    76366
 10         74001    74041
 11             0
 11             0
 11             0
 11             0
 11             0
 11             0
 11         74001
 12      TOXAPHENE
 13         0.000    1.000    1.000    0.000    1.000    0.000    3.000    0.000    0.000
 14           0.4      0.0   0.1000   0.0014   4000.0
 10         74042    74108
 11         74042
 12      FLUOMETURON
 13         1.500   10.000    1.000    0.000    1.000    0.000    0.000    0.000    0.000
 14          90.0      0.0   0.1000   0.0600      2.0
 11             0
 11             0
 11             0
 11             0
 11             0
 11             0
 10         74109    74148
 11             0
 11         74109
 12      TRIFLURALIN
 13         1.000   10.000    1.000    0.000    1.000    0.000    0.000    0.000    0.000
 14           1.0      0.0   0.1000   0.0700    200.0
 11             0
 11             0
 11             0
 11             0
 11             0
 10         74149    74153
 11             0
 11             0
 11         74149
 12      MSMA
 13         0.500    1.000    1.000    0.000    1.000    0.000    0.000    0.000    0.000
 14      100000.0      0.0   0.1000   0.0700   4000.0
 11             0
 11             0
 11             0
 11             0
 10         74154    74171
 11             0
 11             0
```

Figure II-56.—Partial list of pesticide input file, Mississippi Delta.

```
11        74154
12      MSMA
13        0.500    1.000    1.000    0.000    1.000    0.000    0.000    0.000    0.000
14      100000.0     0.0   0.1000   0.0700   4000.0
11            0
11            0
11            0
11            0
10        74172    74181
11            0
11            0
11        74172
12      MSMA
13        0.500    1.000    1.000    0.000    1.000    0.000    0.000    0.000    0.000
14      100000.0     0.0   0.1000   0.0700   4000.0
11            0
11            0
11            0
11            0
10        74182    74197
11            0
11            0
11            0
11        74182
12      DIURON
13        0.200    1.000    1.000    0.000    1.000    0.000    0.000    0.000    0.000
14         42.0      0.0   0.1000   0.1850     15.0
11            0
11            0
11            0
10        74198    74202
11            0
11            0
11            0
11            0
11        74198
12      METHYL PARATHION
13        0.500    1.000    1.000    0.400    0.100    0.000    0.000    0.650    0.200
14         60.0      3.0   0.1000   0.1400     10.0
11        74198
12      EPN
13        0.500    1.000    1.000    0.400    0.100    0.000    0.000    0.650    0.200
14          0.5      5.0   0.1000   0.1400    200.0
11            0
10        74203    74209
11            0
11            0
11            0
11            0
11        74203
12      METHYL PARATHION
13        0.500    1.000    1.000    0.400    0.100    0.000    0.000    0.650    0.200
14         60.0      3.0   0.1000   0.1400     10.0
11        74203
12      EPN
13        0.500    1.000    1.000    0.400    0.100    0.000    0.000    0.650    0.200
14          0.5      5.0   0.1000   0.1400    200.0
11            0
```

Figure II-56.—Partial list of pesticide input file, Mississippi Delta--continued.

Interpretation of Results

Annual summaries are given in table II-53 for the pesticide model output of the Mississippi Delta application. The model output is not discussed in detail. A few aspects are explained, however, as examples of the information contained. In terms of the percent of the amount applied, the greatest predicted pesticide loss was 6.46% for MSMA in 1974. These high losses were caused by an unusually large amount of rainfall shortly after pesticide application. About half of the total MSMA loss for the year occurred in a single storm of 6.96 cm one day after pesticide was applied; 3.37 cm became runoff and soil loss was 3,183 kg/ha. The model output for this storm is in table II-54. Over 98% of the total storm pesticide loss for MSMA was transported by sediment.

Since rainfall, runoff, and soil loss were less in 1975 and 1976, pesticide runoff losses were reduced.

Toxaphene also is transported primarily by sediment (table II-54). Losses were significantly higher in 1974 for several reasons. Soil loss in 1974 was much greater than in following years. Since no additional toxaphene was applied, the pesticide residue available to enter runoff declined because of surface depletion by runoff and pesticide decomposition. In an actual situation, the toxaphene residue at the soil surface would be replaced partially during major tillage operations that bring soil with higher toxaphene concentration to the soil surface. The initial soil residue could have been updated at the time of major tillage operations to partially replace the toxaphene residue available to enter runoff.

Although it is not the purpose of this example to actually compare different management practices and their effects on pesticide runoff potential, it is obvious that practices that limit soil loss will reduce losses of MSMA and toxaphene. Reduction of soil loss has much less effect on the other pesticides. For relatively nonpersistent pesticides, application timing with rainfall/runoff occurrence is a dominant factor in relation to time of pesticide application.

Table II-53.—Summary of pesticide model output for years 1974-76, Mississippi Delta, management practice 1

Pesticide name	Annual rainfall	Annual runoff	Annual sediment yield	Maximum pesticide concentration in runoff water	Maximum pesticide concentration in sediment phase	Total annual pesticide loss	
	(cm)	(cm)	(kg/ha)	(µg/ml)	(µg/g)	(g/ha)	(Percentage of application)
1974	178.4	33.2	38,800				
Fluometuron				0.0202	0.168	0.87	0.06
Trifluralin				.0012	.964	5.30	.53
MSMA				.0014	15.618	96.97	6.46
Diuron				.011	.435	4.40	2.20
Methyl parathion				.0727	3.021	29.62	.59
EPN				.0087	7.232	11.49	.23
Toxaphene				.0007	12.20	1 64.48	---- 1/
1975	145.6	13.7	5,600				
Fluometuron				.0334	.246	3.68	.25
Trifluralin				.0018	1.415	.66	.07
MSMA				.0013	21.062	5.50	.37
Diuron				.0096	.6012	.49	.24
Methyl parathion				.0067	2.713	22.74	.45
EPN				.0112	8.981	8.44	.17
Toxaphene				.0002	3.932	7.08	---- 1/
1976	88.4	6.4	2,220				
Fluometuron				.0205	.1581	3.36	.22
Trifluralin				.0022	1.320	.02	<.01
MSMA				.0010	17.139	.84	.06
Diuron				.0346	2.157	.62	.31
Methyl parathion				.0357	1.353	2.95	.06
EPN				.0052	3.946	.90	.02
Toxaphene				.0001	2.195	1.20	---- 1/

1/ No toxaphene applied, only residue from past applications

Table II-54.—Pesticide model output for a single storm occurring one day after MSMA application

STORM INPUTS

Date	74155	Julian date
Rainfall	6.96	cm
Runoff volume	3.73	cm
Soil loss	3183.74	kg/ha
Enrichment ratio	2.76	
Percolation	1.76	cm
Average temperature	25.00	degrees C
Average soil water	0.29	vol/vol
Accumulated ET	0.33	cm
Potential ET	0.33	cm

Quantity of pesticide in runoff values for storm 74155

Pesticide	Concentration available in residue	Concentration in water	Mass in water	Concentration in sediment	Mass in sediment	Total mass	Remaini residu
	(µg/g)	(µg/ml)	(g/ha)	(µg/g)	(g/ha)	(g/ha)	(µg/g
Fluometuron	0.00	0.0000	0.0000	0.0000	0.0000	0.0000	0.00
Trifluralin	.02	.0001	.0303	.0449	.1430	.1733	.02
MSMA	5.66	.0014	.5272	15.6183	49.7247	50.2519	5.33
Toxaphene	1.61	.0004	.1500	4.4443	14.1495	14.2996	1.52

REFERENCES

(1) Holtan, H. N., C. B. England, G. P. Lawless, and G. A. Schumaker.
1968. Moisture-tension data for selected soils on experimental watersheds. U.S. Department of Agriculture, Agricultural Research Service, ARS 41-144, 609 pp. (Series discontinued; Agricultural Research Service is now Science and Education Administration-Agricultural Research.)

(2) Lund, Z. F., and L. L. Loftin.
1960. Physical characteristics of some representative Louisiana soils. U.S. Department of Agriculture, Agricultural Research Service, ARS 4133, 83 pp. (Series discontinued; Agricultural Research Service is now Science and Education Administration-Agricultural Research.)

(3) Ritchie, J. T.
1972. A model for predicting evaporation from a row crop with complete cover. Water Resources Research 8(5):1204-1213.

(4) Smith, C. N., R. A. Leonard, G. W. Langdale, and G. W. Bailey.
1978. Transport of agricultural chemicals from small upland Piedmont watersheds. EPA-600/3-78-056. Environmental Protection Agency, Office of Research and Development, U.S. Government Printing Office, Washington, D.C. 364 pp.

(5) U.S. Department of Agriculture, Soil Conservation Service.
1972. National Engineering Handbook. Section 4, Hydrology. U.S. Government Printing Office, Washington, D.C. 548 pp.

(6) U.S. Department of Commerce.
1968. Climatic Atlas of the United States. U.S. Government Printing Office, Washington, D.C.

CREAMS

A Field Scale Model for
Chemicals, **R**unoff, and **E**rosion From
Agricultural **M**anagement **S**ystems

VOLUME III. SUPPORTING DOCUMENTATION

CONTENTS

CONTENTS

Chapter 1. TIME DISTRIBUTION OF CLOCK HOUR RAINFALL

A. T. Hjelmfelt[1/]

INTRODUCTION

Simulation of runoff from rainfall events requires rainfall in time increments not greater than the time to equilibrium for the watershed. For small watersheds the time to equilibrium will be measured in minutes, whereas the easily available rainfall will be available at clock-hour intervals. Time distribution must be estimated within the clock hour.

CLOCK HOUR AND 60-MINUTE RAINFALL

A rainfall event indicated as occurring in a single clock hour will have an actual duration of not more than 60 min. Hershfield (2) indicates that the 60-min equivalent is 1.13 times the clock hour rainfall. This is an average of a value that varies irregularly and unpredictably. The validity of the 1.13 value is indicated in Hershfield's (2) graph of 2-yr events shown in figure 1.

TIME DISTRIBUTION OF RAINFALL

Design Storms

Design storms of a particular return period are generated from intensity-duration-frequency studies. Results of these studies do not include a time distribution. The rainfall increments should be arranged to maximize peak discharge but maintain a reasonable sequence within the storm. To achieve this, early portions of the storm will be subject to interception and depression storage losses and to higher infiltration losses.

Williams (8) suggests placing the maximum intensity at a point between one-third and one-half of the storm duration. The other storm increments are grouped around this value. Hjelmfelt and Cassidy (4) recommend this procedure, which Kent (6) formalized. Increments of 30 min. were used to form a 24-hr storm. Maximum intensity was placed near the midpoint of the storm, and remaining increments were placed around this value. An average of the resulting distributions formed the type II distribution for the major portion of the United States. Another distribution, type I, was generated by a similar method for the rest of the United States. Type I and type II distributions are shown in table 1.

1/ Hydraulic engineer, USDA-SEA-AR, Columbia, Mo.

Table 1.--Type I and type II rainfall distributions

Time (hr)	P_x / P_{24} [1/] Type I	Type II
2.0	0.035	0.022
4.0	.076	.048
6.0	.125	.080
7.0	.156	----
8.0	.194	.120
8.5	.219	----
9.0	.254	.147
9.5	.303	.163
9.75	.362	----
10.0	.515	.181
10.5	.583	.204
11.0	.624	.235
11.5	.654	.283
11.75	----	.387
12.0	.682	.663
12.5	----	.735
13.0	.727	.772
13.5	----	.799
14.0	.767	.820
16.0	.830	.880
20.0	.926	.952
24.0	1.000	1.000

1/ Ratio of accumulated rainfall to total.

Source: Kent (6).

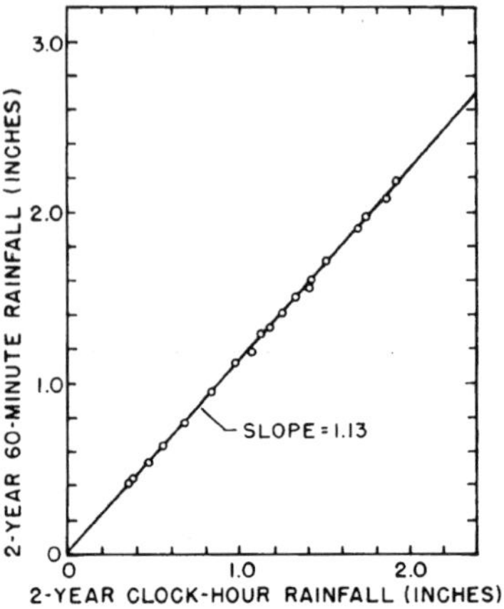

Figure 1.--Relation between 2-yr 60-min rainfall and 2-yr clock-hr rainfall.

Long Duration

Hershfield (3) studied 300 storms to determine the average time distribution of rainfall within 6-, 12-, 18-, and 24-hr storms. The average distribution could be displayed as one curve as shown in figure 2. The type II storm of Kent (6) also is shown for comparison.

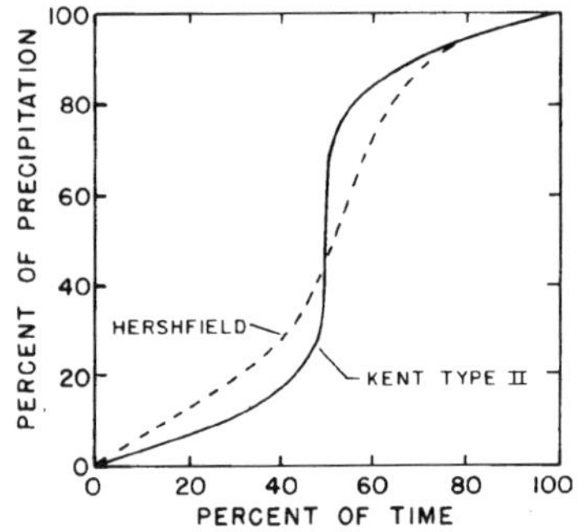

Figure 2.--Time distribution of rainfall for storms of long duration as determined by Hershfield (2) and type II distribution of Kent (6).

In discussing the average curve, Hershfield (3) indicates, "The obvious limitation of such a curve is that it conceals the wide variations in the time distribution and gives no indication of the distribution from an individual storm. Therefore, it would not be unreasonable to refashion the curve by rearranging either the duration or storm magnitude increments, because almost any order is realistic." The Bureau of Reclamation (1) places the most intense 6-hr at the beginning of a probable maximum storm.

Short Events

In an intensive study of thunderstorms, the U.S. Weather Bureau (7) developed a series of curves describing the time distribution of rainfall in 1-hr storms.[2/] The results are shown in figure 3. In these storms the greatest intensity occurs at the beginning. The mass curve of rainfall is the same as the depth-duration curve.

The curve for 2.01 to 3.00 in/hr storm is decribed by the relation

$$\frac{P}{P_{Total}} = \frac{1.37 \quad T/T_{Total}}{0.37 + T/T_{Total}} \tag{1}$$

Points computed using equation 1 are shown on figure 3.

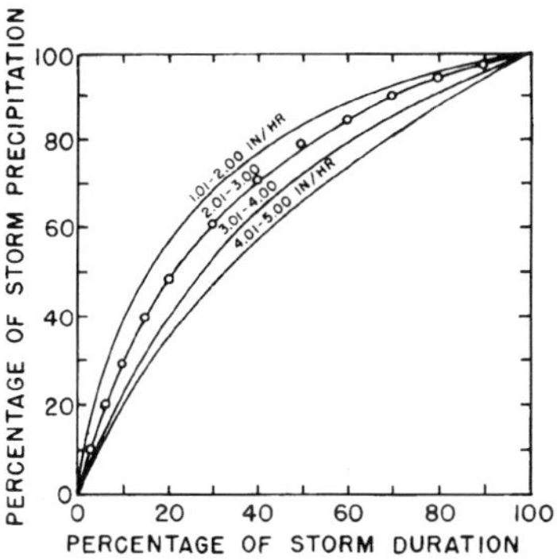

Figure. 3--Time distribution of 1-hr storms of various intensities (7). Points indicate values calculated using equation 1.

TIME DISTRIBUTION OF RAINFALL IN HEAVY STORMS

Huff (5) published the result of a detailed analysis of storm patterns in Illinois. These storms are less than 12 hr, 12 to 24 hr, and greater than 24 hr. They are divided by the timing of the occurrence of the most intense rainfall. Thus, the storms are grouped by most intense portion in the 1st, 2d, 3d, or 4th quartile. Time distributions are expressed in probabilities. The 30% probability can be interpreted that only 30% of the storms will have this distribution or one of the more steep distributions. The results of Huff (5) are shown in figures 4, 5, 6, and 7. The median, 50%, line is probably the most useful.

Storms with the most intense portion in the 1st quartile were commonly of short duration, whereas storms with most intense portion in the 4th quartile were commonly storms of duration greater than 24 hrs. The median distribution from each quartile storm is compared with the 1-hr storm distributions of the National Weather Service

2/ In 1970, the U.S. Weather Bureau became the National Weather Service

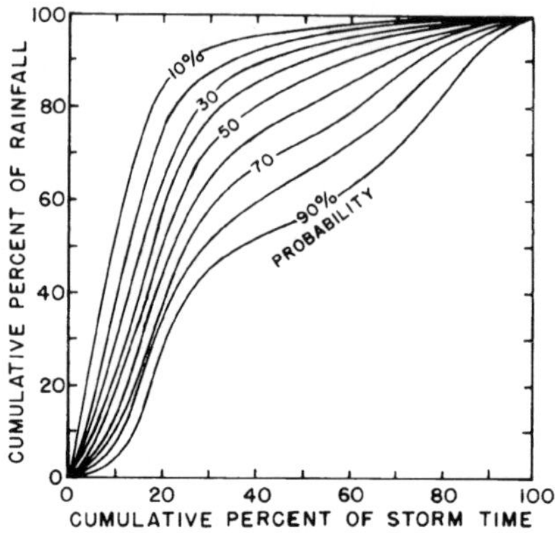

Figure 4.--Time distribution of rainfall with maximum intensity in 1st quartile (5).

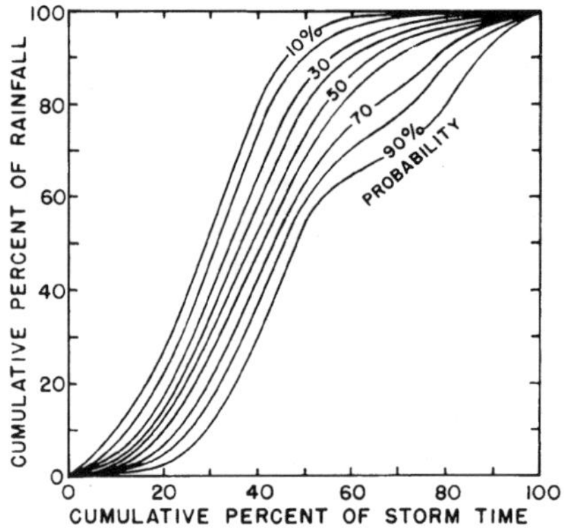

Figure 5.--Time distribution of rainfall with maximum intensity in 2nd quartile (5).

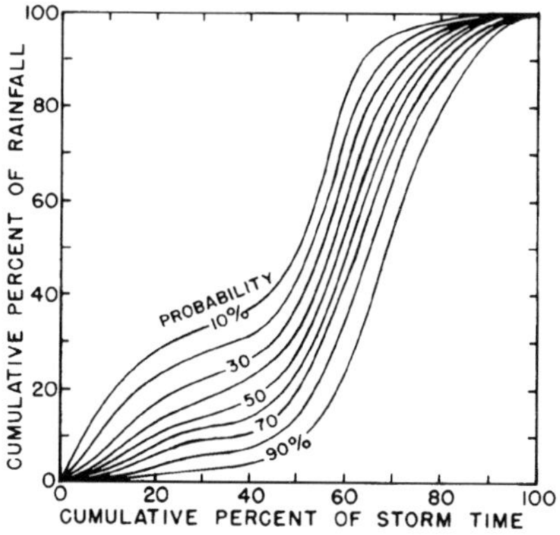

Figure 6.--Time distribution of rainfall with maximum intensity in 3rd quartile (5).

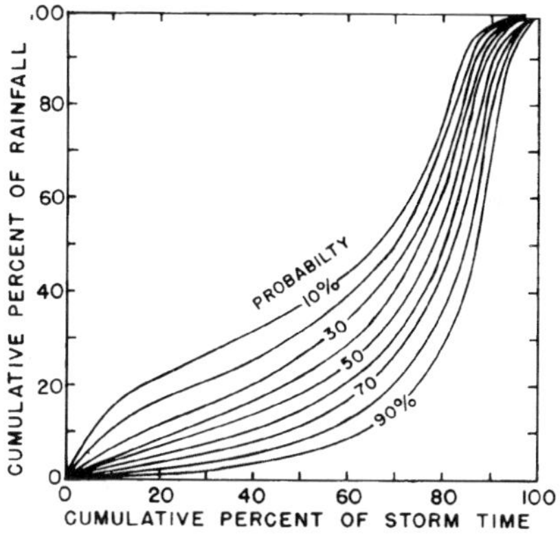

Figure 7.--Time distribution of rain-with maximum intensity in 4th quartile (5).

and with Hershfield's distribution for long duration storms in figure 8.

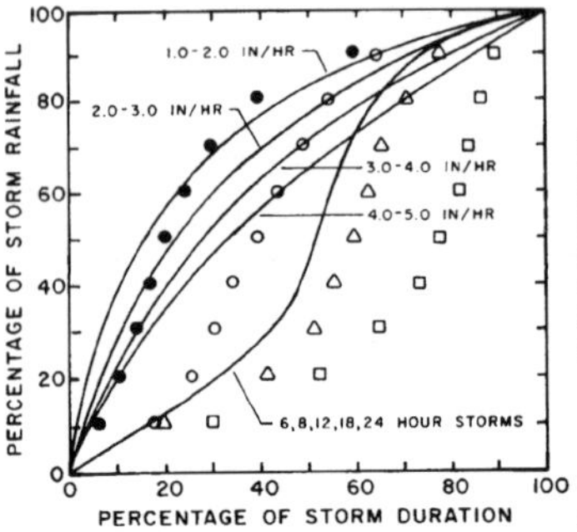

Figure 8.--Comparison of median distributions of Huff (5) with distributions of the Weather Bureau (7) and Hershfield (3).

RECOMMENDED PROCEDURE

When determining time distribution of rainfall from clock hour quantities, the great variability in events must be recognized. At this time there seems little advantage in applying an average distribution to any storm with values recorded in 3 clock-hours or more. The average intensity for each hour is probably the best estimate.

STORMS OCCURRING IN 1 CLOCK-HOUR

Most storms occurring in 1 clock-hour will last less than 60 min. Using the total hour as the duration will underestimate the peak discharge. Multiplying the clock hour value by 1.13 should result in the equivalent, on a return period basis, 60-min catch on the average.

The equivalent 60-min catch be can used with equation 1 to obtain the time distribution for peak discharge estimates. Thus,

$$\frac{P}{P_{equiv}} = \frac{1.37\ (T/60\ \text{min})}{0.37 + (T/60\ \text{min})} \qquad (2)$$

in which T is in min.

STORMS OCCURRING IN 2 CLOCK-HOURS

If the storm occurs in 2 clock-hours, a little more information is available. Equation 1 can be rewritten

$$\frac{T}{T_{Total}} = \frac{0.37\ \ P/P_{Total}}{1.37 - P/P_{Total}}\ . \qquad (3)$$

Let the first-hr catch be p_1 and the 1 second-hr catch be p_2. Then

$$\frac{T_1}{T_{Total}} = \frac{(0.37)\ p_1/(p_1 + p_2)}{1.37 - p_1(p_1 + p_2)}\ . \qquad (4)$$

The result is the fraction of the total duration represented by the first-hr catch. An additional assumption is needed at this point. If the ratio is

0.50 or greater, one can set T_1 equal to 60 min and solve the total time T_{Total}, which total will be 120 min or less. If the ratio is less than 0.5, this process will yield total durations in excess of 2 hr. Recognize that

$$\frac{T_1}{T_{Total}} = \frac{T_1}{T_{Total}} \quad (5)$$

and set T_1 equal to 60 min to determine T_{Total}.

REFERENCES

(1) Bureau of Reclamation.
1973. Design of small dams. U.S. Department of Interior.

(2) Hershfield, D. M.
1961. Rainfall frequency atlas of the United States. U.S. Department of Commerce, Weather Bureau Technical Publication 40. (U.S. Weather Bureau now National Weather Service)

(3) ------
1962. Extreme rainfall relationships. Journal of Hydraulics Division, American Society of Civil Engineers Proceedings 88(HY6):73-92.

(4) Hjelmfelt, A. T., and J. J. Cassidy.
1975. Hydrology for engineers and planners. Iowa State Press.

(5) Huff, F. A.
1967. Time distribution of rainfall in heavy storms. Water Resources Research, Vol. 3, 4th quarter. pp. 1007-1019.

(6) Kent, K. M.
1973. A method for estimating volume and rate of runoff in small watersheds. Department of Agriculture, Soil Conservation Service SCS-TP-149 (revised).

(7) U.S. Weather Bureau.
1947. Thunderstorm rainfall. Hydrometeorological Report 5.

(8) Williams, G. R.
1950. Hydrology. In Hydraulic Engineering (Proceedings of the Fourth Hydraulics Conference), Iowa Institute of Hydraulic Research, H. Rouse, (ed.) John Wiley and Sons, Inc.

Chapter 2. EROSIVITY "R" FOR INDIVIDUAL DESIGN STORMS

Keith R. Cooley[1/]

INTRODUCTION

In assessing nonpoint source pollution as outlined in Section 208 of the Federal Water Pollution Control Act Amendments of 1972, Federal and State agencies need to estimate erosion for individual storms since sediments themselves are pollutants and carry other chemical pollutants. The pollution hazard of many chemicals applied to agricultural lands is restricted to a short period immediately after application because most chemicals deteriorate rather rapidly. The first few runoff-producing storms after chemical applications are, therefore, much more important for assessing possible pollution damage than are later, possibly more intense, storms. Annual runoff calculations are almost meaningless for chemical pollutants.

Maps of the erosivity "R" values normally used in the Universal Soil Loss Equation (USLE) (11) for annual values of erosion do not apply to individual storms, and actual storm hyetographs often are unavailable. Specifying some precipitation frequency, or return period, on which to base estimates frequently is desirable for designers. The method presented here provides R-values for individual storm events of any selected standard design frequency and duration (7, 8) for any of the four types of storms defined by the Soil Conservation Service (SCS) (9). A general equation relating maximum 30-min intensity for storms of any duration and volume of total precipitation also is presented for each type of storm.

Procedure

Although any rainfall distribution could be used, the SCS storm types I, IA, II, and IIA rainfall distributions (9) were used because they are probably the most common. By normalizing the time axis of these four rainfall distribution plots, a table of normalized time vs. normalized rainfall was developed for each type of storm. The rainfall distribution within any selected frequency of design storm was determined by multiplying total storm rainfall by the fractional rainfall increments, corresponding to selected uniform time increments, for the SCS type storm desired. Intensity in inches per hour (1 in/hr = 0.007 mm/s) was calculated for each increment by dividing the rainfall occurring during that increment by the incremental time value.

Energy per inch (1 in = 25.4 mm) for each rainfall increment was calculated according to the relationship:

1/ Hydrologist, USDA, SEA-AR, U.S. Water Conservation Laboratory, Phoenix, Ariz.

$$E = 919 + 331 \log_{10} I \tag{1}$$

where E = energy in foot tons per acre per inch (1 ft-ton/acre-in = 26.38 J/m^3), and I = incremental rainfall intensity in inches per hour (1 in/ hr = 0.007 mm/s), as calculated previously (10). The energy per increment was determined as the product of each energy-per-inch value and the corresponding increment of rainfall in inches (1 in = 25.4 mm). The product of the sum of the individual energy-per-increment values and the maximum 30-min rainfall intensity, divided by 100, provides the erosivity factor "R" for the type and frequency of design storm selected. The maximum 30-min intensity is a function of storm type, and expressed by a general equation. Maximum 30-min intensity in inches per hour (1 in/hr = 0.007 mm/s) equals

$$Imax = (P)\ (\alpha D^{\beta}) \tag{2}$$

where P = total storm rainfall in inches (1 in = 25.4 mm), D = storm duration in hours, and α and β are constants for any given storm type. The values of α and β are presented in table 1 for each of the four SCS storm types used.

Using a similar approach, Ateshian (1) presented a method of determining R values for individual storms of any duration, and 24-hr storms, for types I and II distributions. His main emphasis, however, was to develop a relationship between 2-yr, 6-hr rainfall and the average annual erosion index R. His general equation for individual storms of any duration was:

$$\frac{EI}{100} = R = \frac{a' P^{2.2}}{D^{b'}} \tag{3}$$

Table 1.-- Values of α and β for use in equation 2 for each type of SCS storm

Type of Storm	Coefficients	
	α	β
IA	1.36	-0.56
I	1.51	.40
II	1.68	.25
IIA	1.82	.136

where P and D are as defined above and a' and b' are constants depending on the type of storm.

In this analysis the general equation was found to be:

$$\frac{EI}{100} = R = \frac{a\ P^{f(D)}}{D^{b}} \tag{4}$$

where P and D are as defined previously and a and b are constants depending on type of storm. The power to which rainfall P is raised also is a function of duration f(D). The function f(D) was evaluated by regression analysis using values from the four storm types and seven storm durations. The best fit relationship for all storm types had a regression coefficient r^2 of 0.98 and was found to be:

$$f(D) = 2.119D^{.0086}. \quad (5)$$

Substitution into equation 4 yields:

$$\frac{EI}{100} = R = \frac{a\ P^{2.119D^{.0086}}}{D^{b}}, \quad (6)$$

which can be used to determine individual storm R values for storms of any duration and total precipitation. Maximum values of total rainfall for the different durations were based on reports from U.S. weather stations (3, 6). Table 2 shows the coefficients a and b in equation 6 for each SCS type storm.

Figure 1 shows the percent difference between R values calculated by the Ateshian method (eq. 3) and the method presented here (eq. 6) as a function of storm duration and total precipitation for the type II storm distribution. A positive difference means that the Ateshian method computes an R value greater than that computed by equation 6. The percent difference is greater for short duration storms of high magnitude and is less than 10% for 12- and 24-hr storms of any storm magnitude shown (fig. 1). Since Ateshian used the 24-hr storm to develop his relationship and rounded the coefficient 2.2 to the nearest tenth, the 24-hr values should be nearly the same as those produced by equation 6.

Table 2.--Values of a and b in equation 5 for each SCS type of storm

Type of Storm	Coefficients	
	a	b
IA	12.98	0.7488
I	15.03	.5780
II	17.90	.4134
IIA	21.51	.2811

Renard (4), in discussing Ateshian's paper (1), superimposed typical short-duration, high-intensity air-mass thunderstorm depth-duration curves for 11 storms on a plot of the type I and II distributions presented by Ateshian. Renard suggested that the type I and II distributions poorly represent these thunderstorm distributions. He did not normalize the time scales, however, and only one or two of the longer duration storms appear similar to the type I and II curves. Using 9 of the 11 storms described by Renard and the 4 SCS curves, plots normalized in both precipitation and time are presented in figures 2A and 2B for Walnut Gulch, Ariz., and Alamogordo Creek, N.Mex. respectively. These plots show that the SCS curves more nearly represent the actual storm distributions when time and precipitation are normalized. Even when the most intense storms are selected (as shown here), distributions vary widely and include several storm types. The plots show that the four SCS curves do not cover all possible distributions. A better measure of their representativeness to erosivity would be to compare actual storm R values with those produced by the SCS type storms.

In a separate discussion of Ateshian's paper (1), Renard and Simanton (5) prepared a table showing the the actual computed R values for the same 11 storms. They compared R values calculated using Ateshian's equations for 24-

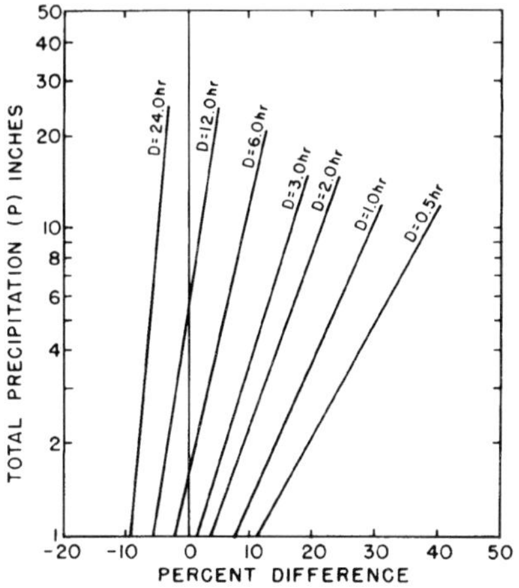

Figure 1.--The percent difference in R value calculated by the Ateshian and the Cooley methods for SCS type II storms of various durations (D) as a function of total storm precipitation (P). A positive difference indicates that the value calculated by the Ateshian method was larger.

hr storms, and for storms of any duration, to the actual R values. As expected, the calculated R values using the equation for 24-hr storms are considerably in error and different from the values calculated by Ateshian's other equation (eq. 3), except for the 25.82-hr storm when both produced essentially the same results since this storm lasted nearly 24 hr.

Using the data of Renard and Simanton (5), table 3 compares actual R values with those determined for the four types of storms, using equation 6. Table 3 also shows the values obtained using Ateshian's method for storms of any duration, the type of storm most nearly matching actual values, and the percent-error for each. As shown, some types of storms may be predominant in an area, but most areas exhibit a large range of variability about this type (2). At Walnut Gulch, and especially Alamogordo Creek (table 3), the type IIA storm is predominant, but essentially the entire range of types is represented. Using equation 6, all types of storms can be considered in a design procedure and which type, or under what set of circumstances, the most critical conditions occur and how often can be determined.

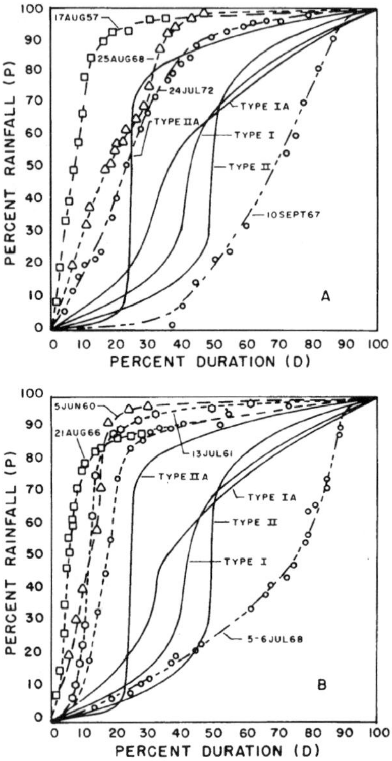

Figure 2.--Plot of normalized rainfall distributions for actual storm data (dashed lines), and SCS type IA, I, II, and IIA normalized storm distributions (solid lines) at (A) Walnut Gulch, Ariz., and (B) Alamogordo Creek, N.Mex.

Table 3.--Actual rainfall erosion index for individual storms compared with values for SCS type IA, I, II, and IIA distributions in the southwestern United States[1]

Watershed and storm data	Duration	Precipitation,	R values					Best fit	Error	R values computed by Ateshian's method	Error
			IA	I	II	IIA	Actual				
	(hr)	(in)	----------(ft-t/ac)----------						(%)	---(ft-t/ac)-----	(%)
Walnut Gulch											
17 Aug 1957	----3.53	2.65	41	58	86	122	135	IIA	-10	91	-33
22 Aug 1964	----5.03	2.54	29	44	68	101	111	IIA	- 9	70	-37
10 Sep 1967	----1.43	3.45	138	170	215	270	193	II	+11	248	+28
25 Aug 1968	----3.98	3.07	51	75	112	162	80	I	- 6	119	+49
24 Jul 1972	----2.28	3.20	84	112	152	204	121	I	- 7	169	+40
Alamogordo Creek											
5 Jun 1960	----3.50	4.07	103	147	216	306	350	IIA	-13	235	-33
13 Jul 1961	----1.80	3.53	123	157	206	268	298	IIA	-10	235	-21
16 Jun 1966	----3.37	3.98	101	143	209	294	259	IIA	+14	228	-12
21 Aug 1966	---13.47	5.46	73	132	242	410	266	II	- 9	239	-10
5-6 Jul 1968	--25.82	3.32	16	31	64	118	17	IA	- 6	59	+247
20 Jul 1972	---17.08	3.73	27	51	97	169	138	IIA	+22	93	-33

1/ Calculated from equation 6.

Note: 1 in = 25.4 mm; 1 ft-ton/acre = 0.67 J/m^2

Table 4 is a similar analysis of 31 storms selected to cover a large range in duration and rainfall from four sites in Hawaii. Type I and IA storms are dominant, although all four types are encountered with type IIA occurring only once. Figure 3 shows the contrast in the distribution of occurrence of each type between the southwestern United States and Hawaii. These data were obtained by combining Walnut Gulch with Alamogordo Creek and by combining the four Hawaiian sites.

RESULTS AND DISCUSSION

In addition to these simple equations, a computer program incorporating the relationships allows one to quickly and easily determine the erosivity factor R for any type and design storm selected (this program is available on request to the author). The effects of types of storms on the erosion potential of any given site also can be determined easily. Using the same design storm of 2-yr frequency and 6-hr duration at Laupahoehoe, Hawaii, for example, the R value ranged from 192 to 677 as type of storm changed. Table 5 shows the computer program for the type IA storm. This table gives the incremental values calculated by the preceding method.

Although some areas of the world are subjected to storms of a predominant type, most areas are subjected to a variety of storms with an annual average approaching one type but with considerable variation occurring about this mean (2). Using the preceding method, the designer can decide what type of storm to use and how much the range in R values will be under his conditions. A type I

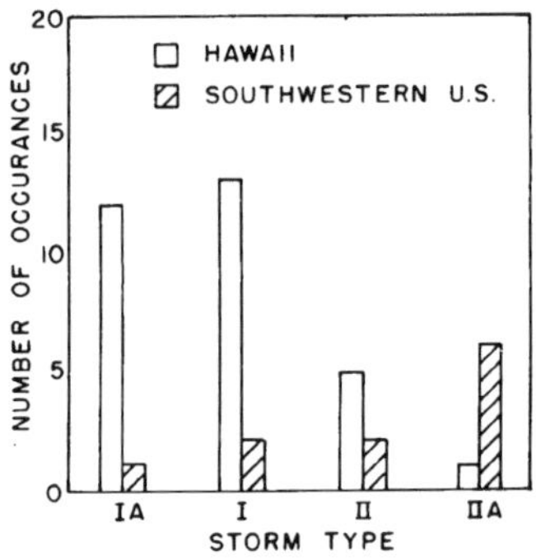

Figure 3.--Distribution of SCS storm types in Hawaii and the southwestern U.S.

Table 4.--Actual rainfall erosion index for individual storms compared with values for SCS type IA, I, II, and IIA distributions in Hawaii[1/]

Watershed and storm data	Duration	Rainfall	Actual R	Calculated R IA	Calculated R I	Calculated R II	Best fit
	(hr)	(in)	(ft-t/ac)				
Laupahoehoe							
342/73	1.83	1.50	26	20	25	33	I
54/72	2.92	1.30	23	10	14	20	II
304/72	3.33	1.50	14	13	18	26	IA
60/76	4.34	1.91	29	17	26	39	I
18/74	5.42	3.40	64	51	79	124	IA
32/74	9.67	3.60	85	38	64	112	I
78/74	12.17	5.40	148	77	137	245	I
51/73	14.75	5.40	52	67	123	228	IA
323/73	19.92	7.20	125	101	195	380	IA
315/73	30.25	17.60	492	528	1094	2283	IA
6/75	75.87	22.28	234	468	1135	2756	IA
Waialua Pineapple							
208/74	1.17	1.50	22	27	32	40	IA
109/74	3.75	2.10	33	24	34	51	I
277/72	5.33	2.00	34	16	25	40	II
77/74	11.00	3.90	75	41	71	126	I
108/74	18.58	15.30	825	546	1042	2007	I
Mililani							
109/74	1.58	1.30	16	16	20	26	IA
184/74	1.83	2.30	49	49	62	82	IA
264/74	2.17	1.80	32	25	34	46	I
262/74	2.75	4.30	234	138	189	266	II
32/74	4.00	2.40	36	30	44	66	IA
38/76	6.91	2.50	59	22	35	58	II
38/76	9.69	2.24	23	14	23	40	I
37/76	13.75	2.66	14	15	28	50	IA
31/75	16.57	4.76	80	47	88	166	I
Kunia							
179/77	5.37	1.37	18	7	11	18	II
132/77	7.83	.85	9	2	3	5	IIA[2/]
337/77	8.31	2.08	11	13	21	36	IA
31/75	19.89	4.11	51	30	58	112	I
37/76	20.67	3.12	29	16	31	61	I
38/76	30.71	5.97	89	49	103	215	I

1/ Calculated by equation 6.
2/ Calculated R value for type IIA storm = 8.

Note: 1 in = 25.4 mm; 1 ft-t/ac = 0.67 J/m^2.

Table 5.--Typical output from computer, program for Laupahoehoe, Hawaii: 2-year, 6-hour type IIA storm[1/]

Time Increment	Rain Increment	Intensity	Energy per acre	Energy per Increment
(hr)	(in)	(in/hr)	(ft-t/ac-in)	(ft-t/ac)
0.125	0.019	0.15	644	12
.125	.019	.15	644	12
.125	.019	.15	644	12
.125	.025	.20	686	17
.125	.025	.20	686	17
.125	.025	.20	686	17
.125	.031	.25	718	23
.125	.044	.35	766	34
.125	.057	.45	802	45
.125	.088	.71	866	76
.125	.372	2.97	1073	399
.125	3.723	29.79	1404	5227
.125	.315	2.52	1049	330
.125	.139	1.11	931	129
.125	.126	1.01	917	116
.125	.101	.81	885	89
.125	.088	.71	866	76
.125	.081	.66	855	70
.125	.075	.60	844	64
.125	.063	.50	818	52
.125	.050	.40	785	40
.125	.050	.40	785	40
.125	.044	.35	766	34
.125	.044	.35	766	34
.125	.044	.35	766	34
.125	.038	.30	744	28
.125	.038	.30	744	28
.125	.038	.30	744	28
.125	.038	.30	744	28
.125	.038	.30	744	28
.125	.038	.30	744	28
.125	.031	.25	718	23
.125	.031	.25	718	23
.125	.031	.25	718	23
.125	.031	.25	718	23
.125	.031	.25	718	23
.125	.025	.20	686	17
.125	.025	.20	686	17
.125	.025	.20	686	17
.125	.025	.20	686	17
.125	.019	.15	644	12
.125	.019	.15	644	12
.125	.019	.15	644	12
.125	.019	.15	644	12
.125	.019	.15	644	12
.125	.019	.15	644	12
.125	.019	.15	644	12
.125	.013	.10	586	7

1/ Depth = 6.3 in; erosivity (R) = 677.05; maximum 30-min intensity = 9.10 in/hr.

storm during winter, when the soil is nearly bare, may be more critical than a type II storm occurring in summer when the soil surface is protected by a full plant canopy.

In areas like Hawaii, where sugarcane and pineapple can be harvested any time during the year, the same type of analysis may help managers determine the best harvest schedule to minimize erosion. Fields most susceptible to erosion because of steepness or soil type, for example, can be harvested during the least critical period, when storms with high erosivity potential are least likely to occur. Figure 4 shows the relationship between individual storm EI or R data and day of the year, using 1974 storm rainfall data from Laupahoehoe, Hawaii. The best harvest date for susceptible fields, as shown in figure 4, would be about day 120.

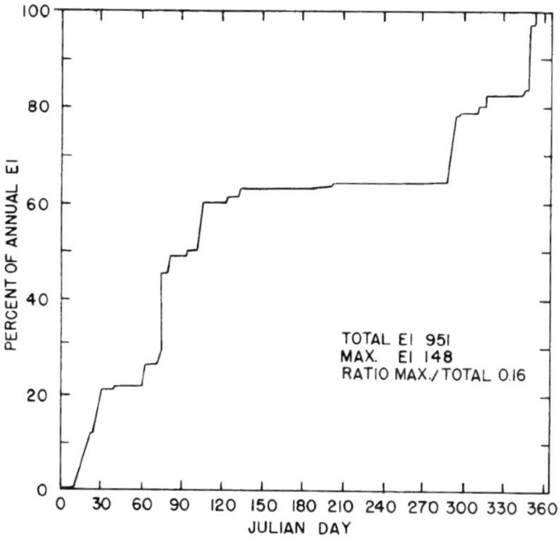

Figure 4.--Relation of individual storm erosivity (EI) with time of year (expressed as Julian day) for a study site near Laupahoehoe, Hawaii, in 1974.

CONCLUSIONS

The method presented here to determine erosivity R values for individual storms provides an easy, rapid procedure to assess erosion or pollution potential for a storm when combined with erosion or chemical transport relationships. This method also allows designers of conservation measures to determine the range of R values that might be encountered and the most critical combination of storm type and soil conditions. Managers in some areas also may use the distribution of erosivity with time to minimize soil losses by proper adjustment of harvesting schedules.

Although the four SCS type curves do not encompass all possible storm distributions, results for two widely different areas (southwestern United States and Hawaii) indicate that R values calculated using these curves would be accurate enough for most designs. The type IIA distribution produced R values close to actual values calculated from selected intense thunderstorms in the southwestern United States.

The author appreciates the assistance of Tom Hansen who wrote the program for this procedure and Dave Larson who did much of the data processing.

REFERENCES

(1) Ateshian, J. K. H.
1974. Estimation of rainfall erosion index. Journal of the Irrigation and Drainage Division. Proceedings of the American Society of Civil Engineers 100(IR3):293-307.

(2) Hershfield, D. M.
1962. Extreme rainfall relationships. Journal of the Hydraulics Division. Proceedings of the American Society of Civil Engineers 88(HY6):73-92.

(3) Niedringhaus, T. E.
1973. Rainfall intensities in the conterminous United States and Hawaii. Special Report ETL-SR-74-3, U.S. Army Engineers, Topographic Laboratories, Fort Belvoir, Va. p. 36.

(4) Renard, K. G.
1975. Discussion of estimation of rainfall erosion index. Journal of the Irrigation and Drainage Division. Proceedings of the American Society of Civil Engineers 101(IR3):240-241.

(5) ------, and J. R. Simanton.
1975. Discussion of estimation of rainfall erosion index. Journal of the Irrigation and Drainage Division. Proceedings of the American Society of Civil Engineers 101(IR3):240-241.

(6) Riordan, P.
1970. Weather extremes around the world. Technical Report 70-45-ES, U.S. Army Natick Laboratories, Natick, Mass. 38 pp.

(7) U. S. Department of Agriculture, Soil Conservation Service.
1970. Estimating peak discharges for watershed evaluation storms and preliminary designs. Technical Service Center Technical Note - Hydrology-PO-2, 6 pp.

(8) U. S. Department of Commerce, Weather Bureau.
1961. Rainfall-frequency atlas of the United States. Technical Paper No. 40, 115 pp.

(9) ------
1962. Rainfall-frequency atlas of the Hawaiian Islands. Technical Paper No. 43, 60 pp.

(10) Wischmeier, W. H., and D. D. Smith.
1958. Rainfall energy and its relationship to soil loss. Transactions of the American Geophysical Union 39(3):285-291.

(11) ------, and D. D. Smith.
1978. Predicting rainfall erosion losses--a guide to conservation planning. U. S. Department of Agriculture, Agriculture Handbook No. 537, 58 pp. (Purdue Agricultural Experiment Station cooperating.)

Chapter 3. ESTIMATING SCS RUNOFF CURVE NUMBERS ON NATIVE GRAZING LANDS

C. L. Hanson, E. L. Neff, and A. D. Nicks[1/]

INTRODUCTION

The United States contains 517 million acres of privately controlled lands classified as native grazing land (rangeland, grazable woodland, and native pasture) (8). This area constitutes 36% of the land area in the United States. Approximately 64% of the total grazing land is on land classified as having an erosion hazard. Conservation treatment and improvement of existing cover are needed on 71% of these lands; brush control and reestablishment of cover are needed on another 10%, and 5% of this land is not suitable for treatment (4).

Water-pollution hazard from native grazing lands is limited mainly to transport of sediments eroded from them. Chemical applications to these grazing lands consist primarily of fertilizer to reestablish cover and herbicides and pesticides to control brush and pests in some areas. Using mathematical models to estimate the best management practices is important, however, because of the large area in grassland that is susceptible to erosion. Estimating the runoff potential of grazing land sites also may be difficult because of the varied soil, cover, and grazing intensity. This chapter gives some methods and data that the planner can use in estimating runoff potential by the Soil Conservation Service (SCS) runoff curve number method. Specifically, the data and procedures are given as a guide to estimate the average curve number parameter required by Hydrology Option One of the CREAMS model.

PROCEDURES OF THE SOIL CONSERVATION SERVICE

The SCS curve number method was used to estimate direct runoff in the report "Control of Water Pollution from Cropland" (3). The procedures used were developed to determine the curve number (CN) for many agricultural practices (5). The following summary of procedures was used by State and Federal agencies to estimate curve numbers for rangeland. Suggested curve numbers for the Northern Great Plains also are given.

The SCS (5) and the Bureau of Land Management (11) have graphs to estimate curve numbers for Pinyon-Juniper and sagebrush cover classes. The Bureau of Land Management (11) also has a graph for grassland. These graphs are based on the hydrologic soil groups and percentage of cover. The percentage of cover

---

1/ Agricultural engineer, USDA-SEA-AR, Northwest Watershed Research Center, Boise, Idaho; hydraulic engineer, USDA-SEA-AR, Northern Plains Soil and Water Research Center, Sidney, Mont.; and agricultural engineer, USDA-SEA-AR, Southern Great Plains Watershed Research Center, Chickasha, Okla.

classifies the cover in poor, fair, and good hydrologic condition. SCS gives two procedures for determining the hydrologic cover condition of rangeland (5, ch. 8). The SCS also developed "Hydrology Technical Note PO-7," a photographic catalog illustrating range sites and hydrologic conditions (6).

The SCS in Arizona and New Mexico developed a figure representing curve numbers for their hydrologic conditions. This figure, based on figures 9.5 and 9.6 in the SCS's National Engineering Handbook, section 4 (5), expresses the runoff curve numbers as a function of cover density and hydrologic soil type for various vegetation types (2, 7). The SCS in Arizona also developed a method of adjusting curve numbers for storm duration (1, 12). The SCS in Wyoming developed a table of soil cover complex numbers derived from range sites and condition of cover (10).

Table 1 lists runoff curve numbers in relation to range sites and condition of cover. These data were adapted from the Wyoming SCS table for use in the Northern Great Plains. The values in table 1 were verified from SEA-AR watershed data in western South Dakota, southeastern Montana, and northeastern Wyoming. They represent antecedent moisture condition I. The range condition classes of fair-good, high-fair, and excellent in the Wyoming SCS table were changed to poor, fair, and good to coincide with tables 8.1 and 8.2 of the National Engineering Handbook (5).

Using a climate index (9) the SCS in Texas developed a procedure to adjust curve numbers. This procedure is for all types of soil-cover complexes and also includes grazing land conditions that are prevalent in large areas of western Texas. Figure 1 shows the adjustment to be made to the curve number of a given soil-cover complex. To use this procedure, select the condition I and condition II curve number from the National Engineering Handbook, section 4 (5). Then select the appropriate isogram values from figure 1 for the range site. Calculate the average curve number, using the isogram and the curve numbers from the SCS table. An example of this calculation is:

$$CN_{ave} = CNI + 0.20\ (CNII-CNI)$$

where CNI and CNII are curve numbers for antecedent moisture condition I and II for the soil-cover complex, and CN_{ave} is the average adjusted curve number for the site.

This procedure is recommended to obtain the average runoff condition curve number for emergency spillway and freeboard hydrographs. No runoff curve number less than 60 is used as a result of this adjustment unless the unadjusted soil-cover complex number is less than 60. When this occurs, no downward adjustment is used.

USDA-SEA-AR WATERSHED CURVE NUMBER

Data from research watershed range sites in the northern and southern Great Plains were used to calculate representative values of average curve numbers. These values are shown for the model user's reference in selecting the average curve number for Hydrology Option One. Values for the northern Plains,

Table 1.--Runoff curve numbers derived from range sites and condition of cover for antecedent moisture condition I.

Range Site	Range condition		
	Poor	Fair	Good
Wetland	95	95	95
Very shallow	95	90	85
Saline subirrigated	90	90	85
Subirrigated	90	90	85
Shale	90	85	80
Dense clay	90	85	80
Alkali clay	90	85	80
Saline upland	90	85	80
Igneous	90	80	75
Shallow clayey	85	80	75
Shallow sandy	80	75	70
Shallow loamy	80	75	70
Shallow igneous	80	75	70
Steep clayey	80	75	70
Clayey	80	75	65
Gravelly loamy	80	75	65
Steep loamy	80	75	65
Overflow	80	70	60
Loamy overflow	80	70	60
Clayey overflow	80	70	60
Coarse upland	80	70	60
Limy upland	80	70	60
Shallow breaks	80	70	60
Stony	80	70	60
Steep stony	80	70	60
Lowland	80	70	60
Saline lowland	80	70	60
Loamy lowland	80	65	55
Loamy	80	65	55
Sandy lowland	75	60	50
Sandy	75	60	50
Gravelly	70	55	45
Sands	70	55	40
Choppy sands	70	55	40

Note: As sites conditions are general, the curve number should be adjusted (interpolated) for each particular site based upon a field investigation.

given in table 2, compare closely with estimates given in table 1. Values for the southern Plains are given in table 3, which compares two types of rangeland conditions that are prevalent in this area. These conditions are native rangeland never cultivated and native rangeland formerly cultivated, abandoned, and left to revert back to native grass. Three watersheds in the table 3 represent

Table 2.--SCS curve number from selected USDA-SEA-AR watersheds in the northern Great Plains

Watershed location	Watershed number	Area	Range type	Range condition	Hydrologic group	SCS curve number		
						Low	Average	High
		(acres)						
Hasting, Nebr.	1H	3.62	Native meadow	Fair	B	40	50	88
Hasting, Nebr.	2H	3.40	Native meadow	Fair	B	41	61	87
Hasting, Nebr.	18H	3.74	Native pasture (heavy grazing).	Fair	B	63	80	96
Ekalaka, Mont.	1	2.00	Saline-upland range site.	Poor	D	86	93	99
Ekalaka, Mont.	2	2.00	Panspots	Poor	D	82	87	98
Ekalaka, Mont.	3	2.00	Panspots	Poor	D	80	89	97
Cottonwood, S. Dak.	4H	8.57	Pierre shale (heavy grazing).	Fair	D	55	71	95
Cottonwood, S. Dak.	5M	8.57	Pierre shale (medium grazing).	Fair	D	57	70	94
Cottonwood, S. Dak.	6L	8.99	Pierre shale (light grazing).	Good	D	53	67	94
Newell, S. Dak.	2	115.00	Medium-textured soils (mixed range sites).	Poor	B	52	70	89
Newell, S. Dak.	55	41.40	Medium-textured soils (mixed range sites).	Fair	B	50	61	94
Newell, S. Dak.	7	160.00	Medium-textured soils (mixed range sites).	Poor	B	55	63	93
Newell, S. Dak.	12	90.00	Fine-textured soils (mixed range sites).	Poor	D	71	89	98
Newell, S. Dak.	13	60.00	Fine-textured soils (mixed range sites).	Poor	D	57	81	96
Newell, S. Dak.	14	35.00	Fine-textured soils (mixed range sites).	Poor	D	66	77	94
Newell, S. Dak.	15	115.00	Fine-textured soils (mixed range sites).	Poor	D	66	77	93
Newell, S. Dak.	51	7.90	Sandy range sites.	Fair	B	52	61	81
Newell, S. Dak.	53	11.30	Sandy range sites.	Fair	B	42	46	86

Table 2.--SCS curve number from selected USDA-SEA-AR watersheds in the northern Great Plains--Continued

Watershed location	Watershed number	Area	Range type	Range condition	Hydrologic group	SCS curve number		
						Low	Average	High
		(acres)						
Newell, S. Dak.	55	16.50	Sandy range sites.	Fair	B	45	50	95
Newell, S. Dak.	P5	8.00	Panspots	Fair	D	64	76	96
Newell, S. Dak.	P6	13.20	Panspots	Fair	D	63	73	90
Newell, S. Dak.	P7	7.25	Panspots	Fair	D	65	81	97
Newell, S. Dak.	P8	6.42	Panspots	Fair	D	72	82	91
Newell, S. Dak.	P9	6.96	Panspots	Fair	D	71	82	95
Aladdin, Wyo.	1	7.70	Silty range site.	Fair	D	61	75	89
Aladdin, Wyo.	2	8.20	Silty range site.	Fair	D	61	74	86
Aladdin, Wyo.	3	11.60	Shallow range site.	Fair	D	71	75	95
Aladdin, Wyo.	4	2.50	Shallow range site.	Fair	D	72	82	95
Reynolds, Idaho	1	205.00	Summit watershed (mixed range site).	Poor	D	74	75	86
Reynolds, Idaho	2	33.00	Lower sheep (mixed range site).	Poor	B	74	74	89
Reynolds, Idaho	3	306.00	Murphy (mixed range site).	Fair	C	69	70	91
Reynolds, Idaho	4	100.00	East Reynolds Mt. (mixed range site).	Fair	C	79	82	88

Table 3.--SCS curve numbers from selected USDA-SEA-AR watersheds in the southern Great Plains

Watershed location	Watershed number	Area	Range type	Range condition	Hydrologic group	SCS curve number		
						Low	Average	High
		(acres)						
Guthrie, Okla.	W-I	2.50	Virgin native grass.	Good	B	33	68	95
Guthrie, Okla.	W-II	5.09	Virgin native grass.	Good	B	32	61	85
Guthrie, Okla.	W-III	9.09	Formerly cultivated; eroded.	Fair	B	56	78	98
Guthrie, Okla.	W-IV	13.40	Formerly cultivated; eroded.	Fair	B	53	78	98
Guthrie, Okla.	W-V	15.70	Formerly cultivated; eroded.	Fair	B	56	76	98
Guthrie, Okla.	PL, L	5.62	Native woodland	Fair	B	30	59	95
Guthrie, Okla.	PL, J	5.28	Severly eroded	Poor	B	53	78	93
Guthrie, Okla.	PL, 15A	3.13	Formerly cultivated; terraced.	Good	B	55	81	96
Guthrie, Okla.	PL, 13	3.21	Gullied; reformed	Good	B	58	81	98
Chickasha, Okla.	R-2	24.08	Sandy range site	Fair	B	45	68	86
Chickasha, Okla.	R-5	23.72	Virgin rangeland site.	Good	D	41	76	98
Chickasha, Okla.	R-7	19.19	Formerly cultivated; treated	Poor	D	52	83	98

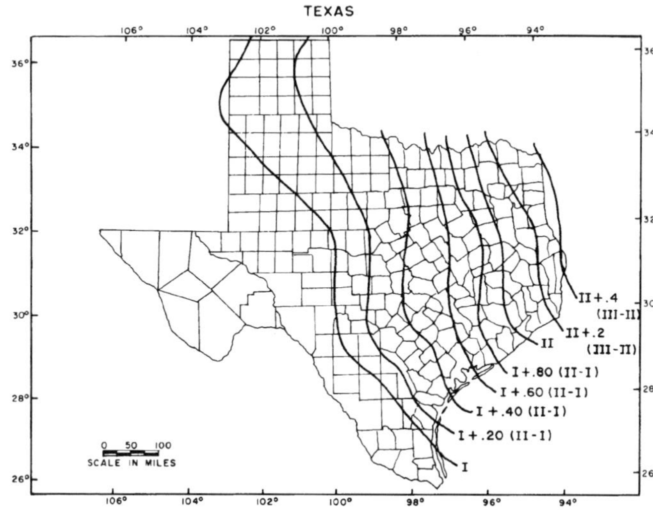

Figure 1.--Adjustments for runoff curve number in Texas (9).

reclaimed eroded, gullied lands (W-III, W-IV, and W-V at Guthrie, Okla.), three represent virgin native conditions (W-I and W-II at Guthrie, and R-5 at Chickasha, Okla.), and two represent no treatment after abandonment with natural reversion to native grass.

REFERENCES

(1) Malone, J.M.
1972. Hydrologic design manual for drainage areas under 25 square miles. [Preliminary draft] U.S. Department of Agriculture, Soil Conservation Service.

(2) Simanton, J.R., K.G. Renard, and N.G. Sutter.
1973. Procedures for identifying parameters affecting storm runoff volumes in a semiarid environment. U.S. Department of Agriculture, Agricultural Research Service, Western Region, ARS-W-1, 12 p. (Series discontinued; Agricultural Research Service is now Science and Educa-

tion Administration-Agricultural Research.)

(3) Stewart, B.A., D.A. Woolhiser, W.H. Wischmeier, J.H. Caro, and M.H. Frere 1976. Control of water pollution from cropland, Vol. II - An overview. U.S. Department of Agriculture, Agricultural Research Service, ARS-H-5-2, 187 pp. (Series discontinued; Agricultural Research Service is now Science and Education Administration-Agricultural Research.)

(4) U.S. Department of Agriculture. 1976. National inventory of soil and water conservation needs. U.S. Department of Agriculture Statistical Bulletin No. 461.

(5) U.S. Department of Agriculture, Soil Conservation Service. 1972. SCS National Engineering Handbook, Section 4, Hydrology. 458 pp.

(6) ------ 1973. Photographic catalog of range sites and their hydrologic condition. U.S. Department of Agriculture, Soil Conservation Service, Hydrology Technical Note PO-7, 23 pp.

(7) ------ 1973. Peak rates of discharge for small watersheds. In Engineering field Manual for conservation practices, ch. 2. (revised 10/73 for New Mexico).

(8) ------ 1976. National Range Handbook Notice-1, Washington, D.C.

(9) ------ 1978. Engineering-Hydrology Memorandum TX-1, Temple, Tex.

(10) ------ 1978. Runoff and yield determination procedures. U.S. Department of Agriculture, Soil Conservation Service, Technical Note Engineering 18, page 16, Table 3. Casper, Wyo.

(11) U.S. Department of Interior, Bureau of Land Management. 1969. Bureau of Land Management Manual 7313, cover.

(12) Woodward, D.E. 1973. Runoff curve numbers for semiarid range and forest conditions. American Society of Agricultural Engineers Paper 73-209, St. Joseph, Mich.

Chapter 4. RESIDUE AND TILLAGE EFFECTS ON SCS RUNOFF CURVE NUMBERS

W. J. Rawls, C. A. Onstad and H. H. Richardson[1/]

ABSTRACT

The effect of conservation tillage on reducing direct runoff ranges from slight to substantial. Procedures of the Soil Conservation Service used today for estimating runoff do not consider the effects of conservation tillage and no-till practices on runoff. To develop the SCS runoff curve numbers for these practices, tillage and crop data were assembled from small watersheds and plots under natural and simulated rainfall from many locations across the country.

The residue left on the ground was chosen as the independent variable to represent the effects of conservation tillage practices. These data were studied to determine runoff curve numbers for single- and double-cropping systems under various conservation tillage practices. These runoff curve numbers can be used with the SCS procedure to evaluate the effect of tillage practices on runoff.

INTRODUCTION

In 1976, about 17 million ha (10% of the farmed cropland in the United States) were farmed with some form of conservation tillage (3). SCS estimates the use of conservation tillage, including no-till, has increased at an average annual rate of 1.2 million ha over the past 10 yr and should continue to increase because of its environmental benefits (3). Conservation tillage is defined as a form of noninversion tillage that retains protective amounts of residue mulch on the surface throughout the year (20). It includes such practices as till plant, chisel plant, no-till, strip tillage, sweep tillage, stubble mulching, chop plant, and other types of noninversion tillage. The relative effectiveness of these practices in controlling runoff (18) can be judged by how much they:

(1) Reduce runoff velocity (10, 11). The velocity of surface runoff water is reduced by decreasing land slope or by increasing surface roughness. Slope usually is decreased by lengthening the flow path of the water. Surface roughness is increased by reducing the number of tillage practices or by increasing vegetative or residue cover. Decreasing runoff velocity usually increases infiltration.

(2) Increase surface storage. Practices that increase surface storage generally reduce the total volume of runoff and increase infiltration.

1/ Hydrologist, USDA-SEA-AR, Beltsville, Md.; agricultural engineer, USDA-SEA-AR, Morris, Minn.; and hydraulic engineer, USDA-SCS, Glenn Dale, Md., respectively.

(3) Increase conductivity and moisture storage. Some practices increase the size and number of soil macropores connected to the soil surface, which can increase greatly infiltration and conductivity. Soil moisture storage can be increased by drainage or by evaporation of moisture in the soil profile.

(4) Reduce raindrop impact. Raindrop impact can form a soil surface crust that is often the limiting factor in infiltration.

Amerman (1) found that comparisons of various types of conservation tillage have resulted in differing conclusions about their effects on runoff. Using a large rainfall simulator (0.36 by 1.05 m) just after planting and cultivation, various authors (2, 7, 8, 9, 10, 11, 12, 15), have shown that, runoff for high intensity storms from areas under conservation tillage was less than that from areas under conventional tillage. Under some conditions, however, which were not defined clearly, the runoff reduction or increase was insignificant (5, 6, 11). Observations have shown that seasonal and annual runoff is consistently less from land under conservation tillage (except no-till) than from land under conventional tillage (4, 5, 13, 14, 16). Research by Mannering and Meyer (9) and Wischmeier (21) indicated that the amount of crop residue left on the ground or the percentage of the ground covered by residue is the best measure for distinguishing among conservation tillage practices. A national method is needed to describe the effect of conservation tillage practices on runoff because of the variable effect of residue on runoff and the desired increase to use residue cover for conservation.

SOIL CONSERVATION SERVICE RUNOFF PROCEDURE

Since the Soil Conservation Service (SCS) has the most frequently used national method for estimating runoff, guides were developed to predict the effect of conservation tillage practices on SCS runoff curve numbers. These guides were based on the fragmentary data available. They will make the SCS procedure useful in evaluating the runoff effects of conservation tillage practices and in comparing these effects with those produced from such conservation measures as contouring, terracing, and structures. Since the amount of residue left on the ground and the percentage of ground covered by residue can be estimated easily from yields and tillage operations or from field inspection, they were used as independent variables to determine how conservation tillage practices affect runoff.

The SCS procedure to predict direct runoff from storm rainfall uses total daily rainfall (19). The runoff curve number (CN) incorporates the effects of infiltration characteristics of the soil, land use, and agricultural practices. Soils are classified into four hydrologic groups according to their minumum infiltration rate, which is obtained for a bare soil after prolonged wetting. The hydrologic soil groups (19) are:

Group A. (Low runoff potential). Soils having high infiltration rates even when thoroughly wetted and consisting chiefly of deep, well-drained to excessively drained sands or gravels. These soils have a high rate of water transmission.

Group B. Soils having moderate infiltration rates when thoroughly wetted and consisting chiefly of moderately deep to deep, moderately well- to well-drained soils with moderately fine to moderately coarse textures. These soils have a moderate rate of water transmission.

Group C. Soils having slow infiltration rates when thoroughly wetted and consisting chiefly of soils with a layer that impedes downward movement of water, or soils with moderately fine to fine texture. These soils have a slow rate of water transmission.

Group D. (High runoff potential). Soils having slow infiltration rates when thoroughly wetted and consisting chiefly of clay soils with a high swelling potential, soils with a permanent high water table, soils with a claypan or clay layer at or near the surface, and shallow soils over nearly impervious material. These soils have a slow rate of water transmission.

The surface conditions of a watershed are described by land use and treatment classes. Land use is defined as the agricultural cover of the watershed, which includes all vegetation, litter and mulch, and fallow, and nonagricultural cover, which includes water surfaces and impervious surfaces. Land treatment includes such agricultural practices as contouring or such terracing and management practices as grazing control or crop rotation. Land use and treatment classes are determined by observing or measuring plant and litter density and extent on sample areas.

The index of the watershed wetness on the day of the storm is described as Antecedent Moisture Condition (AMC), determined by the total rainfall in the 5-day period preceding the storm. Three levels of AMC are used:

AMC-I. Lowest runoff potential. The watershed soils are dry enough for satisfactory plowing or cultivation.

AMC-II. Average condition.

AMC-III. Highest runoff potential. The watershed is practically saturated from antecedent rains.

The antecedent moisture (AMC groupings I, II, and III) was established to explain the variation in event runoff curve number (RCN) that exists at a site. AMC II is the event RCN for a site as determined by fitting an average curve to a rainfall runoff plot. To explain the variation on either side of the average RCN curve, enveloping RCN lines were determined. The lower enveloping RCN curve represents AMC I while the upper enveloping RCN represents AMC III.

The boundaries between the AMC groups vary, depending on the time of the year (table 1). Table 2 gives average runoff curve numbers for several hydrologic soil-cover complexes. Figure 1 shows the relationship between the three AMC levels and runoff.

Table 1.--Seasonal rainfall limits for antecedent moisture conditions

AMC group	Total 5-day antecedent rainfall	
	Dormant season	Growing season
	Inches[1/]	Inches
I	Less than 0.5	Less than 1.4
II	0.5 to 1.1	1.4 to 2.1
III	Over 1.1	Over 2.1

1/ To convert inches to centimeters, multiply by 2.54.

Source: U.S. Department of Agriculture, Soil Conservation Service (19).

Table 2.--Runoff curve numbers for hydrologic soil-cover complexes, antecedent moisture condition II, and $I_a = 0.2\ S$

Cover			Runoff curve number for hydrologic soil groups			
Land Use	Treatment or practice	Hydrologic condition	A	B	C	D
Fallow	Straight row		77	86	91	94
Row Crops	Straight row	Poor	72	81	88	91
	Straight row	Good	67	87	85	89
	Contoured	Poor	70	79	84	88
	Contoured	Good	65	75	82	86
	Contoured and terraced	Poor	66	74	80	82
	Contoured and terraced	Good	62	71	78	81
Small Grain	Straight row	Poor	65	76	84	88
	Straight row	Good	63	75	83	87
	Contoured	Poor	63	74	82	85
	Contoured	Good	61	73	81	84
	Contoured and terraced	Poor	61	72	79	82
	Contoured and terraced	Good	59	70	78	81

Source: U.S. Deaprtment of Agriculture (19, table 9.1).

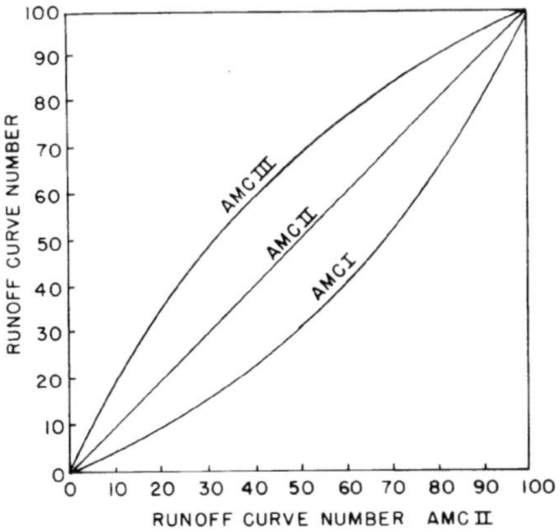

Figure 1.--Relationship between antecedent moisture conditions and curve numbers (19, table 10.1).

The average curve numbers (CN) in table 2 apply to average crop conditions for a growing season. If seasonal variations in the CN are needed, growth stages of the particular crop indicate how much and when to modify the average CN.

For cultivated crops grown during a normal growing season, the CN at plowing or planting time is the same as the CN for fallow. Midway between planting and harvest or cutting times, the CN is the average in table 2. At the time of normal peak growth or height (usually before harvest), the CN is determined with the following equation:

$$CN_{\text{normal peak growth}} = 2\left(CN_{\text{average}}\right) - \left(CN_{\text{fallow}}\right).$$

The Soil Conservation Service (19) suggests that after harvest the CN varies between that for fallow and normal peak growth, depending on the effectiveness of the plant residue ground cover. In general, if two-thirds of the soil surface is exposed, the fallow CN applies; if one-third is exposed, the average CN applies; and if practically none is exposed, the normal peak growth CN applies.

The watershed and plot data assembled for this study are from USDA's Science and Education Administration-Agricultural Research (SEA-AR), universities, and State agricultural experiment stations. The data for each storm event or daily event include (1) date; (2) runoff; (3) precipitation; (4) previous 5-day precipitation; (5) soil series and SCS hydrologic soil group classification; (6) detailed tillage history including type of crop, dates of tillage and harvest, crop rotation, and crop yields; (7) residue amount on the ground for both the growing and dormant season; and (8) size and slope of the watershed or plot. These data were divided into results under simulated rainfall and under natural rainfall because the surface conditions were well defined for the simulated rainfall events but had to be estimated for most of the natural events.

Simulated Rainfall Events

Table 3 summarizes the location and characteristics of the simulated rainfall areas used in this chapter. Before the simulated rainfall events, the amount of residue on the surface and the percentage of the surface covered was measured. These events normally consisted of three rainulator runs, totaling 12.7 cm of applied rain at a uniform intensity of 6.35 cm/hr. The initial 1-hr run began at existing soil moisture. On the following day two 30-min runs were made, separated by a 15-min rainless interval. The last two runs were called the wet and very wet runs, respectively. Curve numbers were determined by using the total precipitation and runoff from the wet and very wet runs because the events had similar antecedent moisture conditions (AMC III). The data set contained 49 events for corn and 22 events for soybeans.

Natural Rainfall Events

Neither residue amount nor the percentage of the surface covered by residue was available for the natural events. The method in table 4 was used, therefore, to calculate the amounts of residue produced by the crops. These residue weights were reduced according to the type and number of tillage operations, using the average residue reductions from tillage operations given in table 5. We assumed no carryover of residue from the previous year's crop. The accuracy of this method in predicting residue varies considerably according to local crop conditions. Laflen and others (6), Sloneker and Moldenhauer (17), and Wischmeier (21) developed curves relating residue to the percentage of the surface covered. These curves show considerable variability for the same crop (fig. 2). Table 6 summarizes the location and characteristics of the natural areas used in this study. To compute the curve numbers for natural events, the precipitation and runoff were used for the maximum precipitation event that caused runoff for the dormant and growing seasons. The maximum precipitation event normally produced the maximum runoff event. All events produced more than 1 in (2.5 cm) of precipitation. If events of similar magnitude occurred during the growing season, the event was used that occurred closest to the middle of the growing season. The data set contained 248 events for corn and 212 events for wheat, oats, sorghum, and soybeans.

ANALYSIS

Preliminary analysis indicated that curve numbers varied considerably among different locations for conventional tillage practices. To eliminate

Table 3.--Location and characteristics of simulated rainfall areas

Location	Soil	SCS hydrologic soil group	Slope	Crop 1/	References
			(%)		
Champaign, Ill.	Catlin silt loam	B	5	Corn, soybeans	(15)
Castana, Iowa	Ida silt loam	B	12	Corn, soybeans	(6)
Ames, Iowa	Clar-Nicol silty clay loam	B	5	Corn, soybeans	(6)
Grinnel, Iowa	Tama silt loam	B	5	Corn	(6)
Waterloo, Iowa	Kenyon silt loam	C	4	Corn	(6)
Lafayette, Ind.	Haskins loam	C	2	Corn, soybeans	(10)
Lafayette, Ind.	Morley clay loam	C	4	Corn, soybeans	(10)
Lafayette, Ind.	Nappanee clay loam	D	1	Corn, soybeans	(10)
Lafayette, Ind.	Hoytville silty clay	D	1	Corn, soybeans	(10)

1/ Type of residue

this effect of location, we converted the curve numbers for the events to a percentage change in curve number. This percentage was calculated from the average of the curve numbers for all events for the specific location, crop, and season that occurred when no residue was on the surface. All curve numbers higher than the calculated average curve number with no residue were set equal to the calculated average curve number. The effect of antecedent moisture also was eliminated by using the SCS procedure (fig. 1) to convert all the curve numbers to an AMC-II condition. Changes in AMC can cause a considerable change in runoff curve numbers (fig. 1). Because the SCS antecedent moisture conditions are a discrete function (table 1), conversion to AMC II to eliminate the effect of antecedent soil moisture can introduce errors of as much as 15% for similar antecedent moisture conditions. If the total 5-day antecedent rainfall were between 0 and 0.49 in (0 and 1.25 cm) during the dormant season, for example, this event would be classed as AMC-I and adjusted to AMC-II condition. If the 5-day antecedent rainfall were 0.51 in (1.29 cm), however, there would be no adjustment. Considerable error can be introduced into the system when events cover a range of antecedent moisture conditions.

Table 4.--Method for converting crop yields to residue[1/]

Crop	Straw/grain ratio	Bushel weight	
		(kg)	(lb)
Barley	1.5	21.8	48
Corn	1.0	25.4	56
Oats	2.0	14.5	32
Rice	1.5	20.4	45
Rye	1.5	25.4	56
Sorghum	1.0	25.4	56
Soybeans	1.5	27.2	60
Winter wheat	1.7	27.2	60
Spring wheat	1.3	27.2	60

1/ Crop residue in lb/acre = (straw/grain ratio) x (bushel weight in lb) x (crop yield in bu/acre).

Table 5.--Residue reduction from tillage operations

Tillage operation	Residue reduction
	(%)
Chisel plow	35
Rod weeder	10
Light disk	30
Heavy disk	70
Moldboard plow	90
Till plant	20
Fluted coulter	10
V sweep	10

Plotting the percent change in runoff curve number vs. the amount of residue on the ground indicated high variability for both the simulated and natural rainfall events. Only 10 to 40% of the variability in the data could be explained by regression analysis; a first order polynomial was the best functional form.

Using regression analysis, the combinination of hydrologic soil groups and seasons did not reduce significantly the explainable variability. The change in curve numbers was computed in such a way that soils and seasons were not expected to give different patterns. Regression analysis also was used to determine whether crops that produced similar amounts of residue could be grouped together. By grouping soybeans, wheat, oats, and sorghum, the variability

Table 6.--Location and characteristics of natural rainfall areas

Location	Soil	SCS hydrologic soil group	Plot or watershed	Slope	Crops[1]
				(%)	
Castana, Iowa	Ida silt loam----------	B	P	6	Corn.
Hastings, Nebr.	Holdrege silt loam.	B	W	5	Wheat, sorghum.
Madison, S. Dak.	Egan-Wentworth silty clay loam.	B	P	6	Corn.
Watkinsville, Ga.	Cecil silt loam.	B	W	3	Corn, soybeans
Akron, Colo.	Sligo loam------------	B,C	W	6	Wheat.
Coshocton, Ohio	Muskingum silt loam.	C	W	20	Corn.
Hollyspring, Miss.	Providence silty clay loam.	C	P	5	Corn, soybeans wheat.
Bushland, Tex.	Pullman silty clay loam.	D	W	3	Sorghum.
McCredie, Mo.	Mexico silt loam.	D	P	3	Corn, soybeans, wheat.

1/ Type of residue.

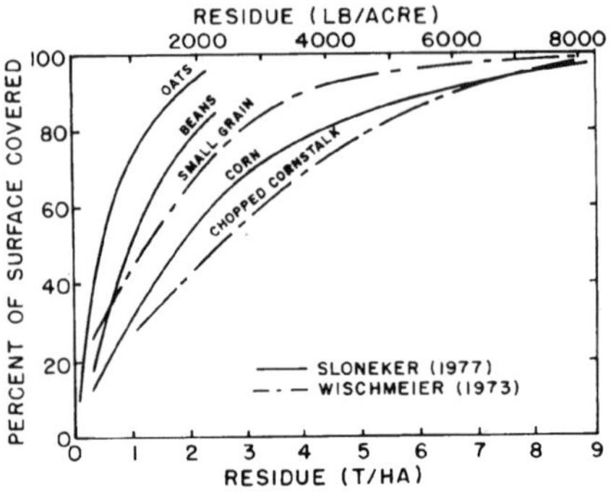

Figure 2.--Relationship of the percentage of surface cover to amount of residue for various crops.

explained by regression analysis was increased only 3%. This grouping differs from the SCS land use classification (19), in which sorghum and soybeans are classified as row crops and wheat and oats as small grain. For residue management, however, classifying crops according to the amount of residue they produce is better. The term "medium residue crops", therefore, includes soybeans, sorghum, oats, and wheat.

As only 10 to 40% of the variability could be explained by equations relating the amount of residue on the ground to the change in runoff curve number, other surface factors affected by tillage (such as soil condition) are influential and must be included in any predictive equation. As data on these factors are not readily available, development of predictive equations by multiple regression techniques was unfeasible. Even so, the available data show that trends between runoff curve number and the amount of residue on the ground or the percentage of the surface covered by residue can still be a useful guide in evaluating conservation tillage practices. Therefore, the data for various residue classes were averaged. After analyzing the distribution of residues, the data were divided into 1,000 lb/acre (1.12 t/ha) residue classes. Table 7 summarizes the number of events for each class. The mean decrease in curve number and mean amount of residue for each class for the natural and simulated rainfall events were used to develop the curves shown in figures 3 and 4 according to crop. Using Wischmeier's (21) curves for small grain and chopped cornstalks (fig. 2), the percentage of the surface covered by residue was calculated (figs. 3 and 4). Between 0 and approximately 1,500 lb/acre (1.68 t/ha) residue, the standard deviation of the data was about 7.2%. For both corn and medium-residued crops, the percent decrease in runoff curve number levels out when about 60% of the surface is covered. The curves for natural and simulated rainfall events were similar. When less than 47% of the surface was covered

Table 7.--Events used in analysis

Residue classes		Simulated rain events		Natural rain events	
		Corn	Other crops[1/]	Corn	Other crops[1/]
(t/ha)	(lb/acre)				
0-1.12	(0-1,000)	18	13	79	76
1.13-2.24	(1,001-2,000)	11	3	40	82
2.25-3.36	(2,001-3,000)	10	6	16	39
3.37-4.48	(3,001-4,000)	6	0	44	15
4.49-up	(4,000-+)	4	0	69	0
Total		49	22	248	212

1/ Other crops include wheat, oats, sorghum, and soybeans.

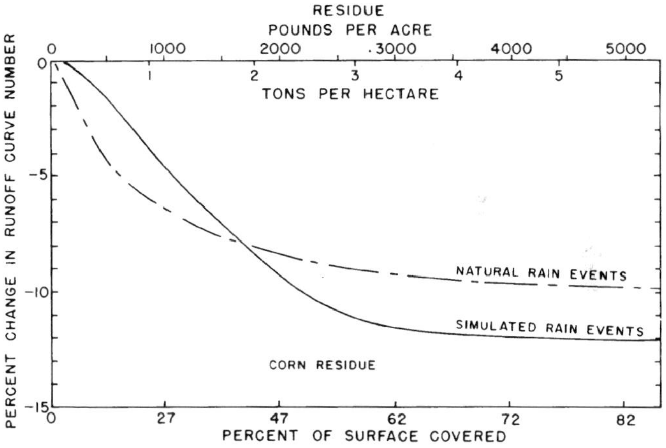

Figure 3.--Relationship between the percent change in runoff curve number and the amount of residue or percentage of the surface covered for corn.

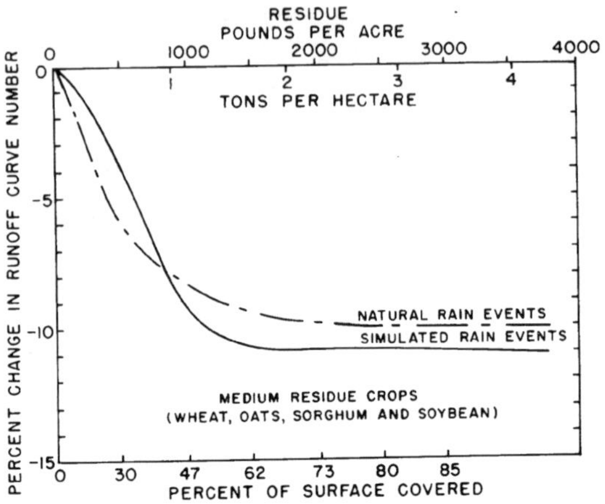

Figure 4.--Relationship between the percent change in runoff curve number and the amount of residue or percentage of the surface covered for small grain.

with residue, however, the curves changed more rapidly for the natural rainfall events than for the simulated rainfall events. This deviation could be a result of the way residue was estimated for the natural events.

The simulated rainfall curves in figures 3 and 4 were compared to the SCS guidelines (19) for estimating the effect of crop residue. These guidelines are based on the percentage of the surface covered by residue. Compared with these guidelines our simulated rainfall curves show that the change in runoff curve numbers are about 5% greater than SCS guidelines until 60% of the surface is covered with residue. At 60% residue, the curves level off, although SCS guidelines suggest a further decrease of about 10%. Figure 5 compares the simulated rainfall curve for corn after harvest and the SCS guidelines. The SCS guidelines imply that the amount of the surface covered by residue during the growing season does not affect curve numbers. We found, however, that residue on the surface had the same percentage effect for the growing season as for the dormant season. The SCS procedure is given in the appendix.

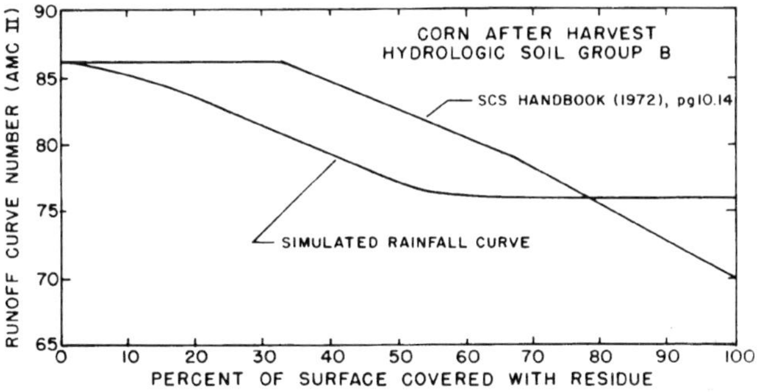

Figure 5.--Comparison between SCS guidelines and the proposed guide.

SUMMARY

There are not enough data to derive an equation for predicting how conservation tillage affects runoff. Data compiled from plots and small watersheds under simulated and natural rainfall for 531 events were used, however, to develop graphs that can be used as practical guides to estimate the decrease in SCS runoff curve numbers due to a certain amount of residue on the ground or a certain percentage of the surface covered with residue. Separate guides were developed for corn and for such medium-residue crops as wheat, oats, sorghum, and soybeans.

ACKNOWLEDGMENTS

The authors gratefully acknowledge the assistance of J. L. McGuinness, Coshocton, Ohio; R. E. Burwell, Columbia, Mo.; O. R. Jones, Bushland, Tex.; K. C. McGregor, Oxford, Miss.; R. H. Michelson, Akron, Colo.; J. V. Mannering, W. Lafayette, Ind.; J. M. Laflen, Ames, Iowa; and J. C. Siemens, Urbana, Ill., for the data they supplied. Without their help this national study never could have been accomplished. We also wish to thank Don Woodward, Leon Kimberlin, and Gerald Darby of the Soil Conservation Service and Jimmy Williams of the Science and Education Administration-Agricultural Research for their critical reviews and helpful comments.

REFERENCES

(1) Amerman, C. R.
1976. Tillage and hydrology in research progress and needs in conservation tillage. U. S. Department of Agriculture, Agricultural Research Service, ARS-NC-57. (Series discontinued; Agricultural Research Service is now Science and Education Administration-Agricultural Research.)

(2) Burwell, R. E., and W. E. Larson.
1969. Infiltration as influenced by tillage-induced random roughness and pore space. Proceedings of the Soil Science Society of America 33(3):449-452.

(3) Davis, R. M.
1977. Soil conservation on agricultural land - the challenge ahead. Journal of Soil and Water Conservation 32(1):5-9.

(4) Free, G. R., and C. E. Bay.
1969. Tillage and slope effects on runoff and erosion. Transactions of the American Society of Agricultural Engineers 12(2): 209-215.

(5) Jones, J. N., J. E. Moody, and J. H. Lillard.
1969. Effects of tillage, no-tillage, and mulch on soil water and plant growth. Agronomy Journal 61:719-721.

(6) Laflen, J.M., J.L. Baker, R.O. Hartwig, W.F. Buchele, and H.P. Johnson.
1978. Soil and water loss from conservation tillage systems. Iowa Agriculture and Home Economics Experiment Station Paper J-8230.

(7) Mannering, J. V., and R. E. Burwell.
1968. Tillage methods to reduce runoff and erosion in the Corn Belt. U. S. Department of Agriculture, Agriculture Information Bulletin 330.

(8) ------, and C. B. Johnson.
1979. Crop sequence and fall tillage effects on soil erosion. Transaction of the American Society of Agricultural Engineers.

(9) ------, and D. L. Meyer.
1963. The effects of various rates of surface mulch on infiltration and erosion. Proceedings of the Soil Science Society of America 27(1):84-86.

(10) ------, D. R. Griffth, and C. B. Richey.
1975. Tillage for moisture conservation. 1975 American Society of Agricultural Engineers Winter meeting, Paper 75-2523.

(11) Meyer, L. D., W. H. Wischmeier, and G. R. Foster.
1970. Mulch rates required for erosion control on steep slopes. Soil Science Society of America Proceedings 34:928-931.

(12) Moldenhauer, W. C., W. G. Lovely, N. P. Swanson, and H. D. Currence.
1971. Effect of row grades and tillage systems on soil and water losses. Journal of Soil and Water Conservation 26(5):192-195.

(13) Onstad, C. A.
[In press] Crop residue effects on runoff. Journal Soil and Water Conservation.

(14) ------
1974. Soil and water losses as affected by tillage practices. Transaction of the American Society of Agricultural Engineers 15(2):287-289.

(15) Siemens, J. C., and W. R. Oschwald.
1976. Corn-soybean tillage systems: erosion control, effects on crop production, costs. 1976 American Society of Agricultural Engineers Winter meeting, Paper 76-2552.

(16) Shanholtz, V. O., and J. H. Lillard.
1969. Tillage system effects on water use efficiency. Journal Soil and Water Conservation 24(5):186-189.

(17) Sloneker, L. L., and W. C. Moldenhauer.
1977. Measuring the amounts of crop residue remaining after tillage. Journal Soil and Water Conservation 32(5):231-236.

(18) Soil Conservation Society of America.
1977. Conservation tillage: Problems and potentials. U. S. Department of Agriculture, Soil Conservation Service Special Publication No. 20.

(19) U. S. Department of Agriculture, Soil Conservation Service.
1972. National engineering handbook, Section 4. Hydrology. U. S. Department of Agriculture, Soil Conservation Service, Washington, D. C.

(20) ------
1977. National handbook on conservation practices. U. S. Department of Agriculture, Soil Conservation Service, Washington, D. C.

(21) Wischmeier, W. H.
1973. Conservation tillage to control water erosion. Conservation Tillage - The Proceedings of a National Conference, Soil Conservation Society of America, Ankeny, Iowa, p. 133.

Appendix PROCEDURE TO ESTIMATE EFFECTS OF CONSERVATION TILLAGE ON REDUCING DIRECT RUNOFF USING THE SCS CURVE NUMBER

Conservation tillage is a form of noninversion tillage that retains protective amounts of residue mulch on the surface throughout the year. Conservation tillage practices include till planting, chisel planting, strip tillage, sweep tillage, chop planting, and no-till. Of these, only no-till has not reduced direct runoff consistently when applied year after year on experimental plots and watersheds. Direct runoff and associated peak discharges are reduced by crop residue cover, which increases infiltration potential through (1) lessening rainfall impact and surface crusting, (2) decreasing runoff velocity by lengthening flow paths and increasing surface roughness, (3) creating additional surface storage, and (4) providing organic matter to improve the soil structure.

Direct runoff is computed using the SCS runoff curve number technique, as described in chapters 9 and 10 of the National Engineering Handbook, Section 4, Hydrology (19). The selected runoff curve number can be reduced by a percentage to account for the effects of conservation tillage and residue management practices. To take advantage of this reduction, conservation tillage and residue management must be continued for the expected life of the engineering practice. The adaptability of the tillage practices to the local soil and crop growth conditions should be checked. This includes drainage limitations of the soils, pest control problems, equipment on hand, and the attitude and abilities of the farmers. No reductions should be used with continuous no-till or similar practices that do not increase infiltration significatly.

Estimating the amount and type of residue cover remaining after harvest is necessary for this procedure. Assumptions incorporated into the procedure are (1) normal decomposition of residue over the dormant season and (2) no carryover of residue from year to year.

The approximate amount of residue can be determined by:

1. Estimating residue for local conditions by experienced personnel.

The SCS State resource conservationist or agronomist can estimate the percentage of the surface presently covered by residue or the amount of residue resulting from specific crops and tillage practices.

2. Estimating residue cover by sampling along a transect.

One technique is to use a cord, 50 ft or longer, that has 100 equally spaced knots or other readily visible markings. This cord is stretched diagonally across several rows, and the knots that contact a piece of mulch are counted. Each knot represents 1% of the sample. This procedure is repeated at randomly selected locations on the field, and the data are averaged to obtain a representative percentage of surface area covered by residue for the field.

3. Estimating residue from empirical data developed from crop and tillage operation records.

a. Estimate the residue produced by the crop in pounds per acre from the estimated crop yield using the equation and data given in table A-1.

b. Compute the amount of residue that will remain on the surface by using the types of tillage practices and remaining residue from table A-2.

The type of residue is classified according to the maximum amount of residue the crop will produce. Medium-residue crops will produce residue amounts up to about 4,000 lb per acre and include wheat, oats, barley, rye, sorghum, and soybeans. A large-residue crop, such as corn, can produce from about 4,000 to 8,000 lb per acre of residue.

The computations of direct runoff from areas using conservation tillage and residue management practices involves five steps:

1. Determine the curve number (CN) for the hydrologic soil group, land use, and treatment according to the procedures given in NEH-4 [table 9.1, (19)].

2. Estimate the percentage of the surface covered by crop residues or the amount (lb per acre) of crop residues to be left on the surface. Any of the three preceding methods can be used.

3. Determine the percentage reduction in runoff curve number caused by conservation tillage practices from table A-3 or

Table A-1.--Method for converting crop yields to residue[1/]

Crop	Straw/grain ratio	Bushel weight
		(lbs)
Barley	1.5	48
Corn	1.0	56
Oats	2.0	32
Rice	1.5	45
Rye	1.5	56
Sorghum	1.0	56
Soybeans	1.5	60
Winter wheat	1.7	60
Spring wheat	1.3	60

[1/]Crop residue = (straw/grain ratio) x (bushel weight in lb/bu) x (crop yield in bu/acre).

Table A-2.--Residue remaining from tillage operations[1/]

Tillage operation	Residue remaining
	(%)
Chisel plow	65
Rod weeder	90
Light disk	70
Heavy disk	30
Moldboard plow	10
Till plant	80
Fluted coulter	90
V Sweep	90

[1/] Crop residue remaining = (crop residue from table 1) x (tillage factor (s)).

or figure A-1. (table A-3 is an adaptation of fig. A-1).

4. Determine the adjusted CN by reducing the CN obtained in step 1 by the percentage obtained in step 3.

5. Obtain the direct runoff from the given rainfall using the curve number obtained in step 4, according to the procedure in NEH-4 [fig. 10.1, (19)]. The adjusted CN also can be used to determine the associated peak discharge.

When conservation tillage and residue management are used in conjunction with contouring or with contouring and terracing, 0 to one-half of the table 3 reduction is needed, based on the type of soil and the increased potential for infiltration. The smaller reduction is applied to the CN for contouring or contouring and terracing. Research data are unavailable to determine the combined effects of residue and these conservation practices to reduce runoff.

Example 1:

A cultivated area in poor hydrologic condition with soils in hydrologic soil group C is farmed in straight-row continuous corn. Corn yields are 90

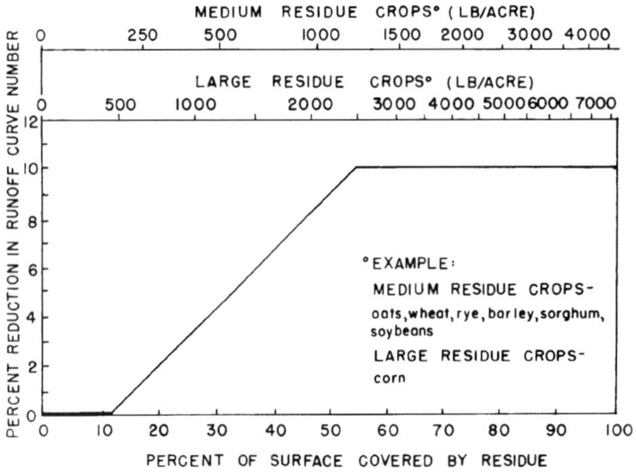

Figure A-1.--Reduction in runoff curve number as a function of crop residue.

Table A-3.--Reduction in runoff curve numbers caused by conservation tillage and residue management

Large residue crop[1/]	Medium residue crop[2/]	Surface covered by residue	Reduction in curve number[3/]
(lb/acre)	(lb/acre)	(%)	(%)
0	0	0	0
400	150	10	0
700	300	19	2
1,100	450	28	4
1,500	700	37	6
2,000	950	46	8
2,500	1,200	55	10
6,200	3,500	90	10

1/ Large-residue crop (corn).
2/ Medium residue crop (wheat, oats, barley, rye, sorghum, soybeans).
3/ Percent reduction in curve numbers can be interpolated linearly. Only apply 0 to 1/2 of these percent reductions to CN's for contouring and terracing practices when they are used in conjunction with conservation tillage.

bu per acre. Conservation tillage operations are estimated to provide a 50% surface coverage with corn residue. Determine the direct runoff from a 3.0 in rainfall in 24-hr.

Step 1. Determine curve number without conservation tillage.
For straight-row, continuous corn, in poor hydrologic condition, in a "C" soil; C = 88 [(19), table 9.1].

Step 2. Determine residue amounts left on surface.
The estimate of surface covered by corn residue was given directly as 50%.

Step 3. Reduce curve number.
Entering table 3 with 50% surface cover; CN reduction = 9%.

Step 4. Adjust curve number for conservation tillage.

$$CN = (\text{CN from step 1})\left(1 - \frac{\text{CN Reduction \%}}{100}\right).$$

$$CN = 88\ (1 - \frac{9}{100}).$$

CN = 80.1, use 80.

Step 5. Determine direct runoff, in inches, with conservation tillage. With 3.0-in rainfall and CN = 80; runoff = 1.3 in [(19), fig. 10.1].

Example 2:

The watershed above a proposed engineering practice is in good hydrologic condition with soils in hydrologic group B and is farmed in a straight-row, corn-soybean rotation. Corn yields are expected to average 100 bu per acre and soybean yields 40 bu per acre. The only tillage operations planned are chisel plowing and heavy disking before planting soybeans and heavy disking only before corn planting. The farmer is committed to these conservation tillage practices, which are suitable for the local conditions. Assume 50% of the cultivated area is in corn and 50% in soybeans in any one year. Determine the direct runoff from a 3.0 in rainfall in 24-hr as follows:

Step 1. Determine curve number without conservation tillage.
For corn: CN = 78 [(19), table 9.1].
For soybeans: CN = 78 [(19), table 9.1].

Step 2. Residue amounts left on surface.
(a) After harvest (from table A-1):
Crop residue = (straw/grain ratio x bushel weight x crop yield).
Corn residue = (1.0 x 56 lb/bu x 100 bu/acre) = 5,600 lb/acre.
Soybean residue = (1.5 x 60 lb/bu x 40 bu/acre) = 3,600 lb/acre.

(b) Reduction crop residue as a result of tillage operations (from table A-2):

Corn residue remaining = (5,600 lb/acre x 0.65 chisel plow x 0.30 heavy disk).
= 1,090 lb/acre

Soybean residue remaining = (3,600 lb/acre x 0.30 heavy disk).
= 1,080 lb/acre.

Step 3. Reduce curve number (from table A-3).
(a) Soybeans following corn, with corn residue = 1,090 lb/acre since corn is a large-residue crop; CN reduction = 4%.

(b) Corn following soybeans, with soybean residue = 1,080 lb/acre since soybeans is a medium-residue crop; CN reduction = 9%.

Step 4. Adjust curve numbers for conservation tillage.

$$CN = (CN\ \text{from step 1})\ (1 - \frac{CN\ \text{reduction}\ \%}{100}).$$

(a) Soybeans following corn (corn residue).

$$CN = 78\ (1 - \frac{4}{100}) = 74.9;\ \text{use } CN = 75.$$

(b) Corn following soybeans (soybean residue).

$$CN = 78\ (1 - \frac{9}{100}) = 71.$$

(c) Average CN for cultivated area, 50% of each crop.

$$CN = (\frac{75\ \text{CN soybeans} + 71\ \text{CN Corn}}{2}).$$

$$CN = 73.$$

Step 5. Direct runoff in inches with conservation tillage with 3.0-in rainfall and CN = 73: runoff = 0.9 in (19, fig. 10.1). Without conservation tillage, CN = 78 (step 1); runoff would be 1.1 in. This amounts to an 18% reduction in runoff.

Chapter 5. SELECTING A FORMULA TO ESTIMATE SEDIMENT TRANSPORT CAPACITY IN NONVEGETATED CHANNELS

Carlos V. Alonso[1/]

INTRODUCTION

The objective of this chapter was to select one or more formulas to predict transport capacity of flows in nonvegetated graded channels. The selected formula should (1) be framed so that it is easy to apply in computer simulation, (2) give the total load of bed material, knowing the hydraulic and geometric properties of the flow, and (3) provide reliable estimates when applied to channels of any size in which sediment particles are transported by the fluid.

Several suitable formulas were selected and tested against a range of available field and flume data. These two phases are discussed in this chapter.

SELECTION OF TRANSPORT FORMULAS

The number of available sediment transport formulas is large (table 1). Many of these formulas, however, have not received extensive application. Others, such as the Toffaletti (30) or the Modified Einstein (10) methods, are too complicated or require knowledge of the concentration of the suspended load, and therefore, are not suitable for hydrologic simulation. For these reasons, the testing of all the existing theories was not considered worthwhile. It was decided, instead, to concentrate on a few simple formulas typical of the different classes of available transport predictors. The formulas are grouped by total and bedload.

Total load formulas include: Ackers and White (1), Engelund and Hansen (14), Yang (39), Laursen (21), and Meyer-Peter, Muller and Einstein (13).

Bedload formulas include: Meyer-Peter and Muller (23), Bagnold (2); and Yalin (37).

The first four total load formulas were selected because they represent modern sediment transport theories.

The Meyer-Peter and Muller formula was considered because it is the best known of the old excess-tractive-force type of equations (11, 26). It also works as well or better than such elaborate methods as the Einstein bedload formula (13), particularly in the range of moderate to large tractive forces

---

1/ Hydraulic engineer, USDA-SEA-AR, USDA Sedimentation Laboratory, Oxford, Miss.

Table 1.--Summary of available sediment transport formulas

Type	Predicted load	Author(s) and reference	Date
Deterministic	Bed	DuBoys, (11)	1879
Deterministic	Bed	Schoklitsh, (26)	1934
Deterministic	Bed	Meyer-Peter, (22)	1934
Deterministic	Bed	Straub, (29)	1935
Deterministic	Bed	Waterways Experiment Station, (31)	1935
Deterministic	Bed	Shields, (28)	1936
Deterministic	Bed	Schoklitsh, (27)	1943
Deterministic	Bed	Kalinske, (19)	1947
Empirical	Total	Inglis, (18)	1947
Deterministic	Bed	Meyer-Peter and Muller, (23)	1948
Stochastic	Bed	Einstein, (13)	1950
Stochastic	Total	Einstein, (13)	1950
Stochastic	Bed	Einstein and Brown, (7)	1950
Stochastic	Total	Colby and Hembree, (10)	1955
Deterministic	Bed	Bagnold, (2)	1956
Deterministic	Total	Egiazaroff, (12)	1957
Deterministic	Total	Bogardi, (6)	1958
Deterministic	Total	Laursen, (21)	1958
Deterministic	Bed	Rottner, (25)	1959
Deterministic	Bed	Yalin, (37)	1963
Empirical	Total	Blench, (5)	1964
Stochastic	Total	Colby, (9)	1964
Stochastic	Total	Bishop, Simons and Richardson, (4)	1965
Deterministic	Total	Bagnold, (3)	1966
Stochastic	Total	Wilson, (36)	1966
Deterministic	Total	Chang, Simons and Richardson, (8)	1967
Deterministic	Total	Engelund and Hansen, (14)	1967
Deterministic	Total	Graf, (15)	1968
Deterministic	Total	Toffaleti, (30)	1968
Deterministic	Total	Ackers and White, (1)	1973
Deterministic	Total	Yang, (39)	1973

(see Yalin (38), p. 139). Thus, the Meyer-Peter and Muller formula represents a class of equations incorporating many of the old deterministic methods, as well as the more recent stochastic methods (7, 15) patterned after the Einstein formula (13).

The MPME method estimates the total load by adding the bedload predicted by the Meyer-Peter and Muller formula to the suspended load computed with the procedure developed by Einstein (13). The formulation of the MPME method (see appendix) is simpler than the Einstein total-load method (13) and, therefore, more suitable for hydrologic simulation.

The Bagnold (2) and Yalin (37) bedload formulas were selected because their derivations are based on assumptions that made them more appropriate for predicting transport rates in shallow flows. These formulas are presented in the appendix in the forms used in this chapter. Where the formulations required graphical solutions (that is, determining threshold conditions from Shields' curve), analytical equivalents (not given in the appendix) have been worked out to facilitate the use of digital computation. The Shields' curve was extended into the range of fine particle sizes by using the data presented by White (32). Theoretical procedures were not used to compute average-velocity versus depth relationships. Measured depths, slopes, and average-velocities were used throughout. This approach makes the comparison between observed and calculated transport rates more meaningful. Since the breadth-to-depth ratio influences bed friction when less than about five, the bed shear-velocity estimates were corrected for sidewall effect in all the flume experiments.

DATA CHARACTERISTICS

Most data in this chapter were acquired from available literature. The data selected were not used in calibrating any of the formulas being tested. The principal characteristics of these data are summarized in table 2. All quoted transport rates represent measured total bed-material discharges.

The data were classified in terms of the following dimensionless parameters (see appendix for definition of symbols):

Dimensionless grain size $D_{gr} = D[g(S-1)/\nu^2]^{1/3}$.

Mobility number $Y = u_*^2/g(S-1)D$.

Dimensionless flow depth $Z = d/D$.

Specific gravity of sediment $S = \gamma_s/\gamma$.

Any mechanical property, particularly the total sediment transport rate, is a function of these four dimensionless quantities.

D_{gr} is a measure of the relative influence of immersed weight of sediment particles and viscous forces acting on those particles. Available data cover

Table 2.--Summary of data

Data source	Tests in set	Water depth	Average velocity	Water surface slope, x10^3	Tempera-ture	Sediment		D_{gr}	Y	Z	Total sediment concentratio
						Spe-cific gra-vity	Size				
		ft	ft/s	ft/ft	^oF		mm				ppm
Field data:											
Hubbell and Matejka (17).	15	0.81-1.22	1.95-3.69	0.93-1.46	35.6-84.9	2.65	0.16-0.24	3.43-6.46	0.926-1.506	1,366-2,021	548-2,440
Colby and Hembree (10).	25	1.36-1.89	2.05-4.17	1.14-1.80	35.1-84.0	2.65	.283	5.21-8.14	1.05-2.222	1,466-2,037	392-2,220
Flume data:											
Willis and others (34).	97	.34-1.24	.52-5.16	.269-2.04	63-87	2.65	.10	2.41-2.94	.357-2.503	1,037-3,782	87-18,600
Willis and Kennedy (35).	32	.34-0.49	1.08-3.93	.831-7.99	51.5-100	2.65	.54	11.57-17.65	.109-0.923	192-277	20-6,670
Williams (33).	37	.094-0.517	1.27-3.49	1.10-22.18	53.4-87.4	2.65	1.35	29.47-37.25	.033-0.405	21.2-117	10-9,218
Kennedy (20).	16	.228-0.252	2.04-2.21	1.98-2.60	54.5-100.4	2.65	.088-0.233	2.39-6.82	.358-0.777	309-818	1,173-4,914
Harmon (16).	43	.041-0.239	.71-2.57	1.3-47.5	69-84	2.65	.41	10.39-11.79	.069-1.042	30.5-178	104-60,900

the interval $2 < D_{gr} < 37$. Thus, the influence of gravitational and viscous effects is well represented within the range of very fine to coarse sands.

About 30% of the data exhibits mobility numbers in the range $Y_c < Y < 0.4$. This criterion usually indicates conditions where the material is transported close to the bed. This tendency is exhibited by most flume data with $Z < 70$.

The dimensionless flow depth ranges from 21 to 3,782, which adequately covers the range of streamflows. No reliable data were found in the range $Z \ll 20$, however, where surface wave effects become as important as viscous effects. This is a range where most transport formulas may fail since their formulations do not account for interactions with free surface waves.

Since all formulas in this chapter predict discharge of bed material, the data were scrutinized to eliminate the measured wash load. That part of the total sediment discharge consisting of particles smaller than 0.062 mm was eliminated from the measured load.

Performance of Selected Transport Formulas

Each formula was applied to every measurement of sediment discharge. The difference between observed and calculated values was denoted by:

$$\text{Discrepancy ratio (DR)} = \frac{\text{calculated transport rate}}{\text{observed transport rate}},$$

where the transport rates are expressed as concentration by dry weight in parts per million. Consistent deviations from DR = 1 are attributed to deficiencies in the formulas in terms of D_{gr}, Y, Z, and S. A detailed error analysis, therefore, should consider the variability of DR within different ranges of those parameters. The extra work required by such an analysis, however, was not considered worthwhile since the basic objective was to determine which formula gives reliable results throughout the entire data range. The data were divided into three sets consisting of field measurements, flume data with $Z \geq 70$, and flume data with $Z < 70$. The latter set included relatively shallow flows in which the sediment probably was transported as bedload. For each data set, the mean discrepancy ratio was calculated with its fiducial limits, the standard deviation, and the percentage of data for which the predictions were between one-half and twice the observed values. These values, presented in table 3, indicate the spread of errors within each data set.

Yang developed the most reliable equation applicable over the entire range of flow conditions. This equation gave predictions that deviated only marginally, with consistently low scatter in all cases.

Both the Ackers and White and the Engelund and Hansen formulas worked reasonably well, without too much scatter. The first formula systematically overestimated the transport rates, however, while the second overestimated the field data but underpredicted the transport in flumes.

Table 3.--Analysis of discrepancy ratio distribution

Formula	No. of tests	Ratio between predicted and measured load				Percentage of tests with ratio between 1/2 and 2
		Mean	95%-confidence limits of the mean		Standard deviation	
						Percent
		Field Data				
Ackers and White	40	1.27	1.05	1.48	0.68	87.8
Engelund and Hansen	40	1.46	1.28	1.64	.56	82.9
Laursen	40	.65	.49	.80	.48	56.1
MPME	40	.83	.50	1.15	1.02	58.5
Yang	40	1.01	.89	1.13	.39	92.7
Bagnold	40	.39	.31	.47	.26	32.0
Meyer-Peter & Muller	40	.24	.22	.27	.09	0
Yalin	40	2.59	2.08	3.11	1.62	46.3
		Flume data with $Z \leq 70$				
Ackers and White	177	1.34	1.24	1.54	1.29	73.0
Engelund and Hansen	177	.73	.63	.83	.68	51.1
Laursen	177	.81	.73	.88	.51	71.4
MPME	177	3.11	2.95	3.52	2.75	42.1
Yang	177	.99	.93	1.08	.60	79.8
Bagnold	177	.85	.81	1.22	2.50	20.8
Meyer-Peter & Muller	177	.40	.39	.47	.49	18.5
Yalin	177	1.62	1.38	2.23	4.08	32.6
		Flume Data with $Z < 70$				
Ackers and White	48	1.12	.93	1.28	.52	89.6
Engelund and Hansen	48	.75	.59	.90	.50	66.7
Laursen	48	1.04	.76	1.32	.99	79.2
MPME	48	1.34	1.04	1.64	1.04	66.7
Yang	48	.90	.79	1.05	.51	85.4
Bagnold	48	1.53	1.46	1.87	1.14	45.8
Meyer-Peter & Muller	48	1.03	1.00	1.27	.83	72.9
Yalin	48	1.92	1.45	2.41	1.65	64.6

The Laursen formula worked fairly well in the flume-data range but gave less satisfactory results for the field data. A possible explanation for this behavior may be that the function $f(u_*/w)$ is not universal, as claimed by Laursen, but rather depends on the dimensionless parameters controlling the transport rates.

Among the bedload formulas, which are all based on excess mobility numbers, the Bagnold and Yalin formulas gave unsatisfactory results, even in the range $Z < 70$. The Meyer-Peter and Muller formula predicted well the total load in the latter range, but underpredicted the loads in the other two data sets. This may indicate that the Meyer-Peter and Muller formula consistently behaves as a bedload predictor. The MPME method always worked poorly and invariably overestimated the suspended load in the flume-data range.

SUMMARY

Eight sediment transport theories were examined with reference to flume and field data. This comparison was based on 40 field measurements and 225 flume experiments. Characteristics of the data were analyzed in terms of four basic dimensionless parameters. The Yang formula was the most reliable over the entire data range. The formulas developed by Ackers and White, Engelund and Hansen, and Laursen, also were found reliable but gave relatively higher errors. Of all the tested formulas, therefore, only the Yang formula can be used with confidence to predict sand-transport capacities over the range of flow conditions usually encountered in nonvegetated channels.

REFERENCES

(1) Ackers, P., and W. R. White.
1973. Sediment transport: New approach and analysis. Journal of Hydraulics Division, American Society of Civil Engineers 99(HY11):2041-2060.

(2) Bagnold, R. A.
1956. The flow of cohesionless grain in fluids. Phil. Transactions of the Royal Society of London, Series A 249(964):235-297.

(3) ------
1966. An approach to the sediment transport problem from general phyics. U.S. Geological Survey Professional Paper 422-1.

(4) Bishop, A. A., D. B. Simons, and E. V. Richardson.
1965. Total bed-material transport. Journal of Hydraulics Division, American Society of Civil Engineers 91(HY2):175-191.

(5) Blench, T.
1966. Mobile-bed fluviology. Department of Technical Services, University of Alberta, Alberta, Canada.

(6) Bogardi, J. L.
1965. European concepts of sediment transportation. Journal of Hydraulics Division, American Society of Civil Engineers 91(HY1):29-54.

(7) Brown, C. B.
1950. Engineering Hydraulics, H. Rouse, ed., John Wiley and Sons, Inc.

(8) Chang, F. M., D. B. Simons, and E. V. Richardson.
1967. Total bed-material discharge in alluvial channels. Proceedings, 12th Congress of the International Association of Hydraulic Research, Vol. I.

(9) Colby, B. R.
1964. Discharge of sands and mean velocity relationships in sand-bed streams. U.S. Geological Survey Professional Paper 462-A.

10) -------, and C. H. Hembree.
1955. Computation of total sediment discharge, Niobrara River near Cody, Nebraska. U.S. Geological Survey Water Supply Paper 1357.

11) DuBoys, M. P.
1879. Etudes de Regime et l'Action Exercee par les Eaux sur un Lit a Fond de Graviers Indefiniment Affouilable. Annales de Ponts et Chausses, Series 5, vol. 18, pp. 141-195.

12) Egiazaroff, J. V.
1965. Calculation of nonuniform sediment concentration. Journal of Hydraulics Division, American Society of Civil Engineers 91(HY4):225-247.

13) Einstein, H. A.
1950. The bed-load function for sediment transportation in open channel flows. U.S. Department of Agriculture Technical Bulletin No. 1026, 71 pp.

14) Engelund, F., and E. Hansen.
1967. A monograph on sediment transport in alluvial streams. Teknisk Vorlag, Copenhagen.

15) Graf, W. H., and E. R. Acarogly.
1968. Sediment transport in conveyance systems-Part 1. Bulletin of the International Association for Scientific Hydrology, XIII Annee, No. 2.

16) Harmon, W. C.
1972. Unpublished experiments. U.S. Department of Agriculture Sedimentation Laboratory. Oxford, Miss.

17) Hubbell, D. W., and D. Q. Matejka.
1959. Investigations of sediment transportation, Middle Loup River at Dunning, Nebraska. U.S. Geological Survey Water Supply Paper 1476.

(18) Inglis, C. C.
1947. Meanders, and their bearing on river training. Institution of Civil Engineers Maritime and Waterways Engineering Division, Paper No. 7.

(19) Kalinske, A. A.
1947. Movement of sediment on bed load in rivers, Transactions, American Geophysical Union 28(4):615-620.

(20) Kennedy, J. F.
1961. Further laboratory studies of the roughness and suspended load of alluvial streams. California Institute of Technology, Report No. KH-R-3, W. M. Keck Laboratory of Hydraulics and Water Resources, 36 pp.

(21) Laursen, E.
1958. The total sediment load of streams. Journal of Hydraulics Division, American Society of Civil Engineers vol. 54, No. HY1, Paper 1530.

(22) Meyer-Peter, E., H. Farre, and A. Einstein.
1934. Neuere Versuchsresultate uber den Geschiebetrieb, Schweizerische Bauzeitung, vol. 103, No. 13, pp. 147-150.

(23) ------and R. Muller.
1948. Formulas for bed-load transport. Proceedings, 2nd Congress of the International Association of Hydraulic Research, pp. 39-64.

(24) Nordin, C. F., Jr.
1964. Aspects of flow resistance and sediment transport-Rio Grande near Bernalillo, New Mexico. U.S. Geological Survey Water Supply Paper 1498-H.

(25) Rottner, J.
1959. A formula for bed load transportation. La Houille Blanche, 4(3):301-307.

(26) Schoklitsh, A.
1934. Geschiebetrieb und die Geschiebefracht, Wasserkraftund Wasserwirtsch, Jgg. 39, Heft 4.

(27) ------
1950. Handbuch der Wasserbaues, Springer-Verlag, Vienna, 2d. ed., vol. I.

(28) Shields, A.
1936. Anwendung der Aehnlichkeitsmechanik und der Turbulentzforschung auf die Geschiebebewegung. Mitteilungen, Preussichen Versuchsanstalt fur Wasserbau und Schiffbau, Helft 26, Berlin.

(29) Straub, L. G.
1935. Missouri River Report. In House Document 238, p. 1135, 73d Congress, 2d. session, U.S. Government Printing Office.

(30) Toffaleti, F. B.
1969. Definitive computations of sand discharge in rivers. Journal of Hydraulics Division, American Society of Civil Engineers 95(HY1): 225-246.

(31) U.S. Waterways Experiment Station.
1935. Studies of river bed materials and their movement. U.S. Waterways Experiment Station, Paper 17.

(32) White, S. J.
1970. Plain bed thresholds for fine grained sediments. Nature 228 (5267):152-153.

(33) Williams, G. P.
1967. Flume experiments on the transport of a coarse sand. U.S. Geological Survey Professional Paper 562-B.

(34) Willis, J. C., N. L. Coleman, and W. M. Ellis.
1972. Laboratory study of transport of fine sand. Journal of Hydraulics Division, American Society of Civil Engineers 98(HY3):489-501.

(35) ------, and J. F. Kennedy.
1977. Sediment discharge of alluvial streams calculated from bed-form statistics. Iowa University, Iowa Institute of Hydraulics Research, Report No. 202, 200 pp.

(36) Wilson, K.C.
1966. Bed load transport at high shear stress. Journal of Hydraulics Division, American Society of Civil Engineers 92(HY6):49-59.

(37) Yalin, M. S.
1963. An expression for the bed-load transportation. Journal of Hydraulics Division, American Society of Civil Engineers 89(HY3):221-250.

(38) ------
1972. Mechanics of Sediment Transport, Pergamon Press.

(39) Yang, C. T.
1973. Incipient motion and sediment transport. Journal of Hydraulics Division, American Society of Civil Engineers 99(HY10):1679-1704.

APPENDIX. TRANSPORT FORMULA

Symbols used in the following formulas are defined at the end of this section.

1. Total load formula of Ackers and White (1):

$$C = 10^6 \frac{S d_{35}}{h} \left(\frac{V}{u_*}\right)^n \left\{\frac{F_{gr}}{A} - 1\right\}^m$$

$$F_{gr} = \frac{u_*^n}{\sqrt{g d_{35}(S-1)}} \left\{\frac{V}{\sqrt{32}\log(10h/d_{35})}\right\}^{1-n}$$

$$n \begin{cases} = 1 - 0.56 \log D_{gr}, & 1 \le D_{gr} \le 60 \\ = 0, & D_{gr} > 60 \end{cases}$$

$$A \begin{cases} = (0.23/\sqrt{D_{gr}}) + 0.14, & 1 \le D_{gr} \le 60 \\ = 0.170, & D_{gr} > 60 \end{cases}$$

$$m \begin{cases} = (9.66/D_{gr}) + 1.34, & 1 \le D_{gr} \le 60 \\ = 1.50 & D_{gr} > 60 \end{cases}$$

$$\log \sigma = 2.86 \log D_{gr} - (\log D_{gr})^{1/2} - 3.53, \quad 1 \le D_{gr} \le 60$$

$$\sigma = 0.25 \qquad D_{gr} > 60$$

2. Total load formula of Engelund and Hansen (14):

$$C = \frac{5 \times 10^4\, S\, V\, u_*^3}{g^2\, h\, d_{50}\, (S-1)^2}$$

3. Total load formula of Yang (39):

$$\log C = 5.435 - 0.286 \log (wd/\nu) - 0.457 \log (u_*/w) +$$

$$[1.799 - 0.409 \log (wd/\nu) - 0.314 \log (u_*/w)] \log (VS_o/w - V_c S_o/w).$$

$$\frac{V_c}{w} \begin{cases} = 2.5/[\log(u_* d/\nu) - 0.06] + 0.06, & 0 < (u_* d/\nu) < 70. \\ = 2.05, & (u_* d/\nu) \geq 70. \end{cases}$$

4. Total load formula of Laursen (21):

$$C = 10^4 \sum_{i=1}^{n} p_i \, (d_{mi}/h)^{7/6} \left\{ V^2 (d_{50})^{1/3} / [58 \, Y_c \, d_{mi} \, g(S-1) h^{1/3}] - 1 \right\} \cdot f(u_*/w).$$

$f(u_*/w)$ is given by Laursen (see 21, fig. 14).

5. MPME formula. (In this method the total load is computed by adding the bedload predicted by the Meyer-Peter and Muller formula (23) to the sus pended load predicted by the Einstein methods (13).)

$$C = C_b \left\{ \frac{A^{E-1}}{11.6(1-A)^E} \left[\left(\frac{V}{u_*} + 2.5 \right) I_1 + 2.5 \, I_2 \right] + 1 \right\}$$

where $A = 2 \, d_{50}/h$, $E = w/0.4 \, u_*$, C_b is the bedload concentration, and I_1 and I_2 are the Einstein integrals.

6. Bedload formula of Meyer-Peter and Muller (23):

$$C = \frac{8 \times 10^6 \, S \, u_*^3}{(S-1) \, V \, h \, g} \left\{ K^{3/2} - \frac{0.047}{Y} \right\}^{3/2}$$

in which $K = (V/u_*) \, (f_b'/8)^{1/2}$. The friction factor associated with the bed skin friction, f_b', is obtained from

$$(f_b')^{-1/2} + 2\log(k_s/4r_b) = 1.14 - 2\log[1 + \frac{37.40 r_b}{k_S R_E (f_B')^{1/2}}], \quad 3 < \frac{u_* k_s}{\nu} \leq 70,$$

$$(f_b')^{-1/2} = 2\log(4r_b/k_s) + 1.14, \qquad \frac{u_* k_s}{\nu} > 70$$

where k_s = 2d, and R_e is the flow Reynolds number.

7. Bedload formula of Bagnold (2):

$$C = [10^6 \text{ f B } Sg^{1/2} (S-1)^{1/2} d_m{}^{3/2}/Vh] \, Y^{1/2}(Y-Y_c)$$

where $f = f(u_* d_m/\nu)$ is the transition velocity function (see Yalin (37), fig. 2.5), and $B = f(d_m)$ is given b Bagnold (2, fig. 12).

8. Bedload formula of Yalin (37):

$$C = (6.35 \times 10^5 \, d_{50} \, u_* \, S\sigma/Vh) \, [1 - (1/a\sigma) \ln (1 + a\sigma)]$$

$$a = 2.45 \, S^{-0.4} \, Y_c{}^{1/2}, \quad = (Y/Y_c) - 1$$

The sediment concentration, C, is related to the sediment transport rate as dry weight per unit width per unit time as follows:

$$g_s = 10^{-6} \, C \, \gamma \, Vh.$$

Symbols used in these formulas are defined as follows:

C	= sediment concentration by dry weight, in parts per million;
d_m	= mean diameter of bed material;
d_{35}, d_{50}	= effective bed-material size;
g	= acceleration of gravity;
h	= mean flow depth;
i	= fraction index;
p	= percentage of bed-material fraction;
r_b	= hydraulic radius of bed;
S	= specific gravity of bed material;

u_* = bed shear velocity;

V = average velocity of flow;

w = settling velocity of sediment particles;

Y_c = mobility number at threshold conditions (Shields);

γ = specific weight of fluid;

γ_s = specific weight of sediment;

ν = kinematic viscosity of fluid.

Chapter 6. CONTOUR FARMING AFFECTS RUNOFF PATTERNS AND SOIL MOVEMENT

L. D. Meyer[1/]

Contouring or contour farming is an erosion-control supporting practice where cropping operations follow a route nearly parallel to contour (equal-elevation) lines rather than up-and-down slope or parallel to the field boundaries. Idealistically, contoured rows each carry all runoff from that row and cause it to flow across the slope at a slight gradient into a grassed waterway or other controlled-erosion channel. Realistically, however, contoured rows often fail to slope continuously toward a waterway or they do not have the capacity to carry all the runoff. Therefore, contouring is usually a quite different erosion-control practice when actually applied on cropland than it is conceptually.

Characteristics that affect the effectiveness of contouring include:

1. Flow cross section--Depends on row width, row microrelief (height and shape of cross section as left by tillage), land slope (perpendicular to the row), irregularity of the row ridges (locations where breakovers might occur), and cross sections of waterways and breakover channels (if they occur).

2. Flow velocity--Depends on flow cross section, surface roughness, row gradient (including the influence of irregularities that may reverse the gradient), row length, and land slope (for waterways and breakover channels).

3. Excess rainfall (runoff) rate--Depends on rainstorm characteristics, infiltration characteristics of the soil, and moisture conditions.

Certain factors, such as soil type and land slope, are characteristics of the field. Others factors may vary from year to year, such as row gradient, row width, and row length. Factors that depend on the farming system include the tillage system, frequency of cultivation, height of row ridges, and roughness of the soil surface. Factors that vary stochastically include rainfall patterns, soil moisture conditions at rainfall, certain surface irregularities, the runoff pattern as determined by whether breakovers previously had occurred, and the size and shape of breakover channels as they enlarge from progressive erosion.

Since contouring is a widely used practice that varies greatly depending upon the individual situation, probably a single evaluation of its effectiveness is inadequate. Instead, various typical contour conditions should be

---

1/ Agricultural engineer, USDA-SEA-AR, Sedimentation Laboratory, Oxford, Miss.

analyzed. The effectiveness of each condition should be indicated as a guide. Variations to be evaluated should include types of tillage that affect row microrelief differently, land slopes, row slopes and their irregularities, row lengths, row widths, cropping systems, and rainstorm patterns.

A limited number of conditions should be tested at first, using techniques incorporated in the model for concentrated flow. Concentrated flow would occur in the rows and waterways until row breakovers developed; thereafter, it would follow the shorter rows and breakover channels. Once breakover occurred, the new flow pattern would be appropriate until new rows with significant microrelief are established by further tillage. Thus, runoff that occurred for a given tillage system, row length, and storm size either would flow down the row to an established waterway or would follow the rows until it reached a breakover channel. This would determine the flow pattern thereafter until a significant change due to subsequent tillage occurred. In both situations, sediment from along the rows plus that from the waterways and breakovers, where present, should be included as appropriate. Figures 1 and 2 illustrate this approach.

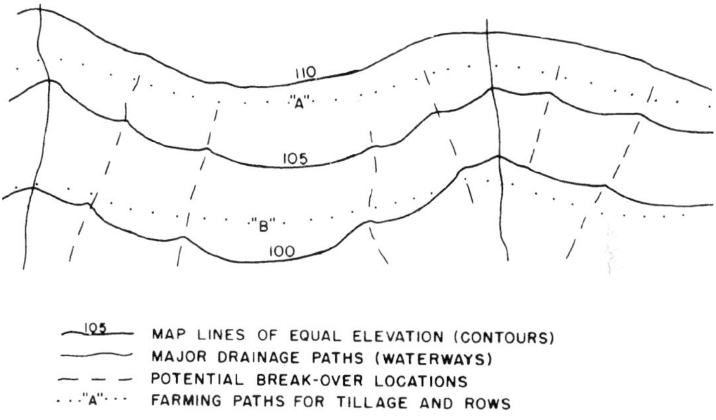

Figure 1.--Typical field topography showing true contour lines and probable tillage paths in fields that are contour farmed.

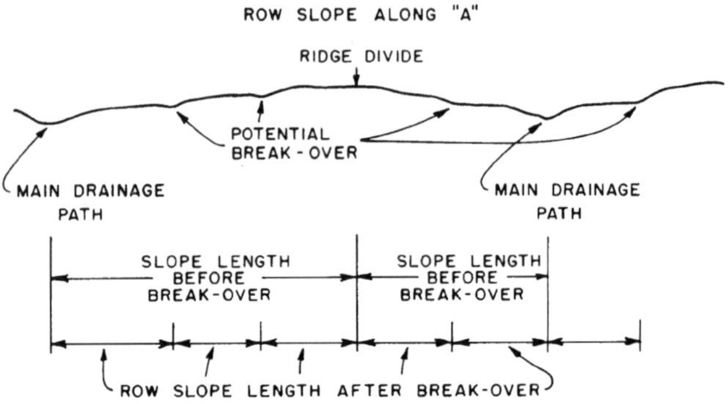

Figure 2.--Changes in runoff pattern and slope length along row "A" when runoff breaks across crop rows of field in Figure 1.

REFERENCES

(1) Beasley, R. P.
1972. Erosion and sediment pollution control. Iowa State University Press, Ames, 320 pp.

(2) Blakely, B. D., and others.
1957. Erosion on cultivated land. In Water. 1957 Yearbook of Agriculture. pp. 290-307.

(3) Diseker, E. G., and R. E. Yoder.
1936. Sheet erosion studies on Cecil clay. Alabama Agricltural Experiment Station Bulletin 245.

(4) Schwab, G. O., and others.
1966. Soil and Water Conservation Engineering. Second edition. John Wiley and Sons, Inc., New York.

(5) Smith, D. D., and others.
1945. Investigations in erosion control and reclamation of eroded Shelby and related soils at the Conservation Experiment Station, Bethany, Missouri, 1930-42. U.S. Department of Agriculture, Technical Bulletin 883.

(6) Stewart, B. A., and others.
1976. Control of water pollution from cropland. U.S. Department of Agriculture, Agriculture Research Service, ARS-H-5-1 (vol. I): pp. 27, 63, 67, 72-76: ARS-H-5-2 (vol. II): pp. 41, 46. (Series discontinued; Agriculture Research Service is now Science and Education Administration-Agricultural Research.)

(7) U.S. Department of Agriculture, Soil Conservation Service.
[Periodically updated]. Specifications for contour farming (National and Mississippi).

(8) Wischmeier, W. H., and D. D. Smith.
1965. Predicting rainfall-erosion losses from cropland east of the Rocky Mountains. U.S. Department of Agriculture, Agriculture Handbook 282, 47 pp.

Chapter 7. ADDING EROSION FROM SNOWMELT TO AN EROSION PREDICTION EQUATION

L. D. Meyer[1/]

Erosion prediction equations evaluate soil losses from rainstorms but may omit losses from snowmelt, as is done with the Universal Soil Loss Equation and this model. The following approach may be used to estimate the additional erosion.

Erosion plot data[2/] from studies at Morris, Minn., for a 10-yr period (1962-71) were:

Treatment	Soil loss from snowmelt: Average annual	Soil loss from snowmelt: Range	Soil loss from rain: Average annual	Soil loss from rain: Range
	(t/acre)	(t/acre)	(t/acre)	(t/acre)
Fallow	1.12	0 to 5.31	15.36	0.8 to 80.4
Continuous Corn	0.25	0 to 1.32	7.10	0.3 to 44.5
Rotation Corn	0	0 to 0.04	3.37	0.2 to 21.1
Rotation Oats	0.14	0 to 0.95	1.80	0 to 16.4
Rotation Hay	0		.01	0 to 0.1

For the fallow plot data with an average annual loss of 16.5 t/acre from rainfall and snowmelt combined, 1.13 t/acre or 7% additional erosion occurred beyond the 15.36 t/acre caused by rainfall. Thus, average annual erosion by rainfall plus snowmelt was 107% of that when only losses due to rainfall were considered. Of this extra 7%, snowmelt erosion records indicated that 5% occurred in March and the remaining 2% in April. Continuous corn and rotation oats averaged an extra 4% and 8% respectively, due to snowmelt. Their small losses and the influence of small plots on snowtrapping make them subject to greater possible error, however.

1/ Agricultural engineer, USDA-SEA-AR, Sedimentation Laboratory, Oxford, Miss.

2/ Mutchler, C. K., R. E. Burwell, and L. C. Stapler. 1976. Runoff and soil losses from Barnes soils in the Northwestern Corn Belt. U. S. Department of Agriculture, Agricultural Research Service, ARS-NC-36, 16 pp. (Series discontinued; Agricultural Research Service is now Science and Education Administration-Agricultural Research.)

The appropriate amount of snowmelt erosion to add to that indicated from rainfall will vary among locations. The time when this snowmelt erosion occurs also will depend on local climatic characteristics. Furthermore, the melting of snow may depend much more on temperature patterns than rainfall patterns. There may not be adequate data to indicate the extent of snowmelt erosion for all areas. Where data are available, however, snowmelt erosion may be added to the predictions as outlined.

Chapter 8. MODELING EROSION AND SEDIMENT YIELD FROM FLATLAND WATERSHEDS

C. K. Mutchler and C. E. Murphree[1/]

INTRODUCTION

Erosion and sediment transport increase in some direct proportion to land slope. A logical question, therefore, is whether sediment yield from flatlands is large enough to be a problem. Research by Murphree and others (5) on a 40-acre watershed in the Mississippi Delta showed that continuous cotton and 0.2% graded slopes allowed an annual soil loss of as much as 13 tons per acre and an average annual loss of 7 tons per acre.

Using data from the 1965 national inventory of soil and water conservation needs (10), over 40% of U.S. cropland was estimated to be on less than 4% slope and about 10% on less than 2% slope. These flatter slopes usually are reserved for row crops, all intensively tilled for weed control. Row crops are most conducive to soil erosion, thus contributing to flatland erosion.

After 7 years of sediment yield research in the Mississippi Delta, erosion from flatlands emerges as an important problem. Erosion from flatlands is also not as well researched as that on steeper slopes. The following discussion will show that:

(1) Flatland erosion and sediment yield problems are quite different from those on steeper slopes.

(2) Use of the Universal Soil Loss Equation (USLE) on flatlands requires extrapolation of its basic relationships beyond much of the available data base. Progress on adjusting and verifying the USLE factors for flatter slopes will be given.

(3) Methodology or techniques are needed to predict sediment deposition in the many types of depositional areas found in flatland watersheds.

EROSIONAL DIFFERENCES

Surface Storage

A depression on flat land will store more water than one with the same dimensions on steeper land just as a bucket held upright holds more water than one tipped at an angle. This principle applies to depressions of all sizes. Thus, surface storage generally will be greater for lower slopes. For depressions created by surface roughness on a plane surface, mean surface water depths will be greater and velocities will be less than on steeper slopes.

1/ Hydraulic engineer and agricultural engineer, respectively, USDA-SEA-AR, USDA Sedimentation Laboratory, Oxford, Miss.

The slope factor of the USLE can be adjusted for nonplanar surfaces using Foster and Wischmeier's (3) coefficient method for gentle irregular slopes. This same irregularity on a flatter mean slope, however, will cause a large depression that may contain some significant depth of water during a rainstorm. Slight depressions or "wet spots" in a flatland area will produce less sediment than other parts of the field because soil detachment by raindrops is eliminated by water depth greater than three raindrop diameters (7). Any runoff from above the depression will be slowed as it passes through, resulting in partial deposition of the sediment load. To the best of our knowledge, this problem has not been researched in the field.

Water Table

Flatter slopes that are farmed intensively usually lie at the bottom of upland slopes or in flood plains. Consequently, they often are affected by water tables caused by intake rates higher than subsurface drainage rates, for example flood plains that are only slightly higher than river levels. Whatever the cause for the particular water table, soil moisture levels may be so high that more rainfall leaves as runoff or remains in depressional areas for longer periods.

Surface Drainage

Surface drainage systems usually limit temporary flooding to 24 hr or less; those fields without a drainage plan may flood for longer than 24 hr. Thus, backwater affects runoff from all large storms and many small ones. This situation is different from that of the free overfall plots used in developing the USLE. Therefore, the drainage system could invalidate the USLE unless some factor is included to reduce the USLE soil loss estimate.

Tillage

Many cultivated flatlands are maintained in beds (rows) of varying heights whereas most steeper slopes in humid regions are tilled flat. These beds reduce the generally high moisture content in the seedbed and enhance surface drainage. Most seeds and plants (rice is a notable exception) cannot survive in saturated soil. The middles between the beds also provide drainage channels that drain only an area one row wide in contrast to a random rill that may carry runoff from an area several rows wide. The steep side-slope of the bed is extremely erodible over the bedded field, however, and the middle provides an efficient channel for transporting detached particles down slope. These tillage practices also contribute to the definable longer slope lengths on flatlands.

Restricted Outlets

Flow restrictions at the ends of slopes along field boundaries and at watershed outlets occur almost everywhere. The restricted outlets in flatlands, however, become more dominant in their effect on sediment yield. Any restriction will cause water to cover a much larger area on a flat slope, and is similar to the situation created by tile outlet terraces. The effects of restricted outlets and temporary ponding must be considered carefully for most flatland watersheds, whereas, they may be neglected or estimated more easily for steeper slopes.

Sediment Source

More sediments from low slopes originate from the surface soil because a rill on 0.2% land would be only 0.8 ft deep at 400 ft from the outlet if the eroded bed was level. In contrast, such a rill (more properly called a gully) would be 16 ft deep on 4% land. The importance of this point is that with similar cropping practices, sediments from low slopes will include a higher proportion of organic matter and farm chemicals than those from steeper slopes that may contain a larger portion of subsurface material.

Surface Runoff Velocities

Although shallow surface velocities may not have the same relationship to slope as deeper channel velocities, they decrease with decreasing slope and thus are lower on flatlands. The effects of backwater also reduce runoff velocities. Surface roughness and more prevalent vegetation also contribute to lower surface velocities. These lower velocities are significant because of their effect on runoff detachment, sediment transport capacity, and the size of sediment particles that runoff can carry. Consequently, sediments from flatlands should be more similar in size to those surface particles detached by rainfall but will be enriched greatly by the finer soil fraction. Smaller total amounts of detached materials should be transported off the field.

USLE FACTORS FOR FLATLANDS

Use of the USLE assumes that the erosion area is a land unit similar to the erosion plot used for the USLE data base. A free outlet from the sediment-producing area is an important requirement for predicting erosion from flatlands. As the erosion plot has a free overfall at the outlet, any conditions that retard runoff will invalidate use of the USLE beyond that point. The following is a discussion of USLE factors that are limited or changed by flatland conditions. All factors of the USLE generally require verification for nearly flat slopes except the R factor, which is independent of slope.

Soil Erodibility Factor (K)

Higher levels of soil moisture of flatlands usually affect soil erodibility. Barnett (1) found K-factors for Commerce silt loam and Sharkey clay soils to be much higher than previously estimated. Soil strength is lower when moisture is high and soil detachment is greater. In other words, the K-factor value for a soil depends upon the moisture content of the soil when the measurement is made. The K-factor determined with an erosion plot under natural rainfall is smoothed by the succession of storm events during annual periods. Thus, it is a single number representative of average annual erodibility. Evaluations made using a single storm value(s), that is, rainulator storms must be weighted and adjusted, however, to account for annual variation of soil moisture (and EI distribution) peculiar to the geographic area. Few field determinations of K-factors for flatland soils have been made because the seriousness of erosion on steeper slopes was recognized more easily. Hence, most of the presently assigned K-factor values are extrapolations that must be used only with caution.

Slope Length Factor (L)

Soil loss is proportional to slope length to some exponent, $SL = \lambda^m$. The exponent, m, in turn, is related to slope steepness. For use in the USLE the length relationship was normalized by the standard plot length of 72.6 ft, resulting in $L = (\lambda/72.6)^m$ where L is the slope length factor and λ is the slope length in feet.

Various values of m have evolved over the years. Zingg (13) proposed m = 0.6 based on published data from Missouri, Iowa, Wisconsin, Oklahoma, and Texas. Musgrave (6) proposed m = 0.35, Smith and Wischmeier (9) proposed m = 0.5 ± 0.1 and Wischmeier and Smith (11) proposed m = 0.6 for slopes steeper than 10%, m = 0.3 for certain slopes less than 0.5%, and 0.5 for all other slopes. In a study of rill and interrill erosion, Foster and Meyer (2) showed that the values of m should be near 0 if interrill erosion was dominant, near 1 if rill erosion was dominant, and between 0 and 1 for combinations of interrill and rill erosion. In the revision of Agriculture Handbook 282, Wischmeier and Smith (12) recommended values of m = 0.2 for slopes less than 1%, 0.3 for slopes between 1 and 3%, 0.4 for slopes 3.5 to 4.5%, and 0.5 for slopes greater than 5%.

In 1977, we obtained m = 0.15 for a slope of 0.2%, using wet runs of a rainfall simulation experiment (8). Using this value, the latest recommended values (2, 12) of $0 < m < 1$, and slope angle, θ, in degrees, the equation was derived as:

$$m = 1.2 (\sin \theta)^{1/3},$$

which is suitable for predicting erosion from average annual conditions on most soils. This equation gives m = 0.15 at 0.2% slope, m = 0.5 at 7% slope, and 0.6 at 12.5% slope. The equation should be appropriate for use on slopes less than 12% which includes slopes used for much of the cultivated agriculture in the United States.

Slope Factor (S)

The present slope relationship in the USLE was derived from data taken from slopes generally above 3%. In 1978, a slope experiment was run on Mississippi Delta soils using standard erosion plots and slopes of 0.1%, 0.2%, 0.5%, 1%, 2%, and 3%. The data from this project have not been analyzed fully. However, preliminary analysis indicates that little change, if any, will be required to use the relationship given by Wischmeier and Smith (12) for slopes of less than 3%. This relationship is $S = 65.41 \sin^2 \theta + 4.56 \sin \theta + 0.065$ where θ is the slope steepness in degrees.

Cropping and Management Factor (C)

The C-factors for flatlands are generally different from those for steeper slopes. Drainage is almost always a problem on low slopes in the humid regions. Therefore, tillage for surface drainage is a dominant requirement. Bedding is used in many areas to provide an unsaturated seedbed and to enhance drainage down the middles between the beds. Evaluation of C-factors under these conditions is still very limited; present values are based primarily on extrapolation of established values for slopes above 3 to 4% and on judgment.

Recent work in Mississippi using the rainulator showed that erosion from a hipped (bedded) plot on a 1% slope was more than twice as great as that from a flat tilled plot. These results have been included in the C-factor tabulations for cotton in (12).

Based on results from a small flatland watershed, Murphree and Mutchler (4) computed C-factors for cotton with conventional tillage. Cropstage C-values ranged from 0.10 for fully grown cotton to over 1.00 for the cool, wet winter period and the spring bedded period. These values of $C>1$ indicate greater erosion than that from the standard flat, fallow plot used to normalize C-factor values.

Other land features that are different or have greater emphasis on low slopes also will affect C-factor values. Surface roughness, slope irregularity, and cover conditions, for instance, may affect erosion from the lower slopes differently. These conditions have yet to be evaluated.

Practice Factors (P)

The P-factor adjusts predicted soil loss for that amount which does not leave the field. The USLE for sloping land considers contour tillage, stripcropping on the contour, and terrace systems (12). Erosion control credit for tilling across the slope becomes very small as slope steepness approaches 0 because the prevailing land slope and row slope approach equality. Since bedded rows on flatlands rarely break over, slope and slope length for the USLE are best taken relative to row direction rather than true slope direction. Terracing is used rarely, but cross-slope ditches that are oriented toward better drainage perform a similar function.

Size of Soil Particles

Researchers and users of the USLE in the past have been little concerned with enrichment of the smaller soil particles in soil loss. Transporting the separate soil fractions must be considered because of low surface velocities. This is particularly important since pollution from farm chemicals in runoff and attached to the finer sediment particles, has been recognized.

Limited research (unpublished) has shown a high enrichment of fines in soil loss from 75-ft erosion plots on flatlands. If the USLE is to be used as a basis of sediment yield, therefore, it must be modified for estimating soil loss by ranges of particle size or a small unit plot (and plot length) must be used for the erosion estimate.

SEDIMENT YIELD

Sediment yield (SY), as used here, is the amount of soil loss estimated by the USLE that is transported beyond the areal limitation of the USLE. Thus, SY = (SDR) (USLE) where SDR is a factor or group of multiplicative factors that represent some portion of the USLE-estimated soil loss that is transported to the watershed outlet. The SDR's are fractions that range from 0 to 1. They also range from best estimates of deposition in the length of transport, as determined by experience, to verified mathematical representations reflecting greater knowledge of the sediment transport processes. The following discussion describes the variety of depositional features in flatlands.

Slow runoff velocities on flatlands require computation of sediment transport by sediment particle size even from small areas. Two modeling approaches for sorting sediment that seem feasible are use of true particle sizes that include aggregates or use only of primary particle sizes. The problem with using aggregate sizes is that aggregate sizes in the soil source are difficult to determine. The problem with using primary particle sizes is that some factor must be used to account for aggregates in the sediment. Further research is needed to determine the best method for modeling sediment transport.

Relationships of sediment transport through a variety of depositional areas in flatlands are needed. A single expression for the SDR probably would be unsuitable for all types of depositional areas in flatlands.

Field depressions vary in area from several hundred square feet to several acres. They pond water from depths of a few inches to perhaps a maximum of 6 in. Such an area can be defined as one that may delay spring tillage but can be farmed along with other parts of the field in most years. Field depressions present a unique modeling problem because they usually have a continuous slope above and below them. This requires an adjustment to the USLE slope-length function or the use of a special modeling function.

Grass filter strips are rarely part of a flatland conservation scheme. They are found in widths of 20 ft or so around fields, however, where tilling is inconvenient. They should be considered more often for conservation, and hence should be included in a field sediment yield model.

Waterways occur in various forms and frequently are used for purposes other than transporting runoff. As turnrows (also used as field roads), they are shallow and wide and usually assume a shape dependent on the equipment available to maintain them. A dirt-scraper often is used to re-shape and deepen the turnrow. The resultant shape is a shallow vee-ditch where the bottom is almost flat. In the past, turnrows were disked to control weeds; however, many farmers now clip and leave volunteer grass for cover. The grade of this type of waterway is invariably flat enough to cause sediment deposition.

On slopes above about 1%, random waterways occur that are common anywhere. Where farmers persist in farming across them, they present a serious channel (gully) erosion problem. Where such ditches occur on lower slopes, they are usually too wet to farm, and weeds, in combination with the lower grade, make them effective depositional areas.

Many fields on flood plains have slopes less than 0.5%, and many are as flat as 0.1%. In humid areas, this topography requires draining the fields before they can be farmed. Surface drainage systems are complex and extensive, and rarely will flow velocities be high enough to degrade field ditches. Hence, all such ditches must be considered as depositional areas. Small ditches with outlets into larger and deeper ones, however, present opportunity for significant amounts of gully type erosion.

Restricted outlets that form major depositional features occur naturally in flatlands. They also are used deliberately to prevent gully type erosion, however. Many types of structures, ranging from a culvert through a ditch spoil bank to more elaborate drop structures, are used to drop runoff into a deeper ditch. Since all are designed with the invert at the field level to prevent field ditch erosion, they must pond water before full capacity is attained. This fact and the economies of size cause large areas of ponding in flatlands.

SUMMARY

The differences in erosion between flatlands and the more sloping lands are dominated by the amount of protective water cover during storm events. This is caused by greater surface storage due to low slope, the bedding type of tillage required for seedbed drainage, higher water tables, generally restricted outlets, and low runoff gradients. The generally higher soil water content and high local slopes of bedding make soil detachment higher and more available for transport, however. A larger part of the sediment probably will originate from the field surface, and fines may be enriched because of lower runoff velocities.

Factors of the USLE generally were derived using data from slopes exceeding 3% steepness. Few K-factor determinations have been made for flatland soils whose erodibility is affected by their higher soil moistures. Research has lowered the exponent of the slope-length factor for slopes of 0.2% to 0.15% and determined the probable validity of the slope steepness relationship. A few C-factor values have been determined that reflect the wetter flatland conditions and the unique tillage required for drainage.

The depositional features of flatlands could be evaluated as sediment delivery ratios to reduce USLE soil loss estimates to sediment yield. Determining soil particle sizes in the USLE-predicted soil loss for determining sediment yield from flatlands is important because of high enrichment of the finer soil fractions in the sediment.

REFERENCES

(1) Barnett, A. P., A. E. Dooley, and G. A. Smith.
1978. Soil erosion and sediment movement under sugar cane culture on the flatlands of southern Louisiana. Transactions of the American Society of Agricultural Engineers 21(6):1144-1150, 1156.

(2) Foster, G. R. and L. D. Meyer.
1975. Mathematical simulation of upland erosion using fundamental erosion mechanics. In Present and prospective technology for predicting sediment yields and sources. Proceedings of the 1972 sediment yield workshop. USDA Sedimentation Lab., Oxford, Miss. U.S. Department of Agriculture, Agricultural Research Service, ARS-S-40, p. 190-207.

(3) ------, and W. H. Wischmeier.
1974. Evaluating irregular slopes for soil loss prediction. Transactions of the American Society of Agricultural Engineers 17:305-309.

(4) Murphree, C. E. and C. K. Muchler.
1979. Cover and management factors for cotton. Accepted for publication, Transactions of the American Society of Agricultural Engineers.

(5) ------, C. K. Mutchler, and L. L. McDowell.
1976. Sediment yields from a Mississippi Delta watershed. In Proceedings of the Third Federal Interagency Sedimentation Conference, Water Resources Council, Denver, Colorado, p. 1-99 to 1-109.

(6) Musgrave, G. W.
1947. The quantitative evaluation of factors in soil erosion, a first approximation. Journal of Soil and Water Conservation 2(3):133-138.

(7) Mutchler, C. K.
1970. Splash of a waterdrop at terminal velocity. Science, Sept. 25, 1970: 169:1311-1312.

(8) ------, and J. D. Greer.
1979. Effect of slope length on erosion. Accepted for publication, Transactions of the American Society of Agricultural Engineers.

(9) Smith, D. D., and W. H. Wischmeier.
1957. Factors affecting sheet and rill erosion. American Geophysical Union 38:889-896.

(10) U.S. Department of Agriculture.
1965. Soil and water conservation needs - a national inventory. U.S. Department of Agriculture, Miscellaneous Publication No. 971. Washington, D.C.

(11) Wischmeier, W. H., and D. D. Smith.
1965. Predicting rainfall-erosion losses from cropland east of the Rocky Mountains - guide for selection of practices for soil and water conservation. U.S. Department of Agriculture, Agriculture Handbook No. 282.

12) ------, and D. D. Smith.
1978. Predicting rainfall erosion losses - a guide to conservation planning. U.S. Department of Agriculture, Agriculture Handbook No. 537.

13) Zingg, A. W.
1940. Degree and length of slope as it affects soil loss in runoff. Agricultural Engineer 21:59-64.

Chapter 9. GULLY EROSION

R. F. Piest and E. H. Grissinger[1]

Types and causes of gullying vary from region to region and are ill-defined in the quantitative sense. Therefore, gully erosion rates will not be estimated in the short-term erosion/sedimentation model and likely will be an "add-on" quantity for the foreseeable future. If gullying is significant in the sediment accounting of a drainage basin, the gully erosion rate is estimated on the basis of SCS-Technical Release No. 32 (13) or considerable judgment or both.

DISCUSSION

Definition and Causes

The concentration of runoff under some circumstances causes waterways to incise and form gullies. The identifying characteristic of an actively eroding gully is an erosional scarp, usually steep-sided and several feet high. Most gullies have cross sections much larger than needed to convey normal flood flows. Gullies can form continuous or intermittent channels. They may occur on the perimeter of upland fields and actively advance into the fields. In the southeastern States, many gullies have eroded nearly to the drainage divide, and erosion is sustained only by rainfall, runoff, and associated weathering forces on the gully itself. By contrast, many valley head gullies of the Great Plains and western United States are located on ephemeral streams that drain large areas. For convenience, an arbitrary separation between gullies and eroding channels sometimes is based on catchment area (6); gullies are limited under this scheme to drainage areas smaller than 1 mi^2.

Runoff is a necessary, but not always sufficient, condition for gully erosion. The size and growth of these incised upland channels often are influenced as much by the progressive erosion of downstream drainageways as by the water flowing through them. In some regions, runoff may describe a basic scouring rate. In localities with the most serious gullying problems, however, a surcharge above this basic rate can be triggered by a random vegetative or soil instability, change in flow gradient or downstream base level, moisture-induced change in soil stress-strength relationships, other weathering forces, or infrequent superstorms. Catastrophic erosion occurrences that initiate an entirely new drainage system or erosion regime probably never can be programmed, especially if they are triggered by an extreme storm or random turbulence. If the "trigger" is a change in gradient or incisement into weakened strata, the consequences might be predictable.

1/ Hydraulic engineer, USDA-SEA-AR, Columbia, Mo., and soil scientist, USDA-SEA-AR, Sedimentation Laboratory, Oxford, Miss.

Measurement, Rates, and Processes

Gully erosion has been measured by using conventional and photogrammetric surveys (1, 3, 13), erosion stakes (17), sampled outflow runoff from a gully (10), or gully inflow to a reservoir (5). Erosion changes commonly are expressed in terms of areal or lineal gully extension or the quantity of soil moved downstream.

Palmer (8) quantified gully erosion rates at several sites in the Connecticut River Basin. At one gully with a 0.75-acre drainage area, an average 246 cubic yards of soil were eroded each year of a 3-yr period, with three-fourths of the total occurring in a single year, 1960. Near Fargo, N. Dak. a 20-in rainfall occurred during June 28-30, 1975 on a single 5-mi^2 watershed, carving a gully one-half mile long where none previously had existed (14), removing 230,000 yd^3 of soil.

Using a grid of erosion stakes on upland gullies in Mississippi, Woodburn (17) found that rainfall and resulting slope wash caused severe gully erosion, averaging 300 t/acre/yr. In a Nebraska study (4), more than 98% (111 acre-ft) of the sediment was eroded from a gully reach in Dry Creek, Frontier County during the first year of the April 1951-April 1956 measurement period.

Typically then, gully erosion rates are highly variable, and have been related to factors both intrinsic to the watershed and drainage system, and superimposed thereon. Intrinsic factors can include profile adjustments in a drainageway and depth and erodibility of soils and subsoils along the flow boundary. Superimposed (extrinsic) factors include climatic and or manmade disturbances, or both. Both types of stimuli, whether external or internal to the drainage system, are usually discontinuous. Yet even a progressive erosion response in a system eventually can precipitate an abrupt change in gullying level if geomorphic thresholds are exceeded (11).

A study of individual storms on a 1.5-acre gully in Oklahoma (16) shows that runoff contributed by the entire catchment was the most significant factor contributing to gully sediment yields. Small storms that produce runoff from only the 1.5-acre gullied area accounted for only a small percentage of the total sediment yield. These erode the gully banks but cannot transport the material out of the system. The material is then available for transport by larger runoff events that originate from the watershed area outside of the gully. Freeze-thaw and wet-dry cycles also produce soil debris that is transported easily. The size of the storm and when it occurs in the sequence of events determine the amount of sediment produced. This explains why runoff, even though one of the better prediction variables, does not account for more of the variation in gully sediment yield. The gully occupies 22% of the watershed area and contributes over 90% of the total sediment yield. Annual gully yields averaged 115 t/acre, with a range of 27 to 284 t/acre. A gully on an adjacent area occupies 2% of the watershed and contributes over 50% of the total sediment yield. Annual yields from this gully have averaged 261 t/acre with a range of 46 to 669 t/acre.

In western Iowa, gully erosion rates were measured by sampling streamflow at locations a short distance above and below each eroding reach of gully being studied. These samples defined within-storm and between-storm differences in gully erosion rates and some processes involved (10). On a storm basis, for example, the gully erosion rate was weakly but significantly (in the statistical sense) correlated with runoff volume. Within-storm movement of soil from the gully varied somewhat with the discharge hydrograph but was chiefly a function of the supply of soil debris derived from slumping banks and headcut. Therefore, movement of sediment often was uncorrelated with runoff rate or runoff tractive force. More specific conclusions were:

(1) The quantity of gully material removed during a runoff event is a function of storm size and prior storm occurrence.

(2) Although loosely correlated with runoff, about 80% of the soil transported from this Iowa gully during a 10-yr period occurred during the months of May and June. May-June storm runoff and rainfall during this time was about 55 and 33%, respectively, of the total.

(3) Quantities of soil removed from gullies, when adjusted for storm runoff size, were greatest for the first spring storm and were reduced greatly for subsequent events.

A sampling of streamflow gives the pattern of soil movement from a gully and defines the cycle of soil debris cleanout by runoff versus renewed weathering processes that supply erodible debris. It only identifies indirectly, however, the instability (supply) processes along the gully boundary that sustain the cycle. Weathering actions during autumn and winter periods of low runoff, seepage forces, and aspects of bank stability must be considered. Some observations are:

(1) Weathering of gully boundaries by frost and freeze-thaw, wet-dry cycles directly increases the supply of gully debris available for transport, although little direct evidence is available. Flaking or spalling of soil from gully banks and some tension cracking along and above the banks and above headcuts (7) have been observed.

(2) Bank wetting or saturation, bank seepage from groundwater, and influent channel seepage from receding runoff have all contributed to gully erosion.

Bank stability is maintained only when the resisting forces within a soil mass exceed the driving forces. Raising the phreatic surface within the potential failure mass decreases this force ratio by: (1) increasing the unit weight of the soil mass, (2) decreasing the shearing strength, and (3) increasing the seepage forces.

Bradford and Piest (2) showed that gravity loading of soil banks and reduction of strength of soil bank by saturation (or reduced effective capillary tension in unsaturated soil) can cause failure of gully banks and headscarps.

In a 10-yr special study (9) whereby subsurface flows and seepage levels along gully banks were increased greatly by the construction of terraces, some deterioration of the banks was observed. The terraces, however, apparently decreased surface runoff rates to levels incapable of channel cleanout, and vegetation became established on weathered debris.

(3) Similar sequences of upstream gully extension and lateral gully enlargement may occur in different regions, but may involve different forces. For example, head cutting in normally dry gullies in Medicine Creek Basin, southwest Nebraska, is enhanced greatly by plunge pool action that undercuts the head cut and causes overburden slumping (3). In western Iowa, the plunge pool turbulence is minimal, but similar head-cutting sequences are attributable to the seepage present in nearly all headcuts, with overburden slumping due to seepage undercutting; gravity failure; and, to some extent, runoff. Many of these mechanisms of head-cutting, along with tension cracking and piping, have been applicable to the southern Piedmont region.

(4) Secondary headscarps form and advance upstream in many incised drainageways. This initial deepening may be a narrow slot that causes bank instabilities from tension cracking. Subsequent runoff widens the channel. Instrumentation on a small drainageway in western Iowa has shown a 6-in streamward displacement of gully banks with scarp passage.

Predicting Gully Erosion

Quantitative information on gully erosion rates have resulted in some empirical prediction procedures. For a severely gullied loessial area of western Iowa, the relationship for predicting gully growth is (1):

$$X_1 = 0.01X_4^{0.0982} \quad X_6^{-0.044} \quad X_8^{0.7954} \quad X_{14}^{-0.2473} \quad e^{-0.0360X_3} \qquad [1]$$

in which X_1 = average annual growth in gully surface area in acres for a given time period; X_3 = deviation of annual precipitation from normal in inches; X_4 = index of surface runoff in inches; X_6 = terraced area of watershed in acres; X_8 = gully length at beginning of period in feet; and X_{14} = length from end of gully to watersheddivide in feet. Some variables, for example X_6, are not universally applicable ($X_1 = 0$ when $X_6 = 0$).

Equations such as [1] illustrate some problems in modeling gully erosion when the state-of-the-art is presently inadequate to define causative processes. These five variables were the most significant of more than a dozen tested. Using a realistic range of values for each variable shows that beginning gully length (X_8) and distance from gully heads to divide (X_{14}) are the dominant indicators of gully growth. These indicators are somewhat arbitrary, however, and only tell us that longer existing gullies and shorter distances from gully head to watershed divide are associated with greater gully erosion rates. In effect, these two variables are related to historic gully growth, and this meager information is used in predicting future erosion rates; the causative processes are not defined.

Variable X_3, which measures the deviation of annual precipitation from normal, shows gullying rates to be higher in dry years than in wet years. For example, gully growth when annual rainfall was 5 in above normal would be only 70% of that for a year with a 5-in annual deficit. This effect is rationalized because extremely low precipitation reduces vegetation density, which in turn, increases runoff. Years of high rainfall would create vegetative growth, thereby reducing storm runoff. If correctly portrayed by the rainfall term of equation [1], this effect apparently overrides such considerations as (1) high rainfall quantities that increase soil moisture near gully boundaries and thereby lower the soil shearing resistance, causing bank caving and scarp movement and (2) high rainfall levels that increase runoff rates to erosive levels.

Gullying processes are not described adequately in equation [1] (or any other relationship derived to date), and it is not possible to determine its range of applicability. Equation [1] is very likely not applicable to wetter climates where the vegetative balance is less critical than in western Iowa or where the runoff is generated from different cropland or management systems.

Although gullying processes are complex, developing a generalized predictive model is impossible unless it is process based. In effect, we need to characterize this complexity. Studies are underway to advance our understanding of surficial stratigraphy along drainageways and to better describe the interaction with hydraulic forces of runoff and soil water.

Another field study of 210 gullies in six scattered locations throughout the eastern United States has resulted in the tentative relationship (14):

$$R = 0.14\ A^{0.49}\ S^{0.14}\ P^{0.74}\ E^{1.00} \qquad [2]$$

in which R = average annual gully head advance in feet; A = drainage area in acres; S = slope of approach channel as a percentage; P = annual summation of rainfall in inches from rains equal to or greater than 0.5 in/24 hr; and E = clay content of eroding soil profile as a percentage by weight.

Seginer (12) found that gully erosion has been related most prominently to size of drainage basin or some intercorrelated variable. He suggested, therefore, that the gully erosion problems of a locality can be evaluated from an equation of the form

$$R = CA^{b} \qquad [3]$$

in which R = average annual lineal gully advance determined by historic or geologic study or calculated; A = area of drainage basin; and C and b = constants. Since the head cuts of all continuous gullies in a drainage system reach their present position by migration from a common origin, C is the same for all gullies of the system. The relative gully advance rate then depends upon b. If b = 0, the average advance rate of all gullies would be the same; nonzero values of b would show the influence of watershed size, which in turn represents the integrated effect of many primary hydrologic variables of the region. The b variation between regions and between watersheds of a given region also could reveal the effect of soils, topography, land use, and management. This procedure would allow conservation planners to place proper emphasis on problem areas.

Field design problems involving gullying were examined by the Soil Conservation Service (15) using the equation

$$R = 1.5\ A^{0.46}\ P^{0.20} \qquad [4]$$

in which R and A are defined as in equation 2 and P = the summation of 24-hr rainfall totals of 0.5 in or more occurring during the time period converted to an average annual basis in inches.

The Soil Conservation Service recognizes that other inadequately defined factors influence the headward advance of gullies, and prefers to account for these factors by using past gullying rates calculated from maps or aerial photos. The actual prediction equation is then

$$R_2 = R_1 \left(\frac{A_2}{A_1}\right)^{0.46} \left(\frac{P_2}{P_1}\right)^{0.20} \qquad [5]$$

in which the subscripts 1 and 2 refer to the past and future, respectively.

The severity of gullying is often a function of variables common to a given region (general topography, depth and character of soils, cropping and climate) as well as site-specific characteristics. Therefore, a knowledge of the overall gully erosion problem of a region can help evaluate a given erosion hazard. Figure 1 shows average gullying rates for Missouri based on measurements available to the Soil Conservation Service. Rates are normalized to the sediment yield to be expected from drainage areas of 1 mi^2. If a single 64-acre field is gullied per mi^2 of land area, on the average, for example, the effective gullying rate for the region is 10 times the value cited in figure 1.

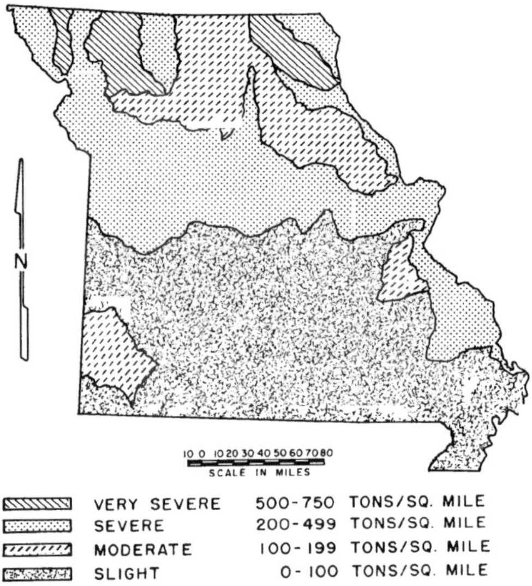

Figure 1.--Missouri gully problem areas.

REFERENCES

(1) Beer, C. E., and H. P. Johnson.
1965. Factors related to gully growth in the deep loess area of western Iowa. In U.S. Department of Agriculture, Miscellaneous Publication No. 970, pp. 37-43.

(2) Bradford, J. M., and R. F. Piest.
1977. Gully wall stability in loess derived alluvium. Journal of Soil Science Society of America 41(1):115-122.

(3) Brice, J. C.
1966. Erosion and deposition in the loess-mantled Great Plains, Medicine Creek Drainage Basin, Nebraska. U.S. Geological Survey Professional Paper 352H, pp. 255-339.

(4) Dvorak, V. I., and H. G. Heinemann.
1967. Cooperative runoff and sediment investigations on Medicine Creek Watershed in Nebraska. U.S. Department of Agriculture, Agricultural Research Service, ARS 41-130. (Series discontinued; Agricultural Research Service now Science and Education Administration-Agricultural Research.)

(5) Glymph, L. M.
1951. Relation of sedimentation to accelerated erosion in the Missouri River Basin. U.S. Department of Agriculture, Soil Conservation Service, SCS-TP-102.

(6) House of Representatives, 91st Congress, Committee on Public Works.
1971. A study of streambank erosion in the United States. Report of the Chief of Engineers to the Secretary of the Army.

(7) Ireland, H. A., C. F. Sharp, and D. H. Eargle.
1939. Principles of gully erosion in the Piedmont of South Carolina. U.S. Department of Agriculture Technical Bulletin 633.

(8) Palmer, R. S.
1965. Causes, control, and prevention of gullies. U.S. Department of Agriculture Conservation Research Report No. 2.

(9) Piest, R. F., J. M. Bradford, and R. G. Spomer.
1975. Mechanisms of erosion and sediment movement from gullies. Proceedings of the Sediment Yield Workshop, Oxford, Mississippi, Nov. 28-30, 1972. U.S. Department of Agriculture, Agricultural Research Service, ARS-S-40, pp. 162-176. (Series discontinued; Agricultural Research Service now Science and Education Administration-Agricultural Research.)

(10) ------, J. M. Bradford, and G. M. Wyatt.
1974. Soil erosion and sediment transport from gullies. American Society of Civil Engineers, Journal of the Hydraulics Division 101 (HY1):65-80, Proceedings Paper No. 11069.

1) Schumm, S. A.
1974. Geomorphic thresholds and complex response of drainage systems. In Proceedings of the Fourth Annual Geomorphology Symposium, Binghamton, New York, pp. 229-310.

2) Seginer, Ido.
1966. Gully development and sediment yield. Research Report No. 13, Israel Ministry of Agriculture, Soil Conservation Service.

3) Spomer, R. G., R. F. Piest, R. L. Poggensee, and J. A. Brophy.
1979. Erosion on the Sheyenne River Delta, North Dakota. Presented at the American Society of Civil Engineers, Irrigation and Drainage Division Specialty Conference, Albuquerque, New Mexico, July 17-21, 1979.

4) Thompson, J. R.
1962. Quantititative effect of watershed variables on the rate of gully head advancement. Paper No. 62-1713, presented at the American Society of Agricultural Engineers Meeting, Chicago, Illinois.

5) U.S. Department of Agriculture, Soil Conservation Service.
1966. Procedures for determining rates of land damage, land depreciation, and volume of sediment produced by gully erosion. U.S. Department of Agriculture, Soil Conservation Service Technical Release No. 32, Geology.

6) Welch, N. H., and D. J. Galindo.
1977. Effect of rainfall and runoff on the sediment yield from a gully. Abstract, American Society of Agronomy, p. 40.

7) Woodburn, R.
1949. Science studies on a gully. Journal of Soil Conservation vol. 15, pp. 11-13, 22.

Chapter 10. SEDIMENT TRANSPORT CAPACITY OF OVERLAND FLOW

W. H. Neibling and G. R. Foster[1/]

Sediment yield from many farm fields is limited by deposition on concave slopes (4). Deposition occurs when the sediment available for transport exceeds the transport capacity of the flow. Therefore, a relationship for sediment transport capacity of overland flow is a key component of a model designed to estimate sediment yield from agricultural fields.

Although extensive literature exists on sediment transport by streamflow (5), little information is available on sediment transport by overland flow. No widely accepted transport equations have been developed for overland flow. Since data are unavailable to develop an equation specifically for overland flow, streamflow sediment transport equations often are used (3, 7, 9). Such equations have not been validated, however, with data from overland flow transport-deposition experiments.

During development of the erosion/sediment yield component for a field scale model for Chemicals, Runoff, and Erosion from Agricultural Management Systems (CREAMS) (vol. I, ch. 3), several streamflow sediment transport equations were examined to determine how well each fit observed overland flow sediment transport data when the equations and their parameters were used directly, without modification or calibration.

EQUATIONS AND DATA

The equations considered in this analysis were Ackers-White, Engelund-Hansen, Laursen, Yang, Yalin, Meyer-Peter and Muller, and Bagnold. They are described in more detail in ch. 5. The last three are bedload equations, while the first four are total load equations. Transport values were computed with the equations using a program supplied by C. V. Alonso, USDA-Sedimentation Laboratory, Oxford, Miss.

The data sets used in the evaluation are identified in table 1. The data are from continuously curving concave slopes where the transport capacity decreased along the slope until sediment available for transport exceeded transport capacity, causing deposition on the toe of the slope. Development of the deposition profile is described in more detail by Foster and Huggins (2). Data sets 1 through 5 were from laboratory studies on a bed 3 m long by 0.6 m wide, which varied continuously in steepness from 20% at the upper end to 1% or less at the lower end (1, 2, 6). Sediment and baseflow were added at controlled rates at the upper end of the bed. Simulated rainfall at 35 mm/h was applied

1/ Agricultural Engineer and hydraulic engineer, respectively, USDA-SEA-AR, W. Lafayette, Ind.

Table 1.--Data sets used to evaluate equations for sediment transport by overland flow

Data set number	Test material	Specific gravity	Tests	Data source1/
			(number)	
1	0.342 mm sand	2.64	22	(2, 6)
2	.265 mm sand	2.64	40	Unpublished.
3	.342 mm sand	2.64	42	(1)
4	.342 mm crushed coal	1.60	31	(1)
5	.156 mm crushed coal	1.67	38	(1)
6	Soil aggregates (>2 mm to <.002 mm)	1.80 2/	3	Unpublished.

1/ Numbers under "Data source" refer to references at the end of this chapter.

2/ Specific gravity for this data set was assumed. It was measured for other data sets.

along the slope in about half the tests. Sediment discharge rate and flow velocity were measured during each test. Data set 6 was from a field study using concave plots 7.0, 8.8, and 10.7 m long by 1.8 m wide on a field soil exposed to 64 mm/h simulated rainfall. The sediment source in this study was from erosion by raindrop impact and runoff on the upper 0 to 7.0 m portion of the plots. The steepness of the 10.7 m plot ranged from 18% at the top to 0% at the bottom. The shape of the other plots was the same as that for the 10.7 m plot except they were terminated at 7.0 and 8.8 m from the top to measure sediment loads at different locations in the deposition area.

The initial slope shape for all data sets was a continuously curving concave surface. During each run, deposition began along the slightly sloping lower section of the bed and then advanced upslope. As deposition continued, the slope of the surface of the deposited material increased, producing a corresponding increase in the transport capacity of the flow passing over the deposited material. Data set 1 was from experiments where deposition was allowed to continue until a uniform slope was formed, with steepness such that transport capacity equaled sediment feed rate (6). In the remaining data sets (1, 2), sediment discharge and profile elevations were recorded before an equilibrium profile was reached and deposition ceased. The transport capacity at the end of the slope was assumed to equal the measured sediment load, that is, no lag occurred between sediment capacity and sediment load (1). This assumption seems valid except for the fine particles in the sediment for data set 6.

For data sets 2 through 5, the slope of the deposited sediment profile over the last 0.3 m at the end of the slope was used to compute flow hydraulic properties required by the transport equations. The slope of the deposition profile at equilibrium was used in computations for data set 1. The flow was laminar, based on Reynolds number, and was assumed to be uniform, that is, friction slope equal to the slope of the deposition profile.

A coefficient of 16, instead of 24, was used in the laminar friction factor - Reynolds number relationship for data sets 1 through 5 to match computed velocities with observed values. A coefficient of 40 for the rougher field plots was required for data set 6. Shear stress was computed using standard equations for uniform, laminar flow.

Since the size range of the particles in data sets 1 through 5 was narrow (1), uniform sized particles were assumed. The measured particle distribution for data set 6 is shown in table 2.

Table 2.--Influence of deposition on a concave slope on the particle size distribution of the sediment

Plot length and slope at end	Particle size distribution of sediment (microns)							
	>2,000	2,000-1,000	1,000-500	500-210	210-50	50-35	35-2	<2
	(%)							
7.0m - 6% 1/	7.6	18.8	16.8	9.3	4.8	3.4	34.0	5.3
8.8m - 3%	6.9	14.0	11.6	7.3	5.1	4.1	44.1	6.9
10.7m - 0%	1.0	1.4	1.0	0.8	1.4	2.3	82.1	10.0

1/ Approximate location where deposition began.

Note: Primary particle distribution of original soil: sand - 19%, silt - 55%, clay - 26% (USDA classification).

Sediment transport was computed in two ways for data set 6: (1) A D50 value was assumed and (2) transport capacity computed for each particle size class was multiplied by the fraction of the total sediment load contained in the class. These products then were summed to obtain a total transport capacity.

TEST RESULTS

Transport capacity was computed for each test and compared with the measured sediment transport for the test. Five error statistics were computed for each data set and are given in table 3. Nothing is shown for the Ackers-White equation because it "blew up" for particles of 0.020 mm diameter and smaller with a 1.8 specific gravity. Particles as small as 0.002 mm were considered in the second method for analyzing data set 6.

The percentage of tests in each data set where the ratio of calculated to observed sediment discharge fell between 0.75 and 1.5 is shown in table 4. Both the Yalin and Meyer-Peter Muller equations fit the data fairly well, with the Yalin equation giving a better overall fit when considering all six data sets. Ratios for the three soil aggregate tests were within the 0.75 to 1.5

Table 3.--Statistics used to evaluate performance for the 6 equations tested

Equation	$\frac{100\ \Sigma(\lvert O-P\rvert/O)}{n}$ 1/	$\frac{[\Sigma(O-P)^2]^{1/2}}{n}$	$\frac{[\Sigma(\ln O - \ln P)^2]^{1/2}}{n}$	$\frac{\Sigma\lvert O - P\rvert}{n}$	$\frac{\Sigma(O/P)}{n}$
		Statistic			
		0.342 mm sand, 22 tests			
Engelund	73.346	0.597	1.597	0.469	0.247
Laursen	1,075.645	9.026	2.297	6.461	11.756
Yalin	34.739	.265	.529	.213	.656
Yang	78.402	.633	1.542	.408	1.023
Meyer	222.008	1.926	1.136	1.233	3.053
Bagnold	44.548	.419	.662	.308	.555
		0.265 mm sand, 40 tests			
Engelund	53.909	.404	.997	.310	.476
Laursen	1,647.730	17.108	2.696	11.167	17.477
Yalin	29.899	.197	.360	.152	.983
Yang	265.450	3.399	1.306	1.940	2.995
Meyer	390.849	3.742	1.514	2.460	4.908
Bagnold	47.178	3.742	1.514	2.460	4.908
		0.342 mm sand, 42 tests			
Engelund	82.853	.535	1.870	.449	.171
Laursen	415.828	2.159	1.625	1.898	5.158
Yalin	37.721	.247	.587	.192	.627
Yang	98.636	.448	2.086	.399	.014
Meyer	26.112	.153	.278	.119	1.109
Bagnold	42.688	.337	.626	.255	.630
		0.342 mm coal, 31 tests			
Engelund	81.871	.578	1.736	.528	.181
Laursen	533.731	3.530	1.841	3.268	6.337
Yalin	19.463	.122	.222	.110	.995
Yang	81.647	.539	1.829	.506	.184
Meyer	25.007	.132	.263	.116	1.230
Bagnold	47.505	.397	.741	.339	.525
		0.156 mm coal, 38 tests			
Engelund	58.518	.434	.961	.382	.415
Laursen	829.721	4.913	2.211	4.661	9.297
Yalin	36.485	.188	.371	.160	1.276
Yang	72.597	.505	.573	.422	1.694
Meyer	76.302	.377	.601	.345	1.758
Bagnold	75.950	.538	1.523	.483	.241
		Soil aggregates, 3 tests, D50 size 2/			
Engelund	57.251	.503	.604	.447	1.205
Laursen	87,681.050	741.395	6.131	551.583	877.810
Yalin	94.815	1.342	.721	.893	1.847
Yang	1.500E06	15,564.031	8.114	10,291.037	14,998.399
Meyer	44.174	.349	.369	.320	1.442
Bagnold	76.157	.528	3.672	.490	.238
		Soil aggregates, 3 tests, 7 sizes 3/			
Engelund	4,920.671	3.186	3.753	2.584	50.207
Laursen	1.879E06	1,467.874	9.281	1,005.365	18,790.018
Yalin	2,438.242	1.621	3.061	1.288	25.382
Yang	1.188E14	9.602E09	24.647	6.405E09	1.188E12
Meyer	4,258.997	3.065	3.510	2.249	43.590
Bagnold	348.888	.271	1.553	.186	4.017

1/ Where O = observed sediment discharge, P = predicted sediment discharge, n = number of tests.

2/ Naturally eroded soil aggregates. The D50 value for the aggregates was used in the sediment transport equations.

3/ Naturally eroded soil aggregates (same data as above). The representative diameter measured for each size class was used in the equations. Transport capacity was calculated for each of the 7 size classes measured and was multiplied by the fraction of each size class. The results were summed to obtain total transport capacity.

range only for the Yalin, Meyer-Peter Muller, and Engelund and Hansen equations. Two in three ratios fell within the specified range for both Yalin and Meyer-Peter Muller.

Lower density coal and soil aggregates in data sets 4 to 6 generally should be more representative of sediment transported from agricultural fields than the higher density sands used in data sets 1 through 3. Both data sets 1 through 3 and 4 through 6 suggest, however, that the Yalin equation best fit the observed data, followed by the Meyer-Peter Muller equation. The Laursen and Yang equations were unacceptable. While the Yang equation gave reasonable results for some tests, it was totally unrealistic for small aggregates.

Figures 1 through 5 show that computed transport rates from the Yalin equation generally fit well. This equation did not have the curvature of the observed data, which could cause error for shear stresses larger than those in this analysis. Tables 3 and 4 and figures 1 through 5 suggest that the Yalin equation, applied to a broad range of particle sizes with aggregate densities characteristic of agricultural soils, will produce more reasonable results for the entire size and density range of particles studied than any of the other six equations.

Table 4.--Percentage of occurrences where the ratio of calculated to observed sediment discharge was between 0.75 and 1.5

Test material	Specific gravity	Tests	Method					
			Engelund	Laursen	Yalin	Yang	Meyer	Bagnold
		(number)	------	------	(%)	------	------	------
0.342 mm sand-----	2.64	22	0	0	23	32	10	10
.265 mm sand-----	2.64	40	10	0	65	23	5	15
.342 mm sand-----	2.64	42	0	0	25	0	78	13
.342 mm coal-----	1.60	31	0	0	87	0	87	16
.156 mm coal-----	1.67	38	3	0	71	29	45	0
Soil aggregates (D50 approach)--	1.8	3	33	0	67	0	67	0
Soil aggregates (7 size classes)--	1.8	3	0	0	67	0	67	0

CONCLUSION

The Yalin equatiom can be used "as is" to compute sediment transport capacities for overland flow. This equation gives reasonable results for the range of sediment sizes and densities characteristic of agricultural fields, in contrast to some streamflow sediment transport equations that "blow up" when extended to overland flow conditions. The Yalin equation is recommended for computing sediment trasport capacity for both overland flow and concentrated flow in channels and waterways on agricultural fields. An equation developed specifically for these conditions is still needed from research, however.

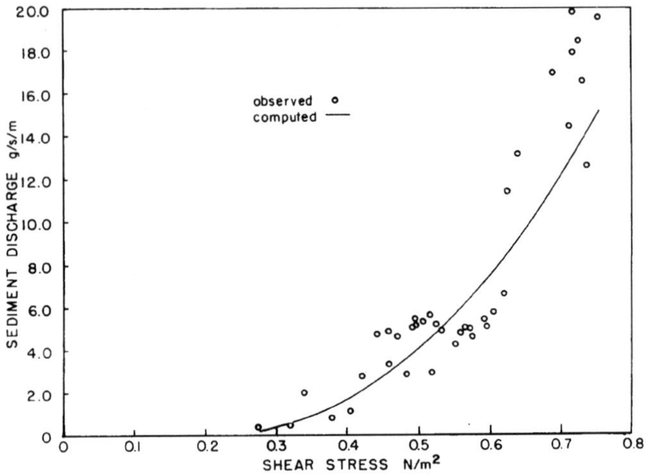

Figure 1.--Yalin equation, 0.265 mm diameter sand, data set No. 2.

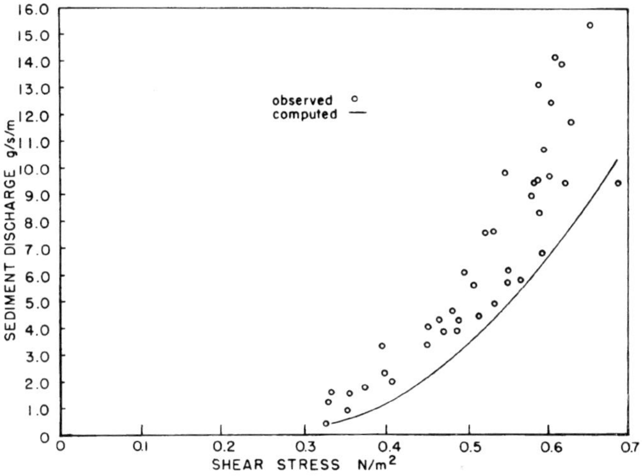

Figure 2.--Yalin equation, 0.342 mm diameter sand, data set No. 3.

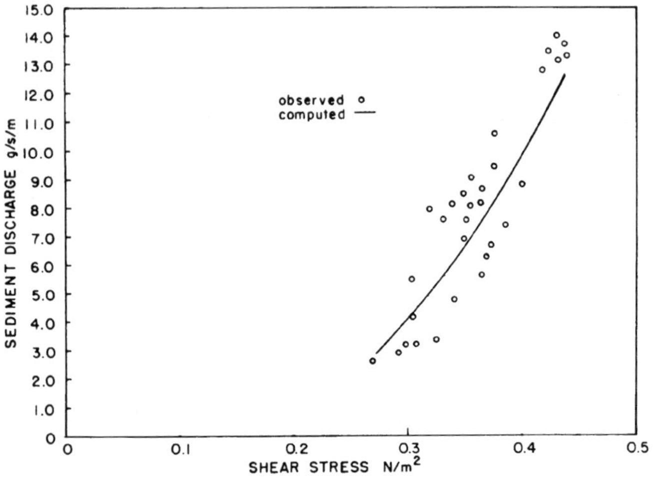

Figure 3.--Yalin equation, 0.342 mm diameter coal, data set No. 4.

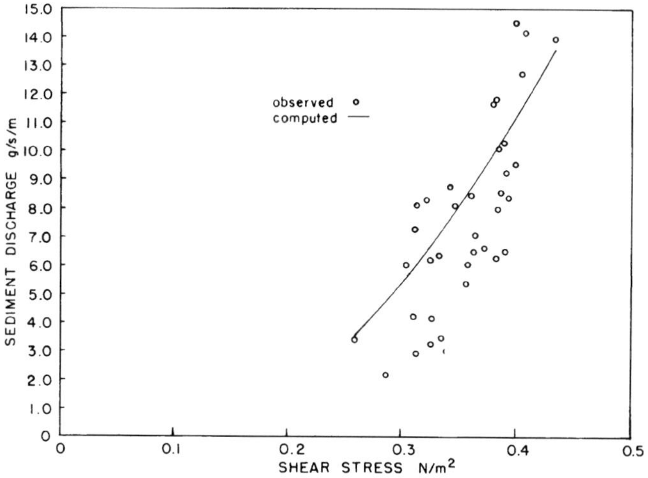

Figure 4.--Yalin equation, 0.156 mm diameter coal, data set No. 5.

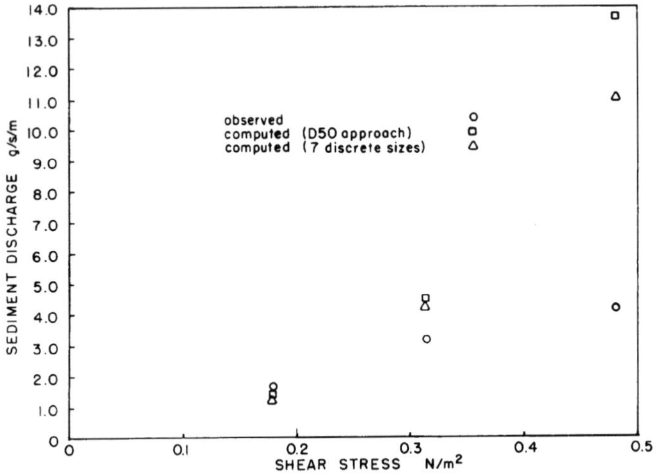

Figure 5.--Yalin equation, naturally eroded soil aggregates, <0.002 to >2.0 mm, data set No. 6.

REFERENCES

(1) Davis, S. S.
1978. Deposition of uniform sediment by overland flow on concave slopes. 137 pp. (Unpublished Master's thesis) Purdue University, Agricultural Engineering Department, W. Lafayette, Indiana.

(2) Foster, G. R., and L. F. Huggins.
1977. Deposition of sediment by overland flow on concave slopes. In Soil Erosion: Prediction and Control. Soil Conservation Society of America. Ankeny, Iowa. pp. 201-210.

(3) ------, and L. D. Meyer.
1972. Transport of soil particles by shallow flow. Transactions of the American Society of Agricultural Engineers 15(1):99-102.

(4) ------, and L. D. Meyer.
1977. Soil erosion and sedimentation. In Proceedings of the National Symposium on Soil Erosion and Sedimentation by Water. American Society of Agricultural Engineers, St. Joseph, Michigan. pp. 1-10.

(5) Graf, W. H.
1971. Hydraulics of sediment transport. 544 pp. McGraw-Hill Book Company, New York, New York.

(6) Karakaplan, S., L. F. Huggins, and G. R. Foster.
1976. Effect of degree of curvature, slope, flow rate, sediment load, sand size and rainfall intensity on deposition and sediment yield. (Unpublished report) Purdue University, Agricultural Engineering Department, W. Lafayette, Indiana.

(7) Li, R. M.
1977. Water and sediment routing from watersheds. In Proceedings of the River Mechanics Institute. Colorado State University, Fort Collins, Colorado. Chapter 9.

(8) Meyer, L. D. and W. H. Wischmeier.
1969. Mathematical simulation of the process of soil erosion by water. Transactions of the American Society of Agricultural Engineers 12(6): 754-758, 762.

(9) Smith, R. E.
1977. Field test of a distributed watershed erosion/sedimentation model. In Soil Erosion: Prediction and Control. Soil Conservation Society of America, Ankeny, Iowa. pp. 201-209.

Chapter 11. CONCENTRATED FLOW RELATIONSHIPS

L. J. Lane and G. R. Foster[1/]

The concentrated flow relationships discussed here are limited to upland areas typical of farm-field situations. Representative situations include (1) erosional channel development in areas of fields where flow is concentrated, such as stream headwaters and terrace channels, (2) small channels as permanent features of the landscape which normally are "tilled over" during cultivation, and (3) temporary channels developing when rows or terraces overtop and flow proceeds cross-contour to the field edge. Specifically excluded are permanent stream channels and active gully systems of a scale larger than described previously. Such large-scale features as permanent stream channels are more in the range of a basin-scale model than the field-scale model. Gully systems are beyond the scope of current field-scale modeling due to our lack of understanding of gully dynamics. An exception to the channel size limitations is developing a final or equilibrium channel width. Relationships developed for field-sized channels apply to larger channels as well.

Hydrologic input to the channel system consists of overland flow hydrographs or volume and peak rate of runoff and a duration of runoff. In the latter case, a characteristic discharge and a time distribution must be chosen with the specific peak, volume, and duration. In this analysis, we assume the peak rate as a characteristic discharge and assume that the temporal distribution of shear stress in the channel is triangular.

SPATIALLY VARIED FLOW

Given the hydrologic input to the channel system, the next step was to compute friction slopes, and, thus, depths of flow and shear stress at each position along a channel segment. The routing equations were simplified by using the peak or characteristic discharge with the assumption of steady flow. This eliminated the unsteady flow equations, leaving spatial variability resulting in steady but spatially varied flow.

In many field-sized flow situations, flow is concentrated in a channel, producing spatially varied flow along the channel length. As described by Chow (2), the dynamic equation for spatially varied flow with increasing discharge is

---

1/ The authors are, respectively, hydrologist, USDA-SEA-AR, Southwest Rangeland Watershed Research Center, Tucson, Ariz.; and hydraulic engineer, USDA-SEA-AR, Purdue University, West Lafayette, Ind.

$$\frac{dy}{dx} = \frac{S_o - S_f - 2\alpha\, Qq_*/gA^2}{1 - Q^2/gA^2\, D} \tag{1}$$

where:

$\frac{dy}{dx}$ = slope of water surface,
S_o = bed slope,
S_f = friction slope,
α = energy coefficient,
Q = discharge at point of interest,
q_* = lateral inflow per unit length of channel,
A = cross-sectional area,
D = hydraulic depth, and
g = acceleration due to gravity.

To avoid solving equation 1 for each runoff event, it was solved under a variety of conditions, and regression equations were used to approximate the solutions (Vol. I, Ch. 3). Given a representative channel and flow conditions, regression equations were derived for the friction slope (approximately the slope of the energy gradeline) as a function of position along the channel. Given the friction slope, $S_f(x)$, the average shear stress along the channel is then

$$\tau(x) = \gamma\, R(x)\, S_f\,(x) \tag{2}$$

where:

$\tau(x)$ = shear stress, force per unit area,
γ = specific gravity of water, and
$R(x)$ = hydraulic radius, length.

In many field situations, outlet controls for a channel have significant influence on sediment detachment and transport capacity. Consequently, backwater or drawdown at the outlet significantly can affect sediment yield. If the outlet rating is known (critical depth or a rating table) for example, then the friction slope at the outlet is

$$S_f = \frac{n^2\, Q^2}{2.22\, A^2\, R^{4/3}} \tag{3}$$

where n is Manning's resistance coefficient and subsequent values of friction slope at positions above the outlet are obtained from the spatially varied flow equation (eq. 1) or the approximating regression equations. This procedure is similar to computation of backwater profiles except that spatially varied flow is considered.

SEDIMENT ROUTING

Sediment load, G, is limited by the lesser of sediment available for transport and sediment transport capacity. The continuity equation for sediment assuming quasi-steady state is:

$$\frac{dG}{dx} = D_L + D_F \tag{4}$$

where:

G = sediment load per unit width per unit time,
x = distance along channel,
D_L = lateral inflow of sediment, and
D_F = detachment or deposition by flow.

Lateral inflow of sediment is from overland flow or contributing channels. The Yalin equation (3) is used to compute the transport capacity, and a critical shear stress equation is used to compute detachment capacity:

$$D = K_{ch}(1.35\,\bar{\tau} - \tau_{cr})^c \tag{5}$$

where:

D = detachment capacity,
K_{ch} = erodibility factor,
$\bar{\tau}$ = average shear stress for the cross section,
τ_{cr} = critical shear stress, and
c = an exponent.

Therefore, the shear stress, τ, obtained from equation 2 is used to compute the shear velocity in the Yalin equation, that is,

$$V_* = \sqrt{gRS_f} = (\tau/\rho)^{1/2} \tag{6}$$

and also to compute detachment capacity in equation 5. Solutions for equations 1-3 are used to derive S and τ, which are then used in equations 4-6 and the Yalin equation to compute sediment load in a channel.

CHANNEL MORPHOLOGY AND SEDIMENT YIELD

As part of the field-scale model development, a simplified channel morphology-erosion model was developed. Objectives of this simplified model included developing equations for erosion in small channels which would:

(1) Remain relatively simple with a minimum number of parameters.
(2) Incorporate what we know of channel hydraulics.
(3) Reproduce observed hydraulic geometry, that is, $W = aQ^b$.
(4) Reproduce observed relations between erosion/sediment yield and time.

Limiting factors in this development include:

(1) Assumption of steady-state discharge.
(2) Occurrence of erosion at potential rate (no deposition-erosion cy-

cles).

(3) Estimation of critical shear stress.

(4) Testing requires data for dynamic developing channel systems. Existing stable channels provide only limited testing.

With these objectives and assumptions, simplified relationships between channel features, discharge, and equilibrium or final eroded channel width were derived and tested. Testing for sediment yield with time was limited to experimental data collected by G. R. Foster, while a large number of equilibrium channel widths (quasi-equilibrium widths corresponding to the principal channel-forming discharge) were obtained from the literature.

Figure 1 schematically represents channel widths developing in homogeneous-erodible material and in material with an erosion resistant or "nonerodible" boundary. The upper part of figure 1 illustrates a channel eroding in homogeneous material, and the lower part illustrates a final eroded width when a nonerodible boundary is reached.

In figure 1, the distance from the water surface around the wetted perimeter is designated as x with the total distance as $L = W_p$, the wetted perimeter. To determine if erosion occurs, and if so, its rate, a distribution of shear stress around the wetted perimeter was represented using a logarithmic-polynomial function with the following conditions:

$$\tau(o) = \tau(L) = o \tag{7}$$

$$\tau(L/2) = \tau_{max} \tag{8}$$

$$\frac{d\tau}{dx} = \begin{cases} o & , x = L/2 \\ >o & , o \leq x < L/2 \\ <o & , L/2 < x \leq L \end{cases} \tag{9}$$

$$\bar{\tau} = \frac{1}{L}\int_0^L \tau(x)dx \tag{10}$$

where:

$\tau(x)$ = shear stress around the channel cross section,
x = distance around the wetted perimeter,
L = total length around the wetted perimeter, and

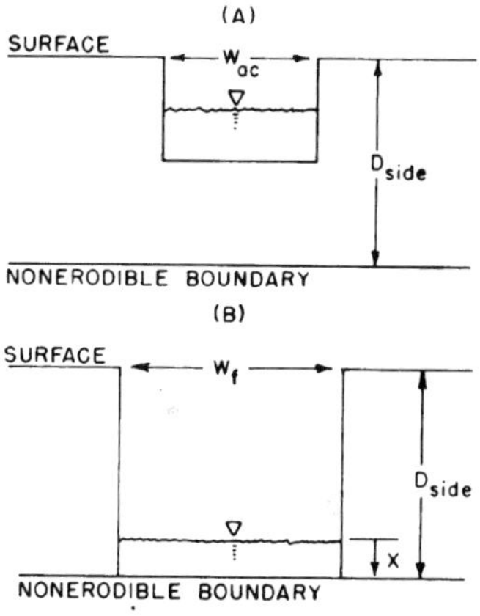

Figure 1.--Definition sketch of eroding channel for (A) width if nonerodible boundary is not reached and (B) final eroded width when nonerodible boundary is reached and channel widens.

$\bar{\tau}$ = average shear stress for the cross section.

Until the channel reaches the nonerodible boundary (fig. 1A), the channel is assumed rectangular with the following relationships between area, hydraulic radius, and depth:

$$A = Wh \tag{11}$$

$$R = \frac{Wh}{W + 2h} \tag{12}$$

Equation 12 is solved for h as

$$h = \frac{RW}{W - 2R} \tag{13}$$

so that with the Manning equation

$$Q = AV = AR^{2/3} \frac{1.49}{n} S^{1/2} \tag{14}$$

which simplifies to

$$\frac{RW^2}{W - 2R} R^{2/3} \frac{1.49}{n} S^{1/2} = Q \tag{15}$$

or

$$\frac{RW^2}{W - 2R} R^{2/3} = \frac{nQ}{1.49 \ S^{1/2}} \tag{16}$$

Now, notice that the wetted perimeter is $W_p = W + 2h$ or

$$W_p = \frac{W^2}{W - 2R} \tag{17}$$

so that equation 16 becomes

$$W_p \ R^{5/3} = \frac{nQ}{1.49 \ S^{1/2}} \tag{18}$$

If $W_\star = W/W_p$ and $R_\star = R/W_p$, then $R^{5/3} = R_\star^{5/3} W_p^{5/3}$, and equation 18 becomes

$$W_p^{8/3} R_\star^{5/3} = \frac{nQ}{1.49 \ S^{1/2}} \tag{19}$$

which is

$$\frac{W}{W_*}^{8/3} = \frac{nQ}{1.49\ S^{1/2}} \quad \frac{1}{R_*^{5/3}} \tag{20}$$

and can be solved for W as,

$$W = W_{ac} = \left(\frac{nQ}{1.49\ S^{1/2}}\right)^{3/8} \frac{W_*}{R_*^{5/8}} \tag{21}$$

In equation 21, W_{ac} is the width of the rectangular channel before the nonerodible boundary is reached. From equations 5 and 21, the erosion rate at this time is

$$E_o = W_{ac}\ K_{ch}(1.35\ \bar{\tau} - \tau_{cr})^c. \tag{22}$$

Once the channel reaches the nonerodible boundary downward movement ceases and if the shear stresses on the channel bank exceed the critical shear stress, widening begins (fig. 1B). The final eroded width, W_f, is obtained when the value of $x = x_{cf}$ is reached such that $(x) = \tau_{cr}$. Following a procedure similar to the derivation of W_{ac}, the final eroded channel width is

$$W_f = \left(\frac{nQ}{1.49\ S^{1/2}}\right)^{3/8} \left(\frac{1 - 2X_{cf}}{X_{cf}^{5/3}}\right)^{3/8} . \tag{23}$$

After the nonerodible boundary is reached, the erosion rate decreases exponentially with time, that is

$$E_o = 2K_{ch}(1.35\tau_B - \tau_{cr})^c/\gamma_{soil} \tag{24}$$

and

$$t_* = t\ E_o/(W_b - W_{ac}) \tag{25}$$

so that the erosion rate as a function of time is

$$E(t) = E_o\ D_{side}\ \gamma_{soil}\ \exp(-dt_*) \tag{26}$$

where τ_B is the shear stress at the boundary, D_{side} is the depth of soil to the nonerodible boundary, γ_{soil} is the specific weight of the soil, and α is a decay constant. Although the depth of flow in the channel is generally less than D_{side} (fig. 1B), we assume that the entire depth of soil erodes uniformly rather than by undercutting and delayed transport of the collapsed bank material.

The process described by equations 21-26 includes an initial width, W_{ac}, and a constant erosion rate, E_o, up until the time t_o when the nonerodible boundary is reached. For $t > t_o$ the width approaches W_f and $E(t)$ decreases exponentially to zero as $t \to \infty$ and $W \to W_f$. If L_c is the length of the channel segment, then the total sediment yield per unit width from the segment as $t \to \infty$ is

$$Q_s = \int_0^{t_o} L_c E_o dt + \int_{t_o}^{\infty} L_c \, E(t)dt, \tag{27}$$

which can be evaluated as

$$Q_s = L_c \, D_{side} \, \gamma_{soil} \tag{28}$$

as the sediment yield per unit width with the total sediment yield equal to

$$Q_{stot} = W_f \, L_c \, D_{side} \, \gamma_{soil} \tag{29}$$

where γ_{soil} is the specific weight of the soil.

Rill Erosion Studies

Data from a rill erosion study conducted by G. R. Foster, USDA-SEA-AR, Lafayette, Ind., were used to validate the model. The experimental procedure was to measure channel (rill) cross sections, flow variables, and sediment yield under controlled conditions. Eight replicated rills were subject to constant discharge introduced at the upstream end of 4.6 m reaches. Discharge rates ranged from 0.00021 ft^3/s to 0.0022 ft^3/s. The simplified model described previously was applied to these rills using measured discharges, velocities, and thus measured Manning's n values. Of these eight experiments, seven reached a nonerodible boundary. Computed and observed sediment yields with time are shown in figure 2. The observed data do not show an initial constant sediment yield with time as predicted by equation 22. The overall fit of the computed curves is very good, however, reproducing the decreases in sediment yield with time. Total sediment yield was given by the area under the curves in figure 2. The total observed and computed sediment yields for the seven rill experiments are shown in figure 3. In general, we slightly underpredicted total sediment yields, but the model accounted for 97% of the variance in sediment yield. As the differences between observed and computed sediment yields shown in figure 3 are well within measurement errors, the model essentially reproduced the observed data.

Width vs. Discharge Relations

In the rill erosion studies, discharge, slope, and Manning's n values were measured. To apply the model to selected discharge-width data from the literature, however, the n values must be estimated. Given the n values, the model can be used to compute final widths, W_f, which can be compared to measured values. Osterkamp (4) selected several streams in the mountains and high plains and related channel width to a characteristic discharge. To apply the model to Osterkamp's data, n values had to be estimated. In similar regions of the West, 17 streams were selected from data by Barnes (1). A regression equation relating channel slope, S, and n values was derived

$$n = 0.143\ S^{0.22} \tag{30}$$

with an R^2 value of 0.73. Equation 30 and Osterkamp's slope data were used to estimate n values for the 32 streams shown in Osterkamp's table 1 (4). These estimated n values were checked and modified when necessary.[2/]

Observed and computed data for these 32 streams and Foster's rill data are summarized in table 1. On the average, the simplified model reproduced the trend in the discharge vs. width data. Widths and discharges were then related by regression of the form

$$W = a\ Q^b \tag{31}$$

where W is channel width in feet and Q is discharge in ft^3/s. These regression results for the observed and computed widths are shown in figure 4. For the rill data, the coefficients and exponents in equation 31 are quite similar. Again these results are for small rills under controlled experimental conditions. For the natural streams, the exponent, b, was larger for the computed widths than for the observed data. The n values were estimated for these natural streams, while the assumed distribution of shear stress (eqs. 7-10) is from the rill study. In wide, natural streams, the distribution of shear stress around the channel cross-section may be "flatter" or more uniform and nearer to the average shear stress over much of the wetted perimeter. Nonetheless, the model

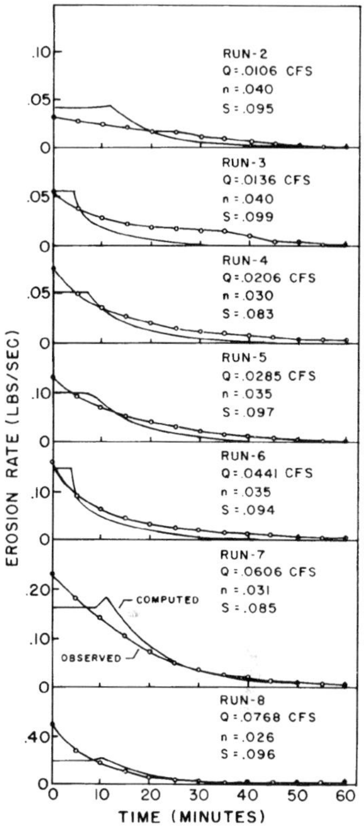

Figure 2.--Observed and computed sediment yield using unpublished rill erosion data by G. R. Foster.

2/ Jarrett, R. 1979. Personal communication

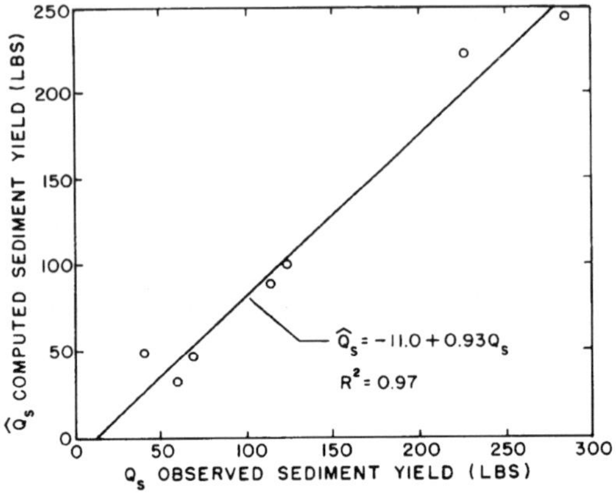

Figure 3.--Relation between observed and computed sediment yield, rill erosion study.

produced a reasonable approximation to the observed width-discharge relationship (fig. 4).

SUMMARY AND CONCLUSIONS

A series of regression equations was developed to estimate friction slope, and, thus, shear stress for spatially varied flow in triangular channels. These equations gave accurate approximations to friction slopes computed under a wide range of conditions.

Once the friction slope is computed, the Yalin sediment transport equation and a critical shear stress equation are used to compute sediment transport, deposition, and detachment for open channel flow in field situations. These equations then are used to route sediment in erodible channels.

A simplified channel morphology-sediment yield model was developed to estimate sediment yields and channel widths for channels in soil with a nonerodible boundary below the surface of the soil. The sediment yield and channel width equations were tested using data from a rill erosion study. The simplified model produced excellent estimates of sediment yield with time, total sediment yield, and final width of the developed rills. The simplified model was applied to natural channels using data from 32 streams. Although the results for the natural channels were not as accurate as for the controlled experiment,

Table 1.--Summary of observed and computed channel discharge-width relationships for natural streams and experimental rill erosion studies

Station No.	Discharge Q (cfs)	Observed width (ft)	Slope (ft/ft)	Estimated Manning's n	Computed final width (ft)
Osterkamp's Natural Streams					
H01	36.6	22.	.0152	.057	24.
H02	48.3	24.	.0158	.057	28.
H03	17.4	19.	.0216	.062	18.
H04	31.9	25.	.0083	.050	18.
H05	20.4	20.	.0179	.059	18.
H06	12.0	17.	.0082	.050	11.
H07	220.0	55.	.0170	.058	66.
H08	146.0	51.	.0190	.060	55.
H09	151.0	49.	.0118	.054	48.
H10	187.0	52.	.0143	.056	58.
H11	12.9	16.	.0238	.063	16.
H12	14.3	16.	.0091	.051	12.
H13	18.5	20.	.0309	.067	21.
H14	22.3	25.	.0123	.054	17.
H15	13.9	18.	.0280	.065	17.
H16	62.5	35.	.0240	.063	37.
H17	10.0	13.	.0082	.050	10.
H18	50.1	34.	.0141	.056	28.
H19	17.7	18.	.0089	.051	13.
H20	54.2	38.	.0357	.050	33.
23	9.0	13.	.0250	.030	9.
H22	138.0	58.	.0136	.050	45.
H23	228.0	73.	.0195	.050	64.
H24	28.1	21.	.0391	.060	26.
H25	54.2	32.	.0231	.062	34.
H26	4.6	8.	.0352	.068	10.
H27	127.0	50.	.0098	.052	41.
H28	12.9	16.	.0087	.050	11.
H29	149.0	39.	.0109	.090	63.
H30	42.2	23.	.0333	.070	34.
H31	62.2	31.	.0260	.090	46.
H32	3.5	9.	.0083	.050	5.
Foster's Rill Study					
F-2	.0106	.27	.0950	.040	.29
F-3	.0136	.37	.0990	.040	.33
F-4	.0206	.43	.0830	.030	.34
F-5	.0285	.44	.0970	.035	.47
F-6	.0441	.58	.0940	.035	.60
F-7	.0606	.68	.0850	.031	.66
F-8	.0768	.78	.0960	.026	.70

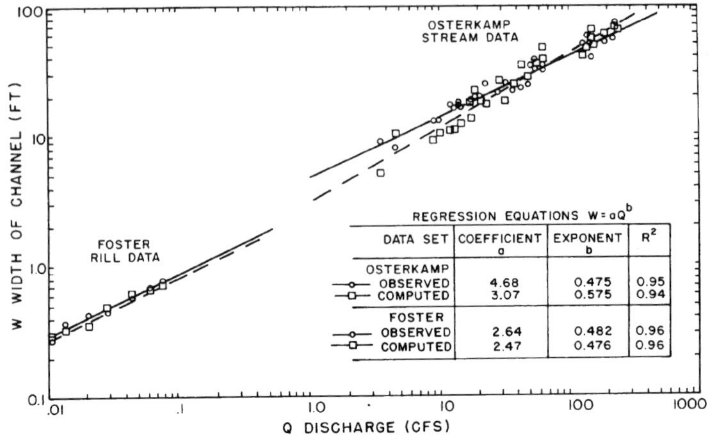

Figure 4.--Relation between channel discharge and width for natural streams and experimental rill systems.

the model predicted a width-discharge relation comparable to that for the observed data.

The concentrated flow-erosion relationships summarized here provide reasonable approximations for erosion in field-sized channels.

Additional research is needed to refine and test these relationships for small channels and to extend them for application in larger channels.

REFERENCES

(1) Barnes, H. H.
1967. Roughness characteristics of natural channels. U.S. Geological Survey Water Supply Paper 1849, 213 pp.

(2) Chow, V. T.
1959. Open-Channel Hydraulics, McGraw-Hill Book Co., Inc., New York, 680 pp.

(3) Foster, G. R., and L. D. Meyer.
1972. Transport of soil particles by shallow flow. Transactions of American Society of Agricultural Engineers 15(1):99-102.

(4) Osterkamp, W. R.
1977. Effect of channel sediment on width-discharge relations, with emphasis on streams in Kansas. Kansas Water Resources Board, Bulletin No. 21, 25 pp.

Chapter 12. ENRICHMENT RATIOS FOR WATER QUALITY MODELING

Ronald G. Menzel[1/]

Eroded soil is usually richer in nutrients than the surface soil of the watershed from which the eroded soil comes. This enrichment results from selective erosion of fine soil particles and organic residues. The degree of enrichment must be known to predict effects of soil erosion and erosion control on water quality.

Degree of enrichment is expressed as an enrichment ratio, which is the concentration of nutrient in the eroded material divided by its concentration in the soil. Enrichment ratios have been long known to increase markedly with a decrease in amount of soil eroded. Massey and Jackson (6) related the logarithm of the enrichment ratio to the logarithms of the sediment discharge and sediment concentration in runoff. They stated, however, that simple correlation coefficients with either discharge or concentration were almost as high as the multiple correlation coefficient.

In this chapter, a general relationship is developed between enrichment ratios and the amount of soil eroded. Data from different soil and climatic areas were examined to detect possible effects of such factors as soil texture, cropping pattern, or runoff amount on the relationship.

METHODS

The basic data were nitrogen, phosphorus, and sediment discharges during individual runoff events. Only the amounts of nitrogen and phosphorus associated with sediment were included in the calculations. Total Kjeldahl nitrogen and total phosphorus were used, except that Massey (5) calculated enrichment ratios with available phosphorus. The concentration of nitrogen or phosphorus in the sediment was divided by the corresponding concentration in soils from the sediment-producing area to get the enrichment ratios. In a few cases where soil analyses were unavailable, a reasonable estimate could be made by assuming no enrichment for the highest amounts of sediment discharge.

After preliminary examination of several relationships with data from Chickasha, Okla., and Riesel, Tex., the logarithmic relationship between enrichment ratios (ER) and sediment discharges (sed) was chosen for comparing all available data:

$$\ln(ER) = a + b \ln(sed). \qquad (1)$$

Massey and Jackson (6) pointed out that logarithmic transformations are applicable if the standard errors are proportional to the values of these variables

---

1/ Soil scientist, USDA-SEA-AR, Water Quality Management Laboratory, Durant, Okla.

as Massey and Jackson considered them to be. The coefficients a and b were computed by the method of least squares. For each regression equation, the correlation coefficient and probability that it would not be exceeded were computed.

RESULTS AND DISCUSSION

All enrichment ratios for which individual storm data were examined are summarized for two groups of studies in table 1. The first group includes 6 studies with at least 20 storm events and a 100-fold range of sediment discharges per event. This group should give the most reliable regression coefficients. The second group includes all other studies. The last two studies listed were carried out with simulated rainfall.

In the first group of studies, the coefficients in the regression equation, $\ln(ER) = a + b \ln(sed)$, were usually $1 < a < 3$ and $-0.3 < b < -0.1$. Although variability was greater in the second group of studies, the values for a and b centered around the same range. Most of the correlation coefficients (r) were significant at the 99% probability level. Less than half the variation in enrichment ratios usually was accounted for by regression on sediment discharge, however.

Similar coefficients were found for nitrogen and phosphorus enrichment ratios, but in the first group of studies the slope "b" was consistently more negative for nitrogen than for phosphorus. A paired "t" test showed that the difference in slopes was significant at the 95% probability level. Mean values for the slopes were -0.26 ± 0.14, and -0.16 ± 0.12 for nitrogen and phosphorus, respectively. Corresponding values for the intercepts were 2.13 ± 0.65 and 1.84 ± 0.62.

Comparing the effects of land use, soil texture, or size of watershed on the enrichment ratios is difficult because these factors generally were not varied independently. Similar land uses and soil textures are grouped to the extent possible in table 1. Few consistent differences appear. Three studies that deviate from the general pattern are those with corn or soybeans in Michigan, terraced corn in Georgia, and fallow in New York.

Close examination of the data from Michigan (3) revealed no reason other than soil texture for the flat slope of the regression line. Nitrogen and phosphorus concentrations in eroded sediment apparently had been estimated for some events with very low sediment discharge. Omitting these events improved the correlation and increased the slope only slightly. Since some other studies with sandy loam (11) and gravelly loam (4) also showed relatively flat slopes, the amount of eroded material may have little effect on enrichment ratios under some conditions on light-textured soils. Nevertheless, enrichment ratios may be quite high with sandy soils. Values for $(ER)_N$ averaged 3.1, 5.0, and 1.4 in the studies by Ellis, Smith, and LeBano, respectively.

The distinctly different slopes of the regression lines for terraced corn and contoured corn in Georgia (11) may offer a clue to the behavior of enrichment ratios on sandy soils. The terraced corn showed little effect of sediment discharge on enrichment ratio, but the contoured corn showed a strong effect.

Table 1.--Summary of enrichment ratios for N and P

Land use	State	Texture	Area	Events	Sediment per event[1/]	ln(ER) = a + b ln(sed)kg/ha						References
						Nitrogen			Phosphorus			
						a[2/]	b	r[3/]	a[2/]	b	r[3/]	
			(ha)	(number)	(kg/ha)							
Corn or soybeans	Mich.	Loamy sand	0.55-0.8	115	0.7-400	1.42	-0.02	0.19*	1.30(?)	-0.01	0.05 n.s.	(3)
Contoured corn	Ga.	Sandy loam	1.3	32	.3-700	2.82	- .34	.56***	2.42	- .18	.50**	(11)
Corn, oats, or hay	Wisc.	Silt loam	.02-0.09	177	5-7,500	2.22	- .25	.77***	1.01	- .05	.20**	(5)
Cotton	Okla.	Silt loam	12-18	137	.6-1,300	1.85	- .20	.67***	1.57	- .13	.50***	(8)
Wheat	Okla.	Silt loam	5.2	121	.5-600	2.95	- .44	.72***	2.40	- .33	.69***	(8)
Range, Chickasha	Okla.	Silt loam	8-11	236	.4-1,500	1.50	- .28	.78***	2.32	- .26	.64***	(8)
Terraced corn	Ga.	Sandy loam	1.4	23	2-300	1.56	- .10	.20 n.s.	1.11	+ .10	.31 n.s.	(11)
Sorghum	Tex.	Clay	7.5	11	50-800	1.84	- .16	.43 n.s.	3.12	- .42	.68*	(8)
Range, El Reno	Okla.	Silt loam	1.6	14	2-200	2.67	- .46	.88***	2.98	- .64	.93***	(8)
Range, Woodward	Okla.	Loam	2.7-5.6	13	5-500	1.81	- .25	.56**	.82	- .06	.19 n.s.	(8)
Cropland and forest	Mo.	Loams & clay loams	1,460	17	50-5,000	----	-----	----	1.40(?)	- .13	.54*	(10)
Burned chaparral	Calif.	Gravelly loam	.0036	5	200-5,000	1.34	- .14	.77 n.s.	.12	- .02	.22 n.s.	(4)
Corn residue	Ind.	Silt loam	.008	8	500-25,000	1.46(?)	- .13	.70 n.s.	1.90(?)	- .19	.79*	(9)
Fallow	N.Y.	?	small plot	43	60-1,800	-0.25	+ .11	.58***	- .64	+ .16	.60***	(1)

1/ About 90% of events fall within this range.
2/ (?) indicates that soil analysis was estimated to get value of a.
3/ n.s. Not significant.
* Significant at 95% level.
** Significant at 99% level.
*** Significant at 99.9% level.

Coarse sand probably was discharged from the contoured corn watershed but was retained on the terraced corn watershed. The enrichment ratio did not change when the coarse sand was retained. This indicates that plant residues either were not selectively eroded with small amounts of sediment or contributed little particulate matter to runoff from these watersheds.

With fallow plots at Cornell University (1), enrichment ratios increased with increasing amounts of sediment discharge. This study was done with three successive applications of simulated rainfall on bare soil. Increasing amounts of sediment were discharged with each successive application. For some reason, the sediment discharged with the first application was low in nitrogen and phosphorus content. When the first application was omitted, there was no significant correlation between enrichment ratio and amount of eroded material. Enrichment ratios averaged 1.6 for both nitrogen and phosphorus.

Enrichment ratios were measured on duplicate watersheds with different land uses and similar soil types near Chickasha, Okla. The logarithmic relationships of enrichment ratio and sediment discharge are plotted in figure 1. An analysis of variance was made on the slopes of the regression equations to test the significance of the differences. The results are shown in table 2.

Table 2.--Slopes of logarithmic regression lines for unit source watersheds at Chickasha, Okla.

Land use	Slope for $(ER)_N$		Slope for $(ER)_P$		Average 1/
	Rep 1	Rep 2	Rep 1	Rep 2	
Wheat- - - - - -	-0.47	-0.40	-0.34	-0.31	-0.38 a
Range, poor condition- - -	- .30	- .29	- 19	- .25	- .26 ab
Cotton - - - - -	- .19	- .21	- .10	- .15	- .16 b
Range, good condition- - -	- .11	- .32	+ .03	- .08	- .12 b
Average- - - - - - - -	-.29 a		-.17 b		

1/ Averages followed by the same letter are not significantly different at 99% level of probability.

As indicated earlier, the slope is more negative for nitrogen than for phosphorus. The slope is also more negative for wheat than for cotton or range in good condition. These differences may be explained if nitrogen enrichment is associated primarily with selective erosion of organic matter, and phosphorus enrichment is associated primarily with selective erosion of fine mineral particles. Further work is needed to establish these associations.

CONCLUSIONS

With few exceptions, the enrichment ratios for nitrogen and phosphorus in eroded soil decrease markedly with amount of eroded soil. A logarithmic

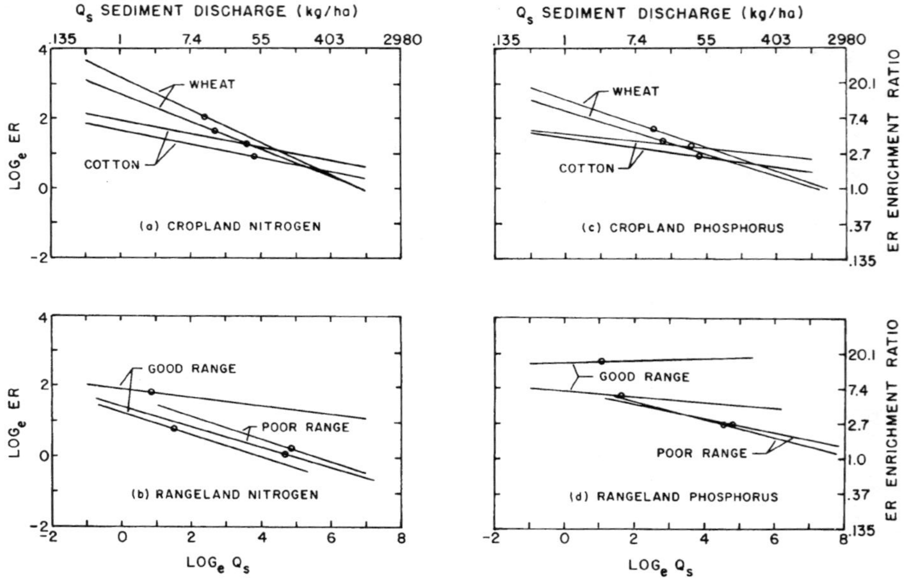

Figure 1.--Relation between sediment yield and enrichment ratios for (a) cropland nitrogen, (b) rangeland nitrogen, (c) cropland phosphorus, and (d) rangeland phosphorus

relationship, ln(ER) = 2-0.2 ln(sed), where (sed) is kilograms per hectare of eroded soil in individual runoff events, appears to hold for a wide range of soil and vegetative conditions. Under these conditions, the enrichment ratio probably could be predicted within a factor of 2 for an annual average, and within a factor of 5 for individual runoff events.

Terraced fields and sandy soils may not follow this relationship. They may have, instead, a nearly constant enrichment ratio, probably greater than 1, and less than 5.

Nitrogen enrichment changes more rapidly with sediment discharge than phosphorus enrichment. Different mechanisms of enrichment may be involved, but not enough information is available to justify separate equations for nitrogen and phosphorus enrichment.

REFERENCES

(1) Cornell University.
1971. Management of nutrients on agricultural land for improved water quality. Cornell University, Deptartment of Agronomy, EPA Report No. 13020DPB08/71.

(2) Duffy, P. D., J. D. Schreiber, D. C. McClurkin, and L. L. McDowell.
1978. Aqueous- and sediment-phase phosphorus yields from five southern pine watersheds. Journal of Environmental Quality 7:45-50.

(3) Ellis, B. G., A. E. Erickson, and A. R. Wolcott.
1978. Nitrate and phosporus runoff from small watersheds in Great Lakes Basin. Report No. EPA 600/3-78028.

(4) LeBano, L. F., and C. E. Conrad.
1976. Nutrients lost in debris and runoff water from a burned chaparral watershed. In Proceedings of the Third Federal Interagency Sedimentation Conference 1976, Water Resources Council. pp. 3-13 to 3-27.

(5) Massey, H. F.
1952. Fertility erosion under different conservation practices on three Wisconsin soils. PhD thesis, University of Wisconsin, Madison, 86 pp.

(6) ------, and M. L. Jackson.
1952. Selective erosion of soil fertility constituents. Soil Science Society of America Proceedings 16:353-356.

(7) McDowell, L. L.
Unpublished data. U.S. Sedimentation Laboratory, Oxford, Miss.

(8) Menzel, R. G.
Unpublished data. U.S. Agricultural Water Quality Management Laboratory, Durant, Okla.

(9) Romkens, M. J. M., D. W. Nelson, and J. V. Mannering.
1973. Nitrogen and phosphorus composition of surface runoff as affected by tillage method. Journal of Environamental Quality 2:292-295.

10) Schreiber, J. D., D. L. Rausch, and L. L. McDowell.
1977. Callahan reservoir. II. Inflow and outflow suspended phosphorus relationship. Transactions of the American Society of Agricultural Engineers 20:285-290.

11) Smith, C. N., G. W. Bailey, R. A. Leonard, and G. W. Langdale.
1978. Transport of agricultural chemicals from small upland Piedmont watersheds. EPA Report No. 600/3-78-056.

Chapter 13. NITRATE PRODUCTION, UPTAKE, AND LEACHING

S. J. Smith, D. E. Kissel, and J. R. Williams[1/]

Nitrate in watershed soils is derived mainly from soil organic matter and applied N fertilizer. Soil microbes convert N from the two sources to nitrate via an ammonium to nitrate conversion pathway. Generally, the conversions follow each other so closely that only nitrate accumulates. The behavior of nitrate in watershed soils is complex, and exact prediction of its fate is neither possible nor practical. The following approach presents a simple, abbreviated way for predicting actual amounts of nitrate produced under field conditions and different options for assessing its fate.

Prediction is on the basis of the soil nitrogen mineralization potential, N_0, which may be considered as the quantity of soil organic N that is susceptible to mineralization according to first-order kinetics (27). Limited evidence from field studies involving fluctuating soil temperatures and water contents tends to support the N_0 approach (19, 24, 31).

The options for assessing the fate of soil and fertilizer-derived nitrate allow for use of different types and amounts of available input data. The "best" option may vary from watershed to watershed.

CALCULATING FIELD NITRATE PRODUCTION

N_0 Determination

N_0 determination can be made in several ways (27, 28, 29, 30), but the most convenient is based on oxidative release of NH_4-N from soil organic matter (29). This new chemical measure of N_0 was developed using 62 soils from 8 soil orders. Amounts of NH_4-N released from 1 g soil samples during 1-hour extractions with 25 ml acid permanganate (0.1 N $KMnO_4$ in 1 N H_2SO_4) were correlated strongly with N_0, for 19 calcareous and 43 noncalcareous soils. The relationship between N_0 (Y) and oxidative release of NH_4-N from soil organic matter (X) is

$$Y = 2.1X + 5.8 \ (r = 0.89, N = 62) \quad (1)$$

The procedure is simple, rapid, and suitable for routine use in soil testing laboratories. N_0 measurements are made for the plow layer (0-15 cm) and a subsurface layer (15-60 cm) based on recommended soil testing depths for nitrate determination (37).

1/ Soil scientist, USDA-SEA-AR, Southern Plains Watershed and Water Quality Management Laboratory, Durant, Okla.; professor of soils, Kansas State University, Manhattan, Kans.; and hydraulic engineer, USDA-SEA-AR, Grassland-Soil and Water Research Laboratory, Temple, Tex.

In the event means are unavailable to determine N_0 by one of the above ways, a less exact estimate can be made from total N or organic C content, utilizing information compiled in table 1.

Table 1.--N_0 as a percentage of total N and organic C in surface soils (0-150 mm) of various soil orders[1/]

Soil order	Soils	N_0	Total N	Organic C	N_0/Total N	N_0/Organic C
	(Number)	(kg/ha)	(kg/ha)	(kg/ha)	(%)	(%)
Alfisol	10	391±167	2,156± 888	23,333±10,444	18.1±3.3	1.74±0.35
Aridisol	7	289±207	1,689± 800	13,111± 7,111	15.3±6.2	2.00±0.77
Entisol	7	342±176	1,844± 800	17,556±10,222	18.2±6.2	2.06±0.83
Inceptisol	1	245	2,267	26,222	10.8	0.94
Mollisol	22	444±195	3,933±2,222	42,178±25,778	12.3±3.9	1.22±0.49
Spodosol	2	600±109	5,822±1,867	75,556±42,667	10.6±1.5	0.89±0.35
Ultisol	12	262±142	1,116± 535	15,178± 7,222	23.9±8.0	1.85±0.76
Vertisol	1	389	2,600	28,600	15.0	1.36
All above	62	373±189	2,622±1,916	28,467±22,844	16.5±6.8	1.60±0.68

1/ See Stanford and Smith (29).

Soil Temperature

Desired information on soil temperature can be obtained from the hydrology compartment or from soil temperatures of the National Weather Service. The hydrology compartment soil temperatures are estimated from air temperatures, whereas the National Weather Service values represent actual readings at selected locations. The National Weather Service currently monitors soil temperature at the 10-cm depth in bare fields at more than 100 locations across the conterminous United States. The average weekly soil temperature isolines are presented on a map in each issue of the "Weekly Weather and Crop Bulletin" during the planting season, and annual daily soil records are available on tape from the Environmental Data Services, Asheville, N.C. By using data from one of these sources, an appropriate soil temperature can be estimated for the watershed.

The relationship between soil temperature and the mineralization rate constant, k, is shown in figure 1. This figure has been constructed using k as 0.0038 day^{-1} for 25°C (34). For 35°C, therefore, k would be 0.0076 day^{-1}, for 15°C, k would be 0.0019 day^{-1}, and so forth. The k value calculated for the 10-cm depth is used for the 0-15 cm layer and the 15-60 cm layer since changes in soil temperature with depth tend to be gradual (9).

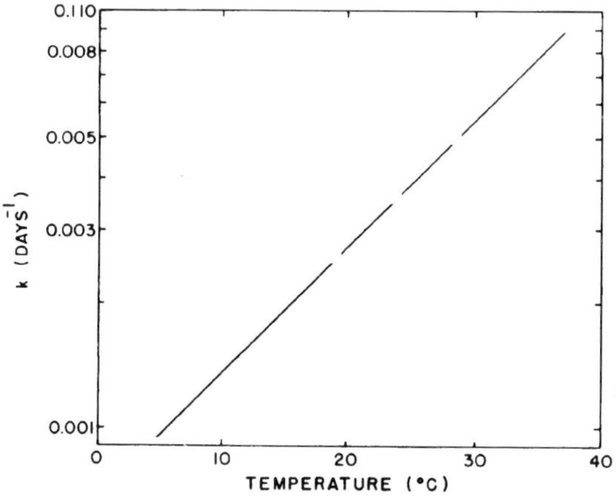

Figure 1.--Relation between nitrogen mineralization reaction rate constant and temperature.

Soil Water Content

Average daily soil water contents for the soil layers are calculated from data available from the hydrology compartment. Relative N mineralization is considered to be a linear function of soil water content, expressed as a percentage of the optimum for biological activity (26). Field capacity (1/3 bar) is used as the optimum soil water content, with no nitrate production assumed during days the water content is more than 5 to 10% above optimum (fig. 2). Excessive moisture does not inhibit conversion of organic N to ammonium, however. When aerobic conditions return, the ammonium readily is converted to nitrate. This "additional" nitrate helps balance any quantities lost by denitrification during excessive moisture conditions.

Soil pH

Soil acidity ($pH < 5.5$) plays an important natural role in minimizing production and subsequent leaching of nitrate. This is because acid soils tend to ammonify rather than nitrify (table 2). Moreover, liming an acid soil often raises the pH only in the surface layers (fig. 3). Consequently, any nitrate prediction technique should consider whether soil N mineralization potentials reflect ammonium or nitrate production. A similar statement holds regarding N fertilizer transformations. With this knowledge, appropriate adjustments can be made in the nitrate prediction technique.

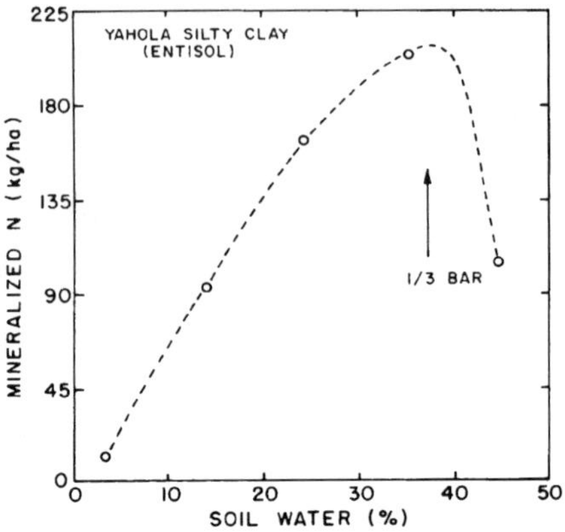

Figure 2.--Two-week aerobic nitrogen mineralization (23), as a function of soil water content.

Table 2.--Relative production of ammonium and nitrate in acid surface soils (0-150 mm) during a 2-week laboratory incubation 1/

Soil	Location	Order	Use	pH	NH_4-N	NO_3-N
					(kg/ha)	(kg/ha)
Caribou silt loam----	Maine	Spodosol	Virgin forest	5.0	220	27.3
Caribou silt loam----	Maine	Spodosol	Potatoes	5.6	94.6	21.3
Cecil sandy loam-----	Georgia	Ultisol	General crops	5.2	44.9	26.4
Goldsboro sandy loam.--------	South Carolina	Ultisol	Corn	5.1	21.8	12.0
Greenville fine, sandy loam.--	Alabama	Ultisol	General crops	5.3	33.3	13.6
Grenada fine, sandy loam.--	Mississippi	Alfisol	Coastal bermudagrass.	5.0	117	14.2
Keene silt loam.---------	Ohio	Alfisol	Virgin forest	4.3	191	21.6
Keene silt loam.---------	Ohio	Alfisol	General crops	6.3	<0.1	144
Norfolk fine, sandy loam.--	South Carolina	Ultisol	Corn	5.5	23.6	13.1
Webster clay loam.---------	Iowa	Mollisol	General crops	4.6	96.4	11.1

1/ Aerobic procedure from Smith and Stanford, (23).

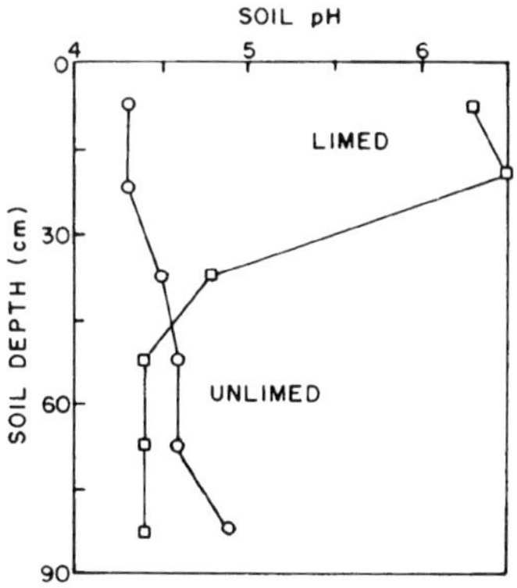

Figure 3.--Soil pH for profiles of limed (general crops) and unlimed (virgin forest), Keene silt loam.

Sample Calculation

The following example illustrates how N_0, soil temperature, and soil water content are used to estimate field nitrate production. Assume that the determined N_0 value for the 0-15 cm layer is 400 kg/ha, and the corresponding first day's soil temperature and water content are 25°C and 75% field capacity, respectively. Then, the amount of nitrate produced during the first day would be 400 x 0.0038 x 0.75, or 1 ppm. For the second day, N_0 is now 399 ppm, and a similar type of calculation is made based on that day's soil temperature and water content. Calculations for the 15-60 cm layer are made in a comparable manner.

For mineralization intervals longer than a day, the integrated expression

$$N = N_0 \, (1 - e^{-kt}) \, M \qquad (2)$$

may be used where N represents the mineralized N; N_0, the mineralization potential; k, the temperature-adjusted rate constant; t, the time; and M, the soil water content expressed as a percentage of field capacity.

Applications of N fertilizer to soils with a pH >5.5 are considered to convert to nitrate within a few days after soil temperature is above 10°C. The fertilizer application can be incorporated in with the preceeding calculation and treated as a broadcast or uniform mixing. Amounts of residual nitrate can be handled similarly.

No direct attempts were made to account for denitrification and immobilization in the preceeding calculations. With increased emphasis on enhancing soil and fertilizer N efficiency, the variable importance of these two factors can be lessened.

Denitrification

If a specific accounting of denitrification is desired, the process may be treated as a first-order reaction, with the rate constant as a function of temperature and organic carbon content (32, 35). Denitrification is considered to occur when the soil water content is 5 to 10% above field capacity and is handled as follows:

$$d(NO_3^-) \ / \ dt = k(NO_3^-)$$

where t = time, and k = the denitrification rate constant. For 30 diverse soils, the relations between k (per hour) and carbon forms X (mg/g soil) were:

Glucose carbon $\quad k_{35^oC} = 0.188 \ X - 0.0093 \ (r^2 = 0.82)$

Total soil carbon $\quad k_{35^oC} = 0.011 \ X + 0.0025 \ (r^2 = 0.69).$

Total soil carbon may be obtained from organic matter content by dividing by 1.724. Over the denitrification temperature range normally encountered in soil, k is considered to have a Q_{10} of 2, with ln k a linear function of temperature.

SIMULATING NITRATE UPTAKE BY CROPS

When the soil supply of available N does not limit plant growth, the accumulation of nitrate by plants is affected by such environmental factors as temperature, light, available soil water, and availability of other nutrients. These environmental factors may influence nitrate uptake by controlling dry matter accumulation of the plant. Any limitation on production of dry matter usually limits nitrate uptake as well. Our quantitative description of how the environment influences dry matter accumulation is incomplete, but realistic computer simulations of dry matter accumulation can be carried out. Option I, which follows, is based on a water use/dry matter simulation, whereas Option II is based on the normal probability curve adjusted for water stress. With both options, nitrate is removed from the root zone, and water use, potential and actual, is calculated in the hydrology model.

Option I

The uptake of nitrate (UN) for any period during the growing season is the difference in the plant nitrogen before and after the period, as indicated by the following equation:

$$UN = c_i G_i - c_{i-1} G_{i-1} \quad (3)$$

where c is the concentration of N in the plant (usually expressed as % N) and G is the accumulative dry matter. Dry matter (G) is computed with the equation

$$G_i = H (YD * K) \quad (4)$$

where YD is the potential economic yield for ideal conditions (for example, this could be expressed as kg of grain per hectare for corn), K is a constant to convert YD to total dry matter (grain + stover + roots), and H is defined as:

$$H = \frac{\Sigma UW_i}{PWU} \quad (5)$$

where ΣUW_i is the total water used by the crop from emergence to day i, and PWU is the potential crop water use for the entire growing season under ideal conditions.

The value of K in equation 4 for corn and grain sorghum can be 2.5 based on an estimate by Stanford (25), considering that roots, stover, and grain were 20, 40, and 40% of the total dry matter, respectively. If roots are assumed to make up 20% of the total dry matter of the other crops, the value of K for wheat, soybeans, and cotton would be 3.7, 4.2, and 6.5, respectively, based on data of Boatwright and Haas (3) for wheat, Hanway and Weber (13) for soybeans, and Bassett and others (2) for cotton. The value of 6.5 for cotton assumes economic yield is described by lint yield.

The concentration of N in the plant (c) declines in plants from early in the season until plant maturity, as shown for grain sorghum in figure 4. To make the concentration of N compatible with the simulation of dry matter, the percentage N of crops is described as a fraction of the total accumulated dry matter (fractional dry matter) for the growing season (GR). The percentage N vs. GR data were fit to log functions. Most crops require two functions, one for early growth and a second as the crops approach maturity. Both equations are evaluated and the smaller value is selected. The resulting equations are given in table 3. The crops considered for simulation of N uptake are corn, grain sorghum, wheat, cotton, and soybeans.

Nitrogen concentration as a function of accumulated dry matter for corn was constructed from data of Hanway (10, 11, 12). These data, summarized in table 4, were plotted with the grain sorghum data in figure 4. The Iowa corn data of Hanway fit very well the curve for the Texas grain sorghum data. Because of the close agreement, the equation developed for grain sorghum (table 3) may be used for corn also.

The equation for wheat was developed from the work of Boatwright and Haas (3) on spring wheat. The percent N of wheat generally is higher than corn or grain sorghum throughout the growth cycle of the crop. By maturity, however, the difference between the crops is not large (1.4% N for wheat and 1.25% for corn with nonlimiting fertility).

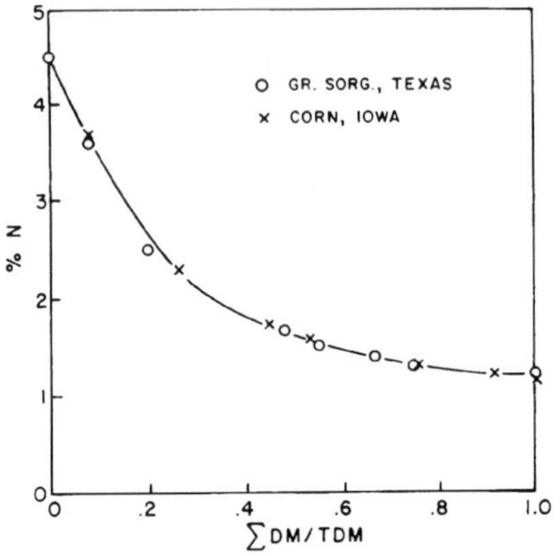

Figure 4.--Relationship between plant nitrogen concentration and accumulated dry matter.

Table 3.--Equations describing nitrogen concentrations as a function of the accumulated dry matter (GR) for grain sorghum and corn, wheat, cotton, and soybeans

General equation:

$$c = \text{minimum } (b_1 \text{ GR}^{b_2} \text{ or } b_3 \text{ GR}^{b_4})$$

	b_1	b_2	b_3	b_4
Grain sorghum and corn-------	0.0209	-0.157	0.0128	-0.415
Wheat------------------------	.0330	- .134	.0134	- .750
Cotton-----------------------	.0177	- .253	.0177	- .253
Soybeans---------------------	.0259	- .104	.0259	- .104

The equation for cotton (Gossypium hirsutum) was developed from the work of Bassett and others (2). They present values of N uptake for irrigated cotton at six locations with different years and planting dates. The mean lint yield for data used is 1,432 kg/ha.

The equation for soybeans was developed from the work of Hanway and Weber (13, 14, 15). The percent N of the whole soybean plants through the growing season is characterized by N contents similar to corn early in the growing season. Near maturity percent N of soybean is about twice as high as corn, however.

Table 4.--Dry matter production, nitrogen accumulation and percent N in the total above ground plant for corn

Date	Dry matter	Fraction dry matter	N uptake	N
	(kg/ha)		(kg/ha)	(%)
July 1	1,241	0.072	-	3.5
July 20	4,582	.267	88.3	2.16
August 1	7,712	.449	119.2	1.73
August 10	9,545	.556	130.9	1.54
August 20	13,057	.760	152.2	1.31
September 10	16,608	.967	178.5	1.20
September 25	17,181	1.00	175	1.14

Source: Hanway (10, 11, 12).

The preceding simulation procedure necessitates the forward estimation of crop yield by the end of the growing season. If the yield at crop maturity is unavailable from another simulation, the yield at crop maturity (YP) can be estimated by

$$YP_i = H + YA\left[1 - \frac{\Sigma\, CU_i}{PWU}\right] \tag{6}$$

where YA is average crop yield and $\Sigma\, CU_i$ is the sum of consumptive water use by the crop for ideal conditions. Both YA and YP are expressed in dimensionless terms and vary between 0 and 1.

The fractional dry matter, GR_i, on any day for determination of nitrogen concentration (c_i) is

$$GR_i = \frac{H}{YP_i}. \tag{7}$$

The GR_i then is used to compute c_i as follows:

$$c_i = \text{minimum } b_1\, GR_i^{\;-b_2} \text{ or } b_3\, GR_i^{\;-b_4} \tag{8}$$

with values of b_1 through b_4 selected from table 3. The resulting c_i is then used in equation (3) to compute UN.

Option II

If data for ideal water use or N content of plants are unavailable, then the following approach can be used. The accumulated nitrogen uptake by plants during the season can be described by the normal probability curve (fig. 5). Water stress during any period of uptake reduces uptake in proportion to the ratio of actual transpiration to potential transpiration.

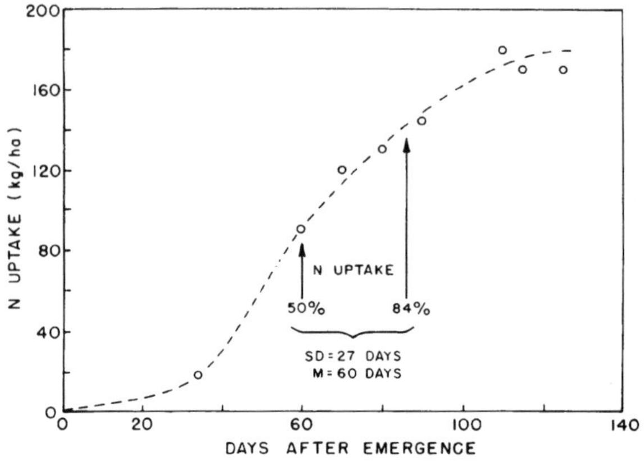

Figure 5.--Nitrogen uptake for corn by days after emergence.

The normal probability curve (1) can be approximated by

$$P = 1 - 1/2\ (S)^{-4} \tag{9}$$

where $$S = 1 + 0.196854X + 0.115194X^2 + 0.000344X^3 + 0.01957X^4 \tag{10}$$

and $$X = (T - M)/SD,\ \text{and if } X < 0,\ P(-X) = 1 - P(X). \tag{11}$$

T is the days since emergence, M is the date of maximum uptake rate or half the N uptake in days since emergence, and SD is one standard deviation or the number of days from 50% uptake to 84% uptake.

This option requires three parameters for the nitrogen uptake pattern (table 5): (1) The potential uptake of N by the whole plant for a year when water was not limiting (maximum yield and percent N are data useful for estimating this parameter); (2) the probability distribution is symmetrical around the midpoint or point of half the N uptake; (3) the standard deviation is the number of days from the midpoint when 16% or 85% of the nitrogen is taken up.

Table 5.--N uptake parameters for Option II

Crop	Potential N uptake per hectare	M (50% uptake)	SD (84% uptake)
	(kg)	(days)	(days)
Corn	150 - 300	60	27
Sorghum	140 - 280	70	15
Wheat	90 - 180	50	8
Cotton	60 - 200	60	28

One additional parameter is needed to account for growth reduction due to water stress. The ratio (TR) of actual transpiration to potential transpiration calculated in the evapotranspiration section of the hydrology model is the stress parameter. The uptake of nitrate (UN) for any period during the growing season is the difference in plant nitrogen before and after that period times the stress parameter as indicated in the following equation:

$$UN = (P_i - P_{i-1}) * TR \quad . \qquad (12)$$

NITRATE LEACHING

The amount of nitrate available for leaching represents accumulated soil nitrate production plus fertilizer-derived nitrate, minus that taken up by the crop. Safeguards against excess nitrate leaching include proper N fertilizer and irrigation management (16, 22, 33). As long as the crop is growing, N uptake by the plant roots tends to lessen the potential for nitrate leaching. Consequently, critical periods for nitrate leaching are often between harvest and the next growing season. Generally, nitrate losses in surface runoff are not great (17, 18, 20, 21, 36), frequently less than that received in precipitation. Exceptions are considered in chapter 15 on surface-runoff losses. Sometimes, also, nitrate in return flow may be too important a factor to ignore. Both return flow and vertical leaching losses can be calculated on the basis of the nitrate concentration in a plant's root zone (table 6), and the amount of water percolating through that zone. The options below assume a maximum root zone. This implies that nitrate leached below an existing root zone can be recovered by additional root growth.

Option I

This option is based on a mathematically linear algorithm that allows the results of individual storms or drainage periods to be added together for monthly, seasonal, or annual totals. The option assumes nitrate in the root zone to be uniformly mixed with all the water in the root zone. This is a rather crude approximation but perhaps adequate on the average. Water moving into dry soil has the highest nitrate concentrations at the wetting front, while water moving through saturated soil may move mainly through the large pores and bypass the smaller pores. Preliminary results indicate this approach provides an upper estimate of leaching.

The algorithm is as follows:

$$N_L = N_R F_L \tag{13}$$

$$F_L = P/(P+S) \tag{14}$$

where N_L is kg nitrate-N/ha leached below the root zone in a storm, N_R is kg nitrate-N/ha in the root zone, F_L is the fraction of the water percolating out of the root zone, P is the mm of water percolating out of the root zone, and S is the mm of soil water remaining in the root zone (field capacity x depth of root zone).

Table 6.--Root zone of various crops

Crop	Potential root zone	Active root zone
	(mm)	(mm)
Corn Sorghum Wheat Alfalfa	1,830 or greater	1,220
Sugar beets Soybeans Potatoes	1,520	910
Pasture	910	610

Source: Doane's Agricultural Service (8).

Option II

This option provides a rapid means of estimating nitrate leaching when the additive feature in Option I is unnecessary. Use is made of a simple, non-linear equation developed recently by Burns (7):

$$f = \left[\frac{10\ P}{10\ P+V_m}\right]^x \tag{15}$$

where f = fraction of nitrate-N leached below a certain soil depth, h,
V_m = field capacity (percent by volume),
P = mm drainage passing through soil,
x = depth function dependent on nitrate-N distribution in the soil profile.

The relationship seems to work best for those situations where the initial soil water content is near field capacity. Assumptions are that leaching is directly proportional to water movement and that leaching occurs only when field capacity is exceeded. The initial distribution of nitrate in the soil profile determines the value for x; x = h, the root zone depth for surface applications, x = h - w/2 for uniform incorporation to a certain depth w cm, and x = h/2 for uniform distribution throughout the whole profile. For simplicity, we assume that the nitrate in the soil is approximately uniform in the top 600 mm of the soil, thus w/2 = 300 and x = (h-300)/10.

Further details regarding development and application of Burns' leaching equation are spelled out in a series of papers (4, 5, 6). Tests were made for a variety of soils and conditions, with good agreement between calculated and observed values.

REFERENCES

(1) Abramowitz, M., and I. A. Stegun.
1964. Handbook of mathmatical functions. National Bureau of Standards Applied Mathematics Series No. 55.

(2) Bassett, D. M., W. D. Anderson, and C. H. E. Werkhoven.
1970. Dry matter production and nutrient uptake in irrigated cotton (Gossypium hirsutum). Agronomy Journal 62:299-303.

(3) Boatwright, G. O., and H. J. Haas.
1961. Development and composition of spring wheat as influenced by nitrogen and phosphorus fertilization. Agronomy Journal 53:33-36.

(4) Burns, I. G.
1974. A model for predicting the redistribution of salts applied to fallow soils after excess rainfall or evaporation. Journal of Soil Science 25:165-178.

(5) ------.
1975. An equation to predict the leaching of surface-applied nitrate. Journal of Agricultural Science Cambridge 85:443-454.

(6) ------.
1976. Equations to predict the leaching of nitrate uniformly incorporated to a known depth or uniformly distributed throughout a soil profile. Journal of Agricultural Science Cambridge 86:305-313.

(7) ------.
1977. Nitrate movement in soil and its agricultural significance. Outlook on Agriculture 9:144-148.

(8) Doane's Agricultural Service.
1977. Report 124.3.

(9) Fluker, B. J.
1958. Soil temperature. Soil Science 86:35-46.

(10) Hanway, J. J.
1962a. Corn growth and composition in relation to soil fertility: I. Growth of different plant parts and relation between leaf weight and grain yield. Agronomy Journal 54:145-148.

(11) Hanway, J. J.
1962b. Corn growth and composition in relation to soil fertility: II. Uptake of N, P, and K in different plant parts in relation to stage of growth. Agronomy Journal 54:217-222.

(12) ------.
1962c. Corn growth and composition in relation to soil fertility: III. Percentages of N, P, and K in different plant parts in relation to stage of growth. Agronomy Journal 54:222-229.

(13) ------, and C. R. Weber.
1971a. Dry matter accumulation in soybean (Glycine max (L.) Merrill) plants as influenced by N, P, and K fertilization. Agronomy Journal 63:263-266.

(14) ------, and C. R. Weber.
1971b. N, P, and K percentages in soybean (Glycine max (L.) Merrill) plant parts. Agronomy Journal 63:286-290.

(15) ------, and C. R. Weber.
1971c. Accumulation of N, P, and K by soybean (Glycine max (L.) Merrill) plants. Agronomy Journal 63:406-408.

(16) Herron, G. M., A. F. Dreier, A. D. Flowerday, W. L. Colville, and R. A. Olson.
1971. Residual mineral N accumulation in soil and its utilization by irrigated corn. Agronomy Journal 63:322-327.

(17) Kissel, D. E., C. W. Richardson, and E. Burnett.
1976. Losses of nitrogen in surface runoff in the Blackland Prairie of Texas. Journal of Environmental Quality 5:288-293.

(18) Olness, A. E., S. J. Smith, E. D. Rhoades, and R. G. Menzel.
1975. Nutrients and sediment discharge from agricultural watersheds in Oklahoma. Journal of Environmental Quality 4:331-336.

(19) Oyanedel, C., and J. Rodriguez.
1977. Estimation of N mineralization in soils. Ciencia E. Investigation Agraria 4:33-44.

(20) Schreiber, J. D., P. D. Duffy, and D. L. McClurkin.
1976. Dissolved nutrient losses in storm runoff from five southern pine watersheds. Journal of Environmental Quality 5:201-205.

(21) Schuman, G. E., R. E. Burwell, R. F. Piest, and R. G. Spomer.
1973. Nitrogen losses in surface runoff from agricultural watersheds on Missouri Valley loess. Journal of Environmental Quality 2:299-302.

(22) Smika, D. E., D. F. Heermann, H. R. Duke, and A. R. Batchelder.
1977. Nitrate N percolation through irrigated sandy soil as affected by water management. Agronomy Journal 69:623-626.

23) Smith, S. J., and G. Stanford.
1971. Evaluation of a chemical index of soil nitrogen availability. Soil Science 111:228-232.

24) ------, L. B. Young, and G. E. Miller.
1977. Evaluation of soil nitrogen mineralization potentials under modified field conditions. Soil Science Society of America Journal 41: 74-76.

25) Stanford, G.
1973. Rationale for optimum nitrogen fertilization in corn production. Journal of Environmental Quality 2:159-166.

26) ------, and E. Epstein.
1974. Nitrogen mineralization-water relations in soils. Soil Science Society of America Proceedings 38:103-107.

27) ------, and S. J. Smith.
1972. Nitrogen mineralization potentials of soils. Soil Science Society of America Proceedings 36:465-472.

28) ------, and S. J. Smith.
1976. Estimating potentially mineralizable soil nitrogen from a chemical index of soil nitrogen availability. Soil Science 122:71-76.

29) ------, and S. J. Smith.
1978. Oxidative release of potentially mineralizable soil nitrogen by acid permanganate extraction. Soil Science 126:210-218.

30) ------, J. N. Carter, and S. J. Smith.
1974. Estimates of potentially mineralizable soil nitrogen based on short-term incubations. Soil Science Society of America Proceedings 38:99-102.

31) ------, J. N. Carter, D. T. Westerman, and J. J. Meisinger.
1977. Residual nitrate and mineralizable soil nitrogen in relation to nitrogen uptake by irrigated sugarbeets. Agronomy Journal 69:303-308.

32) ------, S. Dzienia, and R. A. Vander Pol.
1975. Effect of temperature on denitrification rate in soils. Soil Science Society of America Proceedings 39:867-870.

33) ------, C. B. England, and A. W. Taylor.
1970. Fertilizer use and water quality. U.S. Department of Agriculture, Agricultural Research Service, ARS 41-168. (Series discontinued; Agricultural Research Service now Science and Education Administration-Agricultural Research).

34) ------, M. H. Frere, and D. E. Schwaniger.
1973. Temperature coefficients of soil nitrogen mineralization. Soil Science 115:321-323.

(35) Stanford, G., R. A. Vander Pol, and S. Dzienia.
1975. Denitrification rates in relation to total and extractable soil carbon. Soil Science Society of America Proceedings 39:284-289.

(36) Taylor, A. W., W. M. Edwards, and E. C. Simpson.
1971. Nutrients in streams draining woodland and farmland near Coshocton, Ohio. Water Resources Research 7:81-89.

(37) Ward, R. C.
1971. NO_3-N soil test-approaches to use and interpretation. Soil Science and Plant Analysis 2:61-71.

Chapter 14. ESTIMATING SOLUBLE (PO_4-P) AND LABILE PHOSPHORUS IN RUNOFF FROM CROPLANDS

L. L. McDowell, J. D. Schreiber, and H. B. Pionke[1/]

INTRODUCTION

The Water Quality and Management Program under Section 208 of P.L. 92-500 requires that each State plan and implement programs to achieve water quality goals by decreasing both point and nonpoint pollution. Mathematical models sometimes are being used to predict the movement of water, sediment, and agricultural chemicals from farmlands. Few models have been validated and verified however, because of inadequate data bases and understanding of the physical, chemical, and biological processes involved in sediment-water-chemical interactions. A primary need for 208 planning is to develop models for assessing the effectiveness of land management practices on sediment and agricultural chemical transport from croplands. Phosphorus (P) is an important agricultural chemical to be estimated in runoff from croplands. Estimating P movement from croplands requires knowledge of soil P, P transport, and eutrophication. These topics will be discussed briefly in this section.

Soil Phosphorus

The most significant characteristics of soil P are its low water solubility and high adsorptivity. Much of the fertilizer P applied to soil rapidly is converted to the slightly soluble calcium, magnesium, iron, and aluminum forms. This greatly decreases the amount of P in soil water whether replenished by either mass flow or diffusion. Some phosphate compounds formed are crystalline and slowly dissolve or are transformed further to progressively less plant-available and water-soluble forms (3). In either case, most of the added P released does not stay in solution but is strongly sorbed by the finely divided soil particles (21). Soluble orthophosphorus (PO_4-P) concentrations in the soil vary with soil pH, phosphate compounds, and soil texture (29, 31).

Saturation extracts of conventionally tilled surface soils usually contain only 0.01 to 0.3 mg/l of soluble PO_4-P. As the soil solution percolates, less fertile subsoil layers sorb much of the soluble PO_4-P, reducing the concentration to even lower levels. The obvious exceptions are (1) on highly permeable soils when low P sorption capacities are exceeded by fertilization and (2) in tile drain outlets, under reducing conditions, where soluble PO_4-P

1/ Soil scientist, USDA-SEA-AR, Sedimentation Laboratory, Oxford, Miss.; soil scientist, USDA-SEA-AR, Sedimentation Laboratory, Oxford, Miss.; and soil scientist, USDA-SEA-AR, Northeast Watershed Research Center, University Park, Pa.

concentrations are controlled by soluble ferrous phosphate (12). Topsoils vary greatly in their total P content, ranging from 200 to 10,000 kg/ha, depending mostly on their inherent fertility and to a lesser degree on their fertilization history. Only a small amount of the total P is available for terrestrial plant growth, perhaps 1 to 10 percent. Applying excess P fertilizers over several years increases the total P content of soils.

Eutrophication and Transport of P in Runoff

Many consider P to be the key nutrient element in eutrophication (19, 20). Much research on the effects of P on eutrophication is based on total P concentration, that is, sediment-plus solution-phase P. Some sediment P is thermodynamically stable, however, and hence, not biologically available to support the growth of algae and larger plants. In some eutrophication studies, only soluble PO_4-P is measured. While soluble PO_4-P is readily available for plant growth, this P form usually underestimates the total quantity of P that is available for aquatic growth. Lee and others, (20) proposed that the biologically available P is approximately equal to the PO_4-P plus 0.2 times the difference between the total P and PO_4-P. Other studies have indicated that 5 to 40 percent of the sediment P is labile (14, 15, 38, 40).

P is transported from farmlands into streams and lakes in solution and in association with sediments. The erosion, transport, and deposition processes are selective. The fine sediments, including organic matter, are eroded preferentially. This selectivity enriches the P-associated sediments in runoff relative to that in the soil. Fine sediments have more chemically active surfaces that may sorb or desorb P, depending on the sediment and chemical environment. Fine sediments generally buffer the solution and suspended sediment P loads, whereas coarse sediments modify the erosive potential of streamflow (23).

Estimating P loadings to streams and impoundments requires measurements of the sediment and water discharges and the distribution of P concentrations between the solution and sorbed phases. The distribution of P between the solution and sediment phases in runoff from a monocropped, unit source watershed is controlled by the same factors operating in the soil but with additional complicating factors. These factors include additional P sources, runoff-soil P extraction processes (desorption kinetics), physical and chemical characteristics of the sediments, lower soil:solution ratios (lower sediment concentrations), and time in transport.

Inorganic P being transported generally can be considered in three categories: In solution (intensity factor); soil (or sediment) labile P (capacity factor) interactive with that in solution; and the soil (or sediment) nonlabile fraction. Estimating the intensity and capacity factors is our major interest. To estimate soluble PO_4-P (intensity factor) in soils and sediments, researchers have used water extracts and equilibrium phosphorus concentration (EPC) values derived from sorption-desorption isotherms (13, 16, 17, 40, 47). Such methods as ^{32}P isotopic dilution (32, 38, 43), algal bioassays (14), and resin and chemical extractants (4, 8, 9, 11, 37) have been used to estimate "available" P (capacity factor) associated with soils and sediments. Few studies have related these soil-based measurements of soluble PO_4-P to those observed in runoff from croplands where sediments; fertilizers; animal wastes; leaching and decomposition of crop residues; leaching of green growing crops, including grassed waterways; organophosphate insecticides; and rainfall are important, complex, and potentially interacting sources of soluble PO_4-P in storm runoff.

The objectives of this chapter are: (1) to evaluate methods for extracting or estimating soluble PO_4-P (intensity) and soil labile P (capacity) associated with surface soils of monocropped research plots and watersheds, (2) to relate soluble PO_4-P in soils to PO_4-P concentrations measured in runoff from these instrumented plots and watersheds, (3) to examine crop residues and green plants as sources of PO_4-P in runoff and (4) to recommend simple test(s) to 208 planners for estimating PO_4-P in runoff from monocropped, unit source watersheds.

MATERIALS AND METHODS

Research Plots and Watersheds

The USDA-SEA-AR research plots and watersheds mentioned in this chapter are located at the North Mississippi Agricultural and Forestry Experiment Station (MAFES), near Holly Springs, in north central Mississippi and in the Mississippi Delta near Clarksdale, Miss. (McWilliams watersheds 802 and 803) and Glen Allan, Miss. (Fisher watersheds 807 and 808).

At the MAFES location, corn (*Zea mays* L.) was grown on highly erodible loessial soils (Providence silt loam; Typic Fragiudalfs) by conventional and no-till practices for silage (stalks removed with little residue left on the soil surface) and for grain (stalks chopped and residue left on the soil surface). In addition, corn was grown for grain by no-till planting with two cultivations during the growing season.

Conventional tillage consisted of moldboard plowing in the spring, disking, harrowing, planting, and followup cultivations as needed to control grass and weeds. Fertilizer (30 kg P/ha) was placed about 8-10 cm deep and 7 cm to the side of the seed with a double-disk opener on the conventional planter. No-till corn was fertilized and planted with a no-till planter. A fluted coulter was used to cut through existing crop residues into the soil. Fertilizer (30 kg P/ha) was placed about 8-10 cm deep with a chisel-boot opener. Knives behind the chisel covered the fertilizer with soil. Seeds were planted about 3 cm deep and above the fertilizer with a double-disk opener preceding a rear press wheel.

Runoff from the small plots (0.01 ha) was measured with 0.15 m H-flumes equipped with FW-1 water level recorders and sampled with Coshocton wheel samplers (5). Runoff from the larger watersheds (WC-1, 1.57 ha; WC-2, 0.59 ha; WC-3, 0.65 ha) was measured with modified Parshall flumes and FW-1 water level recorders and was sampled with traversing slot samplers (7).

At the Delta locations, all watersheds were in continuous cotton (*Gossypium hirsutum* L.). These soils are inherently fertile in P; no phosphate fertilizer was applied since cotton does not respond to P fertilization. McWilliams watershed 802 is a 15.6-ha Sharkey silty clay soil (Vertic Haplaquepts). McWilliams watershed 803 is a 30.3-ha watershed consisting of Bruin, Commerce, and Tunica soils (Fluvaquentic Eutrochrepts, Aeric Fluvaquents, and Vertic Haplaquepts, respectively). Fisher watersheds 807 and 808 are about 3 ha each; the soil is classified as Morganfield silt loam (Typic Udifluvents). Runoff from these watersheds was measured with Parshall flumes and sampled with PS-69 automatic pumping samplers (27).

Soil Sampling

Soils were sampled before fertilized in the spring. A grid system was used in sampling all plots and watersheds. Soil samples (plugs) were collected with a ring sampler 2.5 cm x 7.2 cm in diameter and then air-dried, ground to pass a 2 mm sieve, and mixed with a V-type mechanical blender. A composite sample for analysis was prepared for each plot and watershed by blending equal weights of each replicate sample. The sampling program is summarized as follows:

Plot or watershed	Replicate samples	Plugs/sample
MAFES plots	3	9
MAFES watersheds	4	25
McWilliams 802 and 803	6	175
Fisher 807 and 808	5	35

Runoff Sample Preparation and Chemical Analysis

Samples were collected, usually after each storm event, and stored at 4°C until analyzed. Suspended sediments were separated from the aqueous phase by filtration through a 0.45 μm Millipore filter. Soluble PO_4-P and available soil test P were analyzed on a Technicon Auto-analyzer by the procedure of Murphy and Riley (28). Sediment total P also was determined by the method of Murphy and Riley (28) after digestion with perchloric acid. Unless otherwise indicated, suspended sediment and PO_4-P concentrations in runoff for all research plots and watersheds are discharge-weighted values for the 1976 water year (WY).

Soluble PO_4 Extractions From Soil (Intensity Factor)

Duplicate soil samples (0 - 2.5 cm layer) from the research plots and watersheds were extracted with double distilled water and 0.01 M $CaCl_2$ on a rotary shaker for 24 hr. Water extracts were conducted at soil concentrations ranging from 500 to 200,000 mg/l to determine water extractable PO_4-P as a function of the soil concentration. Extractions in 0.01 M $CaCl_2$ were made at 100,000 mg/l. Following extraction, samples were filtered through a 0.45 μm filter and analyzed for PO_4-P.

Soil P Extractions (Capacity Factor)

Duplicate soil samples (0 - 2.5 cm layer) were extracted with 0.5 M $NaHCO_3$ and Bray's No. 1 solution as outlined by Olsen and Dean (32). Soils were extracted by the Mississippi soil test procedure at the recommended soil: solution ratio of 1:5. The Mississippi State soil test consists of a double acid extraction using HCl followed by an acetic-malic-malonic acid solution containing AlF_3. Duplicate resin extractions were made at soil:solution ratios of 1:10 using 10 g of Dowex 1-X8. Before the resin extractions, care was taken to sieve the resin to eliminate sizes < 0.50 mm, saturate the resin exchange sites with Cl^- ions, and wash free of any excess. Following extraction for 24 hr on a rotary shaker, the resin was separated from the soil with a 0.25 mm sieve. P sorbed by the resin was extracted with five 20 ml portions 1 M Na_2SO_4.

Soil P Sorption-Desorption Isotherms - Soluble PO_4 (Intensity Factor) and Soil Labile P (Capacity Factor)

Duplicate soil samples (0 - 2.5 cm layer) were shaken for 24 hr on a rotary shaker with standard solutions containing different PO_4-P concentrations to determine (1) the EPC, that is, the concentration of PO_4-P in solution at which the soil neither sorbs nor desorbs PO_4-P and (2) the soil labile P (P_{sl}) by extrapolating the isotherm to zero PO_4-P concentration (16) or by solving the isotherm equations for Y = 0 and X at some finite PO_4-P concentration, for example, 0.01 mg/l (24). EPC values also were determined on soils from conventional and no-till corn (grain) plots after 2 hr of shaking and at soil concentrations of 2,000, 10,000, and 100,000 mg/l.

Soil concentrations used were 10,000 mg/l based on the recommendation by White (46) that the equilibrium phosphate potential determined at 20,000 mg/l (or smaller) soil concentrations will measure adequately the mean potential of nonequilibrium soils, that is, soils that had P recently added or removed. Phosphate nonequilibrium conditions prevail for many field soils.

Concentrations of PO_4-P in the reacting solutions were 0.0, 0.05, 0.1, 0.2, 0.3, 0.5, 0.8, 1.0, and 2.0 mg/l as KH_2PO_4. After shaking, the samples were filtered (0.45 μm), and the filtrate was analyzed for soluble PO_4-P. To determine the EPC, the PO_4-P sorbed or desorbed (μg P/g soil) was regressed on the PO_4-P in solution (mg/l) after reaction.

Leaching of Soluble PO_4-P From Green Plants

A small multiple-intensity rainfall simulator (25) was used to rain on cotton plants with subsequent measurement of soluble PO_4-P in washoff from the 100% cotton canopy. Simulated rain was applied at 2.5 cm/hr for 3 hr to mature cotton plants having 1.8 and 2.3 m^2 of leaf surface area. Rain at 2.5 cm/hr for 2.3 hr has a 2-year return period at Meridian, Miss. Runoff from the cotton canopy was sampled throughout the hydrograph, using collection pans under the canopy to intercept and route the flow. The experiment was conducted late in the growing season; plant senescence and defoliation were rapidly approaching. No organophosphate insecticides were applied to the cotton during the growing season.

RESULTS AND DISCUSSION

Many mathematical models now use the P sorption-desorption isotherm to estimate soluble PO_4-P and sediment labile P in transport (26). Isotherms may be linear, logarithmic, or power functions (Freundlich) depending on the type of phosphate compounds, sediment characteristics, and soluble PO_4-P concentration range. The commonly used Freundlich equation is $Y = KX^n$, where

Y = µg P sorbed or desorbed/g sediment,
K = distribution coefficient, usually assumed constant,
X = PO_4-P concentration in mg P/l, and
n = slope, usually less than 1.

The calculations often are simplified by assuming linearity (26). This sorption equation, together with other equations relating P uptake by plants, release of P from the organic P pool, and so forth, can be used to simulate the soil P status and estimate the soluble PO_4-P or sediment labile P movement, or both, on a storm runoff basis.

Sorption-desorption isotherms have been used by Taylor and Kunishi (40) and Kunishi and Taylor (16) to distribute P between runoff and sediment at chemical equilibrium and to estimate P losses in the water and sediment. The isotherms (phosphate potential curves) (2) provide measures of the EPC, P_{S1}, and buffer capacity of the soil (fig. 1). These curves are often linear or approximately so over much of the soluble PO_4-P concentration range of interest. The slopes of the curves indicate the buffer capacity of the soil; curves with steep slopes indicate a large capacity to sorb soluble PO_4-P per unit of PO_4-P. Using the isotherm, soluble PO_4-P and P_{S1} concentrations can be calculated (16) and expressed as a function of the soil concentration. Knowing the amount of soil and sediment responsible for sorption-desorption of PO_4-P and assuming equilibrium condition, the soluble PO_4-P and P_{S1} in runoff at the edge-of-field can be calculated on a per storm basis. This approach requires soil sampling and analysis to determine the relationships.

Much of this investigation was to consider easier methods or the use of available data for constructing a sorption-desorption isotherm (buffer curve), thus removing or simplifying some of the analytical requirements. If we could assume (1) a linear isotherm or approximate linearity through the soluble PO_4-P concentration range of interest or (2) the error introduced by assuming linearity is less than that introduced by the hydrology or erosion calculations, the working curve would be defined by (EPC, P_{S1}). Using figure 1, these would be (0.18, 0), (0, -28).

Using regionally and numerically limited soil and runoff data, we attempted to relate water extractable PO_4-P and available soil test P data to the EPC and P_{S1}, respectively. In the later sections, we compare the annual discharge-weighted soluble PO_4-P observed in runoff from plots and watersheds

Table 1.--Comparison of soil PO_4-P intensity values with discharge-weighted soluble PO_4-P concentrations measured in runoff (1976 WY) from research plots and watersheds

Plots and watersheds	Extractants[1/] 0.01 M $CaCl_2$	Extractants[1/] Water	Isotherm[2/] EPC	In runoff PO_4-P	In runoff Sediment concentration
	mg/l				
Plots-MAFES[3/]					
No-till-grain C-4	0.06	0.10	0.18	0.66	170
No-till-grain C-9	.07	.12	.22	.48	350
No-till-silage C-6	.03	.08	.15	.32	300
No-till-silage C-12	.05	.08	.15	.52	370
No-till-grain-2 cult. C-3	.06	.19	.50	.64	800
No-till-grain-2 cult. C-11	.03	.03	.07	.24	880
Conventional-grain C-2	.02	.01	.04	.13	6,600
Conventional-grain C-8	.02	.01	.04	.09	5,500
Conventional-silage C-1	.02	.005	.01	.06	8,700
Conventional-silage C-10	.03	.005	.02	.04	4,800
Watersheds-MAFES[3/]					
No-till-silage WC-1	.03	.12	.34	.99	130
No-till-grain WC-2	.04	.11	.19	1.20	130
No-till-grain-2 cult. WC-3	.04	.08	.17	.97	380
Watersheds-Delta					
McWilliams-802	.07	.14	.25	.20	3,600
McWilliams-803	.03	.09	.14	.15	3,800
Fisher-807	.05	.12	.28	.21	1,500
Fisher-808	.03	.13	.21	.24	1,500

1/ Soil concentrations were 100,000 mg/l.
2/ Soil concentrations were 10,000 mg/l.
3/ North Mississippi Agricultural and Forestry Experiment Station, Holly Springs

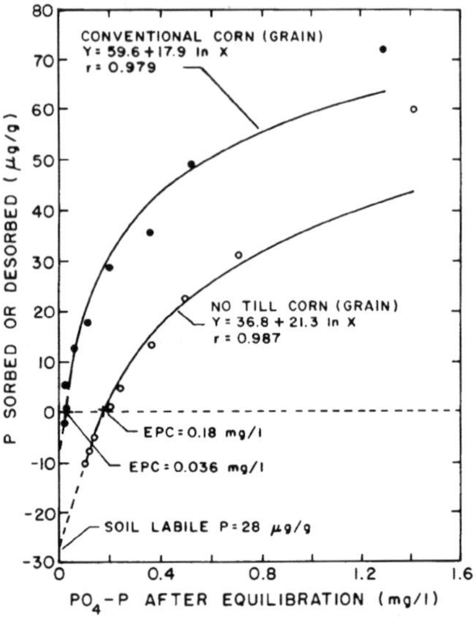

Figure 1.--Phosphorus sorption-desorption isotherms of MAFES soils.

to estimates of soluble PO_4-P and P_{s1} based on soil analysis. We also identify and describe the approximate expected magnitude of major nonsoil sources of soluble PO_4-P and then examine the variability of soluble PO_4-P in runoff from storm to storm.

Evaluation of Methods for Estimating Soluble PO_4-P (Intensity)

Soil samples from the plots and watersheds were extracted with water and 0.01 M $CaCl_2$ to determine if these simple tests might predict soluble PO_4-P in runoff (table 1). Soil concentrations used were much higher than the suspended sediment concentrations observed in runoff because Hope and Syers (13) found that at a given time before equilibrium was reached, the smaller solution:soil ratios were closer to equilibrium than were the larger ratios. Hope and Syers (13) used solution:soil ratios corresponding to soil concentrations of about 25,000, 100,000, and 200,000 mg/l. In this study, water extractable PO_4-P was independent of the soil:solution ratio at soil concentrations > 20,000 mg/l (fig. 2).

As expected, water extractions underestimated the annual discharge-weighted soluble PO_4-P in runoff from all plots and watersheds (table 1). The PO_4-P concentrations in runoff from the MAFES plots and watersheds were 3 to 13 times greater than concentrations measured in water extracts. In contrast, annual discharge-weighted PO_4-P concentrations measured in runoff from the Delta watersheds were only about twice the concentrations measured in water extracts. Water extractable PO_4-P (Y) from all soils was related significantly (0.01 level of probability) to the annual discharge-weighted PO_4-P in runoff (X) as a ln transformed power function equation ($r^2 = 0.61$), $\ln Y = \ln 1.8 + 0.64 \ln X$ (data not shown).

The PO_4-P extracted with 0.01 M $CaCl_2$ generally paralleled but was substantially less than that extracted with water. It greatly underpredicted the PO_4-P observed in runoff from either the Delta or MAFES watersheds. The $CaCl_2$ extractant has been reported to increase sorption or decrease PO_4-P desorption by soil systems (10, 30, 33, 36, 37). Although unclear, the mechanism of $CaCl_2$ influence on PO_4-P sorption in soil systems may be related to the increased ionic strength and the specific influence of the Ca^{++} ion in increasing the positive charge of the sorbing surface (36) (R. A. Leonard, personal communication). Kinetic studies of PO_4-P sorption on soil in $CaCl_2$

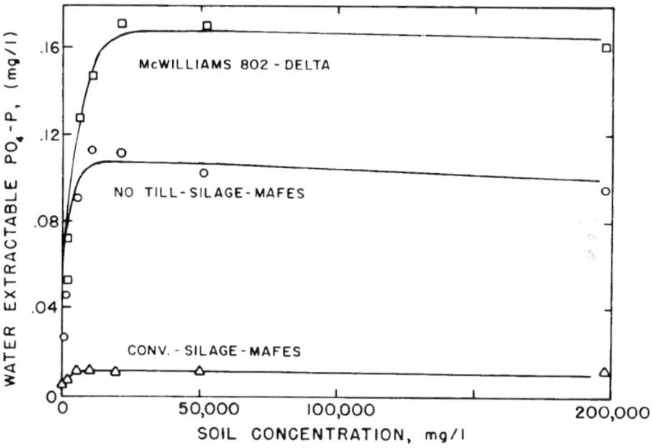

Figure 2.--Water extractable PO_4-P concentrations as related to soil concentration.

systems suggest that $CaCl_2$ increases the rate of PO_4-P sorption (30, 36). Indeed, some kinetic studies have shown that when PO_4-P sorption data are extrapolated to infinite time, the equilibrium PO_4-P concentration is independent of both ionic strength and cation species (36). For either soil or runoff systems, the equilibrium condition is seldom known.

EPC values appear to give good estimates of the annual discharge-weighted soluble PO_4-P in runoff (table 1) for the Delta watersheds. Soluble PO_4-P concentrations in runoff from the MAFES plots were underestimated by a factor of 3 for duplicate plots, but sediment concentrations in runoff from no-till practices were insufficient to buffer the soluble PO_4-P. The decomposition of crop residues on the no-till plots also may have produced organic acids that can form stable complexes with Fe and Al, thereby blocking P sorption (39). Except for the MAFES watersheds, EPC values were higher for soils from the grain plots, suggesting that crop residues were a source of PO_4-P. The annual discharge-weighted PO_4-P concentrations (Y) in runoff from all plots and watersheds were related significantly (0.01 level of probability) to EPC values (X) as a ln transformed power function equation ($r^2 = 0.61$), $\ln Y = \ln 1.4 + 0.75 \ln X$ (data not shown).

EPC values were determined on soils from conventional and no-till corn (grain) plots at soil concentrations of 2,000, 10,000, and 100,000 mg/l after 2 and 24 hr of shaking (table 2). EPC values determined at a soil concentration of 100,000 mg/l were significantly lower (0.05 level of probability) than values determined at 2,000 and 10,000 mg/l for both time periods. Values determined at 2,000 and 10,000 mg/l were not significantly different for both time periods. Using contact periods of less than 4 hr, White (46) and Taylor and Kunishi (40) found that EPC values increased as the soil concentration increased for nonequilibrium soils. For "equilibrium" soils, EPC values were independent of the soil:solution ratio (40, 47). Equilibrium conditions, however, probably are not attained in runoff-transport processes, particularly at low sediment concentrations.

Table 2.--Equilibrium phosphorus concentrations (EPC) determined on soils from conventional and no-till corn (grain) plots at different soil concentrations after 2 and 24 hours of shaking

Soil concentration	Conventional		No-till	
	2 hours	24 hours	2 hours	24 hours
(mg/l)	EPC (mg/l)			
2,000	0.054 ± 0.014	0.037 ± 0.009	0.195 ± 0.012	0.212 ± 0.016
10,000	.046 ± .009	.040 ± .006	.206 ± .011	.184 ± .010
100,000	.030 ± .001	.015 ± .002	.164 ± .002	.110 ± .003

In this study, microbial uptake of inorganic P (45) may have decreased EPC values at a soil concentration of 100,000 mg/l (18) (table 2). Using mercuric chloride as a sterilant, Hope and Syers (13) concluded that the effect of solution:soil ratio on P sorption was controlled kinetically up to 146 hr and that sorption isotherms at equilibrium would be coincident, irrespective of the soil concentration. These findings are controversial, however. Barrow and Shaw (1) questioned the validity of the approach used by Ryden and Syers (36) and Hope and Syers (13) of extrapolating from short sections of reciprocal-time graphs to estimate equilibrium. In determining EPC values, a contact time of less than 4 hr is recommended (H. M. Kunishi, personal communication), particularly at high soil concentrations, to reduce the potential microbial uptake of P. The use of a sterilant may prevent microbial uptake, but the effect of the sterilant on P sorption by the soil should be evaluated carefully.

EPC values were related to water extractable PO_4-P (fig. 3) with a coefficient of determination, r^2, of 0.87 (r^2 = 0.92 from two-way linear regression, Y = a + bx). Water extractable PO_4-P values were about 46% of the EPC values. This close agreement is not surprising since the two methods are

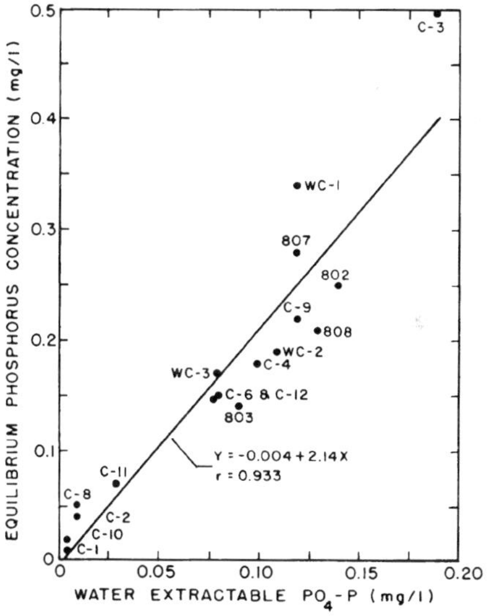

Figure 3.--Equilibrium phosphorus concentrations of soils as a function of water-extractable PO_4-P concentrations.

somewhat similar. In the extraction of several New Zealand soils, the PO_4-P removed in two one-hour distilled water extractions was related to the amount of physically sorbed P whereas $NaHCO_3$ extracted more than the physically sorbed P from noncalcareous soils (37).

Evaluation of Methods for Estimating Soil Labile P (Capacity Factor)

Soil samples from the research plots and watersheds were extracted using 0.5 M $NaHCO_3$, Dowex 1-X8 resin, the Mississippi soil test, and Bray's No. 1 solution to determine if any of these simple tests might predict the isotherm P_{s1} (table 3). Equations relating the isotherm P_{s1} to the soil test and resin extractable P values are given in table 4. The P_{s1} values were related to the $NaHCO_3$ and resin extractable P values with less than 20% of the error in P_{s1} unexplained. State soil tests using the $NaHCO_3$, Mississippi, and Bray methods were designed to measure the amount of soil P available for plant growth (32). These empirical tests were developed by correlating available P with crop growth in field plot and greenhouse studies. Hence, these tests are related to the soil P capacity factor so it is not surprising that these soil test P values correlate with the isotherm P_{s1} values.

P Intensity-Capacity Relationships

The complex interactions between the intensity and capacity factors are not emphasized in this chapter but do exist. The basic problem is that soils can have greatly different P capacities for the same P intensity as indicated by changes in the slope of the sorption-desorption isotherm. Thus, correlations between EPC values and soil test values may be of little potential predictive value unless limited to a common group of soils or an area.

In this chapter, EPC values (table 3) are expressed as a function of the Mississippi soil test values for plant available P in figure 4. A linear equation ($r^2 = 0.77$) applies to all plot and watershed soils with only 23% of the error in EPC unexplained. Similar results were obtained with Bray's No. 1 extracting solution ($r^2 = 0.67$) for plant available P. The EPC values from some Ohio soils (22) and the soluble PO_4-P in runoff from fertilized Indiana soils (35) were related to the quantity of Bray's No. 1 extractable P in their respective sediment systems. While this approach to estimating EPC values from plant available P soil tests may be locally useful, the relationship must be calibrated for that locality. Moreover, the approach must be tested and verified for application to another area.

Estimating soluble PO_4-P and sediment labile P in runoff by any of the methods in tables 1 and 3 obviously is based on two major assumptions: (1) suspended sediment in runoff is not enriched in P relative to the soil, and (2) suspended sediment is the only source of soluble PO_4-P. Sediments may be enriched in P relative to the soil, and this enrichment can be estimated from soil P, flow and sediment characteristics, and/or by calibration data from specific sites. Release of soluble PO_4-P from noneroding soil, fertilizers, crop residues, and green plants may contribute significant quantities of soluble PO_4-P to runoff.

Table 3.--Equilibrium phosphorus concentrations (EPC); soil labile P (P_{sl}); and 0.5 M $NaHCO_3$, resin, and Bray's extractable P from plot and watershed soils at the North Mississippi Agricultural and Forestry Experiment Station (MAFES) and Mississippi Delta locations

Plots and watersheds	EPC	Soil labile P P_{sl} 1/	P_{sl} 2/	Extractable P $NaHCO_3$ 3/	Resin 4/	Bray 5/
	mg/l	µg/g				
Plots-MAFES						
No-till-grain C-4	0.18	28	61	20	27	39
No-till-grain C-9	.22	26	94	22	29	48
No-till-silage C-6	.15	17	62	16	22	53
No-till-silage C-12	.15	21	69	18	24	46
No-till-grain-2 cult. C-3	.50	33	91	21	33	93
No-till-grain-2 cult. C-11	.07	12	40	10	18	26
Conventional-grain C-2	.04	10	23	9	11	22
Conventional-grain C-8	.04	11	34	13	11	33
Conventional-silage C-1	.01	8	6	7	6	17
Conventional-silage C-10	.02	7	8	5	2	15
Watersheds-MAFES						
No-till-silage WC-1	.34	22	56	23	27	60
No-till-grain WC-2	.19	22	64	21	35	41
No-till-grain-2 cult. WC-3	.17	29	80	22	37	53
Watersheds-Delta						
McWilliams-802	.25	37	86	29	56	45
McWilliams-803	.14	20	53	18	36	29
Fisher-807	.28	20	35	12	24	26
Fisher-808	.21	15	29	13	24	26

1/ Determined by extrapolating the soil P sorption-desorption isotherm to the ordinate (zero PO_4-P concentration).

2/ Calculated from logarithmic isotherm equations using 0.01 mg/l soluble PO_4-P concentration.

3/ Soil:solution ratios were 1:20.

4/ Soil:solution ratios were 1:10.

5/ Soil:solution ratios were 1:7.

Leaching of Crop Residues as Sources of PO_4-P

The release of soluble nutrients from crop residues is the result of several factors, including microbial decomposition, leaching by rainfall (amount, intensity, and distribution) and freezing and thawing (42) (see also vol. III, ch. 15). In general, the release of PO_4-P from crop residues is greatest during the fall (44) and total quantities released throughout the year may constitute an appreciable portion of the annual PO_4-P yield in runoff (41).

In this chapter, annual discharge-weighted soluble PO_4-P concentrations in runoff from conventional-till grain plots were greater than from conventional-till silage plots having similar discharge-weighted seasonal runoff, and soil and sediment total P losses (table 5). The higher PO_4-P concentrations in runoff from the grain plots occurred during the fall and winter months (October to March) and are attributed, in part, to the release of PO_4-P from crop residues (fig. 5). Greater contributions of PO_4-P from accumulated and unincorporated residues on the no-till practices would be expected as residues from conventional-till practices were incorporrated into the soil each spring.

Table 4.--Equations relating soil isotherm labile P (Y, µg/g) to P extracted (X, µg/g) from MAFES and Delta soils by different extractants

Extractant	Equation[1/]	r^2
0.5 M $NaHCO_3$	$Y = -0.44 + 1.24X$	0.84
Resin	$Y = 4.93 + 0.60X$	.81
Mississippi soil test	$Y = -2.59 + .42X$	.70
Bray's No. 1	$Y = 6.87 + .33X$	.52

1/ Soil labile P was determined from soil P sorption-desorption isotherms at 0 soluble PO_4-P concentration.

Residues on no-till plots would be expected to increase the EPC of surface soils. In laboratory studies, several PO_4-P sources, including chemical, chemical plus straw, straw, manure, and other organic residues markedly increased the EPC values of several soils (34, 39). EPC values from table 3 are shown in table 6 together with PO_4-P concentration data derived from cumulative-storm mean-concentration frequency-distribution curves. The EPC, C_{50}, DW, and L_{50} concentration values all decrease in the order: No-till grain > no-till silage > conventional-till grain > conventional-till silage. The EPC value is a measure of the higher P intensity status of the 0 - 2.5 cm layer of surface soil in no-till.

In general, PO_4-P concentrations for all plots (table 6) increased in the order EPC < C_{50} < DW < L_{50} emphasizing that the EPC method underestimates PO_4-P inputs other than the P intensity status of the soil. Sediment concentrations in runoff from no-till practices were low (table 1) and insufficient to buffer the soluble PO_4-P. At much higher sediment concentrations, the soluble PO_4-P would be a function of equilibrium between sediment labile P and soluble PO_4-P in runoff.

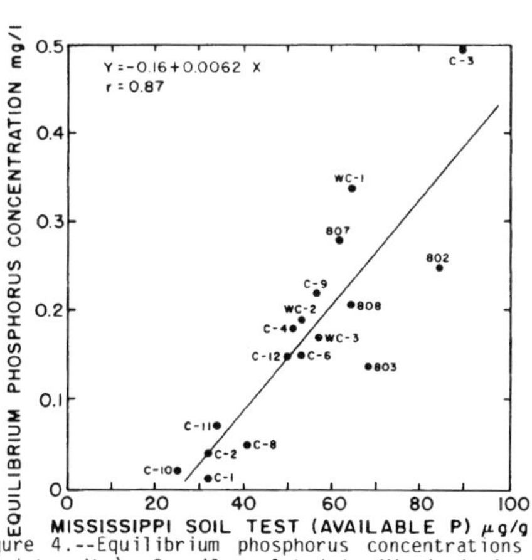

Figure 4.--Equilibrium phosphorus concentrations (P intensity) of soils related to Mississippi soil test values for plant available P (P capacity).

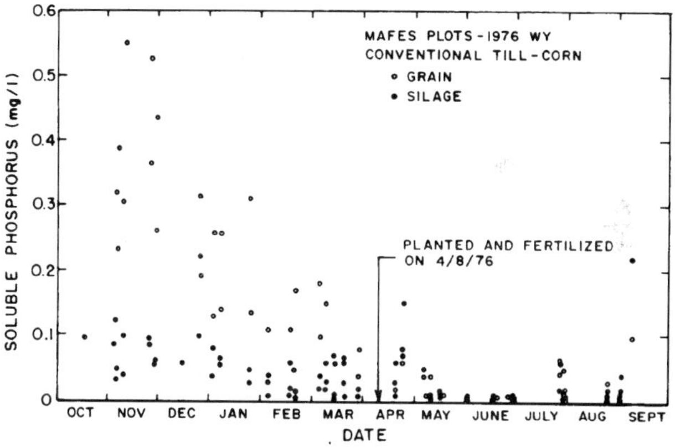

Figure 5.--Temporal distribution of storm discharge-weighted soluble PO_4-P concentrations in runoff from conventionally-tilled corn grown for grain and silage.

Table 5.--Discharge-weighted seasonal runoff, and soil, soluble PO_4-P, and sediment total P losses frc conventional-till grain and silage plots at North Mississippi Agricultural and Forestry Experiment Station, Holly Springs

Plots[1/]	Run-off	Soil loss	Sediment concentration[3/]	Soluble PO_4-P loss	Soluble PO_4-P concentration[3/]	Sediment total P loss	Sediment total P concentratior
	(cm)	(t/ha[2/])	(mg/l)	(g/ha)	(mg/l)	(kg/ha)	(mg/l)
				October-December			
Grain	5.3	0.16	300	201	0.38	0.30	0.57
Silage	6.3	.19	300	50	.08	.27	.43
				January-March			
Grain	14.7	1.33	910	163	.11	1.73	1.18
Silage	14.6	1.99	1,360	37	.03	1.96	1.34
				April-June			
Grain	13.3	15.05	11,300	40	.03	10.89	8.19
Silage	12.6	14.64	11,600	50	.04	9.84	7.81
				July-September			
Grain	5.7	7.24	12,700	22	.04	4.69	8.23
Silage	5.8	7.92	13,660	58	.10	5.36	9.24
				Annual			
Grain	39.0	23.8	6,100	426	.11	17.6	4.5
Silage	39.3	24.7	6,280	195	.05	17.4	4.4

1/ Duplicate plots.
2/ Metric tons/hectare.
3/ Discharge-weighted for 1976 WY

Table 6.--Comparison of EPC values and annual discharge-weighted PO_4-P concentrations in runoff from duplicate corn plots at North Mississippi Agricultural and Forestry Experiment Station, Holly Springs.

Plots	PO_4-P concentration			
	EPC 1/	C_{50} 2/	DW 3/	L_{50} 4/
	(mg/l)			
No-till grain	0.20	0.52	0.57	0.60
No-till silage	.15	.30	.39	.37
Conventional-till grain	.04	.07	.11	.22
Conventional-till silage	.02	.03	.06	.09

1/ Equilibrium phosphorus concentration determined from soil P sorption-desorption isotherms, taken from table 3.

2/ Median concentration in runoff determined from cumulative frequency distribution curves.

3/ Discharge-weighted concentration for 1976 WY, that is, annual PO_4-P loss/annual runoff.

4/ 50% of annual PO_4-P loss was produced by concentrations greater than this value as determined from cumulative frequency distribution curves.

Leaching of Green Crops as Sources of PO_4-P

The leaching and washoff of green plants by rainfall may be an additional significant source of soluble PO_4-P to runoff during the growing season and may contribute to the storm-to-storm variability in PO_4-P concentrations. For example, significant quantities of PO_4-P leached from prairie vegetation during the month of April was attributed to water soluble PO_4-P extracted from growing plants by precipitation (44). Similarly, moisture retained on leaves after a rainstorm contained significant quantities of soluble PO_4-P (6). Data in figure 6 are from a preliminary study on the leaching of soluble PO_4-P from green cotton plants by artificial rainfall. Rain was applied at 2.5 cm/hr for a 3-hr period. During the first hr of rainfall, the plant runoff-weighted PO_4-P concentration was about 0.15 mg/l, a relatively high value, and decreased to 0.05 mg/l during the last 2 hr. All data have been corrected for background PO_4-P in the water supply (rainfall), which amounted to 0.01 mg/l throughout the event. Research is needed to evaluate this potential PO_4-P input as functions of the probability of rainfall and runoff, crop growth stage (leaf area), and plant PO_4-P recovery time. Under natural rainfall, PO_4-P leached from the growing plant becomes part of the soil-water P matrix and is sorbed by the soil, leached from the runoff extraction zone, and/or transported in runoff.

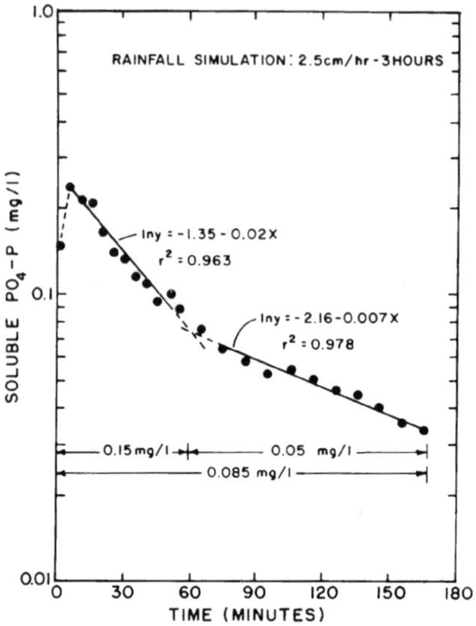

Figure 6.--Soluble PO_4-P concentrations in cotton canopy runoff as a function of time.

Variability of PO_4-P Concentrations in Runoff from Storm to Storm

Because of the diffuse nature of PO_4-P sources and the nonequilibrium between sediment and soluble PO_4-P, considerable variation in PO_4-P concentrations is observed in runoff from storm to storm (fig. 5). Thus, errors are to be expected when soluble PO_4-P concentrations are estimated on a storm basis by empirical methods such as those examined herein, even if a reasonable estimate of the annual discharge-weighted concentration can be made. For example, consider the McWilliams Delta watershed 802 for which 2 yr data are shown as a cumulative frequency distribution curve in figure 7. The storm discharge-weighted PO_4-P concentrations are ln normally distributed with a geometric mean concentration of 0.21 mg/l and a geometric standard deviation of 1.62 mg/l. Thus, for this ln normal distribution the PO_4-P concentration ranges (mg/l) are 0.13 to 0.34 within one geometric standard deviation, and 0.08 to 0.55 within two geometric standard deviations. The PO_4-P concentration in a

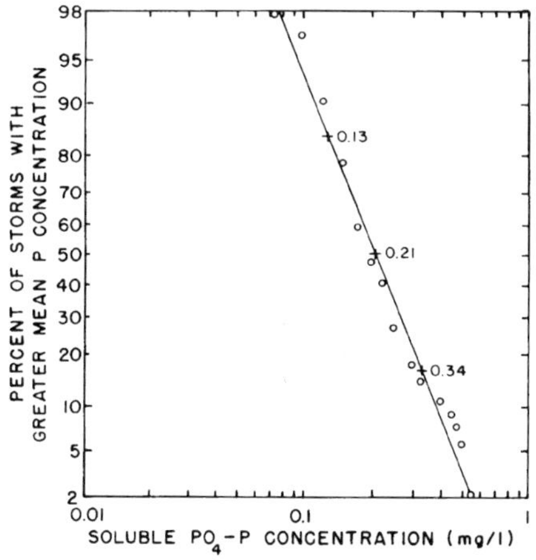

Figure 7.--Cumulative frequency distribution of storm discharge-weighted soluble PO_4-P in runoff from McWilliams watershed 802 (logarithmic probability).

water extract of the soil was about 0.14 mg/l (table 1) but about 75% of the storm discharge-weighted PO_4-P concentrations measured in runoff exceeded this simple laboratory estimate (fig. 7). Storm-to-storm variations in soluble PO_4-P remain unexplained, but the EPC value (table 1) provides a good estimate of the geometric mean (fig. 7) even though other potential PO_4-P inputs are ignored, including organophosphate insecticides applied for bollworm [Heliothis zea (Boddie)] control.

Soluble PO_4-P concentrations estimated from sorption-desorption isotherms (phosphate potential curves) may not provide good estimates of storm PO_4-P concentrations in runoff from fallow soils where other P inputs are excluded. Runoff-extraction processes may produce higher PO_4-P concentrations under field conditions than measured by the pseudoequilibrium EPC method. Research is needed to clarify this issue. It is premature to conclude that the difference between measured storm PO_4-P concentrations and classical equilibrium values is due solely to inputs by crops, crop residues, fertilizers, and so forth.

SUMMARY AND CONCLUSIONS

Soil P intensity values appear useful for estimating potentially soluble PO_4-P in runoff from croplands. Intensity values measured from PO_4-P sorption-desorption isotherms and water extracts of soils were used to estimate the annual discharge-weighted soluble PO_4-P concentrations in runoff from plots and watersheds differing greatly in soils, area, fertility, crops, and management. Equilibrium phosphorus concentration (EPC) values better estimated PO_4-P in runoff, even when PO_4-P from crop residues was an important source. EPC values of soils (Typic Fragiudalfs) from the North Mississippi Agricultural and Forestry Experiment Station (MAFES) underestimated by a factor of 3 the annual discharge-weighted PO_4-P concentrations in runoff from conventional and no-till practices, but sediment concentrations in runoff from no-till practices were very low and insufficient to buffer the soluble PO_4-P in runoff. Runoff-extraction processes also may produce higher PO_4-P concentrations under field conditions than measured by the pseudoequilibrium EPC method. EPC values of Mississippi Delta soils closely approximated the annual discharge-weighted PO_4-P concentrations in runoff. Water extractable PO_4-P concentrations (measured at soil concentrations of 100,000 mg/l) were about 46% of the EPC values (r^2 = 0.87) of MAFES and Delta soils. This correlation is expected since the two methods are somewhat similar.

Soil P extracted by 0.5 M $NaHCO_3$, Dowex 1-X8 resin, and the Mississippi soil test was related significantly (r^2 = 0.84, 0.81, and 0.70, respectively) to soil labile P (capacity) determined from the PO_4-P sorption-desorption isotherm method. This is not surprising since the soil tests for plant available P are related to the soil P capacity.

EPC values were correlated significantly with the Mississippi (r^2 = 0.77) and Bray (r^2 = 0.67) soil tests. This approach to estimating EPC values from plant available P tests is not theoretically sound, however, and if used, must be calibrated and verified for application to an area. Soils can have greatly different P capacities for the same P intensity depending on the buffer capacity of the soil, that is, the slope of the sorption-desorption isotherm.

Soil EPC values derived from sorption-desorption isotherms are recommended for estimating the annual discharge-weighted PO_4-P concentration in runoff. Soil labile P also can be determined from the isotherm, but with a greater uncertainty. Simpler tests, such as water-extractable PO_4-P and plant available P, can be used to estimate the EPC and the soil labile P, respectively. They must be tested and verified for application to an area, however, by correlation with EPC and soil labile P values derived from soil P isotherms. This requires sampling and analysis to determine the relationships. Soils should be sampled before fertilization and in accordance with recommended State procedures.

PO_4-P concentrations measured in runoff may vary considerably from storm to storm. Thus, even though EPC values may estimate the annual discharge-weighted PO_4-P concentration in runoff within a factor of 2 - 3, considerable variation about this estimated value can be expected on a storm basis. Crop residues; fertilizers, depending on method of incorporation and time of runoff after application; and possible leaching of PO_4-P from green plants can contribute significant and variable quantities of soluble PO_4-P.

The observed storm to storm variability of PO_4-P concentrations in runoff demonstrates another point. Models chosen today by the user cannot and need not do better than this variability. This variability, which greatly reduces the precision with which P loss estimates can be made, will remain until the sources of variability are better defined.

Measuring P in runoff from research plots and watersheds will continue to provide (1) a needed data base for model testing and verification, and (2) general guidelines on P sources and temporal distributions, but will not adequately quantify the individual P inputs, their variability, and their relative significance. Similarly, continuing measurements of EPC and distribution coefficients, K, in the laboratory may refine procedures for estimating P intensity and capacity factors but will provide little insight as to their reliability in the field until tested and verified.

ACKNOWLEDGMENTS

We thank E. Gholston and J. D. Lancaster, Mississippi State University, for conducting plant available phosphorus analyses on selected soils. We also thank H. M. Kunishi, USDA-SEA-AR, for confirming the method of labile P calculations and for reviewing the manuscript.

REFERENCES

(1) Barrow, N. J., and T. C. Shaw.
1979. Effects of solution:soil ratio and vigour of shaking on the rate of phosphate adsorption by soil. Journal of Soil Science 30:67-76.

(2) Beckett, P. H. T., and R. E. White.
1964. Studies on the phosphate potentials of soils. Part III. The pool of labile inorganic phosphate. Plant and Soil 21:253-282.

(3) Black, C. A.
1970. Behavior of soil and fertilizer phosphorus in relation to water pollution. In Agricultural Practices and Water Quality, T. L. Willrich and G. E. Smith (eds.), Iowa State University Press, Ames, Iowa. pp. 72-93.

(4) Brewster, J. L., A. N. Gancheva, and P. H. Nye.
1975. The determination of desorption isotherms for soil phosphate using low volumes of solution and an anion exchange resin. Journal of Soil Science 26:364-377.

(5) Carter, C. E., and D. A. Parsons.
1967. Field tests on the Coshocton-type wheel runoff sampler. Transactions of the American Society of Agricultural Engineers 10:133-135

(6) Cowen, W. F., and G. F. Lee.
1973. Leaves as source of phosphorus. Environmental Science and Technology 7:853-854.

(7) Dendy, F. E.
1973. Traversing-slot runoff sampler for small watersheds. U.S. Department of Agriculture, Agricultural Research Service, ARS-S-15. (Series discontinued. Agricultural Research Service is now Science and Education Administration-Agricultural Research.)

(8) Ekpete, D. M.
1976. Evaluation of chemical methods for the determination of available phosphorus in waterlogged soils. Soil Science 121:217-221.

(9) Evans, R. L., and J. J. Jurinak.
1976. Kinetics of phosphate release from a desert soil. Soil Science 121:205-211.

(10) Heylar, K. R., D. N. Munns, and R. G. Burau.
1976. Adsorption of phosphate by gibbsite. I. Effects of neutral chloride salts of calcium, magnesium, sodium, and potassium. Journal of Soil Science 27:307-314.

(11) Hislop, J., and I. J. Cooke.
1968. Anion exchange resin as a means of assessing soil phosphate status: A laboratory technique. ·Soil Science 105:811.

(12) Holt, R. F., H. P. Johnson, and L. L. McDowell.
1973. Surface water quality. Proceedings National Conservation Tillage Conference, Des Moines, Iowa. pp. 141-156.

(13) Hope, G. D., and J. K. Syers.
1976. Effects of solution:soil ratio on phosphate sorption by soils. Journal of Soil Science 27:301-306.

(14) Huettl, P. J., R. C. Wendt, and R. B. Corey.
1979. Prediction of algal-available phosphorus in runoff suspensions. Journal of Environmental Quality 8:130-132.

(15) Johnson, A. H., D. R. Bouldin, E. A. Goyette, and A. M. Hedges.
1976. Phosphorus loss by stream transport from a rural watershed: Quantities, processes, and sources. Journal of Environmental Quality 5:148-157.

(16) Kunishi, H. M., and A. W. Taylor.
1977. Predicting pollution potential of phosphorus at heavy application rates. Proceedings International Seminar on Soil Environment and Fertility Management in Intensive Agriculture, Tokyo, Japan. pp. 350-356.

(17) ------, A. W. Taylor, W. R. Heald, W. J. Gburek, and R. N. Weaver.
1972. Phosphate movement from an agricultural watershed during two rainfall periods. Journal Agricultural and Food Chemistry 20:900-905.

(18) Larsen, S., and A. E. Widdowson.
1964. Effect of soil/solution ratio on determining the chemical potentials of phosphate ions in soil solutions. Nature, (London) 203: 942.

(19) Lee, G. F.
1973. Role of phosphorus in eutrophication and diffuse source control. Water Research 7:111-128.

(20) ------, W. Rast, and R. A. Jones.
1978. Eutrophication of water bodies: Insights for an age-old problem. Environmental Science and Technology 12:900-908.

(21) Marshall, C. E.
1964. The physical chemistry and mineralogy of soils. Vol. I. Soil materials. 388 pp. John Wiley and Sons, Inc., New York, NY.

(22) McCallister, D. L., and T. J. Logan.
1978. Phosphate adsorption-desorption characteristics of soils and bottom sediments in the Maumee River basin of Ohio. Journal of Environmental Quality 7:87-92.

(23) McDowell, L. L., and E. H. Grissinger.
1976. Erosion and water quality. In Proceedings 23d National Watershed Congress, Biloxi, Miss., pp. 40-56.

(24) ------, and K. C. McGregor.
1980. Nitrogen and phosphorus losses in runoff from no-till soybeans. Transactions of the American Society of Agricultural Engineers (In press).

(25) Meyer, L. D., and W. C. Harmon.
1979. Multiple-intensity rainfall simulator for erosion research on row sideslopes. Transactions of the American Society of Agricultural Engineers 22:100-103.

(26) Mulkey, L. A., and J. W. Falco.
1977. Sedimentation and erosion control implications for water quality management. In Proceedings National Symposium on Soil Erosion and Sedimentation by Water. American Society of Agricultural Engineers, St. Joseph, Mich. pp. 69-90.

(27) Murphree, C. E., C. K. Mutchler, and L. L. McDowell.
1976. Sediment yields from a Mississippi Delta watershed. In Proceedings 3d Federal Interagency Sedimentation Conference, Denver, Colo. pp. 1-99 to 1-109.

(28) Murphy, J., and J. P. Riley.
1962. A modified single solution method for the determination of phosphate in natural waters. Analytica Chimica Acta. 27:31-36.

(29) Murrmann, R. P., and M. Peech.
1969. Relative significance of labile and crystalline phosphates in soil. Soil Science 107:249-255.

(30) Novak, L. T., and D. C. Adriano.
1975. Phosphorus movement in soils: Soil-orthophosphate reaction kinetics. Journal of Environmental Quality 4:261-266.

(31) Olsen, S. R., R. A. Bowman, and F. S. Watanabe.
1977. Behavior of phosphorus in the soil and interactions with other nutrients. In Phosphorus in Agriculture, No. 70. Proceedings ISMA Symposium; Paris, France. pp. 31-46.

(32) ------, and L. A. Dean.
1965. Phosphorus. In Methods of Soil Analysis. Part 2. Chemical and Microbiological Properties. C. A. Black (ed.) American Society Agronomy, Madison, Wis. pp. 1035-1049.

(33) Rajan, S. S. S., and R. L. Fox.
1972. Phosphate adsorption by soils. 1. Influence of time and ionic environment on phosphate adsorption. Communications in Soil Science and Plant Analysis 3:493-504.

(34) Reddy, G. Y., E. O. McLean, G. D. Hoyt, and T. J. Logan.
1978. Effects of soil, cover crop, and nutrient source on amounts and forms of phosphorus movement under simulated rainfall conditions. Journal of Environmental Quality 7:50-54.

(35) Romkens, M. J. M., and D. W. Nelson.
1974. Phosphorus relationships in runoff from fertilized soils. Journal of Environmental Quality 3:10-13.

(36) Ryden, J. C., and J. K. Syers.
1975. Rationalization of ionic strength and cation effects on phosphate sorption by soils. Journal of Soil Science 26:395-406.

(37) ------.
1977. Origin of the labile phosphate pool in soils. Soil Science 123:353-361.

(38) Schreiber, J. D., D. L. Rausch, and L. L. McDowell.
1977. Callahan Reservoir: II. Inflow and outflow suspended sediment phosphorus relationships. Transactions of the American Society of Agricultural Engineers 20:285-290.

(39) Singh, B. B., and J. P. Jones.
1976. Phosphorus sorption and desorption characteristics of soil as affected by organic residues. Soil Science Society of America Journal 40:389-394

(40) Taylor, A. W., and H. M. Kunishi.
1971. Phosphate equilibria on stream sediment and soil in a watershed draining an agricultural region. Journal Agricultural and Food Chemistry 19:827-831.

(41) Timmons, D. R., and R. F. Holt.
1977. Nutrient losses in surface runoff from a native prairie. Journal of Environmental Quality 6:369-373.

(42) ------, R. F. Holt, and J. J. Latterell.
1970. Leaching of crop residues as a source of nutrients in surface runoff water. Water Resources Research 6:1367-1375.

(43) Weir, C. C., and R. J. Soper.
1962. Adsorption and exchange studies of phosphorus in some Manitoba soils. Canadian Journal of Soil Science 42:31-42.

(44) White, E. M.
1973. Overwinter changes in the percent Ca, Mg, K, P, and N in vegetation and mulch in an eastern South Dakota prairie. Agronomy Journal 65:680-681.

(45) White, R. E.
1964. Studies on the phosphate potentials of soils. Part II. Microbial effects. Plant and Soil 20:184-193.

(46) ------
1966. Studies on the phosphate potentials of soils. IV. The mechanism of the "soil/solution ratio effect". Australian Journal Soil Research 4:77-85.

(47) ------, and P. H. T. Beckett.
1964. Studies on the phosphate potentials of soils. Part I - The measurement of phosphate potential. Plant and Soil 20:1-15.

Chapter 15. SOLUBLE N AND P CONCENTRATIONS IN SURFACE RUNOFF WATER

D. R. Timmons and R. F. Holt[1]

The loss of soluble nutrients in surface runoff water for a given runoff event is a function of runoff water volume and soluble nutrient concentration. Assuming these soluble nutrients are dissolved in water saturating the upper 1 cm of the soil and are extracted imperfectly from this layer by overland flow, the nutrient concentrations are influenced by an extraction coefficient. Since existing values for extraction coefficients are limited, flow-weighted concentrations from published research were compared to standard treatments for evaluating the effects of various factors on soluble nutrient concentrations.

The flow-weighted concentration for soluble N or P was estimated as the ratio of the weight of nutrient lost to the weight of runoff. Generally, data for flow-weighted concentrations from runoff studies using natural precipitation have been reported on an annual basis, whereas those from simulated rainfall studies have been reported for a series of applied rainstorms.

Soluble N and P concentrations in runoff water result from the interactions of factors which include precipitation, vegetation, fertilizer, tillage, plant residue, and/or soil. Based on existing research data, estimates on how these parameters affect soluble N and P concentrations are discussed in the following sections.

VEGETATION

The vegetative cover on a soil can vary from native species, such as grasslands and forests, to intensely managed cropland. Soluble nutrients can be influenced by the type of vegetation, the stage of physiological development, and the climatic factors preceding the runoff event (5, 6, 18). The leaching of vegetation by precipitation or runoff, or both, is a mechanism that contributes nutrients to surface runoff.

Tables 1 and 2 show the influence of vegetation on soluble N and P concentrations. Compared with fallow (table 1), the concentration of inorganic N (NH_4-N + NO_3-N) changed by 0.9 to 1.1 times for corn, oats, and alfalfa at the two locations in the northwest part of the Corn Belt. Both soluble P concentrations in runoff from vegetated land were higher than runoff from fallow land; ortho-P concentrations ranged from 1 to 4 times and total P from 1.8 to

---

1/ Soil scientists, USDA,SEA-AR, North Central Soil Conservation Research Laboratory, Morris, Minn.

3.3 times. Since fallow is not a cultural practice in the eastern United States, the change in soluble N and P concentrations for different crops also can be based on a single widely grown crop.

Table 2 shows the change in soluble N and P concentrations for different crops compared to corn (except for Oklahoma). Soluble NH_4-N concentrations were about the same for oats, generally, but increased for alfalfa and wheat; soluble NO_3-N concentrations increased slightly for oats and soybeans but decreased for alfalfa and wheat, and soluble P concentrations were inconsistent between locations. These data reflect medium to high fertility and conventional tillage practices.

In Oklahoma, the concentrations of nitrogen compounds in runoff from a continuously grazed upland watershed were a little higher than in runoff from a better managed watershed with rotation grazing, while phosphorus concentrations were a little lower. Runoff from cotton and wheat grown on more fertile bottomland soil had a slightly higher nitrogen concentration but considerably higher phosphorus concentration. Alfalfa, also grown on the fertile bottomland, provided the highest nutrient concentrations in runoff.

FERTILIZER

The concentrations of soluble N and P in surface runoff water will be influenced by (1) rate of application, (2) method of application, (3) form of fertilizer, (4) time of application, and (5) precipitation after fertilization. Considerable data have been reported for studies using simulated rainfall, and these are useful for determining the relative effects of various treatments.

Table 3 shows the effects of fertilizer rate on soluble N and P concentrations in surface runoff water. The data show that soluble N and P concentrations increased with higher fertilizer rates. Studies using simulated rainfall show that inorganic N concentration increased from about 2 to over 15 times compared to unfertilized land, ortho-P concentrations increased from about 3 to 6 times, and total soluble P concentrations increased 4 to 6 times. This large inorganic N concentration increase occurred on a steep slope (13%) for fertilizer broadcast on sod (table 3). The studies using natural precipitation showed inorganic N concentration increased about 1 to 5 times, and ortho-P concentration increased from about 1 to 4 times compared to less fertilized corn (table 3). Bluegrass watersheds in North Carolina showed increases of 1.5, 3.3, and 1.2 times for soluble NH_4-N, NO_3-N, and ortho-P concentrations, respectively, when the N and P applied were quadrupled. Results from Louisiana showed that a greater percentage of applied N and P fertilizer was lost in surface runoff from forage as the amounts applied increased (table 3).

The effect of fertilizer placement on soluble N and P in surface runoff is presented in table 4. Limited data show that incorporation of applied fertilizer can minimize its effect on soluble N and P concentrations since only negligible increases occurred when broadcast fertilizer was plowed-down (compared to no fertilizer). The concentration of soluble NO_3-N in runoff water changes the least because much of the NO_3-N in fertilizer dissolves rapidly and is transported into the soil by infiltrating water before surface runoff starts.

Table 1.--Effect of vegetation on soluble N and P concentrations in runoff water compared with fallow

Location	Soil	Slope	Type of precipitation	Crop	Fertilizer applied	Change in flow-weighted concentration						Reference
						Basis	NH_4-N	NO_3-N	Inorg. N	Ortho. P	Tot. P	
		(%)			(lb/acre)							
Minnesota[1/]	Barnes sil	6	P[2/]	Cont. corn	100 N, 26 P	Annual	1.5	0.9	1.0	3.6	3.2	(2)
				Rot. corn	50 N, 26 P	"	1.7	0.6	0.9	4.0	3.2	(2)
				Rot. oats	16 N, 27 P	"	1.4	1.1	1.1	2.2	2.0	(2)
				Rot. alfa.	0 N, 0 P	"	3.0	0.3	0.9	3.8	3.3	(2)
South Dakota[3/]	Agriboroll	5	P	Corn		Avg. concen.	0.8	1.4	1.0	2.0	2.0	(20)
				Oats		"	0.9	1.8	1.1	2.0	1.8	(20)
				Alfalfa		"	1.2	0.7	1.1	1.0	2.8	(20)
Indiana[4/]	Zanesville sil	13	S[5/]	Sod	0 N	3 storms			0.7			(10)
				Sod	200 N	"			10.8			(10)

1/ N as 33-0-0 and P as 0-46-0 broadcast and disked in before planting.
2/ P stands for natural precipitation.
3/ Based on average chemical composition of runoff water for 1970 and 1971.
4/ Based on flow-weighted concentrations for 5 in of water applied in 3 storms and 33-0-0 fertilizer broadcast on surface.
5/ S stands for simulated rainfall.

Table 2.--Effect of vegetation on soluble N and P concentrations in runoff water compared with corn

Location	Soil	Slope	Type of precipitation	Crop	Fertilizer applied	Change in flow-weighted concentration: Basis	NH_4-N	NO_3-N	Inorg. N	Ortho. P	Tot. P	Reference
		(%)			(lb/acre)							
Minnesota[1/]	Barnes sil	6	P	Rot. corn	50 N, 26 P	Annual	1.2	0.7	0.8	1.1	1.0	(2)
				Rot. oats	16 N, 27 P	"	0.9	1.2	1.1	0.6	0.6	(2)
				Rot. alfa.	0 N, 0 P	"	2.0	0.4	0.9	1.0	1.0	(2)
South Dakota[2/]	Agriborolls	5	P	Oats		Avg. concen.	1.0	1.2	1.1	1.0	0.9	(20)
				Alfalfa		"	1.4	0.5	1.1	0.5	1.4	(20)
New York[3/]	Lima-Kendaia sil		P	Soybeans-		Annual	0.9	1.3	1.3	0.9		(8)
			P	Winter wheat								
				Wheat		"	1.7	0.4	0.7	1.1		(8)
Michigan[4/]	Spinks fsl		P	Soybeans	56 N, 93 P	Annual	1.2	1.1	1.1	0.9	1.0	(4)
Vermont[5/]	Cabot sil	6	P	Alfalfa		9 sampling		0.1		1.1		(1)
				Hay	200 of	dates.		0.3		1.3		(1)
				Pasture	15-10-10							
Oklahoma[6/]	Renfro sil and	3	P	Pasture-		Median						
	Kingfisher sil,			Cont. graze		concen.	1.0	1.6	1.4	0.5	0.8	(12)
	Mclain sil,	<1	"	Alfalfa		"	4.2	5.2	4.9	81.0	17.3	(12)
	Mclain sicl,	<1	"	Wheat		"	0.8	1.7	1.4	20.0	4.7	(12)
	and Reinach sil	<1	"	Dryland cotton		"	0.5	2.1	1.6	38.0	7.6	(12)
		<1	"	Irrig. cotton		"	0.5	3.4	2.4	64.0	13.5	(12)

1/ Compared to continuous corn.
2/ Based on average chemical composition of runoff water for 1970 and 1971.
3/ Based on average annual flow-weighted concentrations for high and moderate fertility with good management.
4/ Based on average flow-weighted concentrations for 1974 and 1975; fertilizer shown is average applied in 1974 and 1975.
5/ Average concentration for 9 sampling dates.
6/ Based on median concentrations and compared to rotation grazed

Table 3.--Effect of fertilizer rate on soluble N and P concentrations in runoff water

Location	Soil	Slope	Type of precipitation	Crop	Fertilizer applied	Change in flow-weighted concentration						Reference
						Basis	NH_4-N	NO_3-N	Inorg. N	Ortho. P	Tot. P	
		(%)			(lb/acre)							
Indiana[1/]	Russel sil	6	S	Fallow	0 P	Avg. concen.				1.0	1.0	(11)(13)
				"	50 P	"				3.4	4.1	(11)(13)
				"	100 P	"				6.3	6.5	(11)(13)
Indiana[2/]	Zanesville sil	13	S	Fallow	0 N	3 storms			1.0			(10)
				Fallow	200 N	"			5.3			(10)
				Sod	0 N	"			1.0			(10)
				Sod	200 N	"			15.6			(10)
Indiana[3/]	Bedford sil	8-12	S	Corn	0 N, 0 P	1 storm	1.0	1.0	1.0	1.0		(14)
				Corn	150 N, 50 P	"	2.2	2.3	2.2	4.0		(14)
Iowa[4/]	Deep loess	2-4	S	Corn	150 N, 35 P	Annual	1.0	1.0	1.0	1.0		(15)(16)
				Corn	400 N, 86 P	"	1.2	1.4	1.3	1.3		(15)(16)
Missouri[5/]	Mexico sil	3	P	Corn	13 N	Annual			1.0			
				Corn	86 N	"			2.1			
				Corn	163 N	"			2.8			
				Corn	223 N	"			4.1			
				Corn	325 N	"			4.4			
New York[6/]	Silt loam	2-4	P	Corn	15 N, 13 P	Annual	1.0	1.0	1.0	1.0		(8)
				Corn	275 N, 66 P	"	3.7	5.7	5.1	4.1		(8)
North Carolina[7/]	Porters and Wautaga	35-40	P	Bluegrass	100 N, 43 P	Annual	1.0	1.0		1.0		(7)
				Pasture	400 N, 171 P	"	1.5	3.3		1.2		(7)
						% Fertilizer lost						
Louisiana[8/]						N		P				
		4-7	P	Ryegrass	100 N, 43 P	0.7		2.4				(3)
				Ryegrass	200 N, 86 P	1.3		2.3				(3)
				Ryegrass	50 N	3.2						(3)
				Ryegrass	100 N	5.2						(3)
				Millet	100 N	2.9						(3)

1/ Based on average composition in runoff for all storms; 0-46-0 broadcast and disked in.
2/ Based on flow-weighted concentrations for 5 in of water applied in 3 storms and 33-0-0 broadcast on surface.
3/ Based on average concentration in runoff for the initial storm.
4/ Based on 3-yr average for contour-planted corn.
5/ Based on conventional planted corn. From personal communication with J. D. Mikulcik.
6/ Based on good management for 1 hydrologic water yr.
7/ Based on 4-yr average.
8/ Based on 3 to 7 week runoff periods.

Table 4.--Effect of fertilizer placement on soluble N and P concentrations in runoff water

Location	Soil	Slope	Type of precipitation	Crop	Fertilizer applied	Change in flow-weighted concentration						Referen
						Basis	NH_4-N	NO_3-N	Inorg. N	Ortho. P	Tot. P	
		(%)			(lb/acre)							
Minnesota[1/]	Barnes sil	6	S	Fallow	Bdct, plowed, disked.	3 storms	1.1	1.2	1.2		1.5	(19)
				"	Bdct and disked.	"	8.8	1.3	3.0		2.7	(19)
				"	Bdct on plowed surface.	"	19.6	1.6	5.7		6.7	(19)
				"	Bdct on disked surface.	"	22.6	1.7	6.4		21.3	(19)
								Percentage of N fertilizer lost				
Louisiana[2/]		4-7	P	Rye	Incorporated into top 4 inches;			0.7				(3)
				"	topdressed.			5.2				(3)
				Millet	Topdressed.			2.9				(3)

1/ Average for 3 storms during 2-yr period, 150 lb N/acre as 33-0-0 and 35 lb P/ac as 0-46-0 applied.
2/ 100 lb N/acre as 8-8-8 incorporated into top 4 in or topdressed as 33-0-0.

Table 5.--Effect of tillage and residues on soluble N and P concentrations in runoff water

Location	Soil	Slope	Type of precipitation	Crop	Fertilizer applied	Change in flow-weighted concentration: Basis	NH_4-N	NO_3-N	Inorg. N	Ortho. P	Tot. P	Reference
		(%)			(lb/acre)							
Indiana[1/]	Bedford sil	8-12	S	Conv. corn	150 N, 50 P	2 storms	1.0	1.0		1.0		(14)
				Till. corn	150 N, 50 P	"	49.0	27.0		28.0		(14)
				Double disk	150 N, 50 P	"	14.0	11.0		19.0		(14)
				Chisel	150 N, 50 P	"	58.0	48.0		50.0		(14)
				Coulter	150 N, 50 P	"	135.0	39.0		1490.0		(14)
Mississippi[2/]	Providence sil		P	Soy-Soy conv.	0 N, 25 P	Annual	1.0	1.0		1.0		(9)
				Soy-Soy no till	0 N, 25 P	"	1.4	1.8		30.0		(9)
				Soy-wheat no till	85 N, 48 P	"	3.7	12.4		83.0		(9)
				Soy-corn no till	0 N, 25 P	"	1.5	3.8		23.0		(9)
				Corn-soy no till	121 N, 18 P	"	2.4	10.8		20.0		(9)
Missouri[3/]	Mexico sil	3	P	Corn	13 N	5-yr total			0.9			
				Corn	86 N	"			1.1			
				Corn	163 N	"			0.9			
				Corn	223 N	"			0.8			
				Corn	325 N	"			1.2			
Ohio[4/]			P	Corn				0.8		0.4		(17)

1/ Based on average flow-weighted concentrations for 2 successive rainstorms.
2/ Based on average flow-weighted concentrations for 2 yr.
3/ Change from conventional corn to no-till corn for each N rate. From personal communication with J. D. Mikulcik.
4/ Change from conventional corn to no-till corn.

Surface applied N and P fertilizer is extremely susceptible to removal by runoff water (table 4). When 33-0-0 was topdressed on ryegrass, the Louisiana researchers indicated that some of the fertilizer was held by the dense vegetation and was washed easily from the plots by precipitation, thus losing nearly 7.5 times as much N compared with soil-incorporated N. They also reported that nearly 50% of the total loss of both the 50- and 100-lb rates of topdressed 33-0-0 occurred for the first two rains following application.

Tillage, fertilizer placement, and plant residue factors are often hard to separate for a specific management system. The conventional tillage system of plowing, disking, and planting has been used widely, but minimum and no-till systems are being encouraged to reduce runoff and erosion. The different tillage systems have a direct effect on incorporation of fertilizer and on the plant residue remaining on the surface. As plant residues left on the soil surface to decrease soil erosion are subject to leaching by precipitation or runoff water or both, they will tend to increase soluble nutrient concentrations, as compared to conventional tillage.

Table 5 presents the effect of tillage and residue on soluble N and P concentrations in surface runoff. Compared with conventional tilled corn, soluble N and P concentrations in runoff from two simulated rainstorms increased substantially for till-plant, double disk and coulter plant, chisel plant, and coulter plant. In contrast, soluble inorganic N concentrations in runoff from Mexico sil (MO) showed little change for no-till corn at five N levels, and Ohio data showed a relative decrease in both soluble N and P concentrations for no-till corn. Data from Mississippi showed increases in soluble N and P concentrations for four no-till management systems compared with conventionally tilled soybeans.

The data presented in these tables represent only a small part of the soil-crop-fertility conditions present throughout the subhumid and humid areas of the United States. Additional research is needed to predict more accurately the concentrations of soluble nutrients in surface runoff water for the various conditions. These data provide factors for estimating the relative changes in soluble concentrations for various management practices, assuming a base concentration for comparison is known or can be approximated. Caution must be exercised however, when using the data for one soil-crop-fertility condition to estimate changes for a different condition.

REFERENCES

(1) Benoit, G. R.
1973. Effect of agricultural management of wet sloping soil on nitrate and phosphorus in surface and subsurface water. Water Resources Research 9:1296-1303.

(2) Burwell, R. E., D. R. Timmons, and R. F. Holt.
1975. Nutrient transport in surface runoff as influenced by soil cover and seasonal periods. Soil Science Society of America Proceedings 39:523-528.

(3) Dunigan, E. P., C. L. Mondart, Jr., R. A. Phelan, and Z. H. Shamsuddin. 1974. Surface runoff losses of fertilizers. Louisiana Agriculture 17(4):4-7.

(4) Ellis, B. G., A. E. Erickson, and A. R. Wolcott. 1978. Nitrate and phosphorus runoff losses from small watersheds in Great Lakes Basins. Ecological Research Series, EPA-600/3-78-028. 84 pp.

(5) Harley, C. P., H. H. Moon, and L. O. Regeimbal. 1951. The release of certain nutrient elements from simulated orchard grass mulch. Proceedings of the American Society for Horticultural Science 57:17-23.

(6) Jones, O. L., and S. M. Bromfield. 1969. Phosphorus changes during the leaching and decomposition of hayed-off pasture plants. Australian Journal of Agricultural Research 20:653-663.

(7) Kilmer, V. J., J. W. Gilliam, J. F. Lutz, R. T. Joyce, and C. D. Eklund. 1974. Nutrient losses from fertilized grassed watersheds in Western North Carolina. Journal of Environmental Quality 3:214-219.

(8) Klausner, S. D., P. J. Zwerman, and D. F. Ellis. 1974. Surface runoff losses of soluble nitrogen and phosphorus under two systems of soil management. Journal of Environmental Quality 3:42-46.

(9) McDowell, L. L., and K. C. McGregor. 1980. Nitrogen and phosphorus losses in runoff from no-till soybeans. Transactions of the American Society of Agricultural Engineers. (In Press.)

(10) Moe, P. G., J. V. Mannering, and C. B. Johnson. 1967. Loss of fertilizer nitrogen in surface runoff water. Soil Science 104:389-394.

(11) Nelson, D. W., and M. J. M. Römkens. 1970. Transport of Phosphorus in surface runoff. Cornell University Conference on Agricultural Waste Management Proceedings. pp. 215-224.

(12) Olness, A. E., S. J. Smith, E. D. Rhodes, and R. G. Menzel. 1975. Nutrient and sediment discharge from agricultural watersheds in Oklahoma. Journal of Environmental Quality 4:331-336.

(13) Römkens, M. J. M., and D. W. Nelson. 1974. Phosphorus relationships in runoff from fertilized soils. Journal of Environmental Quality 3:10-13.

(14) ------, D. W. Nelson, and J. V. Mannering. 1973. Nitrogen and phosphorus composition of surface runoff as affected by tillage methods. Journal of Environmental Quality 2:292-295.

(15) Schuman, G. E., R. G. Spomer, and R. F. Piest.
1973. Phosphorus losses from four agricultural watersheds on Missouri Valley loess. Soil Science Society of America Proceedings 37:424-427.

(16) ------, R. E. Burwell, R. F. Piest, and R. G. Spomer.
1973. Nitrogen losses in surface runoff from agricultural watersheds on Missouri Valley loess. Journal of Environmental Quality 2:299-302.

(17) Schwab, G.O., E.O. McLean, A.C. Waldron, R.K. White, and D.W. Michener.
1973. Quality of drainage water from a heavy-textured soil. Transactions of the America Society of Agricultural Engineers 16:1104-1107.

(18) Timmons, D. R., R. F. Holt, and J. J. Latterell.
1970. Leaching of crop residues as a source of nutrients in surface runoff water. Water Resources Research 6:1367-1375.

(19) ------, R. E. Burwell, and R. F. Holt.
1973. Nitrogen and phosphorus losses in surface runoff from agricultural land as influenced by placement of broadcast fertilizer. Water Resources Research 9:658-667.

(20) White, E. M., and E. J. Williamson.
1973. Plant nutrient concentrations in runoff from fertilized cultivated erosion plots and prairie in eastern South Dakota. Journal of Environmental Quality 2:453-455.

Chapter 16. PESTICIDE CONCENTRATIONS IN AGRICULTURAL RUNOFF: AVAILABLE DATA AND AN APPROXIMATION FORMULA

R. D. Wauchope and R. A. Leonard1/

Data on pesticide concentrations in runoff2/ that have been reported in scientific literature are reviewed in this chapter, and the most important factors affecting pesticide concentration are summarized. These factors are combined into a rough mathematical formula containing two empirical constants, which are calibrated using selected data on individual runoff "events." The formula is crude but general and complements the more conceptual, mechanistic calculations described in the pesticide submodel in volumes I and II. The equation described in this chapter should be studied because it provides a broad perspective on factors that effect pesticide losses, and it should allow first-pass screening of many situations from further consideration as potential problems. The reader also should refer to an extensive review on pesticide losses in runoff published previously (33).

OBSERVED CONCENTRATIONS OF PESTICIDES IN RUNOFF

Pesticide concentrations in runoff may vary by an order of magnitude (factor of 10) during a single runoff event. Event averages for a given chemical vary widely, depending on field site, storm timing, application rate, application target, and formulation.

Looking for generalizations in the wide array of available data is simplified by (1) knowing maximum probable concentrations, since these are the best measure of pollution potential, and (2) examining total concentrations per unit volume of runoff, irrespective of whether the pesticide is really in solution or adsorbed onto sediment. We do this because most available data are reported in this form, and we seldom know if the observed sediment-water partitioning obtains above or below the runoff sampling point. If partitioning information is needed, the techniques in volumes I and II should be used (vol. III, Ch. 15).

MAJOR FACTORS AFFECTING RUNOFF LOSSES OF PESTICIDES

Despite the complexity of data in the literature, four factors explain most results of experiments. These are physical-chemical properties of the

---

1/ Research chemist, USDA-SEA-AR, U.S. Delta States Agricultural Research Center, Stoneville, Miss., and soil scientist, USDA-SEA-AR, Southeast Watershed Research Laboratory, Athens, Ga.

2/ In this chapter, the term "runoff" means both the water and its associated sediment lost from the surfaces of fields.

formulated pesticide; application rate; location of application residue (foliage, soil surface, soil subsurface when incorporated or subsurface applied); and time elapsed between application and runoff storms.

Application Rate

Studies have shown that runoff concentrations are roughly proportional to application rate, all other factors being equal. We will assume this proportionality always holds, but, fortunately, application rates tend to fall within rather narrow ranges. Single applications of 1-5 kg/ha are typical for most herbicides and multiple applications totaling 10-20 kg/ha [3/] for insecticides. Thus, the proportionality assumption will not have to cover a wide range of rates.

Storm Timing

Regardless of pesticide or site, runoff experiments usually show that runoff concentrations decline drastically as the time elapsed between application and runoff increases. Typically, runoff concentrations decrease by a factor of two for every 2-7 days of delay. This decline is generally much more rapid than pesticide persistence values reported in field soils (33) or even on the surface of soils (see vol. III, ch. 17). Declines in runoff concentrations are, in fact, so rapid that a single quasi-exponential function describes runoff concentrations fairly well for all pesticides.

Physical Properties of Formulated Pesticide

Many pesticides are applied as simple aqueous solutions, perhaps with a surfactant or spray drift control agent. When these pesticides reach their target and dry, the resultant residue should have properties similar to the pure chemical.

Many pesticides are insoluble, however, and are formulated with complex carriers. They are sprayed adsorbed onto solid granules or in nonpolar solvents or emulsions. This drastically will affect the availability of a pesticide for washoff, even after drying. Soil surface dust may act as a carrier if pesticides are strongly adsorbed to it on application (33).

Site of Application

Application of highly soluble pesticides to foliage results in greater runoff concentrations than applications to bare soil. At the other extreme, some pesticides are applied incorporated into subsurface soil which, of course, leads to low runoff concentrations. Four distinct "targets" are dry soil surface, wet soil surface, foliage, and soil subsurface. Apparently, water-soluble pesticides are the most affected by application target.

3/ Recent developments involving precise application of herbicides to target weeds may use 1/10 or less of these amounts. Some sophisticated new insecticides also require small application rates--sometimes < 0.1 kg/ha.

These last two factors--pesticide properties and effect of target site--require complicated modeling at the conceptual level. The way they are handled in this chapter is to factor out application rate and storm timing effects, then attempt to lump empirically pesticide property-target site situations into classes characterized by a single "availability index" parameter, which crudely accounts for variations in concentration from these factors.

AVAILABLE DATA

A selection from the data compiled by Wauchope (33) is presented in table 1, along with calculated half-lives based on simple semilog plots, for experiments in which several runoff events were analyzed from the same application.

Generally, an exponential decay crudely describes runoff concentrations, that is, a plot of log (runoff concentration) vs. time of runoff event after application often will be linear for points taken over a few weeks. In figure 1, the "half-lives" of table 1 are shown vs. the midpoint of the time ranges for the data sets. Even for the same chemical, "half-lives" tend to increase with the time range of observation in different experiments--a result of the more rapid decay of the most available residues. Generally, therefore, the decay curve for a long experiment is only approximately exponential.

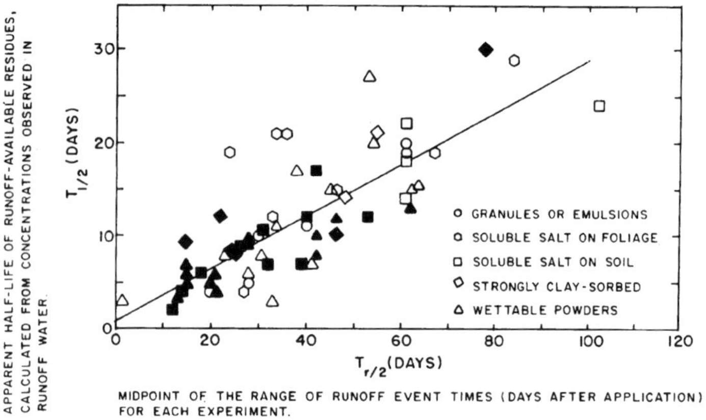

Figure 1.--Half-life data obtained by plotting log (bulk concentration in runoff) vs. elapsed time between application and runoff event. Each point is the result of such a plot for different experiments, plotted vs. the midpoint of the time range of the events in each experiment. Open symbols represent calculations based on few events or scattered data.

To correct for this, we use the fact that figure 1 indicates an approximately linear increase in half-life with time; the data in figure 1 fit the equation

$$t_{1/2}\ (\text{days}) = 1 + 0.28\ t_{r/2}(\text{days}) \quad (1)$$

where $t_{1/2}$ = half-life, $t_{r/2}$ = time range midpoint. The correlation coefficient of this fit is 0.82, a remarkable correlation considering the variety of chemicals and application sites represented. Manipulation and integration of equation (1) lead to

$$C_t = C_o\ (1 + 0.28t)^{-2.5} \quad (2)$$

where C_t is the extrapolated runoff concentration, which would be observed if runoff occurred at time zero--immediately after application.

If we assume that C_o is proportional, for a given formulation/site situation, to application rate R, that is, C_o = AR where A may be called an "availability index," we have

$$C_t = AR\ (1 + 0.28t)^{-2.5} \quad (3)$$

Equation (3) gives the dependence of runoff on storm times and application rates directly. To evaluate formulation/site effects, that is, to calibrate A, we used some 373 runoff events from the literature, some of which are shown in table 1. These data included most common pesticide types and application situations and included runoff times from 0 to 300 days, application rates from 0.29 to 22.5 kg/ha, and runoff volumes from < 0.1 to 17 area-cm. For these events, A values so obtained ranged from 10^1 to 10^6 but tended to cluster into four groups with ranges within a group of about two orders of magnitude (fig. 2). These four groupings of pesticide application situations--called classes I-IV--are given in table 2, along with a single, rounded value of A, which is assigned to each class.

We recalculated C_t for the actual times of runoff of the 373 events, using the assigned A's, to see how well actual observations are predicted. Results for all these events are shown in figure 2 and also are given with the examples in table 1. The correlation coefficient for the log-scale plot in figure 3 is 0.73, that is, we accounted for about 50% of the variance in all runoff concentrations with equation (3).

The average and standard deviation in the logarithm of the ratio C_t(calculated)/C_t(observed) in figure 2 is 0.16 ± 0.86, that is, predicted concentrations, on average, exceeded observed concentration by a factor of < 1.5, but the standard error of the ratios gives a range from overprediction by a factor of 10 to underprediction by a factor of 5. This may seem crude, but we are trying to cover data which ranges over five orders of magnitude.

The gross average for all observed data in figure 1 is interesting. The "average" pesticide runoff event resulted from 5 cm of precipitation occurring 32 days after application; 2 cm of runoff resulted, containing 519 ppb pesticide. These averages reflect the high-intensity "storms" from many simulated runoff experiments in the data. The selection of A values was a somewhat

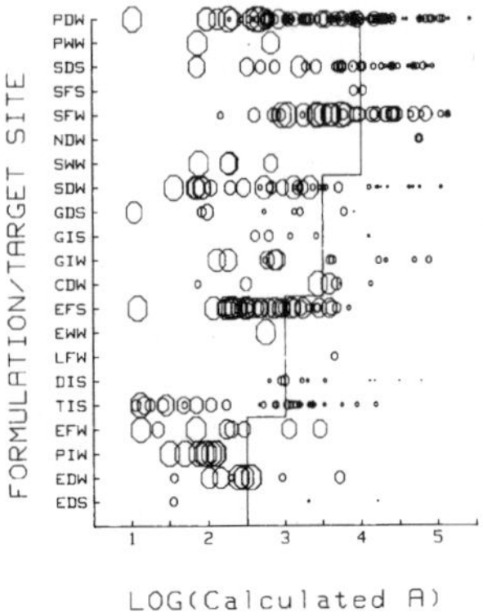

Figure 2.--Calculated values of $A = C_t(1 + 0.28t)^{2.5}/R$ for 373 individual runoff events, for different formulation/target/runoff situations. Three-letter code indicates application situation: First letter indicates formulation; P = wettable powder, S = aqueous solution, N = nonpolar solvent solution, G = granules, C = water-soluble pellets, E = emulsion, L = dispersible liquid, D = dieldrin emulsion, T = trifluralin emulsion. Second letter indicates application target: D = dry soil surface; W = wet soil surface; F = plant foliage; I = incorporated into soil. Third letter indicates mode of transport: S = sediment phase; W = water phase. Size of circle indicates "reliability" of value, that is, larger circles are for A's calculated from a short time (t near 0) extrapolation.

arbitrary process, and we deliberately selected slightly high values for A to minimize the possibility of underpredicting runoff concentrations. This resulted in the slight bias in fit (average prediction slightly high), which, together with somewhat high storm intensities for the selected events, means that equation (3), with A values from table 2, will be unlikely to underpredict pesticide losses by more than one order of magnitude.

A sample calculation will illustrate both the strengths and limitations of equation (3). Consider a 1 kg/ha pesticide application in a class I pesticide/situation combination (table 2). If rainfall occurs immediately afterward, a runoff concentration of 10 ppm is predicted--about the upper limit observed in experiments (33). Only 1 cm of runoff (not an exceptional amount) at that concentration would remove the entire amount of pesticide from the field, however,

Table 1.--Selected runoff concentrations and apparent half-lives of runoff-available pesticide residues reported in the literature, with assigned availability index A (ppb/kg/ha) and concentrations predicted by equation (3)

Compound	Application rate	Rainfall/runoff 1/	Time of runoff 2/	Bulk concentration 3/ Obs	Bulk concentration 3/ Calculated	Observed half-life of available residues 4/	Time range of observed half-life	Reference
	(kg/ha)	(cm)	(days)	(ppb)		(days)	(days)	
				Wettable powders applied to bare soil surface A = 10,000				
Atrazine	0.6	1.30/0.15	24	400	36	13	24-100	(20)
Do.-------------	1.1	1.30/0.15	24	800	66	(10-20)	24-100	(20)
Do.-------------	2.2	1.30/0.20	24	800	130	(10-20)	24-100	(20)
Do.-------------	4.5	1.30/0.25	24	1950	270	8	24-60	(20)
Do.-------------	6.7	1.30/0.15	24	3300	410	10	24-60	(20)
Do.-------------	9.0	1.30/0.20	24	4700	540	10	24-60	(20)
Do.-------------	2.24	1.90/0.10	5	2300	2500	6	5-25	(20)
Do.-------------	4.48	1.90/0.10	5	4600	5000	7	4-25	(20)
Do.-------------	1.12	--	15	180	180	(27)	30-75	(12)
Do.-------------	4.48	--	0	850	45000	(17)	0-75	(12)
Do.-------------	3.36	1.22/0.01	8	200	1800	6	8-33	(25)
Do.-------------	3.36	1.12/0.01	12	160	850	4	8-33	(25)
Do.-------------	3.81	1.57/0.02	6	1900	3200	(3)	6-60	(25)
Do.-------------	4.03	6.86/0.19	24	330	240	(7)	24-59	(25)
Do.-------------	3.36	6.35/3.91	0	1460	34000	(3)	0-3	(30)
Do.-------------	3.36	6.35/3.91	3	627	7300			(30)
Cyanazine	1.61	3.61/0.40	10	193	570	5	10-29	(25)
Do.-------------	2.24		2	101-275	7400			(3)
Diphenamid	3.36	1.90/1.36	0	1645	34000	3.5	0-25	(25)
Do.-------------	3.36	1.90/0.33	2	642	11000	(11)	2-66	(25)
Do.-------------	3.36	6.37/2.62	23	63	220	(6)	23-32	(25)
Do.-------------	3.36	2.54/0.79	27	123	160	12	27-65	(25)
Do.-------------	3.52	1.27/0.18	22	229	260			(25)
Fluometuron	2.8	- /0.65	0	32,44	28000			(5)
Linuron	2.24	0.58/0.01	32	28	72			(32)
Do.-------------	2.24	1.63/0.36	38	124	48			(32)
Do.-------------	2.24	2.29/0.15	49	14	27	(20)	38-69	(32)
Metribuzin	0.56	2.60/1.77	4	53	860	5	4-25	(28)
Do.-------------	0.56	2.40/1.11	9	15	240			(28)

'able 1.--Selected runoff concentrations and apparent half-lives of runoff-available pesticide residues reported in the literatu with assigned availability index A (ppb/kg/ha) and concentrations predicted by equation (3)--Continued

Compound	Application rate	Rainfall/ runoff 1/	Time of runoff 2/	Bulk concentration 3/ Obs	Calculated	Observed half-life of available residues 4/	Time range of observed half-life	Referenc
	(kg/ha)	(cm)	(days)	(ppb)		(days)	(days)	
Wettable powders applied to bare soil surface A = 10,000								
Prometryne--------	2.8	-- /1.00	1	90-150	15000			(4)
Propachlor--------	6.72	-- /1.02	8	1702	3600	(8)	8-37	(22)
Propazine---------	1.66	4.79/2.76	9	401	710	(8)	9-52	(25)
Simazine----------	2.24	-- / --	15	300	360	(15)	15-75	(12)
Terbuthyalazine	2.24	1.40/0.20	14	900	420	9	5-50	(19)
Do.------------	4.48	1.90/0.10	5	1800	5000	10	5-50	(19)
Paraquat and MSMA (aqueous solutions; bind to clay) A = 10,000								
Paraquat	1.53	1.12/0.10	12	200	390	8	12-33	(25)
Do.------------	1.53	1.90/0.58	12	308	390	8	8-40	(25)
Do.------------	1.53	6.37/2.62	23	254	100	(21)	23-86	(25)
Do.------------	1.75	3.56/0.30	17	221	220			(25)
Do.------------	1.84	2.54/0.61	14	193	340			(25)
Do.------------	1.93	3.61/0.40	10	390	690			(25)
Do.------------	1.94	3.30/1.09	28	203	83			(25)
Do.------------	2.12	1.27/0.18	22	1012	150	14	22-73	(25)
Do.------------	2.45	1.57/0.02	6	7406	2100			(25)
Do.-----------	15.34	1.90/0.33	2	7965	50000	12	2-41	(25)
Do.-----------	15.34	2.54/0.79	27	3501	720	10	27-65	(25)
MSMA	8.96	2.60/1.77	4	450	14000	9	4-25	(27)
Do.------------	8.96	3.40/2.82	15	280	1500			(27)
Do.------------	2.24	-- /0.17	2	105	7400			(27)
Aqueous solutions applied to foliage A = 10,000								
\rsenic acid	3.3	-- /0.10	14	500	610	30	16-140	(21)
Do.------------	3.3	-- /1.70	16	500	470			(21)
?,4-D salt	5/0.99	3.96/ --	7	1610	640			(15)
Do.----------	5/1.8	3.96/ --	7	3240	1200			(15)
Do.----------	5/3.0	3.96/ --	7	4210	2000			(15)

Table 1.--Selected runoff concentrations and apparent half-lives of runoff-available pesticide residues reported in the literature with assigned availability index A (ppb/kg/ha) and concentrations predicted by equation (3)--Continued

Compound	Application rate	Rainfall/ runoff 1/	Time of runoff 2/	Bulk concentration 3/ Obs	Bulk concentration 3/ Calculated	Observed half-life of available residues 4/	Time range of observed half-life	Reference
	(kg/ha)	(cm)	(days)	(ppb)		(days)	(days)	
				Aqueous solutions applied to foliage A = 10,000				
2,4-D salt-------	5/ 4.0	3.96/ --	7	4170	2700			(15)
Dicamba salt------	2.24	1.27/0.11	1	4810	12000			(26)
Picloram salt	0.56	6.89/ --	27	285,353	26	(11)	27-52	(23)
Do.------------	0.56	3.80/ --	0	520	5600	(5)	0-56	(8)
Do.-----------	5/ 0.56	3.96/ --	7	1082	370	(10)	9-50	(15)
Do.-----------	5/ 0.7	3.96/ --	7	3240	460			(15)
Do.-----------	5/ 1.3	3.96/ --	7	4170	860			(15)
Do.-----------	5/ 2.0	3.96/ --	7	2716	1300			(15)
Do.------------	1.12	1.27/0.11	1	1980,3810	6000			(26)
Do.------------	2.24	1.27/0.11	1	2170-5210	12000	(19)	1-120	(26)
Do.------------	2.24	-- /0.01	30	15	83			(16)
2,4,5-T salt	0.56	6.89/ --	27	117,160	26			(23)
Do.------------	0.56	3.80/ --	0	618	5600	(4)	0-40	(8)
Do.------------	0.49	3.96/ --	7	850	330			(15)
Do.------------	1.5	3.96/ --	7	1380	1000			(15)
Do.------------	1.8	3.96/ --	7	1010	1200			(15)
Do.------------	2.0	3.96/ --	7	1260	13000			(15)
Do.------------	2.24	1.27/0.11	1	3300	12000	(20)	1-120	(26)
				Aqueous solutions applied to bare soil A = 3,000				
2,4-D acid--------	0.56	12.7/6.3	1	21	910			(1)
2,4-D salt	0.56	19.10/0.53	20	8	15			(29)
Do.------------	0.56	16.41/2.74	27	6	8			(29)
Do.------------	0.56	19.56/0.48	34	2.5	4.7			(29)
Do.------------	0.56	8.25/3.40	1	25	190	6	1-35	(29)
Do.------------	2.46	12.29/8.94	0	90	7400			(6)
Do.------------	2.46	13.72/8.97	2	70	2400			(6)
Do.------------	2.46	13.66/8.15	4	110	1100			(6)
Dicamba salt	2.24	1.27/0.11	1	1600	3600	(18)	1-120	(26)

with assigned availability index A (ppb/kg/ha) and concentrations predicted by equation (3)--Continued

Compound	Application rate	Rainfall/ runoff 1/	Time of runoff 2/	Bulk concentration 3/ Obs	Bulk concentration 3/ Calculated	Observed half-life of available residues 4/	Time range of observed half-life	Reference
	(kg/ha)	(cm)	(days)	(ppb)		(days)	(days)	
Aqueous solutions applied to bare soil A = 3,000								
Fenac salt	3.36	7.37/2.88	8	310	530	9	1-50	(32)
Do.------------	3.36	2.16/0.01	18	24	110			(32)
Picloram salt	0.56	-- /0.05	72	12	0.8			(8)
Do.------------	1.1	7.60/ --	14	105	60	12	14-91	(18)
Do.------------	1.12	-- / --	8	10	180	10	9-50	(15)
Do.------------	1.12	-- / --	9	28	140			(15)
Do.------------	1.12	2.70/ --	6	90	290	12	2-78	(7)
Do.------------	1.12	4.20/ --	30	14	12			(7)
Do.------------	1.12	4.20/ --	52	1.6	3.5			(7)
Do.------------	1.3	3.96/ --	7	4170	260			(15)
Do.------------	2.24	1.27/0.11	1	650	3600	(22)	1-120	(26)
2,4,5-T salt	2.24	1.27/0.11	1	2600	3600	(14)	1-120	(26)
Do.------------	11.2	-- /0.05	22	380	240	(24)	6-143	(13)
Granular and pelleted pesticides on soil surface or incorporated A = 3,000								
Carbaryl	5.03	-- /0.32	17	248	190	(4)	17-37	(11)
Carbofuran	5.41	-- /0.12	2	1400	5300	(12)	1-65	(10)
Do.------------	4.16	-- /0.04	2	1000	4100			(10)
Diazinon	1.12	-- / --	4	100	510			(22)
Fonofas	1.12	21/15	11-35	60-10	100-9			(3)
Picloram	2.24	5.16/2.62	0	6098	6700			
Do.------------	2.24	4.70/2.15	24	1.0	41			
Emulsions applied to soil or foliage A = 1,000								
DDT 6/	0.73	-- / --	1	83	390	(10-13)	1-46	(14)
Do.---------------	0.73	-- / --	1	67	390	(10-13)	1-46	(14)
Do.---------------	1.12	1.55/0.44	2	725	390	(10)	1-50	(24)
Do.---------------	1.12	1.07/0.16	2	450	370	(9)	1-50	(24)
Diuron 6/	2.06	-- / --	21	74	17	15	20-70	(17)

Table 1.--Selected runoff concentrations and apparent half-lives of runoff-available pesticide residues reported in the literature with assigned availability index A (ppb/kg/ha) and concentrations predicted by equation (3)--Continued

Compound	Application rate	Rainfall/runoff 1/	Time of runoff 2/	Bulk concentration 3/ Obs	Bulk concentration 3/ Calculated	Observed half-life of available residues 4/	Time range of observed half-life	Reference
	(kg/ha)	(cm)	(days)	(ppb)		(days)	(days)	
			Emulsions applied to soil or foliage A = 1,000					
Endosulfan 6/	0.351	-- / --	4	18	54	5	3-20	(14)
Endrin 6/	0.289	-- / --	1	48,49	160	6	1-20	(14)
Endrin 5/	0.289	-- / --	4	33	44			(14)
Methylparathion 5/	1.12	0.94/0.19	0	15	340	(3)	0-4	(24)
Methoxychlor	22.50	-- /0.08	18	9	75			(13)
Toxaphene 5/	2.24	1.07/0.16	2	228,543	740			(24)
Do.------------	2.24	2.42/1.13	3	239	490			(24)
			Persistent and incorporated A = 1,000					
Dieldrin	5.6	-- /7.58	13	120	120	(19)	13-120	(9)
Do.------------	5.6	-- /0.06	55	3.2	5.1			(6)
Do.------------	5.6	-- /0.26	58	15.2	4.5	(29)	58-110	(9)
Trifluralin	1.12	2.01/1.04	28	7.3	1.4	(15)	27-65	(25)
Do.------------	1.40	7.37/2.86	8	1.9	22			(32)
Do.------------	1.40	0.58/0.01	56	0.5	0.4			(32)
			Incorporated emulsions or wettable powders A = 3,000					
Atrazine	1.60	13.20/8.64	0	119	480			(2)
Do.------------	1.66	13.90/10.20	0	203	500			(2)
Do.------------	3.22	14.60/10.50	0	408	970			(2)
Do.------------	3.32	13.50/9.37	0	361	1000			(2)
Dichlobenil	6.7	13.20/8.64	0	211	2000			(2)
Do.------------	6.7	14.60/10.50	0	634	2000			(2)
Do.------------	6.7	13.50/9.37	0	330	2000			(2)
Do.------------	6.97	13.90/10.20	0	612	2100			(2)
Trifluralin	0.84	3.40/2.82	15	2.3	4.1			(28)
Do.------------	1.12	6.78/1.25	5	8.8	38	(21)	4-67	(24)

Table 1.--Selected runoff concentrations and apparent half-lives of runoff-available pesticide residues reported in the literatur with assigned availability index A (ppb/kg/ha) and concentrations predicted by equation (3)--Continued

Compound	Application rate	Rainfall/runoff 1/	Time of runoff 2/	Bulk concentration 3/ Obs	Bulk concentration 3/ Calculated	Observed half-life of available residues 4/	Time range of observed half-life	Reference
	(kg/ha)	(cm)	(days)	(ppb)		(days)	(days)	
			Incorporated emulsions or wettable powers A = 3,000					
Trifluralin	1.12	2.87/1.23	15	7.7	5.4	(21)	4-67	(24)
Do.------------	1.12	5.21/1.15	46	13.6	0.5			(24)
Do.------------	1.12	1.90/1.36	0	15	340	(19)	0-47	(25)
Do.------------	1.12	6.61/0.01	5	15	38	(21)	2-65	(25)
Do.------------	1.12	6.37/2.62	23	8.8	2.2			(25)
			Emulsion applied to dry soil A = 3,000					
Alachlor	2.24	20-22/6-16	2	75-184	220			(3)
2,4-D ester	2.46	6.60/4.11	0	870	740			(6)
Do.------------	4.93	6.70/4.78	0	1780	1500			(6)
Do.------------	1.68	3.61/0.40	10	309	18	(5)	10-21	(25)
Do.------------	1.55	3.56/0.30	17	1	6			(25)

1/ Dash indicates value not reported.
2/ Days elapsed between application and time of runoff.
3/ Concentration in µg pesticide/l runoff-sediment mixture; average for runoff event.
4/ Values in parentheses are only suggested by the data.
5/ Only part of watershed was treated; application rate given has been corrected for this.
6/ Several applications were made in experiment; application rate is for last application before runoff

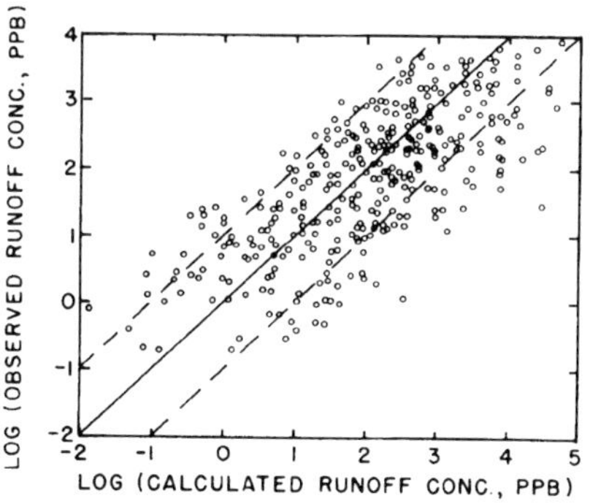

Figure 3.--Observed versus calculated runoff concentrations calculated by equation (3). Dashed lines show order of magnitude limits.

Table 2.--Classification of pesticides by availability index A--the ratio between application rate (kg/ha) and runoff concentration (ppb) if runoff occurs immediately after application

Class	Assigned value of A	Properties of pesticide or application situation	Pesticide data used in calculation of A common name 1/
	(ppb.ha/kg)		
I	10,000	Wettable powders applied to soil surface.	Cyanazine, prometryne, fluometuron, simazine, atrazine, terbuthylazine diphenamid, propazine, propachlor, metribuzin, linuron.
		Soluble salts applied to soil and strongly bound to clay particles.	MSMA, paraquat.
		Soluble salts applied to foliage or wet soil.	Arsenic acid 2/, 2,4,5-T, 2,4-D, picloram, dicamba.
		Nonionic pesticides applied in diesel oil.	2,4,5-T ester.

Table 2.--Classification of pesticides by availability index A--the ratio between application rate (kg/ha) and runoff concentration (ppb) if runoff occurs immediately after application--Continued

Class	Assigned value of A	Properties of pesticide or application situation	Pesticide data used in calculation of A common name 1/
	(ppb.ha/kg)		
II	3,000	Soluble salts applied to dry soil.	2,4-D, picloram, 2,4,5-T, dicamba, fenac.
		Granular and pelleted pesticides, regardless of solubility--even if incorporated.	Picloram (8), endrin, fonofos, dieldrin, carbaryl, carbofuran, diazinon.
III	1,000	Insoluble pesticides applied to foliage.	Endosulfan, endrin, DDT toxaphene, diuron, methoxychlor.
		Persistent and incorporated.	Dieldrin.
		Insoluble; applied to wet soil.	2,4-D ester.3/
IV	300	Insoluble pesticides applied to dry soil.	Alachlor, 2,4-D ester, methoxychlor.
		Incorporated wettable powders.	Atrazine, dichlobenil.
		Incorporated, insoluble nonpersistent.	Trifluralin
		Foliar, insoluble, nonpersistent.	Parathion.

1/ See (33) for pesticide nomenclature, details of experiments, and original references, except where noted.

2/ Arsenic acid was used as a defoliant, rather than as a pesticide, in this experiment.

3/ Limited data available.

and 100% losses in a single event never are observed experimentally even under the most extreme simulated conditions (33). Concentrations on the order of 10 ppm are observed only right at the field edge, never in streams or lakes away from the application site. Equation (3), therefore, will predict relative losses and the worst concentrations at the field edge. Computing an absolute pesticide load in an event and predicting downstream concentrations will require more information. We suggest, however, that the worst total losses (loads) of pesticides should be approximately proportional to the A values in table 1: A ÷ 300 will give about the maximum single event loss in terms of a percentage of that applied.

REFERENCES

(1) Asmussen, L. E., A. W. White, E. W. Hauser, and J. M. Sheridan. 1977. Movement of 2,4-D in a vegetated waterway. Journal of Environmental Quality 6:159-162.

(2) Bailey, G. W., A. P. Barnett, W. R. Payne, Jr., and C. N. Smith. 1974. Herbicide runoff from four coastal plain soil types. U.S. Environmental Protection Agency, Environmental Protection Technology Series EPA-660/2-74-017. 98 pp.

(3) Baker, J. L., H. P. Johnson, and J. M. Laflen. 1976. Completion report-- effect of tillage systems on runoff losses of pesticides: A simulated rainfall study. Iowa State Water Resources Research Institute Report No. ISWRRI-71, Ames, Iowa. 96 pp.

(4) Baldwin, F. L., P. W. Santelmann, and J. M. Davidson. 1975a. Prometryne movement across and through soil. Weed Science 23:285-288.

(5) ------, P. W. Santelmann, and J. M. Davidson. 1975b. Movement of fluometuron across and through the soil. Journal of Environmental Quality 4:191-194.

(6) Barnett, A. P., E. W. Hauser, A. W. White, and J. H. Holladay. 1967. Loss of 2,4-D in washoff from cultivated fallow land. Weeds 15:133-137.

(7) Baur, J. R., R. W. Bovey, and M. G. Merkle. 1972. Concentration of picloram in runoff water. Weed Science 20:309-313.

(8) Bovey, R. W., E. Burnett, C. W. Richardson, and others. 1974. Occurrence of 2,4,5-T and picloram in surface runoff water in the Blacklands of Texas. Journal of Environmental Quality 3:61-64.

(9) Caro, J. H., W. M. Edwards, B. L. Glass, and M. H. Frere. 1972. Dieldrin in runoff from treated watersheds. In Proceedings of the Symposium for Interdisciplinary Aspects of Watershed Management, American Society of Civil Engineers, New York. pp. 141-160.

(10) ------, H. P. Freeman, D. W. Glotfelty, B. J. Turner, and W. M. Edwards. 1973. Dissipation of soil-incorporated carbofuran in the field. Journal of Agricultural and Food Chemistry 21:1010-1015.

(11) ------, H. P. Freeman, and B. C. Turner.
1974. Persistence in soil and losses in runoff of soil-incorporated carbaryl in a small watershed. Journal of Agricultural and Food Chemistry 22:860-863.

(12) Edwards, W. M.
1972. Agricultural chemical pollution as affected by reduced tillage systems. In Proceedings of the No-Tillage Systems Symposium, Feb 21, 1972, Ohio State University. pp. 30-40.

(13) ------, and B. L. Glass.
1971. Methoxychlor and 2,4,5-T in lysimeter percolation and runoff water. Bulletin of Environmental Contamination and Toxicology 6:81-84.

(14) Epstein, E., and W. J. Grant.
1968. Chlorinated insecticides in runoff water as affected by crop rotation. Soil Science Society of America Proceedings 32:423-426.

(15) Evans, J. O., and D. R. Duseja.
1973. Herbicide contamination of surface runoff waters. U.S. Environmental Protection Agency, Environmental Protection Technology Series EPA-Rβ-73-266. 99 pp.

(16) Glass, B. L., and W. M. Edwards.
1974. Picloram in lysimeter runoff and percolation water. Bulletin of Environmental Contamination and Toxicology 11:109-112.

(17) Green, R. E., K. P. Goswami, M. Mukhtar, and H. Y. Young.
1977. Herbicides from cropped watersheds in stream and estuarine sediments in Hawaii. Journal of Environmental Quality 6:145-154.

(18) Haas, R. H., C. J. Scifres, M. G. Merkle, R. R. Hahn, and G. O. Hoffman.
1971. Occurrence and persistence of picloram in grassland water sources. Weed Research 11:54-62.

(19) Hall, J. K.
1974. Erosional losses of s-triazine herbicides. Journal of Environmental Quality 3:174-180.

(20) ------, M. Pawlus, and E. R. Higgins.
1972. Losses of atrazine in runoff water and soil sediment. Journal of Environmental Quality 1:172-176.

(21) Richardson, C. W., J. D. Price, and E. Burnett.
1977. Arsenic concentrations in surface runoff from small watersheds in Texas. Journal of Environmental Quality. (In press)

(22) Ritter, W. F., H. P. Johnson, W. G. Lovely, and M. Molnau.
1974. Atrazine, propachlor, and diazinon residues on small agricultural watersheds. Environmental Science Technology 8:37-42.

(23) Scifres, C. J., H. G. McCall, R. Maxey, and H. Tai.
1977. Residual properties of 2,4,5-T and picloram in sandy rangeland soils. Journal of environmental Quality. (In press)

24) Sheets, T. J., J. R. Bradley, Jr., and M. D. Jackson.
1972. Contamination of surface and ground water with pesticides applied to cotton. University of North Carolina Water Resources Research Institute Report No. 60, Chapel Hill, N.C. 63 pp.

25) Smith, C. N., R. A. Leonard, G. W. Langdale, and G. W. Bailey.
1978. Transport of agricultural chemicals from small upland Piedmont watersheds. U.S. Environmental Protection Agency, Environmental Protection Technology Series EPA-600/3-78-056. 364 pp.

26) Trichell, D. W., H. L. Morton, and M. G. Merkle.
1968. Loss of herbicides in runoff water. Weed Science 16:447-449.

27) Wauchope, R. D., L. E. Asmussen, and E. W. Hauser.
1977. MSMA losses in runoff from small watersheds in the Mississippi Delta and Georgia Coastal Plain. Proceedings of the Southern Weed Science Society 30:395.

28) ------, K. E. Savage, and E. B. Hollingsworth.
1977. Trifluralin, metribuzin, and MSMA in surface water from a Mississippi Delta field. Abstract of the Weed Science Society of America 1977:60.

29) White, A. W., L. E. Asmussen, E. W. Hauser, and J. W. Turnbull.
1976. Loss of 2,4-D in runoff from plots receiving simulated rainfall and from a small agricultural watershed. Journal of Environmental Quality 5:487-490.

30) ------, A. P. Barnett, B. G. Wright, and J. H. Holladay.
1967. Atrazine losses from fallow land caused by runoff and erosion. Environmental Science Technology 1:740-744.

31) Willis, G. H. and R. A. Hamilton.
1973. Agricultural chemicals in surface runoff, ground water, and soil: endrin. Journal of Environmental Quality 2:463-466.

32) ------, R. L. Rogers, and L. M. Southwick.
1975. Losses of diuron, linuron, fenac, and trifluralin in surface drainage water. Journal of Environmental Quality 4:399-402.

33) Wauchope, R. D.
1978. The pesticides content of surface water draining from agricultural fields - a review. Journal of Environmental Quality 7:459-472.

Chapter 17. DISSIPATION RATE OF PESTICIDES FROM SOILS

Ralph G. Nash[1/]

INTRODUCTION

Determining the amount of a pesticide on the soil surface without direct measurement at any given time is not an exact science. First, the amount of pesticide that reaches the soil depends upon the amount applied, application method, ground cover (plants or mulch), and the amount that washes off of plants after application. Second, the dissipation rate (or persistence) of a pesticide varies because of the characteristics of the pesticide itself and soil and climatic factors.

This chapter will be concerned primarily with pesticide dissipation from the soil surface described by a lumped parameter, k_s, a rate constant for the combined processes that affect the persistence of the pesticide in the soil surface. A finite pesticide amount on the soil surface is assumed to be known either from direct application or from direct measurement at time zero (C_o). For this discussion, the following limitations will be assumed or imposed: (1) method of application will not be considered; (2) the pesticide is assumed to be uniform over the soil surface; (3) vapor diffusion from the plants to the soil or from deeper soil depths to the soil surface is considered negligible; (4) movement with water from deeper soil depths to the soil surface is considered negligible; and (5) washoff from plants onto the soil is not considered because this is presented in a separate chapter.

DISCUSSION

The persistence or dissipation of a pesticide on soil depends upon the characteristics of the pesticide itself and may depend upon one or more of the following soil surface factors: type, pH, organic matter and moisture content, temperature, air velocity, and solar radiation.

Two methods for determining the amount of a pesticide on the soil surface at any given time are: (1) sampling and direct measurement and (2) estimating from a disappearance formula or curve when given a known amount, C_o, at t_o. The first method is superior and preferred, especially for pesticide residues in late fall or winter, but often the value obtained is valid, for only a few hours or days because of the continual pesticide concentration decrease as a result of numerous factors. The second method is the concern of this chapter because it is not always feasible to measure the pesticide directly, hence, an estimate is needed.

1/ Soil scientist, USDA-SEA-AR, Pesticide Degradation Laboratory, Beltsville Agricultural Research Center, Beltsville, Md.

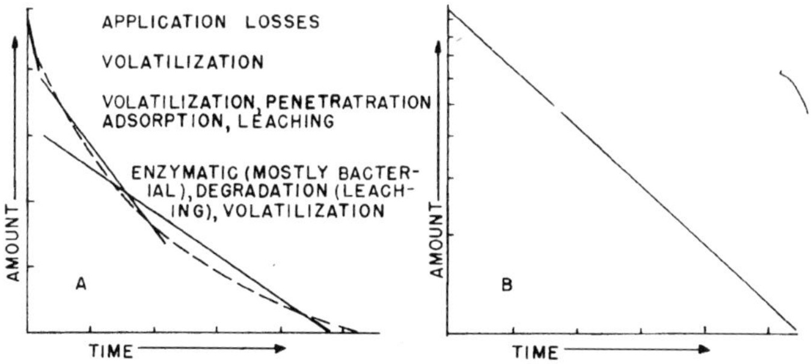

Figure 1.--Theoretical dissipation and loss pathways of pesticides from soil: (A) linear concentration of pesticide on/in soil with time and (B) logarithmic concentration of pesticide on/in soil with time (adapted from (14)).

Theoretical Considerations

Edwards (14) proposed a series of plotted theoretical dissipation lines for pesticides applied to a field (fig. 1A). Initially, losses occur during application. Under certain circumstances, application losses may be large, and considerable engineering effort has been conducted to limit these losses. Losses during application occur over a short period, perhaps of <5 min.

Volatilization apparently occurs for almost all pesticides, primarily because of their distribution over a large surface area. Volatilization losses are governed by the pesticide itself, method of application (surface vs. soil incorporated), formulation, temperature, relative humidity, windspeed, and soil surface. Volatilization losses may occur over a few hours for those pesticides that degrade rapidly or may occur for several months for the more persistent pesticides (53). Most data indicate volatilization losses are reduced greatly after a few hours, or a few days at the most, even for the most persistent pesticides (53, 71).

During the third period, penetration, adsorption, or leaching, or all, become dominant factors that govern the dissipation of a pesticide. Finally, degradation of the pesticide becomes paramount in the disappearance of most pesticides.

The series of four lines in figure 1A can be constructed into a continuous curve (dashed curve). If the pesticide logarithm of the amount is plotted versus time, a linear relationship is possible (fig. 1B).

Expressed mathematically, the line in figure 1B becomes log amount at C_t = log amount at $C_o - k_s t$, which is recognized as the logarithmic expression of $y = b + m x$ (where y = amount, b = intercept of y or vertical axis, x = horizontal axis, and m = positive slope). For convenience, we express the log amount vs. time in natural logarithms or $\ln C_t = \ln C_o - k_s t$ (where C_t = concentration at time t, C_o = concentration at t_o, and k_s = negative slope or a constant).

The gross dissipation of a pesticide from a soil surface or within the soil is assumed to follow a first-order rate equation and should not be considered first order kinetics, decay, or reaction. This is considered a valid assumption (16) for most pesticides as the vast number of k_s values in tables 1 through 4 would indicate. For example, two-thirds of the k_s values given in table 1 were obtained from first-order rate equations with correlation coefficients (r) > 0.9.

The dissipation from soils of many pesticides (especially the chlorinated hydrocarbon insecticides and certain herbicides) follows that of a first-order equation, which is demonstrated further by the data of Edwards (14), Hamaker (24), Nash and Woolson (54), and Walker (82, 83, 84, 85), as well as many others. Similarly, Mill and others[2/] have suggested that the overall rate of disappearance of a pollutant in laboratory aquatic studies was the sum of the first-order rate losses for the individual loss processes.

The first-order rate loss for pesticides from an environmental component sometimes has been referred to as half-life, like that obtained from first-order kinetics. To avoid implying that pesticide loss from an environmental component is explicitly first-order kinetics, this first-order relationship for pesticide loss may be expressed as half-concentration time. By definition, half-concentration time ($C_{1/2}$) refers to the time required to reduce the maximum chemical concentration by one-half in an environmental component.

Hamaker (23) perhaps has studied the dissipation rates of pesticides from soils more than anyone else. He suggests that the rate laws are of two basic types: rate = kc^n and rate = $k_1c/(k_2 + c)$, where c is concentration, n is the order of equation, k is the rate constant, k_1 is a maximum rate approached with increasing concentration, and k_2 is a pseudoequilibrium constant. The former is a "power-rate model." He found that the "power-rate model" usually was the most useful in pesticide dissipation rates, perhaps because not enough observations are made, and that too many complex parameters affect dissipation of pesticides from soils.

Hamaker (23) calculated the apparent order from the "power-rate model" for several compounds found in the literature. A range of apparent orders was from -5.97 to 13.38, with an average of 1.7 ± 2.8. Even for one pesticide, amitrole, the order of the equation varied from 0.307 to 13.38. Hence, assigning a specific k_s value (calculated from the first-order equation or any order

---

2/ Environmental Exposure Assessment Using Laboratory Measurements of Environmental Processes; T. Mill, J. H. Smith, W. Makey, B. Holt, N. Bohonos, S. S. Lee, D. Bomberger, and T. W. Chow; Stanford Research Institute, Menlo Park, Calif. 94025

Table 1.--Values of k_s for dissipation of pesticides from soil surfaces

Pesticide	Soil: Type	pH	OM	Crop or conditions	Application rate	k_s	Reference
			(%)		(kg/ha)		
				FUNGICIDES			
Maneb-------------	Keyport sil 1/	---	---	--	--	0.0126	(58)
Maneb-------------	Galestown	6.7	5.2	Tomato	2	2/ .0278	(51)
Zineb-------------	Galestown	6.7	5.2	Tomato	2	3/ .0512	(57)
				HERBICIDES			
Asulam------------	Regina c	7.7	4.2	--	--	.0141-	(68)
						.1174	
Benefin-----------	Taloka sl	---	1	--	1.7	.3349	(31)
Butralin----------	Taloka sl	---	1	--	2.24	.1077	(31)
2,4-D isooctyl----	Galestown	6.7	5.2	Bluegrass	2.8	.0923	(52)
2,4-D isooctyl----	Acadia sil	5.8	2.6	--	5.7	3/ .0183	(75)
2,4-D-------------	Acadia sil	5.8	2.6	--	7.8	.0788	(75)
2,4-D isooctyl----	Acadia	5.8	2.6	--	15.7	.0486	(75)
2,4-D amine-------	Acadia	5.8	2.6	--	5.6	.0522	(75)
2,4-D amine-------	Acadia	5.8	2.6	--	11.2	.0139	(75)
2,4-D amine-------	Acadia	5.8	2.6	--	22.4	.0108	(75)
2,4-D-------------	Acadia	---	---	Arid range	1.03	.1634	(34)
2,4-D		---	---	Arid range	--	.1036	(34)
2,4-D amine-------	Acadia	5.8	2.6	--	5.7	.0352	(75)
2,4-D isooctyl----	Acadia	5.8	2.6	Ester per se	31.4	.2603	(75)
Dicamba-----------	Acadia	5.8	2.6	--	1.1-4.3	.0151	(75)
Diflubenzuron-----	Lufkin fsl	---	---	Over winter	---	.0040	(9)
Dinitramine-------	Taloka sl	5.8	2.6	--	.4	.0856	(31)
Diuron------------	New Ya'ar (dry)	---	---	--	4	4/ .136	(28)
Diuron------------	New Ya'ar (wet)	---	---	--	4	4/ .214	(28)
Fluchloralin------	Taloka sl	---	---	--	.56	.0169	(31)
Fluometuron-------	New Ya'ar (dry)	---	---	--	.56	4/ .043	(28)
Fluometuron-------	New Ya'ar (wet)	---	---	--	.56	4/ .077	(28)
Isopropalin-------	Taloka sil	---	---	--	1.68	.1948	(31)
Nitralin----------	Taloka sil	---	---	--	.56	.1042	(31)
Oryzalin----------	Taloka	---	---	--	.84	.0284	(31)
Oxmyl-------------	Keyport sil	---	---	--	6.7	.0646	(25)
Oxmyl-------------	Cecil ls	---	---	--	6.7	.0354	(25)
Oxmyl-------------	Leon Immokalee fs	---	---	--	6.7	.0448	(25)
Pendimethalin-----	Taloka sil	---	---	--	.56	.1695	(31)
				HERBICIDES			
Picloram				Arid range	1.05	.2689	(75)
Picloram						.0712	(34)
Profluralin-------	Taloka sil	---	---	--	.74	.2434	(31)
Prometryne--------	Taloka sl	7.2	2	--	--	.0127	(83)
Pronamide---------	Taloka sl	6.1	2	--	4.5	2/ .0203	(82)
						.0603	
Pronamide---------	Newport sl	---	---	--	4.5	3/ .0173	(81)
Propham-----------	New Ya'ar c (dry)	---	---	--	.56	4/ .025	(28)
Propham-----------	New Ya'ar (wet)	---	---	--	.56	4/ .279	(28)
Silvex (spray)----	Galestown sl	6.7	5.2	Bluegrass	2	.0213	(52)
Silvex (granules)-	Galestown	6.7	5.2	Bluegrass	1.3	.0346	(52)
Simazine----------	Galestown sl	7	2	--	--	.0089	(83)
2,4,5-T (isooctyl)		5.8	2.6	--	1.1-	.0266-	(75)
		---	---	--	4.3	.075	(75)
2,4,5-T		---	---	--		.0674	(34)
2,4,5-T		---	---	Arid range	1.07	.1323	(34)
Trifluralin-------	Bosket sl	5.8	2.6	Bluegrass	.86-	.0748	(61)
					1.12	4/ .0681	
Trifluralin-------	(Bushland) cl	---	---	Cotton	.56	4/ .0299	(69)
Trifluralin-------	(Lubbuck)	---	---	Cotton	.56	4/ .0599	(69)
Trifluralin-------	Taloka sl	---	1.0	--	.56	.1729	(31)
Trifluralin-------	Sheep pens (dry)	6.3	2.0	--	--	2/ .0071	(87)
Trifluralin-------	Sheep pens (wet)	---	2.0	--	--	.0956	(87)

Table 1.--Values of k_s for dissipation of pesticides from soil surfaces--Continued

Pesticide	Soil Type	Soil pH	Soil OM	Crop or conditions	Application rate	k_s	Reference
			(%)		(kg/ha)		
				INSECTICIDES			
Aldrin	Coachella fs	---	---	--	20.2	0.2406	(49)
Aldrin (+dieldrin) (granules)	Carrington sil	---	---	--	5.6	2/ .0045	(40)
Azinphosmethyl (spray)	Orchard sl	6.6-7.8	3.4	--	--	4/ .0486	(73)
Azinphosmethyl (spray)	Orchard sl	6.6-7.8	3.4	--	--	4/ .0434	(73)
Carbofuran	Alluvial	6.2	1.6	--	--	.0075	(80)
Carbofuran	Ritzville sil	7.8	1.0	--	2.0	.0690	(19)
Carbofuran	Chehalis c	6.2	7.2	--	2.0	.0180	(19)
Carbofuran	Organic	5.9	4.0	--	2.0	.0048	(19)
Carbofuran	Sultan sil	4.3	3.0	--	2.0	.034	(19)
Carbofuran	Sultan sil	6.0	3.0	--	2.0	.086	(19)
Carbofuran	Sultan sil	6.0	3.0	--	2.0	.0040	(19)
Carbofuran	Sultan sil	6.8	3.0	--	2.0	.0059	(19)
Carbofuran	Sultan sil	7.8	3.0	--	2.0	.0132	(19)
Chlordane	Gullatin Valley	---	---	Alfalfa	2.0	4/ .0101	(6)
Chlordane	Gullatin Valley	---	---	Alfalfa	>2	4/ .007	(6)
DDT	sl	---	---	--	83	2/ .004	(91)
DDT	Coachella fs	---	---	--	22.4	2/ .053	(49)
DDT	Houston c	>7	---	--	--	.0060	(2)
DDT	Pima sic	>7	---	--	--	.0049	(2)
DDT	Pinal gl	>7	---	--	--	.0060	(2)
DDT	Blackwater River	<7	---	Forest	1.12	3/ .00015	(55)
DDT	Pollard Mountain	<7	---	Forest	1.12	.000023	(55)
DDT	Mosquito Bnk Pod	<7	---	Forest	1.12	.00040	(55)
DDT	Route 11	<7	---	Forest	1.12	4/ .00014	(55)
DDT	West Oxbow	<7	---	Forest	1.12	4/ .00024	(55)
DDT	Beach Mountain	<7	---	Forest	1.12	.00044	(55)
Diazinon	Reaville shl	---	---	--	15.7	3/ .1422	(45)
Endrin	Mhoon sicl	6.0	1.2	Sugarcane	--	.0110	(92)
Endrin	Coachella fs	---	---	--	5.4	.2436	(49)
Ethion	Vineyard	---	---	Grapes	1.12	2/ .0647	(37)
Ethion	Vineyard	---	---	Grapes	1.4	.0702	(37)
Ethyl parathion	San Joaquin l	7.2	1.2	Citrus	2.24	.0332 5/ (.0725)	(70)
Ethyl parathion	San Joaquin l	7.2	1.2	Citrus	4.48	.0328 5/ (.0084)	(70)
Ethyl parathion	San Joaquin l	7.2	1.2	Citrus	8.96	.0282 5/ (.0114)	(70)
Hexachlorobenzene	Chewada sl	---	---	Zoysia	--	3/ .050	(4)
Methyl parathion	Houston c	>7	---	--	--	.0165	(2)
Methyl parathion	Pima sic	>7	---	--	--	.0153	(2)
Methyl parathion	Pinal gl	>7	---	--	--	.0147	(2)
Parathion	sl	4.7-7.3	1.0-3.4	--	--	3/ .0058	(94)
Toxaphene	Galestown sl	6.7	5.2	Cotton	(6x2.7)	.0046	(53)

1/ si = silt; s = san; sh = shale; c = clay; l = loam; f = fine; g = gravelly.
2/ Correlation coefficient r = <-0.8.
3/ Correlation coefficient r = <-0.9.
4/ Correlation coefficient r = unknown.
5/ Paraoxin.

Table 2.--Effect of soil type, organic matter content, moisture content, temperature, and pH on dissipation (k_s values) of pesticides in soils

Pesticide	Soil Type	pH	OM[1/]	Temperature 5°C SM[2/]	5°C k_s	10°C SM	10°C k_s	14-15°C SM	14-15°C k_s	20°C SM	20°C k_s	25°C SM	25°C k_s	28-30°C SM	28-30°C k_s	40°C SM	40°C k_s
			(%)	(%)		(%)		(%)		(%)		(%)		(%)		(%)	
Pronamide	Gravel Pits sl	6.1	1.8					11.7	0.0091			4.1	0.0116				
Pronamide	Little Cherry sl	6.9	1.4					12.0	.0099			4.0	.0096				
Pronamide	Soakwaters cl	6.1	1.4					17.8	.0110			6.4	.0124				
Pronamide	Gallas Ley sc	6.9	2.9					16.7	.0062			6.2	.0074				
Pronamide	Water Meadows cl	7.2	5.3					29.5	.0082			14.8	.0142				
Pronamide	Gravel Pits sl	6.1	1.8									11.9	.0217				
Pronamide	Little Cherry sl	6.9	1.4									12.1	.0231				
Pronamide	Soakwaters cl	6.1	1.4									16.3	.0301				
Pronamide	Gallas Ley sl	6.9	2.9									16.7	.0165				
Pronamide	Water Meadows cl	7.2	5.3									29.1	.0198				
Pronamide	Little Cherry sl	6.1	2.0					7.5	.0062					3.5	0.0077		
Pronamide	Little Cherry sl	6.1	2.0											7.5	.0110		
Pronamide	Little Cherry sl	6.1	2.0											10.0	.0127		
Pronamide	Little Cherry sl	6.1	2.0														
Asulam	Regina c	7.7	7.2			20	0.0141	20	.0261	20	0.0440	20	.0590	20	.582	20	0.0774
Asulam	Regina c	7.7	7.2			26	.0290	26	.0397	26	.0564	26	.0723	26	.0596	26	.0792
Asulam	Regina c	7.7	7.2			34	.0358	34	.0504	34	.0981	34	.0957	34	.1052	34	.0814
Asulam	Regina c	7.7	7.2			40	.0386	40	.0525	40	.1127	40	.0970	40	.1174	40	.0868
Linuron	Gravel Pits sl	7.0	2.0	4.0	0.0065	4.0	.0083	4.0	.0104	4.0	.0122	4.0	.0154	4.0	.0293		
Linuron	Gravel Pits sl	7.0	2.0									8.0	.0165				
Linuron	Gravel Pits sl	7.0	2.0	12.0	.0084	12.0	.0110	12.0	.0131	12.0	.0162	12.0	.0187	12.0	.0193		
Linuron	Gravel Pits sl	7.0	2.0									16.0	.0204				
Simazine	Gravel Pits sl	7.0	2.0	4.0	.0033	4.0	.0050	4.0	.0075	4.0	.0116	4.0	.0173	4.0	.0239		
Simazine	Gravel Pits sl	7.0	2.0									8.0	.0248				
Simazine	Gravel Pits sl	7.0	2.0	12.0	.0056	12.0	.0087	12.0	.0131	12.0	.0204	12.0	.0267	12.0	.0433		
Simazine	Gravel Pits sl	7.0	2.0									16.0	.0301				
Simazine	Gravel Pits sl	7.0	2.0									4.8	.0082				
Simazine	Gravel Pits sl	7.0	2.0					6.6	.0030			6.0	.0091				
Simazine	Gravel Pits sl	7.0	2.0									7.9	.0109				
Simazine	Gravel Pits sl	7.0	2.0									9.7	.0131				
Simazine	Gravel Pits sl	7.0	2.0									10.7	.0149				
Prometryn	Gravel Pits sl	7.0	2.0									11.2	.0176				
Prometryn	Gravel Pits sl	7.0	2.0					11.5	.0062			11.4	.0191				
Prometryn	Gravel Pits sl	7.0	2.0									13.2	.0188				
Prometryn	Gravel Pits sl	7.0	2.0									4.8	.0012				

Table 2.--Effect of soil type, organic matter content, moisture content, temperature, and pH on dissipation (k_s values) of pesticides in soils--Continued

Pesticide	Soil			Temperature													
	Type	pH	OM[1/]	5°C		10°C		14-15°C		20°C		25°C		28-30°C		40°C	
			%	SM[2/]	k_s	SM	k_s	SM	k_s	SM	k_s	SM	k_s	SM	k_s	SM	k_s
				(%)		(%)		(%)		(%)		(%)		(%)		(%)	
Prometryn	Gravel Pits sl	7.0	2.0					6.6	0.0014			6.0	0.0029				
Prometryn	Gravel Pits sl	7.0	2.0									7.9	.0084				
Prometryn	Gravel Pits sl	7.0	2.0									9.7	.0152				
Prometryn	Gravel Pits sl	7.0	2.0									10.7	.0167				
Premetryn	Gravel Pits sl	7.0	2.0									11.2	.0234				
Premetryn	Gravel Pits sl	7.0	2.0					11.5	.0062			11.4	.0276				
Premetryn	Gravel Pits sl	7.0	2.0									13.2	.0229				
Carbofuran	sil	4.3	---							--	0.0234						
Carbofuran	sil	6.0	---							--	.0261						
Carbofuran	sil	6.8	---							--	.0421						
Carbofuran	sil	7.8	---							--	.1020						

1/ Organic matter.
2/ Soil moisture.

Source: (19, 68, 82, 83, 84, 85).

Table 3.--Effect of replication and time of year (temperature) on dissipation rates (k_s) of organophosphorus insecticides in soil

Pesticide	Laboratory Replication 1 k_s	Laboratory Replication 2 k_s	Field June-July k_s	Field Aug-Sept k_s	Field Oct-Nov k_s
Aphidan	0.0283	0.0210	0.0480	0.0162	0.0020
Bromophos	.0078	.0124	.0175	.0137	.0111
Chlorfenvinphos	.0083	.0051	.0050	.0097	.0035
Diazinon	.0387	.0403	.0322	.0134	.0082
Dichlofenthion	.0012	.0046	.0037	.0023	.0102
Dimethoate	.0189	.0239	.0663	-	-
Mecarbam	.0080	-	.0532	.0094	.0017
Trichloronate	.0064	.0039	.0048	.0083	.0037

Source: Bro-Rasmussen, Noddegaard, and Voldum-Clausen (8).

equation) for any pesticide would be extremely dangerous. A k_s value from the first-order equation would be a good estimate of the dissipation rate, however, when it is used in a general sense to indicate the dissipation rate of the pesticide from soil.

k_s Values

The weakest link in determining the concentration of a pesticide at time, t, from the equation $C_t = C_0 e^{-k_s t}$, is the k_s value of the pesticide. (C_t is the pesticide concentration at the time under consideration, and C_0 is the pesticide concentration at time 0.) K_s values are constants that measure the rate of dissipation of a pesticide (fig. 1B--slope). To be accurate, k_s must be independent of all edaphic (soil) and climatic factors except temperature. Innumerable edaphic and climatic factors do affect the dissipation rate of a pesticide, however, the k_s values listed in tables 1 through 4 are, at best, only estimates.

Effect of Pesticide Class

The single most important factor affecting the k_s value is the pesticide itself. Some pesticides dissipate from the soil rapidly, and others remain for a long time. Benefin herbicide (table 1), for example, has a k_s value of 0.3349, and the k_s value of fluchloralin is 0.0169, both for the same soil. Their respective first-order half-concentration time ($C_{1/2}$) would be 2 and 41 days. Other examples include ethion insecticide versus parathion, which give $C_{1/2}$'s of 408 and 5 days, respectively.

Table 4.--Values of k_s for dissipation of pesticides in soil

Pesticide	Soil Type		pH	OM[1/]	Crop or conditions	Application rate	k_s	Reference
				(%)		(kg/ha)		
					FUNGICIDES			
BAS 3460F	Potting Soil				Agonis Flexuosa		0.0822	(3)
Benomyl	Potting Soil				Agonis Flexuosa		.1486	(3)
Benomyl		sl					.0058	(35)
Benomyl		l					.0023	(35)
					HERBICIDES			
Alachlor							.0384	(89)
Amitrole							.0768	(89)
Arsenic acid							<.0064	(89)
Asulam	Regina	c	7.7	4.2	14 May		.0986	(86)
Asulam	Regina	c			12 July		.0519	(86)
Asulam	Regina	c	7.7	4.2	30 July		.0310	(86)
Atrazine		sl	4.8	1.0			.0131	(22)
Atrazine	Regina	c	6.5	2.0			.0063	(22)
Atrazine							.0064	(86)
Atrazine	Norfolk	sl	6.8			∿5.2	.0133	(47)
Atrazine	Decatur	cl	6.4			∿5.2	.0149	(47)
Benefin							.0053-.0077	(76)
Benefin							.0077-.0070	(76)
Bifenox	Potting soil mixture				Various	1.7	.142	(36)
Butralin							.0128	(76)
Butralin							.0077	(76)
Cyanazine							.0064	(86)
Di-Allate	Weyburn	l	6.5	6.5	None	2.8	.0138	(66)
Di-Allate	Weyburn	l	7.0	4.5	Laboratory	.65	.0248	(67)
Di-Allate	Regina	c	7.5	4.0	Laboratory	.65	.0180	(67)
Di-Allate	Regina	c	7.8	4.2	None	2.2	.0110	(66)
Dicamba	Asquithse		7.5	3.2	5% moisture	.25	.0197	(65)
Dicamba	Asquithse		7.5	3.2	10% moisture	.25	[2/] .2140	(65)
Dicamba	Meltfort	sic	5.2	11.7	Various moisture	.25	[2/] >.2140	(65)
Dicamba	Regina	c	7.7	4.2	25% moisture	.25	.0486	(65)
Dicamba	Regina	c	7.7	4.2	35% moisture	.25	.0902	(66)
Dicamba	Ouachita	cl		3.3	Forest	.3	.0217	(1)
					HERBICIDES			
Dicamba	Ouachita	cl		2.8	Grass	.3	.0407	(1)
Dicamba	Cross Timbers	l		3.8	Forest	.3	.0267	(1)
2,4-D	Cross Timbers	l		3.8	Forest	.6	.1733	(1)
2,4-D acid	Cross Timbers	l					>.0768	(1)
2,4-D	Cross Timbers	l		3.3	Forest	.6	.1386	(1)
2,4-D salt							.0768	(1)
2,4-D	Ouachita	cl		2.8	Grass	.6	.1733	(1)
2,4-D ester							>.0768	(1)
2,4-D isooctyl ester.	Naff	sil		3.2	Laboratory 30°C	15	.2546	(93)
2,4-D	Naff	sil		3.2	Laboratory 10°C	15	.2731	(93)
2,4-D amine	Naff	sil		3.2	Laboratory carboxyl-^{14}C	12.5	.1457	(93)
2,4-D amine	Naff	sil		3.2	Laboratory carboxyl-^{14}C	12.5	.1008	(93)
2,4-D isooctyl ester α amine.	Naff	sil		3.2	Laboratory carboxyl-^{14}C	12.5	.0951	(93)
2,4-D isooctyl ester α amine.	Naff			3.2	Laboratory carboxyl-^{14}C	12.5	.0555	(93)
2,4-D isooctyl ester α amine.				3.2	Laboratory Ring^{14}C	12.5	.0852	(93)

Table 4.--Values of k_s for dissipation of pesticides in soil--Continued

Pesticide	Soil Type	pH	OM 1/	Crop or conditions	Application rate	k_s	Reference
			(%)		(kg/ha)		
				HERBICIDES			
2,4-D isooctyl ester α amine.			3.2	Laboratory Ring^{14}C	12.5	0.0257	(93)
Dichlorprop	Ouachita sl		3.3	Forest	.6	.0578	(93)
Dichlorprop	Ouachita sl		2.8	Grass	.6	.0866	(93)
Dichlorprop	Cross Timbers l		3.8	Forest	.6	.0693	(93)
Dinitramine						.0193	(76)
Dinitramine						.0193	(76)
Diuron	Norfolk cl	6.8			∿5.2	.0064	(47)
Diuron	Decatur cl	6.4			∿5.2	.0072	(47)
EPTC	Regina c	7.5	4.0	Laboratory	.65	.0220	(67)
EPTC	Weyburn l	7.0	4.5	Laboratory	.65	.0248	(67)
Fluchloralin						2/ .0070	(76)
Fluchloralin						2/ .0045	(76)
Isopropalin	Daummer sic	6.7	5.1	Various		.0023-	(76)
Isopropalin	Eisne sil	7.2	1.6			.0036	(76)
						2/ .0054	(76)
						.0040	(76)
Isopropalin	Ochley sil	4.7	2.9	Sorghum	1.68	.0304	(59)
Isopropalin	Ochley sil	4.7	2.9	Sorghum	3.36	.0214	(59)
Isopropalin	Bloomfield fs	6.3	.6	Sorghum	1.12	.0275	(59)
Karbutilate	cl	6.3	2.2	Rangeland		.0057- .0282	(50)
Karbutilate	lc	6.2	1.1	Rangeland		.0118	(50)
Linuron				Cropped		3/ .0104- .0231	(84)
Linuron	ls			Non-cropped		.0047	(84)
Linuron	cl	7.0	2.0	Carrots	.85	3/ .0280	(17)
Linuron	cl	7.0	2.0	None	3.4	3/ .0039	(17)
Linuron	0-5 cm	7.0	2.0	None	3.4	3/ .0061	(17)
MCPA	Coarse cl	7.0	2.0	Barley	1.7	3/ .1221	(17)
MCPA	Coarse cl	7.0	2.0	None	3.4	4/ .1070	(17)
Metribuzin						4/ .0298	(86)
Metobromuron	sl	4.8	1.0			.0231	(22)
Metobromuron	sl	6.5	2.0			.0248	(22)
Monolinuron						.00216	(62)
Monuron	Romona sl			Various	2.24	.0060	(63)
Monuron	Romona sl				4.48	.0075	(63)
Neburon	Romona sl			Various	2.24	.0073	(63)
Neburon	Romona sl			Various	4.48	.0059	(63)
Nitralin						.0062- .0086	(76)
Nitralin						.0096- .0086	(76)
Nitralin	Ochley sil	4.7	2.9	Sorghum	1.12	.0110	(59)
Nitralin	Ochley sil	4.7	2.9	Sorghum	1.12	.0079	(59)
Nitralin	Ochley sil	4.7	2.9	Sorghum	2.24	.0090	(59)
Nitralin	Ochley sil	4.7	2.9	Sorghum	2.24	.0024	(59)
Nitralin	Bloomfield fs	6.3	.6	Sorghum	.56	.0155	(59)
Nitralin	Bloomfield fs	6.3	.6	Sorghum	1.12	.0091	(59)
Oryzalin	Bloomfield fs					2/ .0054-	(76)
						.0083	(76)
Oryzalin						.0144-	(76)
						.0056	(76)
Pebulate	Regina c	7.5	4.0	Laboratory	.65	.0396	(67)
Pebulate	Weyburn l	7.0	4.5	Laboratory	.65	.0396	(67)
Picloram	Scot l, oxbows cl			Various		.0025	(32)
						2/ .0083	(76)
						2/ .0056	(76)

Table 4.--Values of k_s for dissipation of pesticides in soil--Continued

Pesticide	Soil Type	Soil pH	Soil OM	Crop or conditions	Application rate	k_s	Reference
			(%)		(kg/ha)		
				HERBICIDES			
Pebulate	Regina c	7.5	4.0	Laboratory	.65	0.0396	(67)
Pebulate	Weyburn l	7.0	4.5	Laboratory	.65	.0396	(67)
Picloram	Scot l, oxbows cl			Various		.0025	(32)
Picloram	Various					2/ .00772	(60)
Picloram	Nova Scotia cl				4.8	.0044	(56)
Picloram	Somerset sl	4.8	2.9	Fallow	4.48	.0050	(56)
Picloram	Farin cl	6.3	1.9	Orchard grass	2.24	.0354	(44)
Picloram	Chandler f sl	5.5	1.7	Orchard grass	2.24	.0258	(44)
Picloram	Chester l	5.8	1.9	Orchard grass	2.24	.0268	(44)
Picloram	Chester l	5.8	1.9	Orchard grass	4.48	.0269	(44)
Picloram	Various	Various			.05	2/ .004	(48)
Picloram	Ouachita cl		3.3	Forest	.6	3/ .0019	(1)
Picloram	Ouachita cl		2.8	Grass	.6	.0048	(1)
Picloram	Cross Timbers l		3.8	Forest	.6	.0028	(1)
Profluralin						2/ .0047	(76)
Profluralin						2/ .0051	(76)
Pometryne	sl	7.0	2.0			.0238	(83)
Propazine	sl	4.8	1.0			.0108	(22)
Propazine		6.5	2.0			.0056	(22)
Propyzamide	c to sl			Lettuce		.0061-4/ .0158	(85)
Silvex	Ouachita sl		3.3	Forest	.6	.0330	(1)
Silvex	Ouachita sl		2.8	Grass	.6	.0495	(1)
Silvex	Cross Timbers l		3.8	Forest	.6	.0462	(1)
Simazine		7	2	None	3.4	4/ .0074	(17)
Simazine	0 to 5 cm	7	2	None	3.4	.0083	(17)
Simazine	sl	4.8	1.0			.0116	(22)
Simazine	sl	6.5	2.0			.0082	(22)
Simazine				Cropped		.0539	(84)
Simazine				Noncropped		.062	(84)
Simazine	sl	7.0	2.0			.0187	(83)
Tebuthiuron	Various			Corn	.025	.0024	(11)
Tebuthiuron	Houston Black c, Udic Pellustert			In surface runoff water.	2.24	.0060	(5)
Tebuthiuron	Houston Black Udic Pellustert			In surface pellets.	2.24	.0427	(5)
Tebuthiuron	Houston Black Udic Pellustert			In surface broadcast spray.	2.24	.0201	(5)
Tebuthiuron	Houston Black Udic Pellustert			In pellets	2.24	.0517	(5)
Tebuthiuron	Houston Black Udic Pellustert			In surface band pellets.	2.24	.0624	(5)
Tebuthiuron	Houston Black Udic Pellustert			Broadcast in soil spray.	2.24	.0069	(5)
Triallate	Regina c	7.5	4.0	Laboratory	.65	.0090	(67)
Triallate	Weyburn l	7.0	4.5	Laboratory	.65	.0110	(67)
Triallate	Coarse sl	7.0	2.0	Barley	1.7	3/ .0144	(17)
Triallate	Coarse sl	7.0	2.0	None	3.4	3/ .0067	(17)
Triallate	Weyburn l	6.5	6.5	None	2.8	.0088	(66)
Triallate	Regina c	7.8	4.2	None	2.2	.0053	(66)
2,4,5-T	Ouachita sl		3.3	Forest	.6	.0289	(1)
2,4,5-T	Ouachita sl		2.8	Grass	.6	.0330	(1)
2,4,5-T	Cross Timbers l		3.8	Forest	.6	.0330	(1)
2,4,5-T	Fanin cl	6.3	1.9	Orchard grass	2.24	3/ .0508	(44)
2,4,5-T	Chandler fls	5.5	1.7	Orchard grass	2.24	.0495	(44)

Table 4.--Values of k_s for dissipation of pesticides in soil--Continued

Pesticide	Soil Type	pH	OM	Crop or conditions	Application rate	k_s	Reference
			(%)		(kg/ha)		
				HERBICIDES			
2,4,5-T-------	Chester l	5.8	1.9	Orchard grass	2.24	3/ 0.0416	(44)
2,4,5-T-------	Chester l	5.8	1.9	Orchard grass	4.48	.0414	(44)
Trifluralin						2/ .0037-.0047	(76)
Trifluralin							(76)
Trifluralin						2/ .0051-.0044	(76)
Trifluralin----	Cecil sl	6.5	.6	Soybeans		.0175	(90)
Trifluralin----	Wet soil			None		.0956	(87)
Trifluralin----	Dry soil			None		4/ .0189	
Trifluralin------	Ochley sil	4.7	2.9	Sorghum	.84	.0145	(59)
Trifluralin------	Ochley sil	4.7	2.9	Sorghum	.84	.0117	(59)
Trifluralin------	Ochley sil	4.7	2.9	Sorghum	1.68	.0104	(59)
Trifluralin------	Ochley sil	4.7	2.9	Sorghum	1.68	.0026	(59)
Trifluralin-----	Bloomfield fs	6.3	.6	Sorghum	.56	.0155	(59)
Trifluralin-----	Bloomfield fs	6.3	.6	Sorghum	1.12	.0091	(59)
Vernolate--------	Regina c	7.5	4.0	Laboratory	.65	.0396	(67)
Vernolate--------	Weyburn l	7.0	4.5	Laboratory	.65	.0396	(67)
				INSECTICIDES			
Aldicarb--------	Beaumont c	5.4			130	.00273	(78)
Aldicarb--------	Houston c	7.8			130	.0087	(78)
Aldicarb--------	Houston cl	7.5	.25	Shendi	.5	3/ .0991	(57)
Aldicarb--------	Houston cl	7.5	.25	Shendi	1.0	.0420	(57)
Aldicarb--------	Houston cl	7.5	.25	Orange	2.8-22.4	.0322-<.0032	(30)
Aldrin							(86)
Aldrin----------	Ulysses sil			Fallow	2.24	.0264	(38)
Aldrin------------	Knox sil			Fallow	2.24	.0259	(38)
Aldrin--------	Celeryville muck			Fallow	2.24	4/ .0014	(38)
Aldrin----------	Marietta sl			Fallow	2.24	.0136	(38)
Aldrin------------	Fox fsl			Fallow	2.24	.0256	(38)
Aldrin----------	Miami sil			Fallow	2.24	.0258	(38)
Aldrin-------------	Muck			Fallow	2.24	4/ .0066	(38)
Aldrin--------	Carrington sil	Nondisked		Fallow	4.5	2/ .0101	(38)
Aldrin--------	Carrington sil	Disked		Fallow	4.5	2/ .0136	(42)
Aldrin----------	Udaipur cl	7.8	1.6	Various	3.0	.0149	(72)
						x̄ of 19	(72)
						x̄ of 19	(72)
Aldrin----------	Jobner	8.6	.26	Various	3.0	.0165	(72)
						x̄ of 19	(72)
Aldrin-------------	Muck				22.4	.0061	(41)
Aldrin----------	Miami sil				22.4	.0096	(41)
Aldrin----------	Composite				22.4	.0038	(14)
Aldrin (Dieldrin)----	Carrington sil	Nondisked			4.5	4/ .0006	(42)
Aldrin (Dieldrin)----	Carrington sil	Disked			4.5	4/ .0008	(42)
Aldrin (Dieldrin)----	Carrington sil	Disked		Granules	5.6	4/ .0012	(40)
Aldrin (Dieldrin)----	Carrington sil			Spray		4/ .0017	(40)
Akton-----------	Sultan sil	6.3	3.4	Corn granules	2.24	.0032	(20)
Azinphosmethyl		8.4				.0239	(22)
Azinphosmethyl--	Orchard sl	6.6-7.8	3.4			4/,5/ .0026	(73)
Azinphosmethyl--	Orchard sl	6.6-7.8	3.4			4/,6/ .0014	(73)
Azinphosmethyl--	Gila sil				.018	.0533	(95)
Azinphosmethyl--	Mocho sil					.0273	(29)
Azinphosmethyl--	Linne c					.0516	(29)
Azinphosmethyl--	Madera sl					.0086	(29)
Azinphosmethyl--	Laveen sl					.0119	(29)
Azinphosmethyl--	Santa Lucia sill					.0235	(29)

Table 4.--Values of k_s for dissipation of pesticides in soil--Continued

Pesticide	Soil			Crop or conditions	Application rate	k_s	Reference
	Type	pH	OM				
			(%)		(kg/ha)		
				INSECTICIDES			
Azinphosmethyl----	Windy l					0.0074	(29)
Azinphosmethyl------	fsl					.0101	(29)
Azinphosmethyl------	sicl					.0458	(29)
Azinphosmethyl-------	c					.0505	(29)
Azinphosmethyl-------	sl					.0211	(29)
BHC						.0021	(86)
BHC-----------------	Udaipur	7.8	1.6	Various	5.0	.0140	(72)
BHC							
						$\bar{x}$ of 19	(72)
BHC----------------	Jobner sl	8.6	.26	Various	5.0	.0098	(72)
						$\bar{x}$ of 19	(72)
BHC alpha----------	Berwick sl			Vegetables	7.4 BHC	5/ .0006	(74)
BHC beta-----------	Berwick sl			Vegetables	7.4 BHC	2/ .00015	(74)
BHC gamma----------	Berwick sl			Vegetables	7.4 BHC	2/ .00042	(74)
BHC delta----------	Berwick sl			Vegetables	7.4 BHC	2/ .00036	(74)
Bromophos----------	Composite					.0198	(77)
Carbaryl						.0768	(86)
Carbaryl-----------	Udaipur cl	7.8	1.6	Various	15.0	.1196	(72)
						$\bar{x}$ of 8	
Carbaryl-----------	Jobner sl	8.6	.26	Various	15.0	.0969	(72)
						$\bar{x}$ of 8	
Carbofuran						.0768	(86)
Carbofuran----------	Take sil	8.5			10.0	.0079	(79)
CGA-12223--------------	sil	4.8	1.0			.0385	(22)
CGA-12223 7/------------	sil	6.5	2.2			.0693	(22)
Chlordane----------	Berwick sl				2.0	2/ .00072	(74)
Chlordane--------	Composite					.0020	(14)
Chlorfenvinphos						.0055	(77)
Diazinon---------	Composite					.0330	(77)
Diazinon---------	Sultan sil	6.7	3.1	25°C		.0151	(18)
Diazinon---------	Sultan sil	6.7	3.1	15°C		.0067	(18)
Diazinon---------	Sultan sil	4.3	3.1			.0242	(18)
Diazinon-------------	sl	4.8	1.0			.0239	(22)
Diazinon-------------	sl	6.5	2.0			.0239	(22)
Diazinon-------------	sl	6.5	2.0			.0248	(22)
Diazinon--------	Puyallup sl	5.0	2.1			.0189	(21)
Diazinon---------	Puget sil	5.4	3.0			.0260	(21)
Diazinon--------	Chehalis cl	5.6	7.2			.0166	(21)
Diazinon----------	Organic	5.4	40			.0171	(21)
Dieldrin-------	Carrington sil	Nondisked		Fallow	4.5	2/ .0142	(42)
Dieldrin-------	Carrington sil	Disked		Fallow	4.5	2/ .0187	(42)
Dieldrin						3/ .0003	(15)
Dieldrin---------	Imperial sc	7.8	1.0		20.0	.0002	(12)
Dieldrin--------	Holtville fsl	7.8	.5		20.0	.0001	(12)
Dieldrin----------	Composite					.0008	(14)
Dioxacarb------------	sl	4.8	1.0			.0248	(22)
Dioxacarb------------	sl	6.5	2.0			.3465	(22)
Dioxathion-----------	fsl					.0156	(29)
Dioxathion----------	sicl					0128	(29)
Dioxathion-----------	c					.0141	(29)
Dioxathion------------	sl					.0229	(29)
p,p'-DDT---------	Ulysses sil	6.9	1.8	Fallow	9.4	3/ .0008	(38)
p,p'-DDT-----------	Knox sil	6.8	.8	Fallow	9.4	3/ .0005	(38)
p,p'-DDT-------	Celeryville muck	4.9	74.5	Fallow	9.4	.0021	(38)
p,p'-DDT----------	Marietta sl	6.0	2.0	Fallow	9.4	3/ .0014	(38)
p,p'-DDT------------	Fox fsl	7.2	.8	Fallow	9.4	.0009	(38)
p,p'-DDT-----------	Miami sil	7.1	3.6	Fallow	9.4	3/ .0004	(38)
p,p'-DDT-------------	Muck	6.8	40.0	Fallow	1.0	.0009	(38)
p,p'-DDT--------	Commerce sil				24.4	.0037	(9)

Table 4.--Values of k_s for dissipation of pesticides in soil--Continued

Pesticide	Soil Type	Soil pH	Soil OM	Crop or conditions	Application rate	k_s	Reference
			(%)		(kg/ha)		
				INSECTICIDES			
p,p'-DDT	Carrington sil	Nondisked			4.5	2/ .0024	(42)
p,p'-DDT	Carrington sil	Disked		Fallow	4.5	2/ .0048	(42)
p,p'-DDT	Miami sil				11.6	.0003	(39)
p,p'-DDT	Carrington sil	Disked/non-disked		Fallow	4.5	4/ .0002	(39)
p,p'-DDT	Muck				11.2	.0011	(41)
p,p'-DDT	Miami sil				11.2	.0029	(41)
p,p'-DDT	Berwick sl				37 DDT	2/ .00016	(74)
p,p'-DDT	Composite					.0007	(14)
o,p'-DDT	Berwick sl					2/ .00029	(74)
Dimethoate	Composite					.0990	(77)
Disulfoton				Various		.1604	(64)
Endosulfan	Various				[1].3	.0162	(46)
Ethion	Mocho sil					.0014	(29)
Ethion	Linne c					.0012	(29)
Ethion	Madera sl					.0009	(29)
Ethion	Laveen sl					.0015	(29)
Ethion	Santa Lucia sil					.0014	(29)
Ethion	Windy l					.0015	(29)
Ethion	fsl					3/ .0009	(29)
Ethion	sicl					.0022	(29)
Ethion	c					.0032	(29)
Ethion	sl					.0025	(29)
Fenitrothion	sl	4.8	1.0			.0578	(22)
Fenitrothion	sl	6.5	2.0			.1155	(22)
Fonofos	Take sil	8.5			10	.0158	(79)
Heptachlor	Composite					.0021	(15)
Heptachlor	Composite					.0025	(14)
Heptachlor	Composite					.0028	(7)
Hexachlorobenzene	Chevada			Zoysia		3/ .0006	(4)
Isobenzan	Composite					.0050	(7)
Lindane	Imperial sc	7.8	1.0		20	.0022	(12)
Lindane	Holtville fsl	7.8	.5		20	.0026	(12)
Lindane	Composite					.0017	(14)
Lindane	Gila sil	7.7	.6	None		2/ .0046	(13)
Lindane	Miami sil				11.6	.0011	(39)
Lindane	Muck				11.2	.0014	(41)
Lindane	Miami sil				11.2	.0048	(41)
Lindane	Ulysses sil			Fallow	1.12	.0147	(38)
Lindane	Knox sil			None	11.2	.0264	(41)
Lindane	Celeryville muck			None	11.2	.0074	(41)
Lindane	Marietta sl			None	11.2	.0263	(41)
Lindane	Fox fsl			None	11.2	.0264	(41)
Lindane	Miami sil			None	11.2	.0139	(41)
Lindane	Muck			None	11.2	3/ .0059	(41)
Malathion	Poygan sicl	7.2		None		2.9173	(33)
Malathion	Kewaune c	6.4		None		2.4618	(33)
Malathion	Ella ls	3.8		None		1.2681	(33)
Malathion	Freestone sl	5.3	1.1	None		.4152	(88)
Malathion	Okolona c	7.4	3.1	None		1.9832	(88)
Malathion	Trinity l	7.2	4.7	None		1.9026	(88)
Mecarbam	Composite					.0495	(77)
Methidathion	sl	4.8	1.0			.0108	(22)
Methidathion	sl	6.5	2.0			.0495	(22)
Methoxychlor		4.8	1.0			.0046	(22)
Methoxychlor		6.5	2.0			.0033	(22)
Methyl Parathion	Carrington l			Radishes	5.6	.2207	(43)
Mevinphos	Sacramento s	5.4	.4		13	.2936	(10)
Parathion	Carrington l			Radishes	5.6	.0248	(43)
Parathion						.056	(16)
Parathion	Mocho sil	>7	.6	None		2/ .0046	(13)
Parathion	Udaipur cl	7.8	1.6	Various	10	.1239	(72)

Table 4.--Values of k_s for dissipation of pesticides in soil--Continued

Pesticide	Soil Type	pH	OM	Crop or conditions	Application rate	k_s	Reference
			(%)		(kg/ha)		
				INSECTICIDES			
						$\bar{x}$ of 8	(72)
Parathion	Jobner sl	8.6	.26	Various	10	0.0727	(72)
						$\bar{x}$ of 7	(72)
Parathion	Mocho sil					3/ .1371	(29)
Parathion	Linne c					.1306	(29)
Parathion	Madera sl					3/ .0944	(29)
Parathion	Laveen sl					3/ .1150	(29)
Parathion	Santa Lucia sil					.0866	(29)
Parathion	fsl	6.8	0.8			.0654	(29)
Parathion	sicl	7.3	2.1			3/ .0891	(29)
Parathion	c	7.3	2.3			.2962	(29)
Parathion	sl	7.6	1.8			.2614	(29)
Phenthoate	fls	6.8	.8			.2865	(29)
Phenthoate	sicl	7.3	2.1			.0156	(29)
Phenthoate	c	7.3	2.3			.0141	(29)
Phenthoate	sl	7.6	1.8			.0229	(29)
Phorate	Sacramento muck				13	.0040	(10)
Phorate	Sacramento peat				13	2/ .0043	(10)
Phorate	Sacramento peat				13	2/ .0051	(10)
Phorate	Take sil	8.5			10	3/ .0363	(79)
Phorate	Sacramento s				13	4/ .0078	(10)
Phorate	Sacramento c				13	.0277	(10)
Zinophos	Sultan sil	6.7	3.1	25°C		3/ .0223	(18)
Zinophos	Sultan sil	6.7	3.1	15°C		.0164	(18)
Zinophos	Sultan sil	5.5	3.1			.0144	(18)
Zinophos	Sultan sil	8.1	3.1			.0244	(18)
Zinophos	Sultan sil					.0096	(21)
Zinophos						.0133	(21)
Zinophos						.0206	(21)
Zinophos						.0075	(21)
				NEMATICIDES			
Dichlofenthion	Composite					.0031	(7)
Trichloronate	Composite					.0050	(77)

1/ Organic matter.
2/ Unknown.
3/ r = <-0.9.
4/ r = <-0.8.
5/ Emulsifiable concentrate formulation.
6/ Wettable powder formulation.
7/ Diethyl (1-iso-propyl-5-chloro-1,2,4-triazolyl-3) phosphorothioate.

Briggs (7) proposed a method of predicting pesticide half-concentration time in soil from the physicochemical properties of the pesticides and soils, primarily the former. He suggested that two criteria needed for predicting pesticide persistence (or, here, dissipation) are the availability (in soil solution) of the pesticides and their degradability. The product of these two criteria, availability (A) and degradability (D), estimates half-concentration time of the pesticide in weeks (persistence in weeks = A x D).

Briggs (7) lists 14 availability (A) classes (table 5). To calculate a pesticide's availability class, the octanol/water distribution coefficient (P) is determined first. Then Q is determined from the equation log Q = 0.52 log P + 0.62 (where Q is the soil organic matter/water distribution coefficient). Finally, a value for K (soil/water distribution coefficient) can be calculated from K = (% organic matter x Q)/100. The K values in table 5 allow one to determine the availability class at 15% moisture. Suppose the octanol/water distribution coefficient was P = 1,500, for example. Then from log Q = 0.52 log P + 0.62, log Q = 0.52 (3.176) + 0.62 or 2.271, and Q = 187. For 2% soil organic matter, K = 2 Q/100 = 2 x 187/100 = 3.7. The availability class would be 6. For 5% soil organic matter, K = 5 x 187/100 = 9.3, or availability class 7.

Degradability (D) of a pesticide may be determined from figure 2. The degradability scale of Briggs (7) is based on the pesticide's functional groups (further information on this approach is contained in Briggs' excellent discussion). A readily degraded functional group on a pesticide is assigned a low number on a scale of 1 to 10, whereas a stable functional group is assigned a high number.

Many functional groups may have variable degradability because of the steric and electronic effects in the rest of the molecule. Steric nonhindered esters (RCOOR') are very short lived in soil, for example, and are assigned a degradability of 1, whereas steric hindered esters are more stable and are assigned a degradability of 3.

Persistence of aromatic carboxylic acids (ArCOOH) highly depends on the other substituents in the aromatic ring. Benzoic acid is degraded in soil in a few hours, but 2,3,6-trichlorobenzoic acid (TBA) is very stable in soil because of both steric and electronic effects of the chlorine substituents. Hence, benzoic acid would be assigned a degradability value of 1, and TBA, 10.

Products of the predicted A x D values (half-concentration time in weeks) for several insecticides and environmental pollutants are listed in table 6 and compared with data obtained from the literature by Briggs (7). From these values, k_s values were calculated. Although trends for the predicted (A x D in weeks) are reasonably good, further refinement of the predictive method of Briggs is necessary before it can be considered reliable. This is especially true for the short-lived pesticides where the error appears rather large. The predicted method of Briggs may be useful for the long-lived pesticides and pollutants, especially if no experimental data are available.

Figure 3 gives another approach by Briggs (7), in which 65 pesticides are arranged by persistence and mobility, according to the classification of Helling (26) in soil. Consequently, the k_s value can be estimated if the relative mobility of the pesticide in soil is known

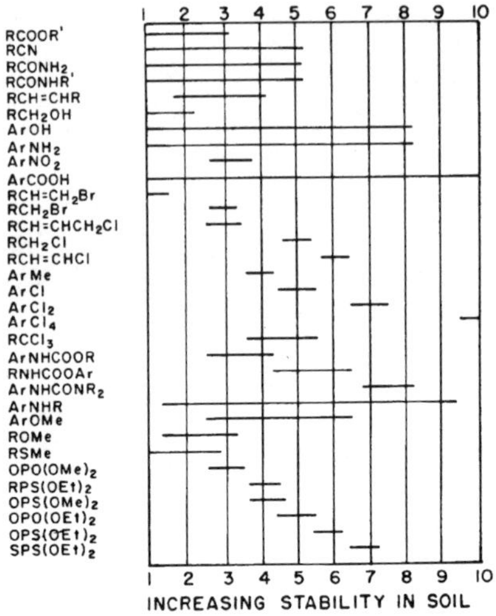

Figure 2.--Degradability of functional groups on a pesticide molecule, based on a 1 to 10 scale. [Adapted from (7)]

A better approach, though empirical, is that of Walker (82, 83, 84, 85). Walker measured the persistence of several herbicides, both in the laboratory and field, at controlled or measured soil temperature and moisture conditions. He then simulated the concentration of the herbicide at any given time t with the aid of a computer program. Walker's persistence in soil data (table 2) for the herbicides propamide, simazine, prometryn, asulam, linuron, and N-(1-ethyl=propyl)-2,6-dinitro-3, 4-e-xylidine is undoubtedly, the most reliable persistence data available (see table 7 for pesticide names).

Effect of Soil Type

Iwata and others (9) determined the persistence of azinphosmethyl insecticide in 10 different soil types in the laboratory. Their data gave a k_s range of values from 0.0074 to 0.0516 (table 1). The $C_{1/2}$'s for azinphosmethyl would range from 13 to 94 days. Likewise, the data of Walker (table 2) show a range of k_s values (0.0074 to 0.0116) as affected by soil type at 25°C, and similar soil moisture content for propyzamide. This range would result in

Table 5.--Availability classes for pesticides in soil containing 15% moisture

Availability	K (%O.M.[1] x Q/100)
1	< .05
2	.15
3	.45
4	1.35
5	2.85
6	5.85
7	14.9
8	29.9
9	59.9
10	150
11	300
12	600
13	1500
14	> 1500

1/ OM = organic matter.

Source: Briggs (7).

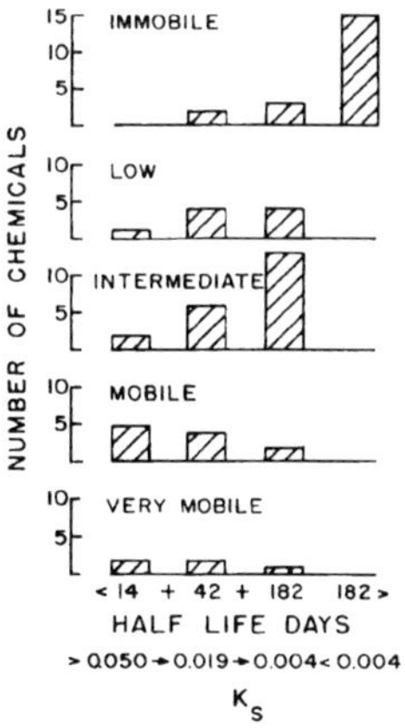

Figure 3.--Values of k_s for 65 pesticides distributed between persistent groups and mobility classes [adapted from (7)].

$C_{1/2}$'s of 60 to 94 days. In a further example, k_s values for the same soil can differ by a wide margin. Table 3 lists replicate samples of the persistence of several insecticides. The k_s values for two replications of bromophos would give $C_{1/2}$ values ranging from 56 to 89 days.

Effect of Soil Temperature

Similarly, temperature or time of year affects the k_s value (tables 2 and 3). Walker (82), may have studied the affect of soil temperature on herbicide persistence more than anyone else. His results show the influence that temperature may have on pesticides that are degraded primarily by microbial action. At 5°C and 12% soil moisture, for example, the $C_{1/2}$ of simazine is 124 days, whereas at 30°C and 12% moisture, the $C_{1/2}$ is reduced to 16 days, just 1/8 the time with a 25°C increase of temperature.

In June and July (table 3) the $C_{1/2}$ for bromophos is 40 days; for August and September, 51 days; and for October and November, 62 days (81). For all practical purposes, when a soil becomes frozen, dissipation of pesticide ceases. When calculating the number of days since application, subtract the number of days the soil was frozen.

Table 6.--Predicted k_s values for several insecticides and environmental pollutants in soil at 15% moisture and 2% organic matter

Compound	Log P 1/	A 2/	D 3/	A x D	$c^{1/2}$	k_s 4/ A x D	k_s 4/ $c^{1/2}$
				(weeks)	(weeks)		
DDT	-	11	6	66	150	0.0015	0.0007
Dieldrin	-	10	6	60	130	.0017	.0032
Lindane	-	10	5	50	60	.0020	.0017
Chlordane	-	10	5	50	50	.0020	.0020
Heptachlor	-	11	4	44	40	.0023	.0025
Heptachlor epoxide	-	9	5	45	35	.0022	.0028
Isobenzan	-	9	4	36	20	.0028	.0050
Aldrin	-	12	3	36	15	.0028	.0066
Dichlofenthion	-	8	6	48	32	.0021	.0031
Trichloronate	-	9	4	36	20	.0028	.0050
Chlorfenvinphos	-	7	5	35	18	.0028	.0055
Bromophos	-	7	3	21	5	.0047	.0198
Diazinon	-	6	6	36	3	.0028	.0330
Mecarbam	-	4	3	12	2	.0083	.0495
Dimethoate	-	3	3	9	1	.0110	.0990
Biphenyl-------------------	4.0	7	3	21	1-2	.0047	.0660
4-Cl-----------------------	4.7	8	3	24	2	.0041	.0495
4,4'-di-Cl-----------------	5.4	9	5	45	30	.0022	.0033
2,4-di-Cl------------------	5.4	9	3	27	4-6	.0037	.0198
3,3',4,4'-tetra-Cl---------	6.8	11	7	77	200	.0013	.0005
Sym hexa-Cl----------------	8.2	14	9	126	>300	.0008	<.0003
Sym octa-Cl----------------	9.6	14	10	140	>300	.0007	<.0003
Terphenyl------------------	5.9	10	3	30	6	.0033	.0765
2,4,5-tri-Cl---------------	8.0	13	5	65	150	.0015	.0007
Pentabromotoluene----------	6.0	10	9	90	>200	.0011	<.0005
Napthalene-----------------	3.4	6	4	24	2	.0041	.0495
hexa-Cl"-------------------	7.6	13	10	130	>300	.008	<.0003

1/ octanol/water distribution coefficient.
2/ availability of pesticide in soil solution.
3/ D = degradability of pesticide.
4/ days.

Source: Briggs (7)

Table 7.--Common and chemical names of pesticides

Common name	Chemical name
	FUNGICIDES
Benomyl	methyl 1-(butylcarbamoyl)-2-benzimidazole-carbamate.
Maneb	manganous ethylenebis[dithiocarbamate].
Zineb	zinc ethylenebis[dithiocarbamate].
	HERBICIDES
Alachlor	2-chloro-2',6'-diethyl-N-(methoxymethyl)acetanilide.
Amitrole	3-amino-s-triazole.
Asulam	methyl sulfanilylcarbamate.
Atrazine	2-chloro-4-(ethylamino)-6-(isopropylamino)-s-triazine
Benefin	N-butyl-N-ethyl-α,α,α-trifluoro-2,6-dinitro-p-toluidine .
Bifenox	methyl 5-(2,4-dichlorophenoxy)-2-nitrobenzoate.
Butralin	4-(1,1-dimethylethyl)-N-(1-methylpropyl)-2,6-dinitroben= zenamine.
Cyanazine	2-[[4-chloro-6-(ethylamino)-s-triazine-2-yl]amino]-2-methylpropionitrile.
2,4-D	(2,4-dichlorophenoxy)acetic acid.
Diallate	S-(2,3-dichloroallyl) diisopropylthiocarbamate.
Dicamba	3,6-dichloro-o-anisic acid.
Dichlorprop	2-(2,4-dichlorophenoxy)propionic acid.
Diflubenzuron	N-[[(4-chlorophenyl)amino]carbonyl]-2,6-difluorobenza mide.
Dinitramine	N^4,N^4-diethyl-α,α,α-trifluoro-3,5-dinitrotoluene-2,4-diamine.
Diuron	3-(3,4-dichlorophenyl)-1,1-dimethylurea.
EPTC	S-ethyl dipropylthiocarbamate.
Fluchloralin	N-(2-chloroethyl)-2,6-dinitro-N-propyl-4-(trifluoromethyl)aniline.

Table 7.--Common and chemical names of pesticides--Continued

Common name	Chemical name
	HERBICIDES
Fluometuron	1,1-dimethyl-3-(α,α,α-trifluoro-m-tolyl)urea.
Isopropalin	2,6-dinitro-N,N-dipropylcumidine.
Karbutilate	tert-butylcarbamic acid ester with 3-(m-hydroxy-phenyl)-1,1-dimethylurea.
Linuron	3-(3,4-dichlorophenyl)-1-methoxy-1-methylurea.
MCPA	[(4-chloro-o-tolyl)oxy]acetic acid.
Metobromuron	3-(p-bromophenyl)-1-methoxy-1-methylurea.
Metribuzin	4-amino-6-tert-butyl-3-(methylthio)-as-triazin-5(4H)-one.
Monolinuron	3-(p-chlorophenyl)-1-methoxy-1-methylurea.
Monuron	3-(p-chlorophenyl)-1,1-dimethylurea.
Neburon	1-butyl-3-(3,4-dichlorophenyl)-1-methylurea.
Nitralin	4-(methylsulfonyl)-2,6-dinitro-N,N-dipropylaniline.
Oryzalin	3,5-dinitro-N^4,N^4-dipropylsulfanilamide.
Oxamyl	methyl N',N'-dimethyl-N-[(methylcarbamoyl)oxy]-1=thiooxamimidate.
Pebulate	S-propyl butylethylthiocarbamate.
Pendimenthalin	N-(1-ethylpropl)-3,4-dimethyl-2,6-dinitrobenzenamine.
Picloram	4-amino-3,5,6-trichloropicolinic acid.
Profluralin	N-(cyclopropylmethyl)-α,α,α-trifluoro-2,6-dinitro-N=propyl-p-toluidine.
Prometryn	2,4-bis(isopropylamino)-6-(methylthio)-s-triazine.
Pronomide	3,5-dichloro-N-(1,1-dimethyl-2-propynyl)benzamide.
Propazine	2-chloro-4,6-bis(isopropylmaino)-s-triazine.
Propham	isopropyl carbanilate.
Simazine	2-chloro-4,6-bis(ethylamino)-s-triazine.

Table 7.--Common and chemical names of pesticides--Continued

Common name	Chemical name
	INSECTICIDES
Aldicarb	2-methyl-2-(methylthio)propionaldehyde O-(methylcarbamoyl)oxime.
Aldrin	1,2,3,4,10,10-hexachloro-1,4,4a,5,8,8a-hexahydro-1,4=endo-exo-5,8-dimethanonaphthalene.
Akton	O-[2-chloro-1-(2,5-dichlorophenyl)vinyl] O,O-diethyl phosphorothioate.
Azinphosmethyl	O,O-dimethyl S-[(4-oxo-1,2,3-benzotriazin-3(4H)-yl) methyl] phosphorodithioate.
BHC	1,2,3,4,5,6-Hexachlorocyclohexane.
BHC alpha	1,2,3,4,5,6-Hexachlorocyclohexane.
BHC beta	1,2,3,4,5,6-Hexachlorocyclohexane.
BHC gamma	1,2,3,4,5,6-Hexachlorocyclohexane.
BHC delta	1,2,3,4,5,6-Hexachlorocyclohexane.
Bromophos	O-(4-bromo-2,5-dichlorophenyl) O,O-dimethyl phosphorothioate.
Carbaryl	1-naphthyl methylcarbamate.
Carbofuran	2,3-dihydro-2,2-dimethyl-7-benzofuranyl methylcarbamate.
CGA-12223	O-(5-chloro-1-(1-methylethyl)-1H-1,2,4-triazol-3-yl] O,O-diethyl phosphorothioate.
Chlordane	1,2,4,5,6,7,8,8-octachloro-3a,4,7,7a-tetrahydro-4,7=methanoindan (60% minimum, and not over 40% of related compounds).
Chlorfenvinphos	2-chloro-1-(2,4-dichlorophenyl)vinyl diethyl phosphate.
o,p'-DDT p,p'-DDT	1,1,1-trichloro-2,2-bis(p-chlorophenyl)ethane (by common usage, an isomeric mexture of dichlorodiphenyltrichloroethane in which the p,p' isomer is not less than 60 to 70%).
Diazinon	O,O-diethyl O-(2-isopropyl-6-methyl-4-pyrimidinyl) phosphorothioate.

Table 7.--Common and chemical names of pesticides--Continued

Common name	Chemical name
	INSECTICIDES
Dieldrin	1,2,3,4,10,10-hexachloro-6,7-epoxy-1,4,4a,5,6,7,8,8a-octahydro-1,4-endo-exo-5,8-dimethano-naphthalene, 85% minimum.
Dimethoate	O,O-dimethyl S-(methylcarbamoylmethyl) phosphorodithioate.
Dioxacarb	o-1,3-dioxolan-2-ylphenyl methylcarbamate.
Dioxathion	S,S'-p-dioxane-2,3-diyl O,O,O',O'-tetraethyl bis (phosphorodithioate).
Disulfoton	O,O-diethyl S-[2-(ethylthio)ethyl] phosphorodithioate
Endosulfan	6,7,8,9,10,10-hexachloro-1,5,5a,6,9,9a-hexahydro-6,9=methano-2,4,3-benzodioxathiepin 3-oxide.
Endrin	1,2,3,4,10,10-hexachloro-6,7-opoxy-1,4,4a,5,6,7,8,8a-octahydro-1,4-endo-endo-5,8-dimethanonaphthalene.
Ethion	O,O,O',O'-tetraethyl S,S'-methylene bis=phosphorodithioate.
Ethyl parathion	O,O-diethyl O-(p-nitrophenyl) phosphorothioate.
Fenitrothion	O,O-dimethyl O-(4-nitro-m-tolyl)phosphorothioate.
Fonofos	O-ethyl S-phenyl ethylphosphonodithioate.
Heptachlor	1,4,5,6,7,8,8-heptachloro-3a,4,7,7a-tetrahydro-4,7=methanoindene.
Hexachlorophene	2,2'-methylenebis[3,4,6-trichlorophenol].
Isobenzan	1,3,4,5,6,7,8,8-octachloro-1,3,3a,4,7,7a-hexahydro-4,7-methanoisobenzofuran.
Lindane	1,2,3,4,5,6-hexachlorocyclohexane, gamma isomer of not less than 99% purity.
Malathion	diethyl mercaptosuccinate S-ester with O,O-dimethyl phosphorodithioate.
Mecarbam	S-[[(ethoxycarbonyl)methylcarbamoyl]methyl]O,O-diethyl phosphorodithioate.

Table 7.--Common and chemical names of pesticides--Continued

Common name	Chemical name
	INSECTICIDES
Methidathion-----------	O,O-dimethyl phosphorodithioate S-ester with 4-(mercaptomethyl)-2-methoxy-Δ^2-1,3,4-thiadiazolin-5-one.
Methoxychlor-----------	1,1,1-trichloro-2,2-bis(p-methoxyphenyl)ethane.
Mevinphos--------------	methyl (E)-3-hydroxycrotonate dimethyl phosphate.
Parathion--------------	O,O-diethyl O-(p-nitrophenyl) phosphorothioate.
Phenthoate-------------	ehtyl mercaptonphenylacetate S-ester with O,O-dimethyl phosphorodithioate.
Phorate----------------	O,O-diethyl S-[(ethylthio)methyl] phosphorodithioate.
Thionazin--------------	O,O-diethyl O-pyrazinyl phosphorothioate.
Toxaphene--------------	chlorinated camphene containing 67-69% chlorine.
	NEMATICIDES
Dichlofenthion---------	O-(2,4-dichlorophenyl) O,O-diethyl phosphorothioate.
Trichloronate----------	O-ethyl O-(2,4,5-trichlorophenyl) ethylphosphonothioate.

Data for picloram dissipation from soils in several locations in the United States and Canada show the effect of temperature (23) also. Figure 4 is a plot of latitude versus $k_{1/2}$ (the half-order equation better describes the dissipation of picloram than the first-order equation). In the northern hemisphere, the $k_{1/2}$ values increase from north to south, which indicates a more rapid dissipation of picloram from north to south. The slope of the curve is described fairly well by the first-order equation.

Effect of soil moisture.--Soil moisture affects the dissipation of most pesticides; dissipation is more rapid from moist than dry soils (23, 31). The extensive work of Walker and coworkers (82) shows the influence of moisture on the dissipation of herbicides: usually, the drier the soil, the smaller the k_s value indicating a longer $C_{1/2}$, the wetter the soil, at least near field capacity (33 kPa soil moisture tension), the bigger the k_s and a shorter $C_{1/2}$. The $C_{1/2}$ of trifluralin was near 100 days on a dry soil, but only 7 days on a moist soil (table 1, 87). The k_s value for propham on dry soil was 0.025, and 0.279 for the wet soil, which gave $C_{1/2}$'s of 27 and 2.5 days, respectively.

Effect of soil pH.--Soil pH affects the dissipation of several pesticides, though no general conclusion can be made (23). Some pesticides are less stable at low soil pH values and some at high pH values, depending on their functional groups. The rate of dissipation of the herbicide carbofuran increased with increased soil pH (19 table 2). At a pH of 4.3, the $C_{1/2}$ of carbofuran was 30 days, whereas at a pH of 7.8 the $C_{1/2}$ was 7 days, 1/4 that of the lower soil pH value.

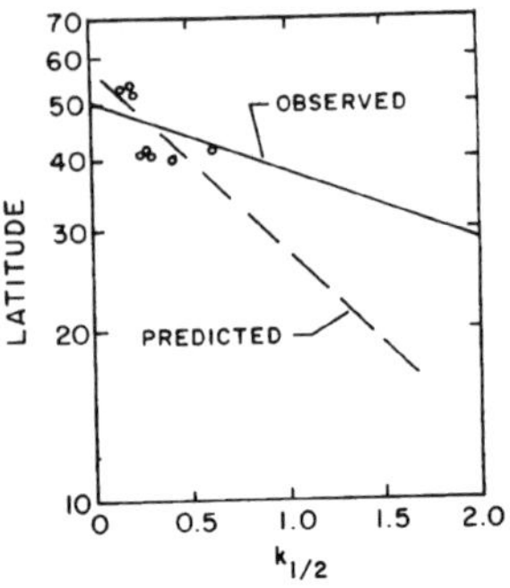

Figure 4.--Relation between latitude and half-order rate constants for the dissipation of picloram in soils at several locations in the United States and Canada (23).

Correlation coefficients (r).--When calculating a first-order equation for logarithmic pesticide concentration vs. time, a correlation coefficient (r) can be determined. The r value describes the goodness-of-fit of the equation. A perfect fit gives an r value of ±1. The sign of the r values indicates positive (+) or negative (-) slope.

Normally, values of r < 0.95 indicate that goodness-of-fit is lacking and that the equation should be used cautiously. This indicates an uncertainty in Y of 10%. An r value of < 0.9, indicating an uncertainty of 20%, usually is interpreted to mean that the equation does not describe adequately the x,y relationship. For this study, an r value of > 0.9 was considered acceptable, however, because variability of the dissipation of pesticides from and in soils is so great. An r value of 0.99 may be obtained in one experiment, and in another experiment, with the same pesticide under similar conditions, the r value may be < 0.9.

If the r value was < 0.9 but > 0.8 (tables 1 through 4), it was identified as such so that the user would be aware of the poor value for k_s. Likewise, r values of < 0.8 were identified. On several occasions, only the beginning and ending concentrations of a pesticide in soil were given. Consequently, an r value could not be determined.

Use of k_s Values

The pesticide submodel (vol. I, ch. 5) computes estimates of runoff-available pesticide residue at the soil surface using the first-order rate equation and k_s values as discussed in this chapter. Table 1 gives a listing of k_s values for several pesticides at the soil surface. This table, rather than table 4, should be the first source of reference.

The amount of data available on the dissipation of pesticides from soil surfaces is small compared with that of pesticides in the soil bulk. Consequently, k_s values for the dissipatian of pesticides in bulk soils are listed in table 4. Literature surveyed for table 4 was to list only a wide range of pesticides under various conditions. Further information on herbicides may be found in the Herbicide Handbook (27) and Weed Science (Weed Science Society of America, 113 N. Neil Street, Champaign, Ill.). The k_s values only indicate the dissipation of a certain pesticide compared with another.

Dissipation of a pesticide generally is less rapid in the soil bulk than at the soil surface, except where the surface soil is dry for a long period (29). The k_s values listed in table 4, therefore, are smaller than those listed in table 1 for pesticides on soil surfaces. Use of k_s values from table 4 in predicting runoff may indicate greater amounts of pesticide than actually remain on a rain or irrigation date, that is, C_4 is too large (see vol. I, ch. 5).

In fields where persistent pesticides have been used for some time, pesticide residue already may be present at the beginning of a model simulation period. The model requires an input value for initial residue. If this value cannot be obtained by direct measurement, the initial residue level may also be estimated if the application history is known by using procedures and k_s values presented in this chapter.

Best-Fit Equations

In several instances, the first-order equation does not describe, or poorly describes, the dissipation rate of a pesticide. This normally can be overcome by determining the best-fit equation of the experimentally obtained data. Although the best-fit equation may be a complicated polynomial expression, it more accurately describes the data. With the availability of computers today, the obstacle of complicated mathematics for many data sets is overcome and the best-fit equation and more accurate descriptions of data can be obtained.

The author has developed a computer storage bank for k_s dissipation values and the best-fit dissipation equation. The parameters (soil type, temperature, and moisture; where pesticide was applied, etc.) associated with the experiment are stored also. This bank of k_s values is more comprehensive and kept up-to-date more than those given in table 1 through 4. The user can obtain a k_s value for dissipation of a pesticide from a specific soil-type surface, if it is in the bank, or of a pesticide from any soil surface. Presently, users should contact the author on use of the bank.

This bank resulted from inadequacies of the k_s values given in tables 1 through 4. Although k_s-means within standard limits generally describe the dissipation of a pesticide from a soil surface, k_s cannot be considered specific. If the user has available the k_s value or the best-fit equation obtained from several controlled experiments, perhaps he could select a k_s or best-fit equation describing data similar to his own, thereby obtaining a more precise estimate of the pesticide concentration at any given time. Hopefully these improvements can be incorporated into future versions of the pesticide model.

REFERENCES

(1) Altom, J. D., and J. F. Stritzke.
1973. Degradation of dicamba, picloram, and four phenoxy herbicides in soils. Weed Science 21:556-560.

(2) Baker, R. D., and H. G. Applegate.
1970. Effect of temperature and ultraviolet radiation on the persistence of methyl parathion and DDT in soils. Agronomy Journal 62:509-512.

(3) Baker, H. J., A. Bertus, and G. F. Courtney.
1976. The uptake of some benzimidazole fungicides from soil into plants and their persistence in soil and plant tissue. Acta Phytopathology Academy of Science in Hungary, 11:331-334.

(4) Beall, M. L., Jr.
1976. Persistence of aerially applied hexachlorobenzene on grass and soil. Journal of Environmental Quality 5:367-369.

(5) Bovey, R. W., E. Burnett, R. E. Meyer, C. Richardson, and A. Lok.
1978. Persistence of tebuthiuron in surface runoff water, soil, and vegetation in the Texas blacklands prairie. Journal of Environmental Quality 7:233-236.

(6) Boyd, J. C.
1971. Field study of a chlordane residue problem: Soil and plant relationships. Bulletin of Environmental Contamination and Toxicology 6:177-182.

(7) Briggs, G. G.
1976. Degradation in soils. In:Procceedings, British Crop Protection Council Symposium: Persistence of Insecticides and Herbicides. Monograph 17. pp. 41-54.

(8) Bro-Rasmussen, F., E. Noddegaard, and K. Voldum-Clausen.
1970. Comparison of the disappearance of eight organophosphorous insecticides from soil in laboratory and in outdoor experiments. Pesticide Science 1:179-182.

(9) Bull, D. L., and G. W. Ivie.
1978. Fate of diflubenzuron in cotton, soil, and rotational crops. Journal of Agricultural Food Chemistry 26:515-520.

(10) Burns, R. G.
1971. Loss of phosdrin and phorate insecticides from a range of soil types. Bulletin of Environmental Contamination Toxicology 6:316-321.

(11) Chang, S. S., and J. F. Stritzke.
1977. Adsorption: Movement and persistence of tebuthiuron in soil. Weed Science 25:184-187.

(12) Cliath, M. M., and W. F. Spencer.
1971. Movement and persistence of dieldrin and lindane in soil as influenced by placement and irrigation. Soil Science Society of America Proceedings 35:791-795.

(13) ------, and W. F. Spencer.
1972. Dissipation of pesticides from soil by volatilization of degradation products. I. Lindane and DDT. Environmental Science and Technology 6:910-914.

(14) Edwards, C. A.
1966. Insecticide residues in soils. Residue Reviews 13:83-132.

(15) Freeman, H. P., H. W. Taylor, and W. M. Edwards.
1975. Heptachlor and dieldrin disappearance from a field soil measured by annual residue determinations. Journal of Agricultural Food Chemistry 23:1101-1105.

(16) Freshe, H.
1976. The perspective of persistence. In: Proceedings, British Crop Protection Council Symposium: Persistence of Insecticides and Herbicides. Monograph 17. pp. 1-39.

(17) Fryer, J. D., and K. Kirkland.
1970. Field experiments to investigate long term effects of repeated applications of MCPA, tri-allate, simazine and linuron: Report after 6 years. Weed Research 10:133-158.

(18) Getzin, L. W.
1968. Persistence of diazinon and zinophos in soil: Effects of autoclaving, temperature, moisture, and acidity. Journal of Economic Entomology 61:1560-1565.

(19) ------
1973. Persistence and degradation of carbofuran in soil. Environmental Entomology 2:461-467.

(20) ------
1977. Persistence of the insecticide Akton in soil. Bulletin of Environmental Contamination Toxicology 18:499-502.

(21) ------, and I. Rosefield.
1966. Persistence of diazinon and zinophos in soils. Journal of Economic Entomology 59:512-516.

(22) Guth, J. A., N. Burkhard, and D. O. Eberle.
1976. Experimental models for studying the persistence of pesticides in soils. In: Proceedings British Crop Protection Council Symposium: Persistence of Insecticides and Herbicides. Monograph 17.

23) Hamaker, J. W.
1972. Decomposition: Quantitative aspects. In: Organic Chemicals in the Soil Environment. C. A. A. Goring and J. W. Hamaker (editors), Marcel Dekker, Inc., New York. 1:253-340.

24) ------
1976. The application of mathematical modeling to the soil persistence and accumulation of pesticides. In: Proceedings, British Crop Protection Council Symposium: Persistence of Insecticides and Herbicides, Monograph 17, pp. 181-199.

25) Harvey, J., Jr., and J. C. Y. Hah.
1978. Decomposition of oxamyl in soil and water. Journal of Agricultural Food Chemistry 26:536-541.

26) Helling, C. S.
1971. Pesticide mobility in soils. II. Applications of soil thin-layer chromatography. Soil Science Society of America Proceedings 35:737-743.

27) Hilton, J. L., R. W. Bovey, H. M. Hull, W. R. Mullison, and R. E. Talbert.
1974. Herbicide Handbook. Weed Science Society of America, 3rd edition. 113 N. Neil St., Champaign, Ill. 61820.

28) Horowitz, M., and G. Herzlinger.
1974. Soil conditions affecting the dissipation of diuron, fluometuron and propham from the soil surface. Weed Research 14:257-259.

29) Iwata, Y., M. E. Dusch, W. E. Westlake, and F. A. Gunther.
1975. Behavior of five organo-phosphorus pesticides in dust derived from several soil types. Bulletin of Environmental Contamination Toxicology 14:49-56.

30) ------, W. E. Westlake, J. H. Barkley, G. E. Carman, and F. A. Gunther.
1977. Aldicarb residues in oranges, citrus by-products, orange leaves, and soil after an aldicarb soil application in an orange grove. Journal of Agricultural Food Chemistry 25:933-937.

31) Kennedy, J. M., and R. E. Talbert.
1977. Comparative persistence of dinitroaniline type herbicides on the soil surface. Weed Science 25:373-381.

32) Keys, C. H., and H. A. Friesen.
1968. Persistence of picloram activity in soil. Weed Science 16:341-343.

33) Konrad, J. G., G. Chesters, and D. E. Armstrong.
1969. Soil degradation of malathion, a phosphorodithioate insecticide. Soil Science Society of America Proceedings 33:259-262.

(34) Lane, L. J., H. L. Morton, D. E. Wallace, R. E. Wilson, and R. D. Martin.
1977. Nonpoint source pollutants to determine runoff source areas. Hydrology and Water Resources in Arizona and the Southwest 7:89-102.

(35) Lauridsen, M. G., K. Voldum-Clausen, M. Dahl, K. E. Hansen, T. Hansen, and K. Bech.
1978. Residues of benomyl in various crops and soils. State Tidsskrift for Planteavl 81:33-48.

(36) Leather, G. R., and C. L. Foy.
1977. Metabolism of bifenox in soil and plants. Pesticide Biochemistry and Physiology 7:437-442.

(37) Leffingwell, J. T., R. C. Spear, and D. Jenkins.
1975. The persistence of ethion and zolone residues on grape foliage in the central valley of Calif. Archives Environmental Contamination 3:40-54.

(38) Lichtenstein, E. P., L. J. De Pew, E. L. Eshbaugh, and J. P. Sleesman.
1960. Persistence of DDT, aldrin, and lindane in some Midwestern soils. Journal of Economic Entomology 54:136-142.

(39) ------, T. W. Fuhremann, and K. R. Schulz.
1971. Persistence and vertical distribution of DDT, lindane, and aldrin residues 10 and 15 years after a single soil application. Journal of Agricultural Food Chemistry 19:718-721.

(40) ------, G. R. Myrdal, and K. R. Schulz.
1964. Effect of formulation and mode of application of aldrin on the loss of aldrin and its epoxide from soils and their translocation into carrots. Journal of Ecomomic Entomology 57:133-136.

(41) ------, and K. R. Schulz.
1959. Persistence of some chlorinated hydrocarbon insecticides as influenced by soil types, rate of application, and temperature. Journal of Economic Entomology 52:124-131.

(42) ------, and K. R. Schulz.
1961. Effect of soil cultivation, soil surface, and water on the persistence of insecticidal residues in soils. Journal of Economic Entomology 54:517-522.

(43) ------, and K. R. Schulz.
1964. The effect of moisture and micro-organisms on the persistence and metabolism of some organophosphorus insecticides in soils, with special emphasis on parathion. Journal of Economic Entomology 57:618-627.

(44) Lutz, J. F., G. E. Byers, and T. J. Sheets.
1973. The persistence and movement of picloran and 2,4,5-T in soils. Journal of Environmental Quality 2:485-488.

45) Malone, C. R., A. G. Winnett, and K. Helrich.
1967. Insecticide induced responses in an old field ecosystem: Persistence of diazinon in the soil. Bulletin of Environmental Contamination Toxicology 2:83-89.

46) Martens, R.
1977. Degradation of Endosulfan-8,9 -^{14}C in soil under different conditions. Environmental Contamination Toxicology 17:438-446.

47) McCormick, L. L., and A. E. Hiltbold.
1966. Microbial decomposition of atrazine and diuron in soil. Weeds 14:77-82.

48) Meikle, R. W., C. R. Youngson, R. T. Hedlund, C. A. A. Goring, J. W. Hamaker, and W. W. Addington.
1973. Measurement and prediction of picloram disappearance rates from soil. Weed Science 21:549-555.

49) Mula, M. S.
1960. Loss of chlorinated hydrocarbon insecticides from soil surface in the field. Journal of Economic Entomology 53:650-654.

50) Mutz, J. L., C. J. Scifner, S. Selim, and R. F. Cook.
1978. Movement and persistence of karbutilate in rangeland soils following application as spheres. Weed Science 26:179-181.

51) Nash, R. G., and M. L. Beall, Jr.
1980. Fate of maneb and zineb fungicides in microagroecosystem chambers. Journal of Agricultural Food Chemistry 28:[accepted].

52) ------
1980. Distribution of silvex, 2,4-D, and TCDD applied to turf in chambers and field plots. Journal of Agricultural Food Chemistry 28:[accepted].

53) ------, M. L. Beall, Jr., and W. G. Harris.
1977. Toxaphene and 1,1,1-trichloro-2,2-bis(p-chlorophenyl) ethane (DDT) losses from cotton in an agro-ecosystem chamber. Journal of Agricultural Food Chemistry 25:336-341.

54) ------, and E. A. Woolson.
1967. Persistence of chlorinated hydrocarbon insecticides in soils. Science 157:924-927.

55) Owen, B., Jr., J. B. Dimond, and A. S. Getchell.
1977. DDT: Persistence in northern spodosols. Journal of Environmental Quality 6:359-360.

56) Ragab, M. T. H.
1975. Residues of picloram in soil and their effects on crops. Canadian Journal of Soil Science 55:55-59.

(57) Rajukkannu, K., R. Reguraj, T. R. Subramaniam, K. K. Krishnamoorthy.
1977. Persistence of Temik (Aldicarb) in soil and its residues in Bhendi fruits. Current Science 46:108-109.

(58) Rhodes, R. C.
1977. Studies with manganese [^{14}C] ethylenebis (dithiocarbamate) ([^{14}C] maneb) fungicide and [^{14}C] ethylenethiourea ([^{14}C] ETU) in plants, soil, and water. Journal of Agricultural Food Chemistry 25:528-533.

(59) Romanowski, R. R., and A. W. Libik.
1978. Soil persistence of isopropalin, nitralin, and trifluralin. Weed Science 26:258-261.

(60) Sanborn, J. R., B. M. Francis, and R. L. Metcalf.
1976 or 1977. The degradation of selected pesticides in soil: A review of the published literature. Project R 803591-01. U.S. Environmental Protection Agency, Cincinnati, Ohio 45268.

(61) Savage, K. E., and W. L. Barrentine.
1969. Trifluralin persistence as affected by depth of soil incorporation. Weed Science 17:349-352.

(62) Schuphan, I.
1974. Zum metabolismus von phenylharnstoffen. II. Abbau und Metabolismus von monolinuron-O-methyl ^{14}C in Boden. Chemosphere 3:127-130.

(63) Shadbolt, C. A., F. L. Whiting, and J. M. Lyons.
1964. Persistence of monuron and neburon under semi-arid field conditions. Weeds 12:304-306.

(64) Singh, K., K. C. Gulati, and R. S. Sewan.
1972. Persistence of disyston residues in soil and plant. Indian Journal of Agricultural Science 42:1135-1138 (Chemistry Abstracts. 79:14421f, 1973).

(65) Smith, A. E.
1973. Degradation of dicamba in prairie soils. Weed Research 13:373-378.

(66) ------
1970. Degradation, adsorption, and volatility of di-allate and tri-allate in prairie soils. Weed Research 10:331-339.

(67) ------, and A. Fitzpatrick.
1970. The loss of five thiolcarbamate herbicides in nonsterile soils and their stability in acidic and basic solutions. Journal of Agricultural Food Chemistry 18:720-722.

(68) ------, and A. Walker.
1977. A quantitative study of asulam in soil. Pesticide Science 8:449-456.

(69) Smith, D. T., and A. F. Wiese.
1973. Delayed incorporation of trifluralin and nitralin. Weed Science 21:163-165.

(70) Spencer, W. F., M. M. Cliath, K. R. Davis, R. C. Spear, and W. J. Popendorf.
1975. Persistence of parathion and its oxidation of paraoxon on the soil surface as related to worker reentry into treated crops. Bulletin of Environmental Contamination Toxicology 14:265-272.

(71) ------, and M. M. Cliath.
1972. Volatility of DDT and related compounds. Journal of Agricultural Food Chemistry 2:645-649.

(72) Srivastava, B. P., and V. S. Kavadia.
1977. Studies on the phytotoxicity and persistence of some soil insecticides and their residues in crops grown on treated soils. Department of Entomology, University Udaipur, Udaipur 313001, India (PL-480 A7-CR-294, Grant No. FG-In-431) 198 pp.

(73) Staiff, D. C., S. W. Comer, J. F. Armstrong, and H. R. Wolfe.
1975. Persistence of azinphosmethyl in soil. Bulletin of Environmental Contamination Toxicology 13:362-368.

(74) Stewart, D. K. R., and D. Chisholm.
1971. Long-term persistence of BHC, DDT, and chlordane in a sandy loam soil. Canadian Journal of Soil Science 51:379-383.

(75) ------, and S. O. Gaul.
1977. Persistence of 2,4-D Dichlorophenoxyacetic acid, 2,4,5-T, and dicamba in a dykeland soil. Bulletin of Environmental Contamination Toxicology 18:210-218.

(76) Stroller, E. W., and L. M. Wax.
1977. Persistence and activity of dinitroaniline herbicides in soil. Journal of Environmental Quality 6:124-127.

(77) Suett, D. L.
1975. Persistence and degradation of chlorfenvinpohos, disulfoton, phorate, and pirimiphos-ethyl following spring and late-summer soil application. Pesticide Science 6:385-393.

(78) Supak, J. R., A. R. Swoboda, and J. B. Dixon.
1977. Volatilization and degradation losses of aldicarb from soils. Journal of Environmental Quality 6:413-417.

(79) Talekar, N. S., L. T. Sun, E. M. Lee, and J. S. Chen.
1977. Persistence of some insecticides in subtropical soil. Journal of Agricultural Food Chemistry 25:348-352.

(80) Venkateswarlu, K., T. K. Gowada, and N. Sethsunathan.
1977. Persistence and biodegradation of carbofuran in flooded soils. Journal of Agricultural Food Chemistry 25:533-536.

81) Walker, A.
1970. Persistence of pronamide in soil. Pesticide Science 1:237-239.

82) ------
1974. A simulation model for prediction of herbicide persistence. Journal of Environmental Quality 3:396-401.

83) ------
1976a. Sumulation of herbicide persistence in soil. I. Simazine and prometryne. Pesticide Science 7:41-49.

84) ------
1976b. Simulation of herbicide persistence in soil. II. Simazine and linuron in long-term experiments. Pesticide Science 7:50-58.

85) ------
1976c. Simulation of herbicide persistence in soil. III. Propyzamide in different soil types. Pesticide Science 7:59-64.

86) ------
1978. Simulation of herbicide persistence in soil. Weed Science Society of America, Dallas, Texas, February 9. Number 209.

87) ------, and W. Bond.
1977. Persistence of the herbicide AC 92,553, N-(ethylpropyl)-2,6-dinitro-3,4-xylidine, in soils. Pesticide Science 8:359-365.

88) Walker, W. W., and B. J. Stojanovic.
1973. Microbial versus chemical degradation of malathion in soil. Journal of Environmental Quality 2:229-232.

89) Wauchope, R. D.
1978. The pesticide content of surface water draining from agricultural fields - A review. Journal of Environmental Quality 7:459-472.

90) White, A. W., Jr., L. A. Harper, R. A. Leonard, and J. W. Turnbull.
1977. Trifluralin volatilization losses from a soybean field. Journal of Environmental Quality 6:105-110.

91) Willis, G. H., J. F. Parr, and S. Smith.
1971. Volatilization of soil-applied DDT and DDD from flooded and non-flooded plots. Pesticide Monitoring Journal 4:204-208.

92) ------, and R. A. Hamilton.
1973. Agricultural chemicals in surface runoff, groundwater, and soil. I. Endrin. Journal of Environmental Quality 2:463-466.

93) Wilson, R. G., Jr., and H. H. Cheng.
1978. Fate of 2,4-D in a Naff silt loam soil. Journal of Environmental Quality 7:281-286.

(94) Wolfe, H. R., D. C. Staiff, J. F. Armstrong, and S. W. Comer. 1973. Persistence of parathion in soil. Bulletin of Environmental Contamination Toxicology 10:1-9.

(95) Yaron, B., B. Heuer, and Y. Bink. 1974. Kinetics of azinphosmethyl losses in the soil environment. Journal of Agricultural Food Chemistry 22:439-441.

Chapter 18. THE INTERCEPTION OF APPLIED PESTICIDES BY FOLIAGE AND THEIR PERSISTENCE AND WASHOFF POTENTIAL

G. H. Willis, W. F. Spencer, and L. L. McDowell[1/]

INTRODUCTION AND BACKGROUND

A pesticide molecule leaving the nozzle of an application device (airplane, tractor, hand applicator, and so forth) may follow any of several pathways, and many factors may affect its fate. Depending on droplet size, falltime, carrier, and climatic conditions, the molecule may volatilize into the atmosphere, drift out of the target zone, or be intercepted by a plant or soil surface in the target area. A molecule intercepted by a plant surface may be adsorbed, absorbed, degraded, volatilized, or removed by the mechanical action of wind or rain.

Several factors may influence the amount of pesticide that is intercepted by plants in a spray target area. The mode of application is important; ground application usually deposits more pesticide on plants than aerial application (11, 36, 38, and Willis and McDowell, unpublished). Nozzle design (34) and spray volume (5, 8, 12) also can be influential. The percentage of soil surface covered by the plant canopy and depth and density of the canopy also directly affect the amount of pesticide intercepted by plant surfaces (35, Willis and McDowell, unpublished). Windspeed and size of spray droplet can have an effect because of their influence on drift and droplet evaporation.

Many researchers have reported that pesticide loss from plant leaf surfaces usually follows a decreasing exponential curve, and that first-order kinetics prevail (1, 6, 7, 15, 20, 21, 40). Some researchers have suggested, however, that such losses may not be truly exponential (1, 6). Wheatley (40) pointed out that residues often decline very rapidly for a brief period, but the loss rate gradually declines so that many residues ultimately persist longer than predicted by first-order kinetics. Hamaker (13) suggested that hyperbolic rate models better describe pesticide disappearance from soil than power rate models which may be equally applicable to pesticide losses from plant leaf surfaces.

Taylor (33) reported that pesticide loss rates from vegetation may be interpreted in terms of the decreasing coverage of leaf surfaces by the layer of pesticide residue. Assuming constant environmental conditions, volatilization rates from leaves should decrease as coverage moves from (1) fully covered to (2) discontinuous patches or "islands" to (3) filled cracks and surface irregularities. Thus, the geometrical distribution of pesticides on leaf surface is

---

1/ Soil scientist, USDA-SEA-AR, Baton Rouge, La., soil scientist, USDA-SEA-AR, Riverside, Calif.; and soil scientist, USDA-SEA-AR, Sedimentation Laboratory, Oxford, Miss.

an important factor affecting the rate of pesticide dissipation.

After deposition on a leaf surface, many factors may influence the rate at which a pesticide dissipates from that site. Several workers have reported that plant species, and sometimes even varieties within species, can influence disappearance rates (3, 6, 8, 12, 21). Differences in form and wettability of leaf cuticles appear important (3). Differences in chemical properties of the pesticide also can affect the disappearance rate (8, 12). The amount of leaf surface penetration and absorption depends primarily on the molecular polarity of the pesticide. Polar compounds do not penetrate waxy, hydrophobic layers as readily as nonpolar and lipophilic compounds (8). Pesticide formulation and combination are additional factors that can affect penetration of the leaf surface (6, 8) or resistance to weathering (6, 8, 12, 15, 20, 37, 40), or both. Pesticides formulated as emulsifiable concentrates generally are more resistant to weathering than those formulated as dusts or wettable powder.

Environmental parameters reported to affect the rate of pesticide disappearance from leaf surfaces include relative humidity, rainfall, wind, temperature, and sunlight. High humidity has been reported (1) to increase (8, 14, 17) pesticide persistence on plants by facilitating foliar absorption through favoring stomatal opening and slowing drying time, and (2) to decrease (6, 19) persistence by favoring volatilization. Most researchers report that rainfall has the most dramatic effect on pesticide residues on plants of all environmental parameters (6, 8, 9, 12, 17, 20, 21, 24, 25, 26, 27). Pesticide removal from leaf surfaces is greatest if rainfall occurs within 24 h after pesticide application (8, 21). Dusts and wettable powders are more susceptible to washoff than emulsions (6, 8, 20). The extent of removal also is related to rainfall amount and intensity (24, 25). Hull (17) pointed out that whether rain enhances the penetration of a pesticide into a leaf or washes it off depends on the quantity of rain, time interval between pesticide application and rainfall, solubility characteristic of the pesticide, and physical nature of the leaf surface.

The effect of wind on pesticide persistence on leaves usually is manifested through mechanical action (leaves rubbing one another and removing pesticide deposits) and turbulent transfer (volatility). The effects of temperature are primarily through enhanced volatility. The main effect of sunlight is through photodegradation. Method of application also has been reported to affect pesticide persistence within a plant canopy (through its effect on depth of penetration into the canopy) (5, 8).

ESTIMATING THE AVAILABILITY OF FOLIAR PESTICIDES FOR WASHOFF

Pesticide Interception by Foliage

The range of values in table 1 represent, but do not include, all data reported in the literature. These data suggest that ground application generally results in higher levels of pesticides on foliage than aerial application. Too little data exist to develop reliable mathematical expressions for the effect of variable amount of ground cover on the percentage of applied pesticide intercepted by foliage. Available data indicates, however, that the percentage intercepted increases as the amount of ground cover increased. Based on past

Table 1.--Interception of pesticides by plants

Pesticide	Application mode	Target	Amount of pesticide applied on target	Reference
			(Percent)	
Organochlorine				
DDT	Ground applicator	Paper sheet 1/	35	Willis and McDowell. 2/
Kelthane	do.	Paper sheet 3/	44	Do.
Toxaphene	do.	Paper sheet 4/	65	Do.
Toxaphene	do.	Paper sheet 5/	62	Do.
DDT	do.	Cotton 6/	12	Do.
Toxaphene	do.	do.	28	Do.
Toxaphene	do.	do.	25	Do.
DDT	do.	Cotton 7/	39	(35)
DDT	do.	Cotton 8/	83	Do.
Toxaphene	Airplane	Cotton 9/	14	Willis and McDowell. 2/
Methoxychlor	do.	Glass plates 10/	47	(36)
Methoxychlor	do.	Alfalfa	49	(38)
Methoxychlor	do.	Cotton	57	(37)
Organophosphate				
Parathion	Air blast	Citrus trees	11/ 22-35	(31)
Parathion	do.	do.	11/ 28-39	Spencer. 2/
Parathion	Oscillating boom	do.	11/ 20-40	(12)
Phenthoate	Oscillating boom and air blast.	do.	11/ 14-26	(18)

1/ Chromatography paper (46 x 57 cm) placed in top of cotton row, 50 cm above soil; 22% applied on soil.

2/ Preliminary unpublished data.

3/ Chromatography paper (46 x 57 cm) placed in top of cotton row, 50 cm above soil; 22% applied on soil.

4/ Chromatography paper (46 x 57 cm) placed in top of cotton row, 50 cm above soil; 35% applied on soil.

5/ Chromatography paper (46 x 57 cm) placed in top of cotton row, 50 cm above soil; 32% applied on soil.

6/ Cotton height, 50 cm; estimated 40% groundcover.

7/ Cotton height, 74 cm; 34% of applied on soil.

8/ Cotton height, 124 cm; 6% of applied on soil.

9/ Cotton height, 165 cm; 100% groundcover; none applied on soil.

10/ Glass plates installed in cotton and alfalfa.

11/ Values are the ranges of percent of applied pesticide found on ground beneath citrus trees. Unknown portion of remainder was intercepted by foliage.

experience and the little data reported in literature, the authors estimate that 75 ± 20 and 50 ± 20% of pesticide spray applied by ground and aerial equipment, respectively, are intercepted by foliage, assuming a nearly complete plant canopy. Most values reported in table 1 are from studies with incomplete plant canopies. Approximately 20 to 40% of the pesticides applied to citrus trees by air blast and ocillating boom sprayers is intercepted by the orchard soil. An unknown portion of the remaining 60 to 80% is intercepted by the foliage.

Pesticide Persistence on Fsoliage

Most pesticide dissipation data indicate a rapid dissipation phase followed by a slower rate of dissipation. Both rates appear first order. The rate constants depend on the pesticide and environmental conditions that affect dissipation rates. Researchers concerned with pesticides on citrus categorize the total leaf residue into dislodgeable (surface) and penetrated fractions. The dislodgeable residues (susceptible to removal by rain) dissipate much more readily from foliage than penetrated residues. The initial rapid loss phase during which much of the dislodgeable fraction is dissipated usually occurs within 5 to 10 days after application. Table 2 presents average literature values for half-lives of several pesticides on foliage. In most reports cited in table 2, no mention is made of rainfall or other parameters that may have affected the listed persistence information. Where rainfall did not occur, volatilization, degradation, and absorption probably would have the greatest effect on the disappearance rate. For calculation, table 3 averages the values in each class (EC and WP combined) given in table 2. These classes have been divided arbitrarily into fast (0 to 5 days) and slow (> 5 days) groups.

Pesticides Removed from Foliage by Rain

Little definitive data are available on the relationship between intensity and duration of rainfall and pesticide washoff from foliage. Of the studies cited in table 4, only that shown for toxaphene was concerned with the effect of rainfall intensity on the amount of pesticide removed. In many of these studies, rainfall was simulated using sprinkling devices that failed to duplicate natural raindrop size distribution and kinetic energy. Rainfall amounts often were not measured or reported. Rain that occurred within a week after pesticide application generally was the most effective in removing pesticide from foliage. The washoff potential of pesticide residues on foliage is important only as long as significant amounts of dislodgeable residues are present, often no more than a week to 10 days after application. The limited data (table 4) suggest that approximately 60±10% of the dislodgeable residue for most pesticides (excluding organochlorine insecticides) is removed by rain. For the organochlorine insecticides (toxaphene), less than 10% of the applied is removed.

Table 2.--Pesticide half-lives on foliage

Pestsicide	Formulation[1/]	Number of observations	Half-life	Reference
			(days)	
	Pesticide Class - Organochlorine			
Aldrin	EC	14	4.6± 2.9	(16)
	WP	8	4.9± 3.6	(16)
	D	1	12.0	(16)
Chlordane	EC	2	12.5± 6.4	(16)
	WP	[illegible]	12.5± 6.4	(16)
	D	2	12.5± 6.4	(16)
DDT	EC	9	8.4± 3.8	(16)
	WP	3	13.9±12.3	(16)
	D	4	3.8± 1.3	(16)
DDT	EC	2	10.6± 0.5	Willis and McDowell.[2/]
Dieldrin	EC	6	4.3± 1.4	(16)
	WP	5	4.9± 3.3	(16)
	D	2	5.8± 0.4	(16)
	G	2	17.0± 5.7	(16)
Endrin	EC	15	8.0± 4.1	(16)
	WP	1	15.0	(16)
	D	2	9.0±11.3	(16)
Ethylan	EC	1	3.4	(32)
Heptachlor	EC	10	5.3± 3.2	(16)
	WP	3	2.7± 0.6	(16)
	D	10	6.4± 3.9	(16)
	G	5	6.3± 4.4	(16)
Lindane	EC	5	5.5± 3.3	(16)
	WP	4	3.9± 2.4	(16)
	D	4	2.1± 0.3	(16)
	G	2	7.5± 0.7	(16)
Methoxychlor	EC	2	2.5± 0.7	(16)
	WP	10	4.9± 2.7	(16)
	D	11	3.5± 2.5	(16)
Toxaphene	EC	13	6.4± 4.8	(16)
	WP	2	2.3± 0.4	(16)
	D	5	2.3± 1.2	(16)
Toxaphene	EC	6	7.1± 3.7	Willis and McDowell.[2/]
Acephate	WP	3	2.8± 0.4	(23)
Azinphosmethyl	EC	3	1.6± 1.0	(16)
	WP	3	7.9± 9.7	(16)
	D	3	5.2± 4.5	(16)
	EC	5	7.4± 2.4	(29)
	WP	1	9.8	(29)

Table 2.--Pesticide half-lives on foliage--Continued

Pestsicide	Formulation[1/]	Number of observations	Half-life	Reference
			(days)	
		Pesticide Class - Organophosphate		
Azinphosmethyl	WP	1	16.0	(41)
Chlorpyrifos-methyl	EC	1	1.0	(32)
	EC	6	3.3± 0.7	(22)
Cyanophenphos	EC	1	2.6	(32)
Demeton	EC	20	5.4± 1.9	(16)
Diazinon	EC	2	3.0± 2.8	(16)
	WP	6	4.0± 3.3	(16)
	EC	1	1.1	(32)
Dimethoate	EC	5	5.1± 1.7	(29)
	WP	1	7.1	(29)
Dipterex	EC	1	2.0	(16)
EPN	WP	5	5.5± 2.1	(16)
Ethion	WP	1	7.5	(16)
	EC	10	5.0± 2.7	(28)
	WP	1	7.4	(41)
Fenitrothion	EC	1	1.5	(32)
Leptophos	EC	1	2.2	(32)
Malathion	EC	22	2.4± 1.5	(16)
	WP	8	4.8± 2.5	(16)
	D	13	2.3± 1.4	(16)
	EC	1	1.8	(32)
Methidathion	EC	1	1.1	(32)
Methyl Parathion	EC	2	3.0± 2.8	(16)
Parathion	EC	8	2.6± 1.7	(16)
	WP	11	4.3± 4.0	(16)
	D	2	2.6± 0.6	(16)
Phorate	G	1	2.0	(16)
Phosalone	EC	1	16.0	(41)
Phosdrin	EC	5	0.9± 0.2	(16)
Phosphamidon	EC	4	2.0± 1.2	(16)
Quinalphos	EC	2	1.6± 0.0	(32)
Salithion	EC	1	0.7	(32)
Tokuthion	EC	1	3.1	(32)
Triazophos	EC	1	3.0	(32)
Trithion	WP	1	2.0	(16)
	D	2	2.5± 0.7	(16)
		Pesticide Class - Carbamate		
Carbaryl	WP	3	6.8± 1.3	(16)
	D	2	1.5± 0.7	(16)
Carbofuran	F	1	1.1	(32)

Table 2.--Pesticide half-lives on foliage--Continued

Pestsicide	Formulation[1/]	Number of observations	Half-life	Reference
			(days)	
		Pesticide Class - Other		
Permethrin	--	1	35.4	(10)
Dicamba	--	1	9.3	L. J. Lane[2/]
2,4-D	--	1	8.9	Do.
2,4,5-T	--	1	9.6	Do.
Picloram	--	1	8.0	Do.

1/ EC = emulsifiable concentrate; WP = wettable powder; D = dust; G = granular; F = flowable.
2/ Unpublished.

Table 3.--Guide to pesticide half-lives on foliage

Class	Group	Mean half-life
		(days)
Organochlorine	Fast (aldrin, dieldrin, ethylan, heptachlor, lindane, methoxychlor).	4.0±1.0
	Slow (chlordane, DDT, endrin, toxaphene).	9.7±3.9
Organophosphate	Fast (acephate, chlorpyrifos-methyl, cyanophenphos, diazinon, dipterex, ethion, fenitrothion, leptophos, malathion, methidathion, methyl parathion, phorate, phosdrin, phosphamidon, quinalphos, alithion, tokuthion, triazophos, trithion).	2.4±.1
	Slow (azinphosmethyl, demeton, dimethoate, EPN, phosalone).	8.2±4.6
Carbamate	Fast (carbofuran)	1.1
	Slow (carbaryl WP)	6.8±1.3

Table 4.--Pesticide removal from foliage by rainfall

Class	Time after pesticide application	Rainfall Amount	Rainfall Method of application	Dislodgeable residues removed	Reference
	(days)	(cm)		(%)	
Organophosphate					
Phenthoate	3	0.28	Oscillating boom sprayer.	68	(18)
Phenthoate	10	.28	Oscillating boom sprayer.	71	(18)
Parathion	48	6.40	Natural	65	(12)
	22	6.40	do.	70	(12)
	3	.38	Oscillating boom sprayer.	60	(12)
Azinphosmethyl	3	.33	do.	60	(12)
Dioxathion	4	--	do.	56	(39)
	10	--	do.	37	(39)
Carbamate					
Benomyl	13	1.34	"Sprinkling" (simulated rain).	55	(4)
Organochlorine					
Toxaphene	2 hr	2.54	Rainfall simulator.	5	McDowell and Willis.1/
Other					
Diflubenzuron	14	7.60	Natural	70	(2)

1/ Unpublished.

REFERENCES

(1) Brooks, G. T.
1974. Chlorinated insecticides. Vol. II. Biological and environmental aspects. CRC Press, Cleveland, Ohio.

(2) Bull, D. L., and G. W. Ivie.
1978. Fate of diflubenzuron in cotton, soil, and rotational crops. Journal of Agricultural and Food Chemistry 26:515-520.

(3) Crafts, A. S., and C. L. Foy.
1962. The chemical and physical nature of plant surfaces in relation to the use of pesticides and to their residues. Residue Reviews 1: 112-139.

(4) Dejonckheere, Willy, Walter Steurbaut, and Rene H. Kips.
1976. Influence of artificial rainfall and washing on the benomyl and thiophanate-methyl residue content on lettuce. Pesticide Science 7: 161-170.

(5) Dorough, H. W., and N. M. Randolph.
1967. Comparative residual nature of certain insecticides applied as low volume concentrate and water emulsion sprays. Bulletin of Environmental Contamination and Toxicology 2:340-348.

(6) Ebeling, Walter.
1963. Analysis of the basic processes involved in the deposition, degradation, persistence, and effectiveness of pesticides. Residue Reviews 3:35-163.

(7) Fahey, J. E., D. W. Hamilton, and R. W. Rings.
1952. Longevity of parathion and related insecticides in spray residues. Journal of Economic Entomology 45:700-703.

(8) Finlayson, D. G., and H. R. MacCarthy.
1973. Pesticide residues in plants. pp. 57-86. In Environmental Pollution by Pesticides. C. A. Edwards (ed.), Plenum Press, New York.

(9) Gaines, J. C., and W. J. Mistric.
1951. Effect of rainfall and other factors on the toxicity of certain insecticides. Journal of Economic Entomology 44:580-585.

(10) Gaughan, Loretta C., and John E. Casida.
1978. Degradation of trans- and cis-permethrin on cotton and bean plants. Journal of Agricultural and Food Chemistry 26:525-528.

(11) Goering, C. E., and B. J. Butler.
1973. Paired field studies of herbicide drift. ASAE Paper No. 73-1575. American Society of Agricultural Engineers, St. Joseph, Mich.

(12) Gunther, F. A., Y. Iwata, G. E. Carman, and C. A. Smith.
1977. The citrus reentry problem: Research on its causes and effects and approaches to its minimization. Residue Reviews 67:1-139.

(13) Hamaker, J. W.
1972. Decomposition: Quantitative aspects. In Organic chemicals in the soil environment, pp. 253-340. C. A. I. Goring and J. W. Hamaker (eds.), vol. I, Marcel Dekker, Inc., New York.

(14) Hopkins, T. L.
1967. Humidity effects on the persistence of malathion leaf-surface residues. Journal of Economic Entomology 60:1167-1168.

(15) Hoskins, W. M.
1962. Some important properties of pesticide deposits on various surfaces. Residue Reviews 1:66-91.

(16) ------
1961. Methods for expressing the persistence of insecticidal residues on plants. Final report to regional project W-45. Department of Entomology and Parasitology, University of California, Berkeley.

(17) Hull, H. L.
1970. Leaf structure as related to absorption of pesticides and other compounds. Residue Reviews 31:1-150.

(18) Iwata, Y., W. E. Westlake, J. H. Barkley, G. E. Carman, and F. A. Gunther.
1977. Behavior of phenthoate (cidial) deposits and residues on and in grapefruits, lemons and lemon leaves, oranges and orange leaves, and in the soil beneath orange trees. Journal of Agricultural and Food Chemistry 25:362-368.

(19) Kalkat, G. S., R. H. Davidson, and C. L. Brass.
1961. The effect of controlled temperature and humidity on the residual life of certain insecticides. Journal of Economic Entomology 54: 1186-1190.

(20) Keiser, I., and C. F. Henderson.
1951. DDT spray formulations for control of white-fringed beetles. Journal of Economic Entomology 44:646-652.

(21) Kuhr, R. J., and H. W. Dorough.
1976. Carbamate insecticides: Chemistry, biochemistry, and toxicology. CRC Press, Cleveland, Ohio.

(22) Leuck, D. B., R. L. Jones, and M. C. Bowman.
1975. Chlorpyrifosmethyl insecticide residues: Their analysis and persistence in coastal Bermuda grass and forage corn. Journal of Economic Entomology 68:287-290.

(23) Lindquist, R. K., and H. R. Krueger.
1975. Application of acephate to greenhouse tomatoes: External vs. internal residues, and vegetable leafminer control. Journal of Economic Entomology 68:122-123.

(24) Linscott, D. L., and R. D. Hagin.
1968. Effects of two environmental factors on removal of 2,4-DB from forage. Weed Science 16:114-116.

(25) Linskens, H. F., W. Heinin, and A. L. Stoffers.
1965. Cuticula of leaves and the residue problem. Residue Reviews 8:136-178.

(26) Mistric, W. J., and J. C. Gains.
1953. Effect of wind and other factors on the toxicity of certain insecticides. Journal of Economic Entomology 46:341-349.

(27) ------, and D. F. Martin.
1956. Effect of sunlight and other factors on the toxicity of certain insecticides. Journal of Economic Entomology 49:757-760.

(28) Nigg, H. N., J. C. Allen, R. F. Brooks, G. J. Edwards, N. P. Thompson, Roy W. King, and A. H. Blagg.
1977. Dislodgeable residues of ethion on Florida citrus and relationship to weather variables. Archives of Environmental Contamination and Toxicology 6:257-267.

(29) Pree, D. J., K. P. Butler, E. R. Kimball, and D. K. R. Stewart.
1976. Persistence of foliar residues of dimethoate and azinphosmethyl and their toxicity to the apple maggot. Journal of Economic Entomology 69:473-478.

(30) Sloan, M. J., W. A. Rawlins, and L. B. Norton.
1951. Factors affecting loss of DDT and parathion residues on lettuce. Journal of Economic Entomology 44:701-708.

(31) Spencer, W. F., M. M. Cliath, K. R. Davis, R. C. Spear, and W. J. Popendorf.
1975. Persistence of parathion and its oxidation to paraoxon on the soil surface as related to worker reentry into treated crops. Bulletin of Environmental Contamination and Toxicology 14:265-272.

(32) Talekar, N. S., L. T. Sun, E. M. Lee, J. S. Chen, T. M. Lee, and S. Lu.
1977. Residual behavior of several insecticides on Chinese cabbage. Journal of Economic Entomology 70:689-692.

(33) Taylor, Alan W.
1978. Post-application volatilization of pesticides under field conditions. Journal of Air Pollution Control Association 28:922-927.

(34) Ware, G. W., W. P. Cahill, and B. J. Estesen.
1975a. Pesticide drift: Aerial applications comparing conventional flooding vs. raindrop nozzles. Journal of Economic Entomology 68: 329-330.

(35) ------, W. P. Cahill, B. J. Estesen, W. C. Kronland, and N. A. Buck.
1975b. Pesticide drift: Deposit efficiency from ground sprays on cotton. Journal of Economic Entomology 68:549

(36) ------, W. P. Cahill, P. D. Gerhardt, and J. M. Witt.
1970. Pesticide drift. IV. On-target deposits from aerial application of insecticides. Journal of Economic Entomology 63:1982-1983.

(37) ------, B. Estesen, and W. P. Cahill.
1975c. Dislodgeable insecticide residues on cotton. Bulletin of Environmental Contamination and Toxicology 14:606-609.

(38) ------, B. J. Estesen, W. P. Cahill, and K. R. Frost.
1972. Pesticide drift. VI. Target and drift deposits vs. time of applications. Journam of Economic Entomology 65:1170-1172.

(39) Westlake, W. E., F. A. Gunther, and G. E. Carman.
1973. Worker environment research: Dioxathion (Delnav) residues on and in orange fruit and leaves, in dislodgeable particulate matter, and in the soil beneath sprayed trees. Archives of Environmental Contamination and Toxicology 1:60-83.

(40) Wheatley, G. A.
1973. Pesticides in the atmosphere. In Environmental pollution by pesticides. pp. 365-408. C. A. Edwards (ed.), Plenum Press, New York.

(41) Winterlin, Wray, Charles Mourer, and J. Blair Bailey.
1974. Degradation of four organophosphate insecticides in grape tissues. Pesticide Monitoring Journal 8:59-65.

Chapter 19. METHOD FOR DISTRIBUTING PESTICIDE LOSS IN FIELD RUNOFF BETWEEN THE SOLUTION AND ADSORBED PHASE

H. B. Pionke and R. J. DeAngelis[1/]

In the pesticide submodel, an estimate of the pesticidal concentration in the aqueous and eroded soil phase is required in addition to an estimate of runoff water volume and eroded soil mass lost from the field (see vol. I, Ch. 5). This chapter provides estimates of pesticidal concentrations in the water and eroded soil phases. Additional information is provided in other chapters of this volume.

GENERAL INTRODUCTION

The distribution of pesticides between soil and water depends on pesticidal and soil properties.

A considerable amount of work has been done to categorize and generalize properties of pesticides as they relate to the inherent adsorptivity of the pesticide. This effort has not been very successful, particularly with respect to generalizing the relationship for chemically diverse pesticides or between pesticidal classes. Perhaps the simplest is the water solubility-adsorptivity concept, which appears to hold within pesticidal families (4, 5, 18, 40), but not generally (4, 5). Other techniques have attempted to quantitate potential adsorptivity directly from elementary properties or such measurements as molar volume calculated directly (46) or modified to include hydrogen bonding (29), or from Hancsch's π constant, which is the distribution ratio of the pesticide between 1-octanol and water (6). These techniques, or similar ones, are designed to estimate the potential adsorptivity of the pesticide.

The primary soil properties affecting adsorptivity are variable, depending on the specific compound. One generalization does appear to hold in the literature. The single most important soil property affecting pesticidal adsorption is the soil organic carbon content. Whether this is due only to the peculiar characteristics of organic carbon or reflects the dominance of organic carbon as the primary source of specific surface in most intact soils is still unanswered. The independent effect of clay in the intact soil appears secondary and sometimes is doubtful because of its intercorrelation with organic matter concentrations. The degree of adsorptivity for many weak acid and basic pesticides is affected by pH. Where the soil surface pH approaches the pK for protonizing some of the basic ($RH + H^+ = RH_2^+$) or the acid ($R^- + H^+ = RH$) pesticides, the pH can be the controlling soil property either by reducing the number of molecules repulsed (R^-) or increasing the number attracted (RH_2^+) to the

1/ Soil scientist and mathematician, respectively, USDA-SEA-AR, Northeast Watershed Research Center, University Park, Pa.

negatively charged soil colloids. The cation exchange capacity has been correlated significantly with the extent of pesticide adsorption on soils for a variety of pesticides but is almost always less well correlated than is organic carbon (1, 4, 32, 40, 41, 47). The strongly cationic pesticides such as paraquat and diquat are very strongly and often "irreversibly adsorbed"; in water-soil systems, the degree of adsorption appears controlled by the cation exchange capacity (CEC) of the soil. The soil moisture content at field moisture capacity, 1/3 or 15 atmospheres soil water tension, often were well correlated but usually less well correlated with pesticide adsorption than was organic carbon (1, 19, 32). Experimentally determined specific surface has been related to pesticide adsorption on soils with varying degrees of success (32, 53).

The soil properties that control pesticide adsorptivity essentially translate the adsorption potential (that inherent to the compound) to the actual amount adsorbed for a specific soil.

The method described herein for estimating pesticide distribution between soil and runoff waters combines an estimate of the dominant soil property affecting adsorption with an estimation of the inherent pesticide adsorptivity. The dominating soil property is either the % organic carbon or the calculated specific surface. The inherent pesticide adsorptivity is estimated from water solubility data for the more strongly adsorbed pesticides and by published R_f values[2/] derived from soil thin layer chromatography for the others. The basic method provides an elementary pesticide distribution that may be usable directly or may need to be modified in accordance with other circumstances. Environmental conditions and soil properties that can dominate adsorptivity under special circumstances are presented and evaluated.

This chapter is essentially presented in three parts as follows. The first introduces and evaluates the basic method without considering the effect of special circumstances; the second introduces and evaluates the effect of special circumstances, and the third states the procedure.

INTRODUCTION TO THE BASIC METHOD FOR DISTRIBUTING PESTICIDE CONCENTRATION BETWEEN RUNOFF WATER AND SOIL

The proposed method estimates pesticide distribution (K_d) between the soil and water phase by mathematically combining the organic carbon content (%OC) or calculated specific surface (SS) of the soil with a measure of the adsorption potential of the pesticide. This potential can be estimated from pesticide solubility or from the relationships of either %OC or SS with K_d. For the less extensively adsorbed pesticides, the slope of the K_d vs. %OC or SS is a measure of this potential and can be estimated by the soil thin-layer chromatography R_f value.

This method was developed by comparing the K_d for 35 pesticides with observed soil properties using published data. The K_d was either taken directly from the literature, or calculated or approximated from the Freundlich K. The Freundlich equation is $x/m = KC^{1/n}$. Where either 1/n approximates 1.0 or C (equilibrium solution concentration) equals 1.0 mg/l, the Freundlich equation

---

2/ R_f = distance traveled by pesticide front divided by the distance traveled by the solvent front, in this case water.

becomes either x/m = KC or x/m = K where x/m is the equilibrium pesticide concentration adsorbed on the suspended phase (μg/gm). Under these conditions, the Freundlich K K_d. Furthermore, the K_d value may apply over a considerable range in C if several conditions are met. If, for example, we set the maximum error in estimating K_d from Freundlich K as 25% and expect the Freundlich 1/n to fall between 0.9 and 1.1 from the literature or our knowledge of the system, the K_d Freundlich K constant over the C range of 10.0 to 0.1 mg/l. The reader can test this by using figure 1. Of course, another major condition is to recognize and not violate the assumptions. In this example, it would be

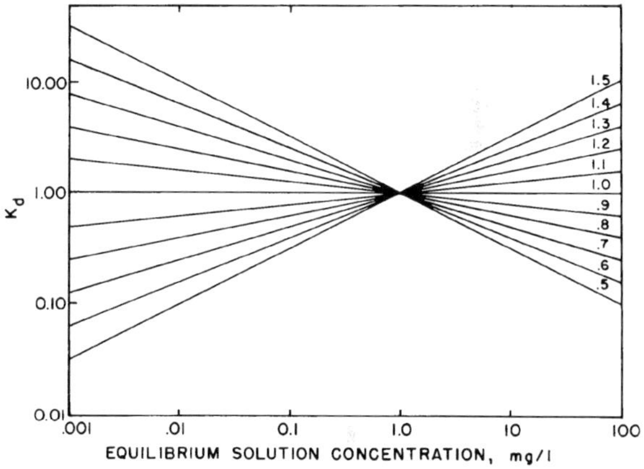

Figure 1.--The effect of Freundlich exponent (1/n) and equilibrium pesticide concentration (C) on changing the pesticide distribution between sorbed and soluble concentrations (K_d = K where 1/n = 1.0 or C = 1.0).

invalid to automatically assume that K_d would remain within comparable error limits at concentrations (C) less than 0.1 mg/l without reevaluating the assumptions. If the 1/n value shifts substantially upon dilution, for example, test calculations using figure 1 show that K_d extrapolation errors could be large. Many K_d values reported in the literature were determined at pesticide concentrations higher than normally occur in the field, sometimes by several orders of magnitude.

Initially, the K_d data for each pesticide, either combined or grouped by author, were plotted against %OC[3/], %clay, calculated SS[4/], analytically determined SS, CEC, pH, [[H^+] x %OC + %OC], and field moisture capacity. The plots indicated that the two soil properties consistently and most highly associated with K_d were %OC and calculated SS. The relationship of the analytically determined specific surface with K_d varied substantially for the two studies for which data were available. Before performing regression and correlation analysis, a deleted data set was established in addition to the complete data set. The deleted data set excluded clusters or groups of high OC or calculated SS values that visually appeared separate from the rest of the data set. Results of the regression analysis of K_d on aforementioned soil properties showed %OC and calculated SS to be the dominant soil properties for estimating K_d. The results for %OC and SS for both the deleted and complete data sets by pesticide are presented in tables 1-4. It should be noted that intercorrelation between calculated SS and %OC is great except for high clay-low OC soils.

Important relationships for each of the 35 pesticides are:

$$K_d = m\ \%OC + b,$$

$$K_d = m\ SS,$$

where the m (slope) and b (intercept) are fitted by regression. The K_d vs. SS relationship was forced through the origin (K_d = 0, SS = 0).

The calculated SS value is the preferred prediction equation because it puts both the organic and inorganic fraction on a common surface area basis. One result is that the curve (K_d vs. SS) theoretically passes through the origin. In contrast, a positive intercept must be theoretically part of the K_d vs. %OC relationship. When %OC goes to zero, there likely is still an inorganic adsorbing surface present. Thus, under conditions of low %OC-high clay content or the predominance of high specific surface clays, the use of SS rather than %OC is more philosophically sound. The degree to which the calculated SS accurately represents the clay contributed SS is open to question, however. Although clay minerals or soil fractions pretreated to remove organic matter and amorphous coating have been demonstrated extensively to be effective pesticide adsorbants, the contribution of clay fraction by type and mineralogy to intact soils is largely unknown. For intact soils, it appears that high specific surface clays have a surprisingly small impact on K_d relative to that expected from studies based on pure or treated clay minerals. Once better defined, the contribution of various clay fractions and type could be included by expanding the %clay term in the SS equation accordingly.

3/ Percent organic carbon is the analytically determined organic carbon or %organic matter ÷ 1.732.

4/ Calculated specific surface = 100 %OC + 2.0 %clay + 0.4 %silt + 0.005 %sand. Where silt and sand were not determined, calculated SS = 100 %OC + 2.0 %clay. Coefficients for %clay, silt, and sand taken from Young and Onstad (76). Coefficient for %OC was based on organic matter values provided by Bailey and White (4) converted to the organic carbon basis (1,000 m^2/g OC), then adjusted to fit the Young and Onstad equation assuming that most clay in the intact soil would exhibit the surface area of kaolinite (20 m^2/g).

Table 1.--Parameters and statistical information provided for the relation: K_d = m %OC + b (using all data, no deletion)[1/]

Pesticide	Slope, m[2/]	Intercept, b[2/]	Coefficient of determination	F value for the relation[2/]	Mean K_d ± one standard deviation	Number of observations	Authors[3/]	Identification number used in figures
	Statistics							
Aldrin- - - - - -	.2650	2.52*	0.802	4.05 ns	2.8 ± .16	3	(75)	
Ametryne- - - - -	8.14**	-8.1**	.961	8.54**	10.4 ± 5.4	37	(17,47)	1
Atrazine- - - - -	1.49**	- .59	.885	378.**	4.3 ± 3.6	51	(3,14,17,20,21,28,31,33, 58,63,72,74)	2
Benefin - - - - -	4.22**	5.07+	.815	35.**	16.7 ± 4.3	10	(40)	
Carbofenthion -	125.	90.	.164	.39 ns	124. ± 81.	4	(43)	
Chloramben- - - -	.164**	.02	.892	66.**	.63 ± .44	10	(33,65)	22
Chlorbromuron -	20.8**	-29.7	.810	21.**	32.9 ± 18.	7	(14,18)	3
Chloroxorun - -	30.4*	44.8	.903	28.*	234. ± 60.	5	(26)	4
Ciodrin - - - - -	.5730	4.31+	.932	13.6 ns	6.0 ± .55	3	(44)	
CIPC- - - - - - -	9.30**	-9.4**	.833	165.**	8.5 ± 8.7	35	(32)	5
DDT - - - - -	2333.*	1690.	.998	444. *	42000. ±4000.	3	(61)	25
Dicamba - - - - -	.040**	-.097**	.945	187.**	.077± .061	13	(7,22)	21
Dieldrin- - - -	59.1**	325.**	.954	124.**	638. ± 101.	8	(9)	24
Dinitramine - - -	2.44**	.64	.727	21.**	7.4 ± 3.2	10	(40)	
Disulfoton- - - -	4.55**	15.0**	.881	111.**	43. ± 13.6	17	(13)	6
Diuron- - - - - -	9.35**	-7.8**	.787	359.**	14.1 ± 21.3	99	(18,20,26,32,47,49,53,58)	7
Ethion- - - - -	70.7	26.2	.416	1.43 ns	45.7 ± 24.	4	(43)	
Fluchloralin- - -	1.78*	4.23	.417	5.7**	9.1 ± 4.5	10	(40)	
Fluometuron - - -	1.12+	-.47	.977	43.+	2. ± .23	3	(17)	8
Lindane - - - -	20.5**	-24.3**	.879	196.**	57.8 ± 33.	29	(1,39,42,48,51,52,62)	
Linuron - - - - -	8.03**	-.83	.907	293.**	30. ± 17.	32	(18,26,28,67)	10
Monuron - - - - -	2.20**	-.02	.964	459.**	16. ± 5.3	19	(18,28,30,31,33,59)	11
Oryzalin- - - - -	1.63**	-.50	.897	70.**	4.0 ± 1.2	10	(40)	
Parathion - - -	31.20**	-18.5	.913	148.**	60.1 ± 36.6	16	(43,59,66)	12
Phorate - - - -	17.5	4.55	.522	2.18 ns	9.4 ± 4.9	4	(43)	
Picloram- - - - -	.241**	-.13	.760	117.**	.71 ± .71	39	(16,24)	13
Profluralin - - -	5.12**	8.38	.610	12.5**	22.4 ± 8.8	10	(40)	
Prometone - - - -	3.12	2.77	.036	1.0 ns	7.2 ± 10.6	29	(31,64)	14
Prometryne- - - -	3.97+	2.79	.115	3.9+	8.7 ± 7.8	32	(17,31,64)	15
Propazine - - - -	1.14**	.48	.548	33.**	2.1 ± .7	29	(31,64)	16
Pyrazon - - - - -	2.49**	-4.70	.975	154.**	22.7 ± 7.9	6	(41)	17
Simazine- - - - -	1.72**	-.32	.677	321.**	3.3 ± 4.5	155	(8,31,32,33,54,60,64,73)	18
Trifluralin - - -	5.09**	5.21	.748	29.0**	19.2 ± 5.7	10	(40,60)	9
2,4-D - - - - - -	.376**	-.65**	.916	283.**	1.36 ± 1.0	28	(17,19,22,24,33,60)	19
2,4,5-T - - - - -	.263**	.65+	.747	32.**	1.7 ± .95	13	(24,56)	20

Table 1.--Parameters and statistical information provided for the relation: $K = m\ \%OC + b$ (using all data, no deletion)--Continued

---

1/ K_d = the distribution coefficient of the pesticide between sediment and water phase where (1) the relationship between adsorbed and equilibrium concentrations for different pesticide concentrations are linear, or (2) the distribution is determined at equilibrium solution concentrations of 1.0 mg/l. Percent OC is organic carbon determined directly or by % organic matter ÷ 1.732, the m and b values are the slope and intercept, respectively, as fitted by regression.

2/ ns or unmarked means not significant; + , *, **, significant at the 10, 5, and 1% level, respectively.

3/ Coding number refers to References.

Table 2.--Parameters and statistical information provided for the relation: $K_d = m \text{ \%OC} + b$ after deleting high K_d values[1/]

Pesticide	Statistics: Slope, m[2/]	Intercept, b[2/]	Coefficient of determination	F value for the relation[2/]	Mean K_d ± one standard deviation	Number of observations	Identification number used in figures
Aldrin - - - - - - -	.2650	2.52*	.802	4.05 ns	2.8 ± .16	3	
Ametryne - - - - - -	2.32*	1.96	.169	6.9*	6.0 ± 3.6	36	1
Atrazine - - - - - -	0.628**	1.33**	.546	58.**	2.9 ± 1.8	50	2
Benefin- - - - - - -	4.22**	5.07+	.815	35.**	16.7 ± 4.3	10	
Carbofenthion- - -	125.	90.	.164	.39 ns	124 ± 81.	4	
Choramben- - - - - -	.020	.00	.000	.000		9	22
Chlorbromuron- - - -	5.52+	5.14	.614	6.4+	18.9 ± 4.7	6	3
Chloroxorun- - - -	30.4	44.8	.903	28.*	234. ± 60.	5	4
Ciodrin- - - - - - -	.5703	4.31+	.932	13.6 ns	60. ± 55.	3	
CIPC - - - - - - - -	2.86**	-.15	.803	122.**	4.3 ± 1.5	32	5
DDT- - - - - - -	2333.	1690.	.998	444.*	42000. ±4000.	3	25
Dicamba- - - - - - -	.040**	-.097**	.945	187.**	.077± .061	13	21
Dieldrin - - - - -	59.1**	325.**	.954	124.**	638. ± 101.	8	24
Dinitramine- - - - -	2.44**	.64	.727	21.**	7.4 ± 3.2	10	
Disulfoton - - - - -	6.25**	8.64**	.907	132.**	36.8 ± 9.4	16	6
Diuron - - - - - - -	3.34**	.44	.426	67.**	5.4 ± 3.6	92	
Ethion - - - - - -	70.7	26.2	.416	1.43 ns	45.7 ± 24.	4	
Fluchloralin - - - -	1.78*	4.23	.417	5.7**	9.1 ± 4.5	10	
Fluometuron- - - - -	1.12+	-.47	.977	43.+	2. ± .23	3	8
Lindane- - - - - - -	6.52**	5.6**	.720	59.**	21.9 ± 6.9	25	
Linuron- - - - - - -	5.57**	5.2	.585	41.**	20.5 ± 15.5	31	10
Monuron- - - - - - -	2.36**	-.48	.711	37.**	7.6 ± 5.4	17	11
Oryzalin - - - - - -	1.63**	-.50	.897	70.**	4.0 ± 1.2	10	
Parathion- - - - -	24.1**	-8.3	.693	29.4**	33.7 ± 33.7	15	12
Phorate- - - - - -	17.5	4.55	.522	2.18 ns	9.4 ± 4.9	4	
Picloram - - - - - -	.123**	.14	.206	8.8**	0.39 ± .45	36	13
Profluralin- - - - -	5.12**	8.38	.610	12.5**	22.4 ± 8.8	10	
Prometone- - - - - -	2.75+	1.6	.115	3.4+	5.5 ± 5.	28	14
Prometryne - - - - -	3.94**	1.72	.283	11.4**	7.6 ± 4.4	30	15
Propazine- - - - - -	1.14**	.48	.548	33.**	2.1 ± .7	29	16
Pyrazon- - - - - - -	.636*	2.11	.770	10.*	4.6 ± 1.6	5	17
Symazine - - - - - -	.636**	1.16	.106	18.**	2.3 ± 2.4	151	18
Trifluralin- - - - -	5.09**	5.21	.784	29.0**	19.2 ± 5.7	10	9
2,4-D- - - - - - - -	.194**	.01	.736	70.**	.74 ± .5	27	19
2,4,5-T- - - - - - -	.606**	.0022	.787	30.**	1.13 ± .65	10	20

Table 2.--Parameters and statistical information provided for the relation: $K_d = m\ \%OC + b$ after deleting high K_d values--Continued

1/ Deleted following visual inspection of scatter diagrams. If the high or dominating K_d values as a group appeared to be of a different statistical population, they were deleted from the analysis.

K_d = the distribution coefficient of the pesticide between sediment and water phase where (1) the relationship between adsorbed and solution equilibrium concentrations as a function of different pesticide concentrations are linear, or (2) the distribution is determined at equilibrium solution concentration of 1.0 mg/l. Percent OC is organic carbon determined directly or by %organic matter ÷ 1.732; the m and b values are the slope and intercept, respectively, as fitted by regression.

References same as those in table 1.

2/ ns or unmarked means not significant; +, *, **, significant at the 10, 5, and 1% level, respectively.

Table 3.--Parameters and statistical information provided for the relation: $K_d = m\,SS$ (using all data; no deletion)[1/]

Pesticide	Slope, m[2/]	Coefficient of determination	F value for the relation[2/]	Mean K_d ± one standard deviation	Number of observations	Authors[3/]	Identification number used in figures
	Statistics						
Aldrin- - - - - - -	.0107	.782	7.2 ns	2.8 ± 1.6	4	(75)	
Ametryne- - - - - -	.0581**	.822	166.**	10.4 ±12.1	38	(17,47)	1
Atrazine- - - - - -	.0127**	.936	496.**	3.0 ± 1.1	36	(17,28,60,64)	2
Benefin - - - - - -	.0379**	.964	243.**	16.7 ± 3.8	11	(40)	
Carbofenthion - -	1.43*	.895	26.*	124. ±52.	5	(43)	
Chlorbromuron - - -	.1071**	.770	20.**	33. ±25.	8	(17,18)	3
Chloxoruron - - - -	.333**	.963	105.**	234. ±59.	6	(26)	4
Ciodrin - - - - - -	.0136+	.872	13.6+	6.0 ± 2.7	4	(44)	
CIPC- - - - - - - -	.055**	.646	62.**	8.5 ±13.6	36	(32)	5
Dicamba - - - - - -	.00030**	.815	53.**	.007± .11	14	(7,22)	21
Dieldrin- - - - -	1.42**	.990	583.**	493. ±56.	8	(9)	24
Dinitramine - - - -	.0177**	.891	73.**	7.4 ± 3.2	11	(40)	
Disulfoton- - - - -	.0676**	.946	264.**	36.8 ±11.4	17.	(13)	6
Diuron- - - - - - -	.078**	.743	275.**	14.4 ±24.7	97	(18,26,32,47,49,53)	7
Ethion- - - - - - -	.544**	.961	74.**	46. ±12.	5	(43)	
Fluchloralin- - - -	.0200**	.865	58.**	9.1 ± 4.1	11	(40)	
Fluometuron - - - -	.0080**	.989	187.**	2.0 ± .3	4	(17)	8
Lindane - - - - - -	.1128**	.772	61.**	33.8 ±27.7	20	(1,42)	
Linuron - - - - - -	.0776**	.926	386.**	30. ±18.	33	(18,26,28,67,)	10
Monuron - - - - - -	.0206**	.943	249.**	12.5 ± 5.5	17	(18,28,30,31)	11
Oryzalin- - - - - -	.00997**	.896	78.**	3.4 ± 1.8	11	(40)	
Parathion - - - - -	.252**	.866	97.**	60. ±49.	17	(43,59,66)	12
Phorate - - - - - -	.114**	.974	112.**	9.4 ± 2.0	5	(43)	
Picloram- - - - - -	.0021**	.798	87.**	.9 ± .9	24	(16,24)	13
Profluralin - - - -	.0502**	.919	102.**	22.5 ± 7.7	11	(40)	
Prometone - - - - -	.0355	.390	17.9**	7.2 ±10.	30	(31,64)	14
Prometryne- - - - -	.0417**	.644	56.**	8.7 ± 7.2	33	(17,31,64)	15
Propazine - - - - -	.0100**	.933	389.**	2.1 ± .6	30	(31,64)	16
Pyrazon - - - - - -	.023**	.969	157.**	23 ± 9.	7	(41)	17
Simazine- - - - - -	.0138**	.378	79.**	2.9 ± 4.5	132	(8,31,32,54,64)	18
Trifluralin - - - -	.0440**	.949	168.**	19.2 ± 5.3	11	(40,60)	9
2,4-D - - - - - - -	.00178**	.825	118.**	.77 ± .5	27	(17,19,22,24,60)	19
2,4,5-T - - - - - -	.0029**	.896	69.**	1.8 ± .9	10	(24)	20

Table 3.--Parameters and statistical information provided for the relation: K_d = m SS (using all data, no deletion)--Continued

1/ K_d= distribution coefficient defined in previous two tables; SS = 100 %OC + 2 %clay + 0.4 %silt + .005 %S or = 100 %OC + 2 %C if sand and silt data not available.

Intercept is 0 in all cases.

2/ ns or unmarked means not significant; +, *, **, significant at the 10, 5, and 1% level, respectively.

3/Coding number refers to References.

Table 4.--Parameters and statistical information provided for the relation: $K_d = m\ SS$ after deleting high K_d values[1/]

Pesticide	Statistics					Identification number used in figures
	Slope, m[2/]	Coefficient of determination	F value for the relation[2/]	Mean K_d ± one standard deviation	Number of observations	
Aldrin- - - - - - - -	.0170	.782	7.2 ns	2.8 ± 1.6	4	
Ametryne- - - - - - -	.0232**	.759	110.**	6. ± 3.6	37	1
Atrazine- - - - - - -	.0127**	.936	532.**	3.0 ± 1.1	36	2
Benefin - - - - - - -	.0379**	.964	243.**	16.7 ± 3.8	11	
Carbofenthion - - -	1.43*	.895	26.*	124. ±52.	5	
Chlorbromuron - - - -	.0593**	.946	88.**	19. ± 5.	7	3
Chloxoruron - - - - -	.333**	.963	105.**	234. ±59.	6	4
Ciodrin - - - - - - -	.0136+	.872	13.6+	6.0 ± 2.7	4	
CIPC- - - - - - - - -	.0205**	.877	220.**	4.3 ± 1.9	33	5
Dicamba - - - - - - -	.00030**	.815	53.**	.077± .11	14	21
Dieldrin- - - - - -	1.42**	.990	583.**	493. ±56.	8	24
Dinitramine - - - - -	.0177**	.891	73.**	7.4 ± 3.2	11	
Disulfoton- - - - - -	.0676**	.946	264.**	36.8 ±11.4	17	6
Diuron- - - - - - - -	.0256**	.752	266.**	5.5 ± 3.7	90	
Ethion- - - - - - - -	.544**	.961	74.**	46. ±12.	5	
Fluchloralin- - - - -	.0200**	.865	58.**	9.1 ± 4.1	11	
Fluometuron - - - - -	.0080**	.989	187.**	2.0 ± .3	4	8
Lindane - - - - - - -	.0686**	.954	348.**	23.4 ± 5.6	19	
Linuron - - - - - - -	.0603**	.771	101.**	20.5 ±15.	32	10
Monuron - - - - - - -	.0210**	.813	61.**	8.4 ± 5.7	16	11
Oryzalin- - - - - - -	.00997**	.896	78.**	3.4 ± 1.8	11	
Parathion - - - - - -	.172**	.697	32.**	34. ±38.	16	12
Phorate - - - - - - -	.114**	.974	112.**	9.4 ± 2.0	5	
Picloram- - - - - - -	.0016**	.556	24.**	.4 ± .4	21	13
Profluralin - - - - -	.0502**	.919	102.**	22.5 ± 7.7	11	
Prometone - - - - - -	.0270**	.628	46.**	5.5 ± 4.7	29	14
Prometryne- - - - - -	.0365**	.817	134.**	7.6 ± 4.0	32	15
Propazine - - - - - -	.0100**	.933	389.**	2.1 ± .6	30	16
Pyrazon - - - - - - -	.0086**	.906	38.6**	4.6 ± 1.8	6	17
Simazine- - - - - - -	.0112**	.503	129.**	2.3 ± 2.4	129	18
Trifluralin - - - - -	.0440**	.949	168.**	19.2 ± 5.3	11	9
2,4-D - - - - - - - -	.00178**	.825	118.**	.77 ± .5	27	19
2,4,5-T - - - - - - -	.0043**	.891	41.**	.85 ± .54	7	20

Table 4.--Parameters and statistical information provided for the relation: K_d = m SS after deleting high K_d values--Continued

1/ K_d = distribution coefficient defined in previous two tables: SS = 100 %OC + 2 %clay + 0.4 %silt + .005 %S or = 100 %OC + 2 %C if sand and silt data not available.

Intercept is 0 in all cases.

Deleted following visual inspection of scatter diagrams. If the high or dominating K_d values as a group appeared to be of a statistically different population, they were deleted from this analysis.

References same as those listed in table 3.

2/ ns or unmarked means not significant; +, *, **, significant at the 10, 5, and 1% level, respectively.

From the nontheoretical or data-based viewpoint, the relationships of %OC + b and %OC with K_d are probably not that different. This is because of data variability. In table 2, note that the intercept, b, is never simultaneously negative and significant at the 5% level. Also, 30 of 35 pesticides do not provide a significant (5% level) nonzero intercept for which the overall relationship was also statistically significant at the 5% level. The deleted (table 2) rather than the total data set (table 1) was used because the high %OC values associated with the total data set statistically could bias the location of the fitted regression line, thereby controlling the intercept.

In comparing tables 1-4, direct statistical comparisons of SS with the %OC relationships are not strictly valid because (1) the SS equation has one more degree of freedom due to no intercept, and (2) there were often less observations for SS than %OC, because the textural analyses were not always published.

The next step was to expand this relationship beyond the 35 study pesticides for which no soil adsorption data are published. From the two equations presented earlier, the slope m value was considered the vehicle for expanding the soil adsorption data base. The m value is predominantly a characteristic of the potential adsorptivity of the pesticide, whereas %OC or SS is primarily a measure of the soil's adsorptivity. Because the m value is derived from soils data, the general applicability of m does assume that the data base is comprised of representative soils exhibiting a reasonable range in the soil properties that cause adsorption.

The m value, which is primarily a pesticidal property, was compared with the R_f values for each pesticide as determined by soil thin layer chromatography or the water solubility of the pesticide. The R_f and water solubility to m relationships are presented in figures 2-5 and 6-8, respectively.

The R_f values for approximately 100 pesticides were determined on Hagerstown sicl originally to estimate pesticide leachability rather than adsorptivity (2, 34, 35, 36, 37, 38). These values are presented in table 5. From the four figures (2-5) that present m versus R_f,the relationship is theoretically reasonable, becoming asymptotic at both limits (R_f = 0,1.0). The relationship for both m's, SS, and %OC, appeared improved when the outlier %OC data were removed (fig. 3 and 5). The selection of a standard soil, Hagerstown sicl, to obtain R_f values provides an adsorptivity measure based on one adsorbent. In this case, the adsorbent also provides the additional advantage of being a reasonably representative soil. The pesticide R_f values for Hagerstown sicl appear to represent reasonably well the average R_f for 13 geographically different soils (fig. 9) and fine-textured soils grouped into different textural classifications (fig. 10) but not coarse-textured soils (fig. 11). The physical, chemical, and mineralogical properties of these geographically different soils are described (table 6). R_f values derived from another soil can be translated to the Hagerstown sicl basis if the user can develop a curve such as figure 11. This assumes that the primary mechanism for adsorbing the pesticide in question on the Hagerstown sicl and other soil is not greatly different.

If the R_f value exceeds 0.9 or is less than 0.1, R_f values become less responsive to changes in the m value as the limits are approached (figs. 2-5). From an adsorption prospective, an R_f greater than 0.9 is not of much interest. In contrast, an R_f less than 0.1 represents an extensively adsorbed pesticide

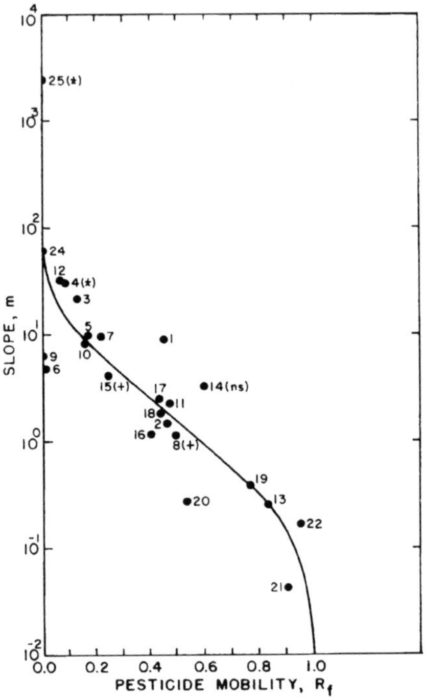

Figure 2.--Relation of slope in equation $K_d = m$ %OC + b, with R_f value as determined by soil-thin layer chromatography for Hagerstown silty clay loam for all data (table 1). Pesticide identification number and m values taken from table 1 (ns = not significant; + = significant at the 10% level; * = significant at the 5% level; remainder significant at the 1% level).

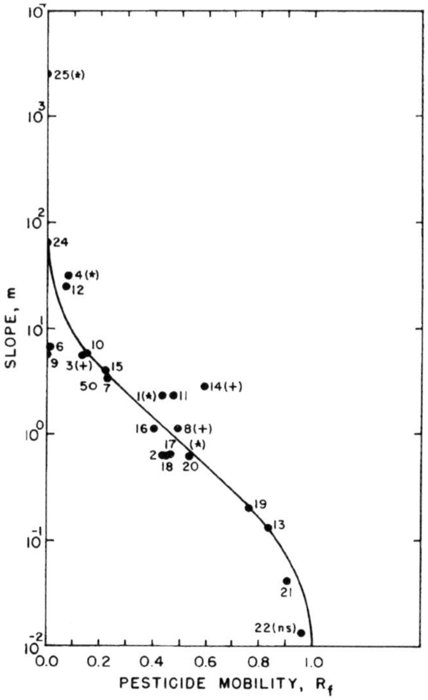

Figure 3.--Relation between slope of equation K_d = m %OC + b with R_f as determined by soil-thin layer chromatography for Hagerstown silty clay loam after high K_d values removed. Identification numbers from table 1 and m values from table 2 (ns = not significant; + = significant at 10% level; * = significant at 5% level; and remainder significant at 1% level).

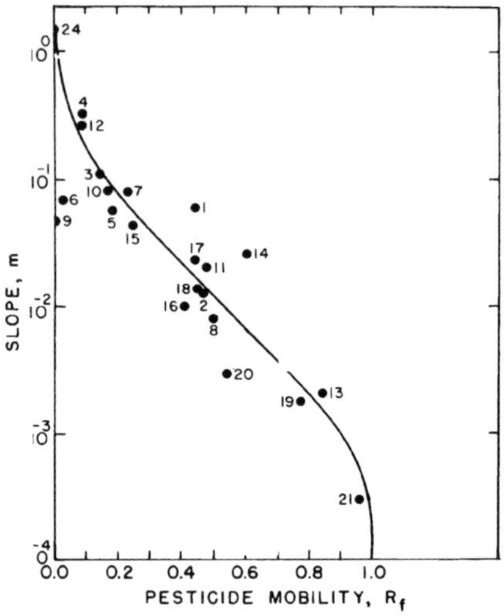

Figure 4.--Relation between slope of equation K_d = m SS with R_f values as determined from soil-thin layer chromatography for Hagerstown silty clay loam for all data (table 3). (All points statistically significant at the 1% level. Pesticide identified by number in table 1.)

and is of major interest. In figures 2 through 5, R_f less than 0.1 includes organophosphorus insecticides, dinitroaniline herbicides, and chlorinated hydrocarbon insecticides. For these pesticides, the m or K_d value needs to be estimated in another way.

The relationship between m values and pesticide solubility in water for the organophosphorus insecticides, dinitroaniline herbicides, and the chlorinated hydrocarbon insecticides are presented in figures 6, 7, and 8, respectively. The mean K_d ± the standard deviation taken from tables 1 through 4 are included. The lines were fitted by eye. The sources of the solubility data are

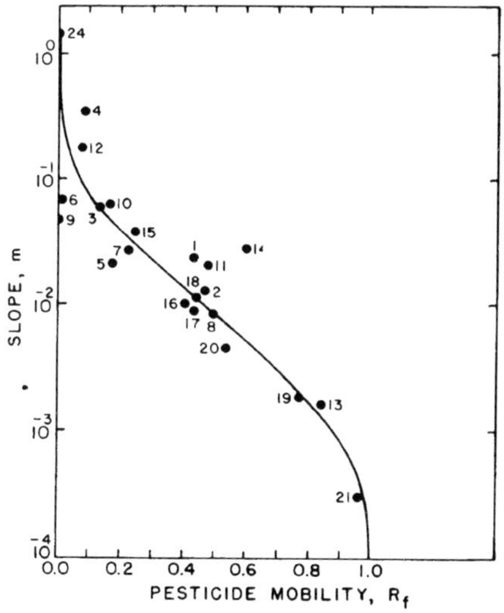

Figure 5.--Relation between slope of equation K_d = m SS with R_f values as determined from soil-thin layer chromatography for Hagerstown silty clay loam after high K_d values removed (table 4). (All points statistically significant at the 1% level. Pesticide identified by number in table 1.)

as follows: dinitroanilines (70), organophosphorus (23), and chlorinated hydrocarbon pesticides (23). For all three classes of pesticides, a relationship existed between m (%OC) and solubility. For two of the three classes, a relationship existed between m (SS) or K_d and solubility. An erratic relationship exists between K_d and water solubility for the organophosphorus insecticides. Within a pesticide class, we used solubility data based on comparable methodologies or data published about the same time. Where old solubility data had been updated greatly for one or several pesticides in a class, the relationship between solubility and m or K_d became erratic.

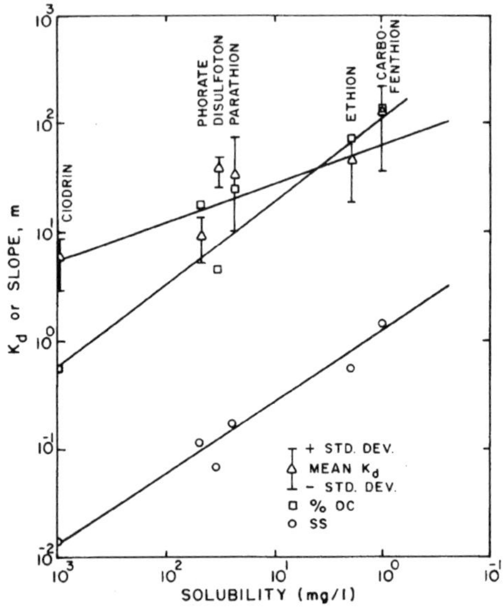

Figure 6.--Water solubility of organophosphorus pesticides as an estimate of K_d or slope of equation for the deleted data sets (tables 2 and 4). Values of m or SS for ciodrin and carbofenthion significant at 10% and 5%, respectively; remaining values significant at the 1% level. Values of m for %OC for ciodrin, ethion, phorate, and carbofenthion not statistically significant; remaining values significant at 1% level. K_d line is for ± one standard deviation. Solubility data are lowest values (23) except for disulfoton and carbofenthion for which it is the average of several low values.

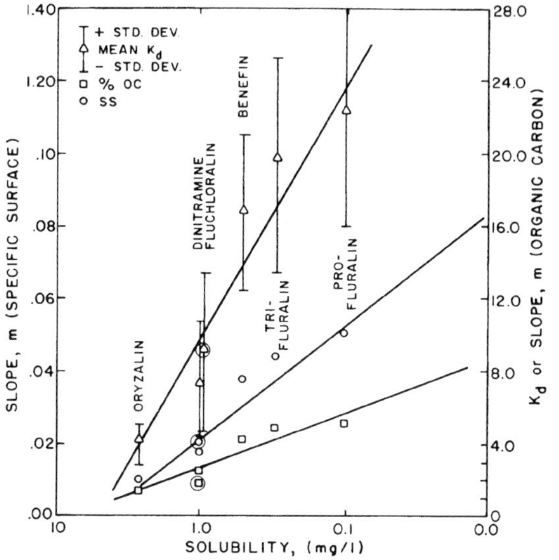

Figure 7.--Water solubility of the dinitroaniline herbicides as an estimate of K_d or the slope of the equation for the deleted data sets (tables 2 and 4). (All slopes statistically significant at 1% level. Circled data points refer to solubility given as less than 1 mg/l.)

ENVIRONMENTAL CONDITIONS AND OTHER SOIL PROPERTIES THAT CAN DOMINATE PESTICIDE ADSORPTIVITY UNDER SPECIAL CIRCUMSTANCES

Once a K_d for a specific pesticide-soil combination has been selected, that is, K_d = m SS or m %OC + b, other environmental factors or soil properties may alter the expected K_d. If the resultant alteration is great, relative to the normal variability introduced by estimating runoff, erosion, and the original K_d, then an appropriate correction may be necessary. Those discussed here include water temperature, salinity, soil-water content, adsorption-desorption hysteresis, soil:solution ratio, and soil pH.

It appears from the literature that the effect of variability introduced by normal changes in water temperature, that is, 0°-25° C (15, 47, 64) or by

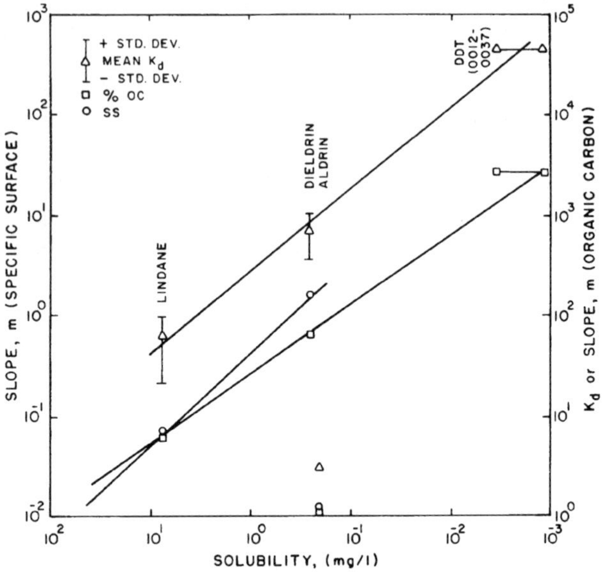

Figure 8.--Water solubility of chlorinated hydrocarbon insecticides as an estimate of K_d or slope of the equation for the deleted data sets (tables 2 and 4). (No DDT data for SS; aldrin m value for SS and OC not significant; m value for DDT in OC plot significant at 5% level; remaining values significant at 1% level, K_d line is ± one standard deviation.)

water salinity variability on the noncationic pesticides (57), is relatively minor compared to the variability introduced by the K_d, hydrologic, and erosion prediction methods. For example, the variabilities in the K_d vs. %OC or SS relationships expressed as standard deviation appear normally to range from ± 50 to ± 100% mean K_d (tables 1-4).

The effect of soil water content is ignored because this methodology is directed only at the runoff event during which the system is water saturated. Adsorption-desorption hysteresis is defined as occurring when the desorption rate is less than the adsorption rate but does not include the effects of "irreversible adsorption." We offer no method for identifying this condition or

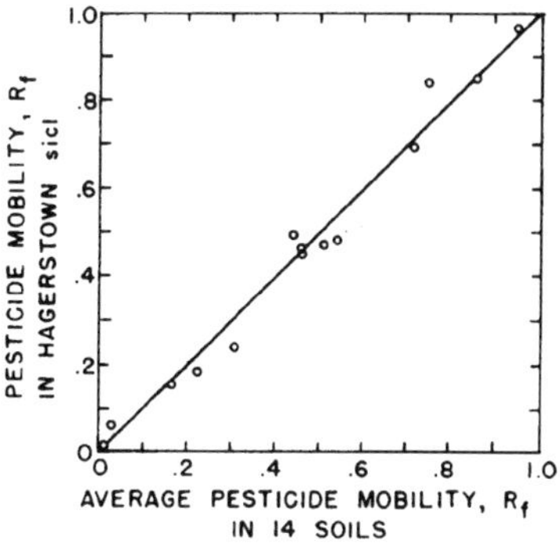

Figure 9.--Mobility of 13 pesticides on Hagerstown silty clay loam compared to the mobility of each pesticide averaged for 14 soils (35).

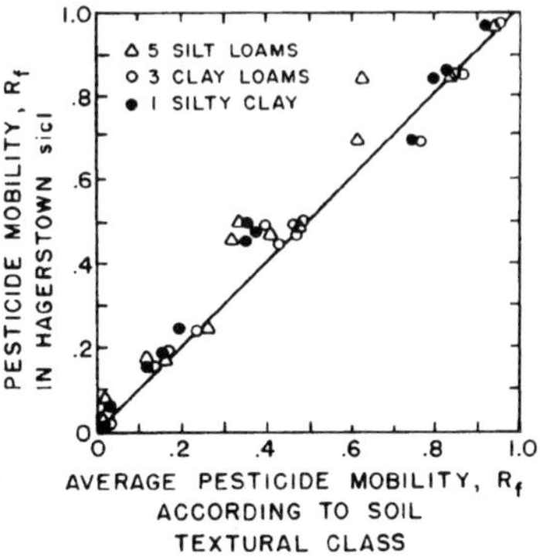

Figure 10.--Mobility of 13 pesticides on Hagerstown silty clay loam compared to the mobility of each pesticide averaged by soil external class (35).

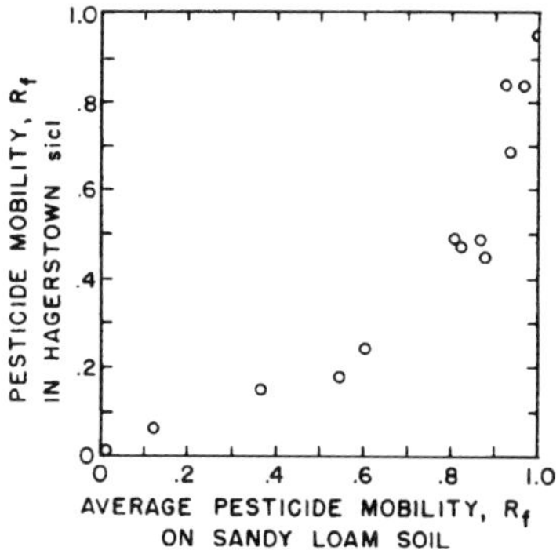

Figure 11.--Mobility of 13 pesticides on Hagerstown silty clay loam compared with the mobility of each pesticide averaged for two sandy loam soils (35).

correcting the K_d value. However, this effect can be ignored (1) when much or most of the desorption that will occur has occurred or (2) where the variability or known error in hydrologic, erosional, or other associated data exceeds the expected or projected error caused by hysteresis. With respect to (1), the desorption versus adsorption rate for some pesticides is sufficiently rapid so that much of the equilibrium concentration is achieved within the time interval associated with most storm or runoff events at a field scale (28). With respect to (2), the suspected or observed error or variability of K_d could be within desired accuracy and precision needs to achieve the modeling objective or within the accuracy and precision of other controlling input data. This must be considered by the user before each specific application.

The soil:solution ratio affects the k_d if the Freundlich 1/n is unequal to 1. This is critical because the soil:solution ratios can change drastically during the period from initial precipitation to first runoff and transport of that runoff to the field edge. Some of this change is inconsequential, for example, the result of deposition or entrainment of sands and coarse materials that are unimportant in the adsorption process. Other changes in runoff or adsorptive material content are critical, and can greatly alter K_d when the 1/n value is substantially different from one. From figure 1, note that dilution of equilibrium solution concentrations (C) from 1.0 to .01 mg/l for a relationship where 1/n = .7 will increase the K_d by 4-times its original value (4.0 ÷ 1.0). In other cases, the 1/n value is not constant but has been observed to shift with changes in C. Thus, a shift in 1/n from 0.8 to 0.6 as C is decreasing from 10 to 0.1 mg/l would alter the K_d to 4-times its original value (2.5 ÷ .62). This decrease in 1/n with decreasing C has been observed for atrazine and monuron (25). It would be ideal if we could experimentally derive the figure 1 relationship for each pesticide of interest. The applicability of the Freundlich relationship has not been tested on a wide range of soil or equilibrium solution concentrations with the possible exception of simazine, atrazine, and monuron (11, 25). Instead, figure 1 may be useful for defining permissible limits of extrapolating K_d data or adsorption relationships. Note that perhaps 90% or more of the observed 1/n values for a variety of pesticides range from 0.8 to 1.0 (11, 25). If we permit a maximum error of -50 to +100% K_d, the following extrapolations hold, assuming a reasonably stable 1/n value where: (1) $0.9 \leq 1/n \leq 1.1$ up to a 1,000-fold dilution in C is permissible, (2) $0.8 \leq 1/n \leq 1.0$, up to a 32-fold dilution in C is permissible, or (3) less than 10-fold dilution in C, the 1/n range can shift from 0.7 to 1.0 and still remain within error limit.

We concluded that, for the purpose of many hydrologically based predictions, the K_d could be extrapolated if these conditions or similar ones were observed. Where the reader wishes to estimate the change in K_d upon projected dilution or change in 1/n, figure 1, or the corresponding equation, $K_{di}/K_d = C\ (C^{1/n})_i / C_i (C^{1/n})$, can be used to make the estimate. If a known error can be tolerated, for example, $0.5 \leq K_{di}/K_d \leq 2.0$, the equation can be solved for specific 1/n or equilibrium concentration (C) limits. Those terms designated with the subscript i refer to the projected or extrapolated terms.

Another soil property that may substantially alter the K_d of weakly acidic and basic pesticides is soil pH. As the soil pH was reduced from 7.0 to 5.0, adsorption of the triazines (45, 47, 54, 55, 64, 68, 69) and of acidic herbicides (24) increased by varying amounts. Below a pH of 5.0, picloram (24) and ametryne (47) adsorption to soil increased substantially. Generally, pH exerts little effect on adsorption of the nonionic or strongly ionic pesticides, except, possibly, where the characteristics of the dispersed adsorbent system are altered substantially. This phenomenon will not be considered here. The pH effect on weakly acidic or basic compounds is attributed to (1) reducing the repulsion of negatively charged acids ($R^- + H^+ = RH$), or (2) increasing the attraction of normally uncharged weak bases ($RH + H^+ = RH_2^+$) to the negatively charged colloids. We assume insignificant anion exchange capacity. The extent of protonation depends on a pesticidal property defined as pK_a (disassociation constant). Thus, as the pH decreases relative to the pK_a, the adsorptivity and K_d of the weakly basic and acidic pesticides generally increase.

Where potentially important, quantitatively generalizing the pH effect on

Table 5.--Pesticide mobility (R_f) in Hagerstown silty clay loam soil

Pesticide	R_f: Autoradiography	R_f: Bioassay corrected for diffusion during incubation
Herbicides		
Ametryne	0.44	0.39
Atrazine	.47	.44
Atratone	-- 1/	.56
Amitrole	.73	--
Aziprotryne	--	.30
Barban	--	.06
Bromacil	.69	--
Dalapon	.96	--
Dichlormate	--	.14
Dichlobenil	.22	--
Dicamba	.96	--
Diphenamid	.49	--
Diuron	.22, .24	.26
Diquat 1/	.06	--
Chloramben (Amiben)	.91	--
Chlorbromuron	--	.14
CDAA	--	.82
Chloroxuron	.09	.09
C-8250 (Triazine)	--	.23
CIPC (Chlorpropham)	.18	--
2,4-D	.69, .85	--
2,4,5-T	.54	--
Fenac	.84	--
Fenuron	--	.69
Fluometuron	.50	.46
Linuron	--	.17
MCPA	.78	--
Metobromuron	--	.31
Monuron	.48	.44
Neburon	--	.07

Pesticides	R_f: Autoradiography	R_f: Bioassay corrected for diffusion during incubation
Herbicides		
TCA	0.96	--
Trietazine	.36	--
Trifluralin	.00	--
Insecticides		
Azinphosmethyl	.15	--
Carbanolate	.47	0.36
Carbaryl	.38	.35
Chlorphenamidine	.04	--
DDT	.00	--
Dieldrin	.00	--
Disulfoton	.01, .02	--
Endrin	.00	--
Formetanate-HCl	--	.79
Formparanate-HCl	--	.78
Methyl Parathion	.14	--
Morestan	.02	.00
Parathion	.08	--
Phosalone	.02	--
Promecarb	--	3/ .34
Fungicides		
ACQN	--	.08
Benomyl	--	.19
Binapacryl	--	4/ .03
Captan	--	.39
Ceresan L	--	5/ .07
Ceresan M	--	.05
Chloranil	--	.00
Chloroneb	--	.01

Table 5.--Pesticide mobility (R_f) in Hagerstown silty clay loam soil--Continued

Pesticide	R_f Autoradiography	R_f Bioassay corrected for diffusion during incubation	Pesticides	R_f Autoradiography	R_f Bioassay corrected for diffusion during incubation
	Herbicides			Fungicides	
Nitrofen	--	0.00	Cycloheximide	--	0.89
Nortron	0.65	--	Cycloheximide Oxime	--	.90
Oxadiazon	.04		D198	--	.51
Paraquat 2/	.00	--	DCNA	--	.01
Phenmedipham	--	--	Dexon	--	.88
Picloram	.84	--	Dichlone	--	.02
Flourodifen	.04	--	Dodine	--	.00
Prometryne	.25	.29	Dyrene	--	.03
Prometone	.60	.53	E-275	--	.09
Propachlor	.63	--	Hexachlorophene	--	.01
Propanil	.24	--	Karathane	--	.13
Propazine	.41	.35	Nabam	6/ 0.22 (.94, .00))	6/ .32 (.98, .77, .02
Propham	.51	--			
Pyrazon	--	.44	Oxycarboxin	--	.73
			Panogen 15	--	7/ .56
Siduron	.30	--	PCNB	--	.00
Simazine	.45	.36	PCP	--	.42
Simatone	--	.45	PCP-(Na)	--	.47
Simetryne	--	.33	TCNA	--	.00
Solan	--	.08	Terrazole	--	.00
			Zineb	--	.01

1/ Dashes denote data not available.

2/ Since adsorption largely considered irreversible, the R_f or K_d is not a useful concept. Assume most of the pesticide is adsorbed irreversibly and moves with the soil colloid fraction regardless of hydrologic dilution.

3/ Promecarb was poorly visualized by bioassay, so R_f is uncertain.

4/ Binapacryl also had a 0.49 R_f component thought to be hydrolysis product.

5/ Ceresan L a mixture of 2 components. The less mobile (0.07) is presumed to be methyl mercuric acetate, whereas the mobile component is presumed to be 2,3-dihydroxypropylmercaptide.

6/ The 0.94 R_f spot is thought to be a breakdown product, ETU. The 0.00 R_f spot may be an unknown breakdown product.

7/ Probable R_f of Panagen 15. Uncertain because of breakdown products

Table 6.--Soil properties of soils used in pesticide mobility (R_f) comparisons[1/]

Soil name	Origin (State)	Organic matter	Clay	Moisture at Field capacity	Moisture at Air-dry	pH	CEC	Dominant clay minerals[2/]
		%	%	%	%		(Meq/100 g)	
Norfolk sl- - - - -	?	0.14	11.3	6.5	0.17	5.1	0.2	Int Vm, Kn
Lakeland sl - - - -	Maryland	0.90	12.0	8.5	0.14	6.4	3.0	Kn, Qz
Christiana l- - - -	Maryland	0.99	24.4	19.7	0.47	4.4	5.6	Kn, Mica
Ascalon scl - - - -	Colorado	1.48	26.6	18.0	1.50	7.3	12.7	Mica, Int Mt
Sterling cl - - - -	Colorado	1.64	30.7	23.8	2.82	7.7	22.5	Mt
Dundee sil- - - - -	Mississippi	1.67	29.0	24.3	0.85	5.0	18.1	Mt
Wehadkee sil- - - -	Maryland	1.93	25.2	23.7	0.82	5.6	10.2	Kn, Vm
Duffield cl - - - -	Maryland	2.20	33.7	21.5	1.38	6.3	10.9	Mica, Vm
Beltsville sil- - -	Maryland	2.42	22.4	22.5	0.58	4.3	4.2	Vm, Kn
Hagerstown sicl [3/] -	Maryland	2.50	39.5	25.8	0.82	6.8	14.7	Vm, Kn
Chillum sil - - - -	Maryland	4.40	22.1	24.6	0.86	4.6	7.6	Vm, Kn
Iredell sil - - - -	?	5.27	23.2	31.3	1.61	5.4	17.0	Mt
Barnes cl - - - - -	Minnesota	6.90	34.4	28.5	2.99	7.4	33.8	Mt
Berkeley sic- - - -	West Virginia	8.02	50.5	30.8	3.23	7.1	33.7	Vm, Kn

1/ Helling, (35)

2/ Abbreviations: Int = Interstratified; Vm = Vermiculite; Kn = Kaolinite; Qz = Quartz; Mt = Montmorillonite.

3/ The R_f values and corresponding pesticides in figures are as follows: 0.96 - dicamba; 0.84 - picloram, fenac; 0.69 - 2,4-D; 0.49 - diphenamid; 0.48 - monuron; 0.47 - atrazine, 0.45 - simazine; 0.24 - diuron; 0.18 - chlorpropham; 0.15 - azinphosmethyl; 0.06 - diquat; 0.00 - trifluralin.

K_d is difficult. One reason is that most adsorption constants from the literature are determined over a range of soil pHs. The published K_d value can be a composite measure of the adsorptivity of both charged and uncharged species. One, perhaps, useful approach is to assume that two K_d limits exist that are defined by approximately 100% of the pesticide being charged or uncharged. The composite K_d would be calculated from

$$K_d = AK_{do} + BK_{d+}$$

where o and + subscripts refer to noncharged and positively charged species, respectively.

If above equation is combined with $[RH]\,[H^+]/[RH_2^+] = K_a$; $A = [RH]/[RH] + [RH_2^+]$; $B = [RH_2^+]/[RH] + [RH_2^+]$ and $[RH] + [RH_2^+]$ constant, then

$$K_d = \frac{10^{-pKa}\,K_{do} + 10^{-pH}K_{d+}}{10^{-pKa} + 10^{-pH}} .$$

For those pesticides where the charged species is adsorbed, less strongly that is, $RH = R^- + H^+$, the coefficients for the K_d values are switched so that

$$K_d = \frac{10^{-pH}\,K_{do} + 10^{-pKa}\,K_{d-}}{10^{-pKa} + 10^{-pH}} .$$

The bracketed terms [ ] refer to concentration in moles/liter; the - subscript refers to negatively charged species. The use of K_d assumes the adsorption isotherm for the charged species or uncharged species to be approximately linear, which is usually true at the lower concentrations of pesticide. Specifically, this means the Freundlich 1/n value and corresponding exponent for the exchange reaction ≈ 1.0. The needed input data to use these equations may be estimated or determined experimentally depending on your objectives. Both the pKa values (table 7) and the K_d, probably weighted toward the K_d of the neutral base or negatively charged acid, are presented in this paper. The K_d of the positively charged base or uncharged acid, usually, is not readily available or logically estimated. The value will probably have to be estimated or determined experimentally on site. The soil pH, often considered a simply measured and straightforward term, requires some interpretation.

The present theory defines soil pH differently for the weakly acidic and basic pesticides. The soil pH for weakly acidic pesticides used in the equations is considered equivalent to that conventionally measured in a distilled water-soil slurry or 1:1 soil to distilled water mix. In contrast, this bulk soil pH measurement for the weakly basic compounds is theorized to be 1 to 2 units greater than the pH at the soil colloid surface (5). Here, the corrected bulk soil pH should be used in the equations. Under dry soil conditions in the field, the corrections to bulk soil pH may require the subtraction of 3.0 or more pH units. Apparently, it is the acidity at the colloid surface which controls the percentage of the base, which is positively charged. Assuming the pH differential to be 2.0, 50% of the prometryne (pKa = 4.0) population is positively charged at a bulk soil pH of 6.0. This theoretically would increase the percentage of prometryne adsorbed over that expected from bulk soil pH values. It should be noted that most acidic and basic pesticides, however, exhibit pK_a

Table 7.--pK_a of acidic and basic pesticides

Common name or designation	Chemical name	pK_a ($R^-/RH = 1$)
2,4-D - - - - - -	2,4-dichlorophenoxyacetic acid	2.80
2,4,5-T - - - - -	2,4,5-trichlorophenoxyacetic acid	2.84
MCPA- - - - - - -	(4-chloro-o-tolyloxy) acetic acid	3.11
MCPB- - - - - - -	4-(4-chloro-o-tolyloxy) butyric acid	4.80
Silvex- - - - - -	2-(2,4,5-trichlorophenoxy) propionic aci d	3.0
Chloramben- - - -	3-amino-2,5-dichlorobenzoic acid	3.40
Dicamba - - - - -	3,6-dichloro-o-anisic acid	1.93
Tricamba- - - - -	3,5,6-trichloro-o-anisic acid	1.5
2,3,6-TBA - - - -	2,3,6-trichlorobenzoic acid	1.5
TIBA- - - - - - -	2,3,5-triiodobenzoic acid	1.5
Fenac - - - - - -	2,3,6-trichlorophenylacetic acid	3.70
Benzadox- - - - -	Benzamidooxyacetic acid	4.7
Picloram- - - - -	4-amino-3,5,6-trichloropicolinic acid	1.90, 5/3.7, 4/4.1
Endothall - - - -	7-oxabicyclo [2.2.1] heptane-2,3-dicarboxylic acid	4.0
Naptalam- - - - -	N-1-naphthylphatalmic acid	4.0
Dalapon - - - - -	2,2-dichloropropionic acid	1.84
TCA - - - - - - -	trichloroacetic acid	.63
DMSA- - - - - - -	N-dimethylamino succinamic acid	4.0
MH- - - - - - - -	1,2-dihydro-3,6-pyridazinedione	4.0
Dinoseb - - - - -	2-sec-butyl-4,6-dinitrophenol	4.40
DNOC- - - - - - -	4,6-dinitro-o-cresol	4.35
Ioxynil - - - - -	4-hydroxy-3,5-diiodobenzonitrile	3.96
Bromoxynil- - - -	3,5-dibromo-4-hydroxybenzonitrile	4.08
Cacodylic acid- -	hydroxydimethylarsine oxide	6.19
Bromacil- - - - -	5-bromo-3-sec-butyl-6-methyluracil	9.1
Isocil- - - - - -	5-bromo-3-isopropyl-6-methyluracil	9.1
Terbacil- - - - -	3-tert-butyl-5-chloro-6-methyluracil	9.0
Oryzalin- - - - -	3,5-dinitro-N,N-dipropylsulfanilamide	8.6
		($RH/RH_2^+ = 1$)
Metribuzin- - - -	4-amino-6-tert-butyl-3(methylthio-as-triazin-5-(4H)-one	2/1.00
Atrazine- - - - -	2-chloro-4-ethylamino-6-isopropylamino-s-triazine	1/1.68
Propazine - - - -	2-chloro-4,6-bis(isopropylamino)-s-triazine	1/1.85
Simazine- - - - -	2-chloro-4,6-bis(ethylamino)-s-triazine	1/1.65
Ametryne- - - - -	2-methythio-4-ethylamino-6-isopropylamino-s-triazine	3/4.0
Prometryne- - - -	2,4,-bis(isopropylamino)-6-methylmercapto-s-triazine	1/4.05
Atratone- - - - -	2-methoxy-4-ethylamino-6-isopropylamino-s-triazine	3/4.20
Prometone - - - -	2-methoxy-4,6-bis(isopropylamino)-s-triazine	1/4.28
Amitrole- - - - -	3-amino-1,2,4-triazole	1/4.17

Source: Weber (70) except for: 1/Weed and Weber (72), 2/Ladlie and others (45), 3/Weber (69), 4/Hamaker and others (26), 5/Hamaker and Thompson (25).

values (table 7) well below the bulk soil pH range of most productive and actively farmed agronomic soils, even when the 2.0 pH unit correction is made.

Figure 12 is an example application of the preceeding equations. The upper and lower K_d limits of prometryne (8.0, 200) and picloram (0.4, 10) correspond to the assumed K_d's of the charged and uncharged prometryne.

In conclusion, the effect of pH on adsorptivity of these pesticides is exceedingly complex but can only be treated simply for our purposes. The user often can ignore pH effects on K_d. Where these effects cannot be ignored, the equation with some assumptions and data provide a useful guide to adjusting the K_d.

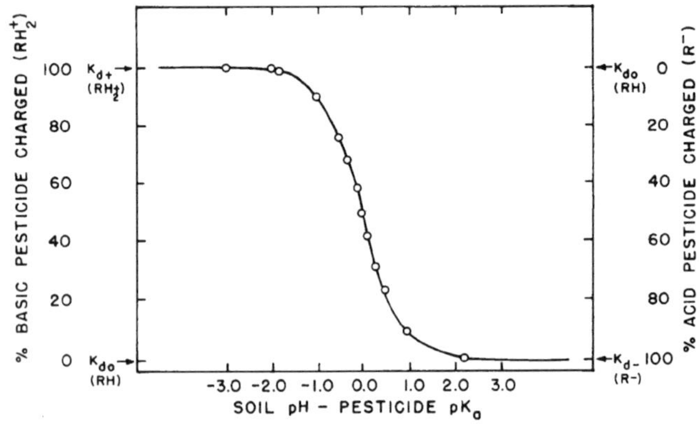

Figure 12.--Relation between K_d and soil pH minus pesticide pKa for weakly basic and acidic pesticides. (Basic pesticide is prometryne with K_{d+} assumed 200, K_{do} = 8, acidic pesticide is picloram with K_{do} assumed 10, K_{d-}= 0.4.) Soil pH = bulk soil pH for acid pesticides; soil pH = bulk soil pH -1.0 for basic pesticides.

METHOD

1. Set up specific objectives of the simulation. Consider whether your objective is to estimate the distributed concentration of a specific pesticide in runoff under present land use/management conditions or to compare the effects of alternative land use/management options, for example, selecting best management practices. The former requires a data base against which

simulation techniques should be tested, calibrated, or recalibrated before making a prediction. The last requires that key processes be explicitly modeled, but the simulation technique need not be calibrated against a data base. This is because the comparison can be relative, choosing one land use/management option as the base.

2. This is an approximate method, the accuracy of which is affected by the data base from which it was derived and the assumptions made in developing the method. Where pesticide adsorptivity data on soils from the site of interest or an analogous site are available or resources are sufficient to determine experimentally the adsorption isotherms, that method should be used.

3. The method estimates K_d [equilibrium pesticide concentration (μg/g) of the adsorbed phase ÷ equilibrium pesticide concentration (μg/g) in solution]. The K_d is calculated from (a) K_d = m %OC, or (b) K_d = m SS. %OC is % organic carbon or % of organic matter ÷ 1.732. The specific surface (SS) is calculated as follows: SS = 100 %OC + 2.0 %clay + 0.4 %silt + 0.005 %sand. The K_d value may be modified for bulk soil pH or the soil:solution ratio. Specific steps are:

   (a) Select the pesticide to be studied.

   (b) Determine the %OC or SS for the soil at the study site. Either equation (a) or (b) can be used, but the SS containing equation is preferred.

   (c) Select figures 3, 5, or 6 through 8 to determine the m value. Figure 3 is used in conjunction with %OC, figure 5 is used with SS. Figures 6 through 8 can be used in conjunction with %OC or SS. For soils containing high concentrations of organic matter or organic sediment, figures 2 and 4, and tables 1 and 3 are preferred.

      (1) If the pesticide is one of those used to derive the relationship in figure 3 or 5, the m value can be taken direttly from the respective figure or corresponding tables (table 2 - %OC; table 4 - SS). The number code used in these figures is identified by the pesticide name on table 1.

      (2) For other pesticides, select the appropriate R_f value from Table 5 if the R_f value is greater than 0.1. Translate the R_f value through the curve in figure 3 (%OC) or 5 (SS) to generate the m value, subject to the following modifications:

         (i) If the soil is medium-textured (silt to a silt loam to clay loam), do not modify; read directly from figures 3 or 5.

         ii) If the soil is coarse-textured (loamy sand to sandy loam), a transformation of the R_f to the Hagerstown silty clay loam basis (standard soil) may be needed. Figure 11 can be used for this purpose, either directly or modified, according to the user's wish. Take, for example, the R_f from table 5, which represents Hagerstown sicl, and plot on the Y-axis (fig. 11); from the Y-axis, extend a line horizontally to where it intersects the new curve (sandy loam); read the new R_f value on the x-axis. Then, this adjusted R_f value can be entered directly into figures 3 or 5 to generate <u>m</u>.

(3) For pesticides with R_f values less than 0.1, select figure 6 for organophosphorus insecticides, figure 7 for dinitroaniline herbicides, or figure 8 for chlorinated hydrocarbon insecticides.

(i) For the specific pesticides used to construct these figures (except aldrin), <u>m</u> values can be taken directly.

(ii) For other pesticides in these three classifications, the <u>m</u> value can be estimated from figures 6 through 8, using the pesticidal water solubility. General sources of water solubility data are Farm Chemicals Handbook (<u>10</u>) and Gunther and others (<u>23</u>). Weber and Monaco (<u>70</u>) provide solubility data for many of the dinitroanilines.

(d) Calculate the K_d using the <u>m</u> value obtained from (c), above, and the estimated or analyzed %OC or SS for the soil or suspended sediment by:

$$K_d = m_{oc}\ \%OC$$

or

$$K_d = m_{ss}\ SS$$

(e) Modify the calculated K_d or limit the applications of the calculated K_d according to the following conditions:

(1) The effect of changes in soil or runoff temperature from 0 to 25°C can be ignored.

(2) The effect of normal changes in indigenous salinity common to inland waters in humid-subhumid climates can be ignored.

(3) The effect of adsorption-desorption hysteresis normally can be ignored on the field-scale, composited runoff event basis. Strongly adsorbed pesticides, such as the cationic diquat or paraquat or the "irreversible adsorbed" portions of other pesticides, should be treated as moving totally within the sediment phase.

(4) Changes in soil:solution ratios in runoff on a field scale can be ignored under certain conditions or within certain limits. The change in K_d with change in soil:solution ratio is primarily a function of the nonlinearity of the Freundlich equation or the instability of the 1/n value. The user must first determine (a) the precision needed in the simulated answer to achieve the simulation objectives or (b) the precision of the final answer based on the combined variability introduced by the prediction methodology and data sets. On this basis, limits in both the allowable hydrologic dilution and 1/n change or deviation from 1/n = 1 can be established. For establishing limits of application or estimating the change in K_d, effects of changing the soil:solution ratio can be estimated from figure 1 or the corresponding equation, $K_{di}/K_d = C(C^{1/n})_i/C_i(C^{1/n})$, where C is the equilibrium pesticide concentra-

tion in solution and 1/n = Freundlich 1/n. The subscript i identifies the projected or extrapolated terms; the present or starting K_d, C, and 1/n values are not subscripted.

(5) The effect of soil or runoff pH can be ignored except for the weakly acidic and basic pesticides under certain conditions. Where the bulk soil or runoff pH is within 2.0 units above the basic pesticide pK or within $\pm$ 1.0 unit of the acid pesticide pK_a, a progressive reduction in solution pH will increase adsorptivity to a maximum at which pK_a = bulk soil pH (bulk soil pH - 1.0 for weakly basic pesticides). Under special conditions, that is, dry antecedent soil moisture for which the rate of chemical requilibration upon wetting during rainfall is much slower than the time to generate the bulk of runoff, the pH differential may increase to 2.0 or 3.0 units. If the pH estimated at the soil surface from the bulk soil pH, as determined above, is one or more units above the pK_a value, the effect of pH on K_d can be ignored. See table 7 for pK_a values of selected pesticides. The pK of most acid pesticides and many basic pesticides is generally well below the observed or corrected bulk soil pH on most productive agronomic soils. Consider modifying the K_d if the bulk soil pH is not outside these limits.

For weakly basic pesticides, the equation is:

$$K_d = \frac{10^{-pKa} K_{do} + 10^{-pH} K_{d+}}{10^{-pKa} +. 10^{pH}} . \qquad (1)$$

For weakly acidic pesticides, the equation is:

$$K_d = \frac{10^{-pH} K_{do} + 10^{-pKa} K_{d-}}{10^{-pKa} + 10^{-pH}} . \qquad (2)$$

K_{do} = K_d of the uncharged species (RH).
K_{d+} = K_d of the positively charged species (RH_2^+).
K_{d-} = K_d of the negatively charged species (R^-).
pH = bulk soil pH for the weakly acidic pesticides; bulk soil pH minus 1.0 for weakly basic pesticides, for example, the triazines. The bulk soil pH may be corrected by subtracting 2.0 or 3.0 if the user feels this is warranted according to special conditions described above.
pKa = negative logarithm of the disassociation constant at chemical equilibrium for reactions. Acidic pesticides are RH_2^+ = RH + H^+; (RH/RH_2^+ = 1); basic pesticides are RH = $\bar{R}$ + H^+; (R^-/RH = 1). Many of these values are provided in table 7.

The K_{d+} values can be estimated or determined experimentally for the site in question. The K_{do} values can be obtained similarly or estimated from the literature (tables 1-4). Most K_d's presented for weakly basic and acidic herbicides were determined on soils predominately outside the critical bulk soil pH ranges. The user, however, should check the specific literature cited in tables 1 through 4 and, if necessary, recalculate the K_d to obtain K_{do} by excluding those soils for which bulk soil pH's are in critical range. K_{d-} appears to be a very low value (≈0), unless soil anion exchange capacity is important.

REFERENCES

(1) Adams, R. S. Jr., and Paulina, Li.
1971. Soil properties influencing sorption and desorption of lindane. Soil Science Society of America Proceedings 35:78-81.

(2) Ambrosi, D., and C. S. Helling.
1977. Leaching of oxadiazon and phosalone in soils. Journal of Agriculture and Food Chemistry 25(1):215-217.

(3) Armstrong, D. E., and G. Chesters.
1968. Adsorption catalyzed chemical hydrolysis of atrazine. Environmental Science Technology 2:683-689.

(4) Bailey, G. W., and J. L. White.
1964. Review of adsorption of organic pesticides by soil colloids, with implications concerning pesticide bioactivity. Journal of Agriculture and Food Chemistry 12:324-332.

(5) ------, J. L. White, and T. Rothberg.
1968. Adsorption of organic herbicides by montmorillonite: Role of pH, and chemical character of adsorbate. Soil Science Society of America Proceedings 32:222-234.

(6) Briggs, G. G.
1969. Molecular structure of herbicides and their sorption by soils. Nature 223:1288.

(7) Burnside, O. C., and T. L. Lavy.
1966. Dissiaption of dicamba. Weeds 14:211-214.

(8) Day, B. E., L. S. Jordan, and V. A. Joliffe.
1968. The influence of soil characteristics on the adsorption and phytotoxicity of simazine. Weeds 16:209-213.

(9) Eye, J. D.
1968. Aqueous transport of dieldrin residues in soils. Journal of Water Pollution Control Federation 40(8):R316-R332.

(10) Farm Chemicals Handbook.
1978. Meister Publishing Co., Willoughby, Ohio.

(11) Farmer, W. J.
1976. Leaching, diffusion, and sorption. In A Literature Survey of Benchmark Pesticides, Chapter 5, pp. 185-238. The George Washington University Medical Center, Washington, DC.

(12) Gerber, H. R., and J. A. Guth.
1973. Short theory, techniques, and practical importance of leaching and adsorption studies, An introductory lecture. Proceedings of the European Weed Research Council Symposium, Herbicides-Soil. pp. 51-69.

13) Graham-Bryce, I. J.
1967. Adsorption of disulfoton by soil. Journal of the Science of Food and Agriculture 18:72-77.

14) Green, R. E., and S. R. Obien.
1969. Herbicide equilibrium in soils in relation to soil water content. Weed Science 17:514-519.

15) Griffin, R. P., and J. S. Dranoff.
1963. The rate of adsorption by ion exchange resins. American Institute of Chemical Engineers, Journal 9(2):283-287.

16) Grover, R.
1971. Adsorption of picloram by soil colloids and various other absorbents. Weed Science 19:417-418.

17) ------
1973. The adsorptive behavior of acid and ester forms of 2,4-D on soils. Weed Research 13:51-58.

18) ------
1975. Adsorption and desorption of urea herbicides on soils. Canadian Journal of Soil Science 55:127-135.

19) ------
1977. Mobility of dicamba, picloram, and 2,4-D in soil columns. Weed Science 25(2):159-162.

20) ------, and R. J. Hance.
1969. Adsorption of some herbicides by soil and roots. Canadian Journal of Plant Science 49:378-380.

21) ------
1970. Effect of ratio of soil to water on adsorption of linuron and atrazine. Soil Science 109:136-138.

22) ------, and A. E. Smith.
1974. Adsorption studies with the acid and dimethylamine forms of 2,4-D and Dicamba. Canadian Journal of Soil Science 54:179-186.

23) Gunther, F. A., W. E. Westlake, and P. S. Jaglan.
1968. Reported solubilities of 733 pesticide chemicals in water. Residue Review 20:1148.

24) Hamaker, J. W., C. A. I. Goring, and C. R. Youngson.
1966. Sorption and leaching of 4-amino-3,5,6-trichloropicolinic acid in soils. Advances in Chemistry Series 60:23-37.

25) ------, and T. M. Thompson.
1972. Adsorption. Chapter 2 In Organic Chemicals in the Soil Environment. pp. 49-143. C. A. I. Goring and J. W. Hamaker, Ed. Marcel Dekker, Inc., New York, New York.

(26) Hance, R. J.
1965. The adsorption of urea and some of its derivatives by a variety of soils. Weed Research 5:98-107.

(27) ------
1965. Observations on the relationship between the adsorption of diuron and the nature of the adsorbent. Weed Research 5:108-114.

(28) ------
1976. The speed of attainment of sorption equilibria in some systems involving herbicides. Weed Research 7:29-36.

(29) ------
1969. An experimental relationship between chemical structure and the sorption of some herbicides in soils. Journal of Agriculture and Food Chemistry 17:667.

(30) ------
1977. The adsorption of atraton and monuron by soils at different water contents. Weed Research 17:197-201.

(31) Harris, C. I.
1966. Adsorption, movement, and phytotoxicity of monuron and s-triazine herbicides in soil. Weeds 14:6-10.

(32) ------, and T. J. Sheets.
1965. Influence of soil properties on adsorption and phytotoxicity of CIPC, diuron, and simazine. Weeds 13:215- 219.

(33) ------, and G. F. Warren.
1964. Adsorption and desorption of herbicides by soil. Weeds 12:120-126.

(34) Helling, C. S.
1971. Pesticide mobility in soils II. Applications of soil thin-layer chromatography. Soil Science Society of America Proceedings 35:737-743.

(35) ------
1971. Pesticide mobility in soils III. Influence of soil properties. Soil Science Society of America Proceedings 35:743-748.

(36) ------, D. G. Dennison, and D. D. Kaufman.
1974. Fungicide movement in soils. Phytopathology 64:1091-1100.

(37) ------, D. D. Kaufman, and C. T. Dieter.
1971. Algae bioassay detection of pesticide mobility in soils. Weed Science 19:685-690.

(38) ------, and B. C. Turner.
1968. Pesticide mobility: Determination of soil thin-layer chromatography. Science 162:562-563.

(39) Huggenberger, F., J. Letey, and W. J. Farmer.
1972. Observed and calculated distribution of lindane in soil columns as influenced by water movement. Soil Science Society of America Proceedings 36:544-548.

(40) Jacques, G. L., and R. G. Harvey.
1978. Adsorption and diffusion of dinitroaniline herbicides in soils. In press.

(41) Jamet, P., and Piedallu, Marie-Andree.
1975. Etude de l'adsorption et de la desorption de la pyrazone (amino-5 chloro-4 phenyl-2 2H pryidazinone 3) par differents type de sols. Weed Research 15:113-121,

(42) Kay, B. D., and D. E. Elrick.
1967. Adsorption and movement of lindane in soils. Soil Science 104: 314-322.

(43) King, P. H., and P. L. McCarty.
1968. A chromatographic model for predicting pesticide migration in soils. Soil Science 106:248-261.

(44) Konrad, J. G., and G. Chesters.
1969. Degradation in soils of ciodrin, an organo-phosphate insecticide. Journal of Agriculture and Food Chemistry 17:226-230.

(45) Ladlie, J. S., W. F. Meggitt, and D. Penner.
1976. Effect of soil pH on microbial degradation, adsorption, and mobility of metribuzin. Weed Science 24:477-481.

(46) Lambert, S. M.
1967. Functional relationship between sorption in soil and chemical structure. Journal of Agriculture and Food Chemistry 15:572-576.

(47) Liu, L. C., H. Cibes-Viade, and F. K. S. Koo.
1970. Adsorption of ametryne and diuron by soils. Weed Science 18: 470-474.

(48) Lotse, E. G., D. A. Graetz, G. Chesters, G. B. Lee, and L. W. Newland.
1968. Lindane adsortion by lake sediments. Environmental Science Technology 2:353-357.

(49) Majka, J. T., and T. L. Lavy.
1977. Adsdorption, mobility, and degradation of cyanizine and diuron in soils. Weed Science 25:401-406.

(50) McCall, P. J.
1978. Desorption kinetics of picloram. 17th ACS National Meeting, March 13-17, 1978, Anaheim, California.

(51) Mills, A. C., and J. W. Biggar.
1969. Adsorption of gamma-BHC from solutions on several selected adsorbents. Journal of Agriculture and Food Chemistry 17:604-608.

(52) ------, J. W. Biggar.
1969. Solubility-temperature effect on the adsorption of gamma- and beta-BHC from aqueous and hexane solution by soil materials. Soil Science Society of America Proceedings 33:210-216.

(53) Mustafa, M. A., and Y. Gamar.
1965. Adsorption and desorption of diuron as a function of soil properties. Soil Science Society of America Proceedings 36:561-565.

(54) Nearpass, D. C.
1965. Effects of soil acidity on the adsorption, penetration, and persistence of sinazine. Weeds 13:341-346.

(55) ------
1967. Effect of the predominating action on the adsorption of simazine and atrazine by Bayboro clay soil. Soil Science 103:177-182.

(56) O'Connor, G. A., and J. U. Anderson.
1970. Soil Factors affecting the adsorption of 2,4,5-T. Soil Science Society of America Proceedings 38:433-436.

(57) Pionke, H. B., and G. Chesters.
1973. Pesticide-sediment-water interactions. Journal of Environmental Quality 3:29-45.

(58) Rhodes, R. C., I. J. Belasco, and H. L. Pease.
1970. Determination of mobility and adsorption of agrichemicals on soils. Journal of Agriculture and Food Chemistry 18:524-528

(59) Saltzman, S., and B. Yaron.
1971. Parathion adsorption from aqueous solutions as influenced by soil components. In Fate of Pesticides in the Environment. A. S. Tohori, Ed. Proceedings 2nd International IUPAC Congress Pesticide Chemistry 6:87-100. Gordon and Breach Science publishers, New York.

(60) Scott, H. D., and R. E. Phillips.
1972. Diffusion of selected herbicides in soil. Soil Science Society of America Proceedings 36:714-719.

(61) Shin, Young-Oh, J. J. Chodan, and A. R. Wolcott.
1970. Adsorption of DDT by soils, soil fractions, and biological materials. Journal of Agriculture and Food Chemistry 18-1129-1133.

(62) Spencer, W. F., and M. M. Cliath.
1970. Desorption of lindane from soil as related to vapor density. Soil Science Society of America Proceedings 34:574-578.

(63) Swanson, R. A., and G. R. Dutt.
1973. Chemical and physical processes that affect atrazine and distribution in soil systems. Soil Science Society of America Proceedings 37:872-876.

(64) Talbert, R. E., and O. H. Fletchall.
1965. The adsorption of some s-triazines in soils. Weeds 13:46-52

(65) ------, R. L. Runyan. and H. R. Baker.
1970. Behavior of amiben and dinoben derivatives in Arkansas soils. Weed Science 18:10-15.

(66) Wahid, P. A., and N. Sethunathan.
1978. Sorption-desorption of parathion in soils. Journal of Agriculture and Food Chemistry 26:101-105.

(67) Walker, W. A.
1973. Availability of linuron to plants in different soils. Pesticide Science 4:665-675.

(68) Weber, J. B.
1970. Mechanisms of adsorption of s-triazines by clay colloids and factors affecting plant availability. Residue Review 32:93-130.

(69) ------
1972. Interaction of organic pesticides with particulate matter in aquatic and soil systems. Advance Chemistry Series 111:55-120.

(70) ------, and T. J. Monaco.
1972. Review of the chemical and physical properties of the substituted dinitroaniline herbicides. Proceedings of the Southern Weed Science Society 25:31-37.

(71) Weed, S. B., and J. B. Weber.
1974. Pesticide-organic matter interactions. In Pesticides in Soil and Water. pp. 39-66. W. D. Guenzi, Ed. Soil Science Society of America, Madison, Wisconsin.

(72) Wildung, R. E., G. Chesters, and D. E. Armstrong.
1968. Chloramben (Amiben) degradation in soil. Weed Research 8:213-225.

(73) Williams, J. D. H.
1968. Adsorption and desorption of simazine by some Rothamsted soils. Weed Research 8:327-335.

(74) Yamane, V. K., and R. E. Green.
1972. Adsorption of ametryne and atrazine on an oxisol, montmorillonite, and charcoal in relation to pH and solubility effects. Soil Science Society of America Proceedings 36:58-64.

(75) Yaron, B., A. R. Swoboda, and G. W. Thomas.
1967. Aldrin adsorption by soils and clays. Journal of Agriculture and Food Chemistry 15:671-675.

(76) Young, R. A., and C. A. Onstad.
1976. Predicting particle-size composition of eroded soil. Transactions of the American Society of Agricultural Engineers. pp. 1071-1075.

*U.S. GOVERNMENT PRINTING OFFICE : 1980 0-310-945/SEA-15

Spotted something that isn't quite right?

Let us know and we'll fix it for future reprints. And as a thank you for your diligence, we'll send you a free Sothis book of your choice.

publisher@sothis.press